# The Heart of Mathematics

*The Cover:* The mirror-faced dodecahedron (a 12-sided regular solid) reflects how mathematics allows us to see and understand our world with greater clarity. It also illustrates how mathematics allows us to abstract nature and see the world in new ways. Finally, the mirrored face in front reflects the power of mathematics to help us see ourselves with greater focus and fresh dimensions.

# The Heart of Mathematics
## An invitation to effective thinking

**Edward B. Burger**
*Williams College*

**Michael Starbird**
*The University of Texas at Austin*

Key College Publishing
Innovators in Higher Education
www.keycollege.com

in cooperation with

Springer

Edward B. Burger  Michael Starbird
Williams College  The University of Texas at Austin
Williamstown, MA 01267  Austin, TX 78712

© 2000 by Edward B. Burger and Michael Starbird
Published by Key College Publishing, Emeryville, California, an imprint of Key Curriculum Press in cooperation with Springer-Verlag New York.

This work consists of a printed book. A manipulative kit and CD-ROM are also available for bundling or for separate purchase. All of these are protected by federal copyright law and international treaty. The book may not be translated or copied in whole or in part without the permission of the publisher, except for brief excerpts in connection with reviews or scholarly analysis. Use in connection with any form of information storage or retrieval, electronic adaptation, computer software, or by similar or dissimilar methodology now known or hereafter developed is forbidden.

The use of general descriptive names, trade names, trademarks, etc., in this publication, even if the former are not especially identified, is not to be taken as a sign that such names, as understood by the Trade Marks and Merchandise Marks Act, may accordingly be used freely by anyone. Where those designations appear in this book and the publisher was aware of a trademark claim, the designations follow the capitalization style used by the manufacturer.

Key College Publishing was founded in 1999 as a division of Key Curriculum Press in cooperation with Springer-Verlag New York. It publishes innovative curriculum materials for undergraduate courses in mathematics, statistics, and mathematics and statistics education.
www.keycollege.com   (510)595-7000
Key College Publishing
1150 65th Street
Emeryville, CA 94608 USA

Publisher: Jeremiah J. Lyons
Production and Manufacturing Manager: Diana Jean Parks
Cover Illustrator: Joan Greenfield
Project and Art Management: GTS Publishing Services
Design: Adrienne Bosworth
Composition and prepress: GTS Graphics
Photo Research: Judy Mason and Billie Porter
All illustrations rendered by GTS Graphics except:
3D imagery on pages xxi, 84, 126, 270, 271, 272, 277, 278, 280, 281, 296, 312, 315, 372, 378, 591, 592, by Jim Carbonetti, 3D CG artist, and Daniel Symmes, DIMENSION 3, Direction/encoding; illustrations on pages 274, 279, 282, 283, 287, 359, 367, 378, by Judy and John Waller; and computer-generated art in chapter 6 by Jennifer Trotts, except where noted in the photo credits.
Photo credits appear on page 645.
Printed and bound by Von Hoffmann Press.

Printed in the United States of America.

10 9 8 7 6 5 4 3     05 04 03 02 01

ISBN 1-55953-407-9   Key College Publishing

# Contents

*Welcome!* xi

*Surfing the book* xiii

**CHAPTER ONE**  *Fun and Games: An introduction to rigorous thought* 2

- 1.1. Silly Stories All Having a Moral:
  *Conundrums that evoke techniques of effective thinking* 4
- 1.2. Nudges:
  *Leading questions and hints for resolving the stories* 13
- 1.3. The Punch Lines:
  *Solutions and further commentary* 17
- 1.4. From Play to Power:
  *Discovering strategies of thought for life* 25

**CHAPTER TWO**  *Number Contemplation* 36

- 2.1. Counting:
  *How the Pigeonhole principle leads to precision through estimation* 38
- 2.2. Numerical Patterns in Nature:
  *Discovering nature's beauty and the Fibonacci numbers* 47
- 2.3. Prime Cuts of Numbers:
  *How the prime numbers are the building blocks of all natural numbers* 64
- 2.4. Crazy Clocks and Checking Out Bars:
  *Checking bar codes on products with cyclical clock arithmetic* 83
- 2.5. Secret Codes and How to Become a Spy:
  *How modular arithmetic and primes lead to secret public codes* 96
- 2.6. The Irrational Side of Numbers:
  *Are there numbers beyond fractions?* 112
- 2.7. Get Real:
  *The point of decimals and pinpointing numbers on the real line* 124

CHAPTER THREE  *Infinity*  138

3.1. Beyond Numbers:
*What does infinity mean?*  140

3.2. Comparing the Infinite:
*Pairing up collections via a one-to-one correspondence*  146

3.3. The Missing Member:
*Georg Cantor answers: Are some infinities larger than others?*  164

3.4. Travels Toward the Stratosphere of Infinities:
*The power set and the question of an infinite galaxy of infinities*  174

3.5. Straightening Up the Circle:
*Exploring the infinite within geometrical objects*  192


CHAPTER FOUR  *Geometric Gems*  208

4.1. Pythagoras and His Hypotenuse:
*How a puzzle leads to the proof of one of the gems of mathematics*  210

4.2. A View of an Art Gallery:
*Using computational geometry to place security cameras in museums*  219

4.3. The Sexiest Rectangle:
*Finding aesthetics in life, art, and math through the Golden Rectangle*  233

4.4. Soothing Symmetry and Spinning Pinwheels:
*Can the floor be tiled without any repeating pattern?*  249

4.5. The Platonic Solids Turn Amorous:
*Discovering the symmetry and interconnections among the Platonic Solids*  268

4.6. The Shape of Reality?:
*How straight lines can bend in non-Euclidean geometries*  288

4.7. The Fourth Dimension:
*Can you see it?*  306

CHAPTER FIVE  *Contortions of Space*  326

5.1. Rubber Sheet Geometry:
*Discovering the topological idea of equivalence by distortion*  328

5.2. The Band That Wouldn't Stop Playing:
*Experimenting with the Möbius Band and Klein Bottle*  344

5.3. Feeling Edgy?:
*Exploring relationships among vertices, edges, and faces*  356

5.4. Knots and Links:
*Untangling ropes and rings*  371

5.5. Fixed Points, Hot Loops, and Rainy Days:
*How the certainty of fixed points implies certain weather phenomena*  385


CHAPTER SIX  *Chaos and Fractals*  398

6.1. Images:
*Viewing a gallery of fractals*  400

6.2. The Dynamics of Change:
*Can change be modeled by repeated applications of simple processes?*  409

6.3. The Infinitely Detailed Beauty of Fractals:
*How to create works of infinite intricacy through repeated processes*  428

6.4. The Mysterious Art of Imaginary Fractals:
*Creating Julia and Mandelbrot Sets by stepping out in the complex plane*  457

6.5. Predetermined Chaos:
*How repeated simple processes result in utter chaos*  481

6.6. Between Dimensions:
*Can the dimensions of fractals fall through the cracks?*  501

vii

CHAPTER SEVEN  **Risky Business**  *512*

7.1. Chance Surprises:
*Some scenarios involving chance that confound our intuition* 514

7.2. Predicting the Future in an Uncertain World:
*How to measure uncertainty through the idea of probability* 520

7.3. Random Thoughts:
*Are coincidences as truly amazing as they first appear?* 538

7.4. Down for the Count:
*Systematically counting all possible outcomes* 551

7.5. Great Expectations:
*Weighing the unknown future through the notion of expected value* 568

7.6. What the Average American Has:
*Peering into the pitfalls of statistics* 583

7.7. Navigating Through a Sea of Data:
*How interpreting data reveals surprising and unintended results* 601

**Farewell** *616*

**Acknowledgments** *620*

**Hints and Solutions** *623*

**Index** *641*

**Credits** *645*

# Welcome!

> *Of course, no one actually reads their math textbooks.*
> —An anonymous math student

We wrote this book to be read. We designed many attractions—a kit for grasping concepts hands-on, jokes (some aren't too lame), 3D pictures & glasses, and a style of presentation that we hope invites you to discover new ideas. Most of all, this book contains intriguing lessons for thinking that can change your life.

*The False Mirror* (1928) by René Magritte. *Discover a new world view.*

## A World of Ideas

Most people do not have an accurate picture of mathematics. For many, mathematics is the torture of tests, homework, and problems, problems, problems. The very word *problems* suggests unpleasantness and anxiety. But mathematics is not "problems."

Hell's library

Some people view mathematics as a set of formulas to be applied to a list of problems at the ends of textbook chapters. Toss that idea into the trash. Formulas in algebra, trigonometry, and calculus are incredibly useful. But, in this book, you will see that mathematics is a network of intriguing ideas—not a dry, formal list of techniques.

We want you to discover what mathematics really is and to

> ... mathematicians are really seeking to behold the things themselves, which can be seen only with the eye of the mind. —Plato

become a fan. However, if you are not intrigued by the romance of the subject, that's fine too, because at least you will have a firmer understanding of what it is you are judging. Mathematics is a living, breathing, changing organism with many facets to its personality. It is creative, powerful, and even artistic.

Mathematics uses penetrating techniques of thought that we can all use to solve problems, analyze situations, and sharpen the way we look at our world. This book emphasizes basic strategies of thought and analysis. These strategies have their greatest value to us in dealing with real-life decisions and situations that are completely outside mathematics. These "life lessons," inspired by mathematical thinking, empower us to better grapple with and conquer the problems and issues that we all face in our lives from love to business, from art to politics. If you can conquer infinity and the fourth dimension, then what can't you do?

> This, therefore, is mathematics: she reminds you of the invisible form of the soul; she gives to her own discoveries; she awakens the mind and purifies the intellect; she brings light to our intrinsic ideas; she abolishes oblivion and ignorance which are ours by birth. —Proclus

> Mathematics seems to endow one with something like a new sense. —Charles Darwin

As you read this book, we hope you discover the beauty and fascination of mathematics, admire its strength, and see its value to your life. We do not have modest goals for this book. We want you to look at your life, your habits of thought, and your perception of the world in a new way. And we hope you enjoy the view.

Part of the power of mathematics lies in its inexorable quest for elegance, symmetry, order, and grace. Seeking pattern, order, and understanding is a transforming process that mathematics can help us develop.

> In mathematics I can report no deficiency, except it be that men do not sufficiently understand the excellent use of Pure Mathematics. —Francis Bacon

## A Mathematical Journey

> The advancement and perfection of mathematics are intimately connected with the prosperity of the State. —Napoleon I

The realm of mathematics contains some of the greatest ideas of humankind—ideas comparable to the works of Shakespeare, Plato, and Michelangelo. These mathematical ideas helped shape history, and they can add texture, beauty, and wonder to our lives.

To make our mathematical excursion as pleasant as possible, we have tried to make it all fun—fun to read, fun to do, and fun to think about. We hope you explore some, learn some, think some, enjoy some, and add

> *It may well be doubted whether, in all the range of science, there is any field so fascinating to the explorer—so rich with hidden treasures—so fruitful in delightful surprises—as Pure Mathematics.* —Lewis Carroll

a new aspect to your view of everything. We hope you laugh at our bad jokes and silly remarks, forgive our sometimes unbridled enthusiasm, but also embrace the profound issues at hand.

The road through this book is not free from perils, bumps, and jolts. Sometimes you will confront issues that start beyond your comprehension, but they won't stay beyond your comprehension. The journey to true understanding can be difficult and frustrating, but stay the course and be patient. There is light at the end of the tunnel—and throughout the journey, too.

What's the point of it all? Well, the bottom line is that mathematics involves profound ideas.

> *If we do not expect the unexpected, we will never find it.* —Heraclitus

Making these ideas our own empowers us with the strength, the techniques, and the confidence to accomplish wonders.

## *Travel Tips—Read the Book*

We have some suggestions about how to use this book:

- *Answer our questions*—We often pose questions in the middle of a section and invite you to give an answer or a guess before continuing. Please attempt to answer these questions. If you don't know an answer for sure, guess. Don't be afraid to make lots of mistakes—that is the only way to learn. It is much better to guess wrong than not to think about the question at all.

- *Think*—This is our main goal. We want you to contemplate some of the greatest and most intriguing creations of human thought. Constantly *stop* and *think*.

- *Be active not passive*—Our wish is for you to be an active participant. Take the concepts and make them your own. Look beyond the mathematical ideas, and don't be satisfied with mere knowledge. Challenge yourself to attain the power to figure things out on your own.

- *Have fun*—We truly believe that the ideas presented in this book are some of the most fascinating and beautiful ones around. We sincerely hope that some, if not all, of the themes will appeal to your intellect. More importantly, this journey of the imagination and of the mind should be fun. Enjoy yourself!

Finally, reading mathematics is much different from reading about many other subjects. Here's how we read mathematics. We read a sentence or two, stop reading, think about what we've read, and then realize we're completely and utterly confused. Usually we discover that we didn't really understand the previous paragraph. But, we don't get frustrated . . .

it's the nature of the beast. Instead, we either reread some previous sections or just reread the previous sentence. The fuzziness slowly begins to fade ever so slightly, and the concept begins to come into focus. Then we attempt to think about the issue and work with it on our own or with friends. It is at this point that we begin to appreciate and understand the ideas presented. One of the great features of mathematics is that once we do understand an idea our grasp of it is completely solid. There is no vagueness or uncertainty. So, adopt high standards for what you view as 'understanding.' Be actively engaged as you read. Draw pictures, explain ideas to friends. Put yourself in the position of the discoverer of each idea. Ask questions, search for answers, and let those answers guide you to still more questions.

> *Shall any gazer with mortal eyes*
> *Or any searcher know with mortal mind—*
> *Veil after veil will lift—but there must be*
> *Veil after veil behind.* —Sir Edwin Arnold

With all good wishes,

*Edward Burger     Michael Starbird*

# Surfing the book

It's too early to get caught up in details. Instead, let's just surf the book and get a quick overview of what's ahead. The whole book revolves around just two basic themes:
- Effective thinking
- Some truly great ideas

What mathematical sites lie ahead? Instead of just starting in with a hot and spicy math topic, we thought it would be more fun to surf the entire book and get a quick, big-picture overview of what is on the horizon. We hope these "home pages" will pique your curiosity and tantalize your intellect. Keep an open mind; forego any previous biases and prejudices toward mathematics; and do not censor any inventive thoughts or sparks of interest you may develop toward the subject. Let's surf.

*The Masterpiece or The Mysteries of the Horizon* (1955) by René Magritte.

Welcome  Games  Number  Infinity  Gems  Space  Chaos  Chance  Farewell

http://www.heartofmath.com/FunandGames

# Fun and Games:
## an introduction to rigorous thought

Can this book help you think more effectively, more inventively, solve life problems more creatively, and analyze issues more logically?

Go to page  3

The short answer is "Yes."

Is there a better way to meet the powerful world of logical thought than through <u>Fun and Games</u>?

Go to page  25

The short answer is "No."

Is this book strange and sometimes over the edge?

Go to page  8

The short answer is "Absolutely."

Just hang with us and see how far we'll go.

. . . are we having fun yet?

This site is an invitation to think and have fun with genies, damsels, and Dodge Ball and, in the process, develop a system of logical inquiry that we will use throughout the book and throughout our lives.

Who can better develop your thinking skills than you? As you resolve the many dilemmas in these crazy stories, you will automatically discover your own path to logical and strategic thought. Don't feel like going at it alone? Get a friend or a roommate to try some with you . . . it's all fun and games.

> . . . *the primary question was not What do we know, but How do we know it.* — Aristotle

| Welcome | Games | Number | Infinity | Gems | Space | Chaos | Chance | Farewell |

http://www.heartofmath.com/NumberContemplation

# Number Contemplation

Worried about balding? How about this one: Are there two hairy people on Earth with exactly the same number of hairs on their bodies? Does Rogaine change the answer?

Go to page **39**

What do the reproductive habits of 13th-century rabbits have in common with the Parthenon?

Go to page **57**

More than you think.

Are art and music branches of mathematics?

Go to page **238**

You betcha Bach!

Don't give up! Think working on really challenging questions that others have tried to solve is fruitless? Ask Andrew Wiles. In 1994 he answered a 350-year-old question—only took him seven years.
Hey, intellectual triumphs happen—just takes tenacity!

Go to page **75**

If 2 can be 1, who is the 1 to become??

xvi ◆ SURFING THE BOOK

> *Wherever there is a number, there is beauty.* — **Proclus**

Can you tell time? If so, then you might have a promising career at decoding the numbers at the bottom of Universal Product Codes. Want to know how?  Go to page **87**

XQE TPS LPBE AX TZ?  Go to page **96**

So numbers are no biggie? In Greece you were thrown from a ship and drowned if you told people about certain numbers. Sounds irrational?  Go to page **117**

How close is 1 to .99999 . . . ? Closer than you might think.  Go to page **132**

Ancient questions about numbers still remain unanswered. Act now . . . mathematicians are standing by.  Go to page **77**

SURFING THE BOOK

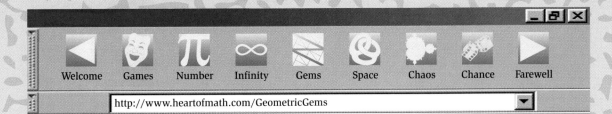

# Geometric Gems

Good at jigsaw puzzles? Check out the <u>Pythagorean Theorem</u>.

Go to page **211**

Want to see a picture of the <u>sexiest rectangle</u>? If you're 18 or over,

Go to page **233**

What kind of attractive <u>patterns</u> can cover our floors and walls? Can special, jumbled-looking patterns have some <u>symmetry</u> that regular checkerboard patterns have? Probably not . . . but hey, you never know.

Go to page **257**

Are straight lines really <u>straight</u>? Does <u>space</u> bend? For a free tour of the universe

Go to page **291**

(not valid in all states)

Is there a <u>fourth dimension</u>? Can you <u>see</u> it? (Warning. . . if you click on this site, you may not be able to return to this page.)

Go to page **315**

> *Might is geometry; joined with art, resistless.* — **Euripides**

Want to get in SHAPE?

SURFING THE BOOK ◆ xix

Welcome  Games  Number  Infinity  Gems  Space  Chaos  Chance  Farewell

http://www.heartofmath.com/ContortionsofSpace

# Contortions of Space

Bend and stretch—sound advice for both aerobics and <u>topology</u>.  Go to page **328**

If you want to take off some, but not all, of your clothes . . .  Go to page **331**

Does every issue have <u>two sides</u>? Answer: "No."  Go to page **349**

<u>Elastic</u> thoughts lead to <u>solid</u> ideas . . . the power of rubber.  Go to page **360**

> *The true spirit of delight . . . is to be found in mathematics as surely as in poetry.*  — Bertrand Russell

Ever thought about a world of rubber?

XX ♦ SURFING THE BOOK

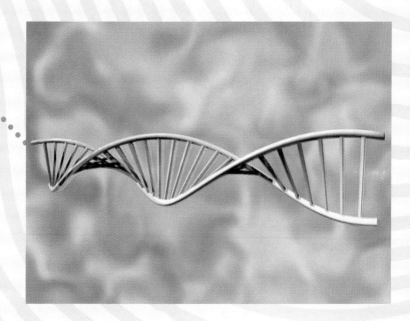

3D dude, put on your
3D glasses and enjoy.

Wondering about the mysteries of life? Want to untangle DNA?  Go to page 371

You've first got to untangle knots . . . good luck skipper!

The weather and rubber—are there two places on Earth  Go to page 393
that are exactly opposite each other and yet have identical
temperatures and pressures?

Either ask your local weather forecaster or . . .

SURFING THE BOOK ◆ xxi

Welcome  Games  Number  Infinity  Gems  Space  Chaos  Chance  Farewell

http://www.heartofmath.com/Chaos&Fractals

# CHAOS & FRACTALS

Can pictures or ideas be infinitely intricate?  Go to page 404

Can we predict the population, the weather, or even the positions of the planets in the future?  Go to page 481

Answer: "No."

. . . details . . . details . . . details . . .

[Fractals](#)—is there anything that is not one? Probably not . . . but what *is* one?   Go to page **444**

A [butterfly](#) flaps its wings in Brazil. Two weeks later there is a tornado in Kansas—kiss [Dorothy](#) good-bye. Are these events related?   Go to page **491**

Nature is sheer and utter [chaos](#). Why bother cleaning your room?   Go to page **486**

Can objects [straddle](#) between two dimensions?   Go to page **501**

> God has put a secret art into the forces of Nature so as to enable it to fashion itself out of chaos into a perfect world system.
> — Immanuel Kant

SURFING THE BOOK ◆ xxiii

Welcome  Games  Number  Infinity  Gems  Space  Chaos  Chance  Farewell

http://www.heartofmath.com/RiskyBusiness

# Risky Business

John and Jim are identical twins who were separated at birth. Both have married women named Jennifer who watch *Seinfeld* and love ice cream. What are the odds?

Go to page **607**

Answer: "Higher than you might think."

Will two people in a room of thirty have the same birth date? How would you bet?

Go to page **528**

Why are *amazing coincidences* nearly certain to happen? Here's one: Take "amazing coincidences" and look at the letters or spaces in the prime positions: 2, 3, (skip 5 because that's the number of fingers on a hand), 7, 11, and 13. What does it spell? *m a g i c*!!! What an amazing coincidence?

Go to page **538**

Surprised?

Buy their book? Or buy the lottery tickets?

Chances are . . .

Consider all graduates of Lakeside School over its whole history. The average net worth of each graduate increased by millions of dollars in 1996.

Amazing . . . or not?

Go to page **594**

Can improved airline safety lead to more accidental deaths?
Answer: "Yes."

Go to page **606**

> *Chance, too, which seems to rush along with slack reins, is bridled and governed by law.* — **Boethius**

Now that we have a sense of what's ahead, let's dig in.

# The Heart of Mathematics

CHAPTER ONE

# Fun and Games: An Introduction to Rigorous Thought

*. . . where the senses fail us, reason must step in.*
—Galileo Galilei

**1.1**

*Silly Stories All Having a Moral*

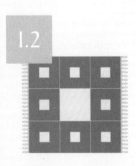

**1.2**

*Nudges*

**1.3**

*The Punch Lines*

**1.4**

*From Play to Power*

Fun and games—rigorous thought: only in mathematics. Who says that deep ideas and important consequences come only from hard work? Sure, we can consider the discipline of mathematics broadly from a philosophical perspective as in the *Welcome* or quickly surf its intriguing, mysterious sites as in *Surfing the Book*. But, when we really get down to it, when we think of mathematics, we think about rigorous thought along with fun and games, and we hope that one day you will too.

We start with two important and fundamental observations:

1. Logical and creative thinking are involved in mathematics.
2. Thinking can be fun.

By grappling with conundrums serious or otherwise, we can discover significant concepts. As we grope for solutions to silly stories, we begin to develop effective strategies for serious thinking.

❖ ❖ ❖ ❖ ❖ ❖ ❖ ❖ ❖

# 1.1 Silly Stories All Having a Moral:
## Conundrums That Evoke Techniques of Effective Thinking

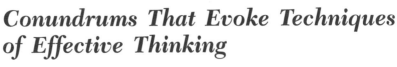

*Beware gentle knight, there is no greater monster than reason.*

—Miguel de Cervantes

You now have a mission. Your mission, should you decide to accept it, is to read the following stories and attempt to answer the questions they raise. The only rules are as follows:

1. Make an earnest attempt to resolve each story.
2. Think about each story, and be creative.
3. Don't give up: When stuck, look at the story in a different way.
4. If you become frustrated, stop working, move on, and then return to the story later.
5. Share these stories with your family, friends, or random people.
6. HAVE FUN!

*The Great Parade* (1954) by Fernand Léger

*A journey of one thousand miles begins with a single step.* —Chinese proverb

What is the fundamental strategy for attacking issues in life that need to be resolved? A critical part of the answer is simply to begin. Think a bit and then take one step forward. Taking that first step is often scary; more often than not, we do not possess a clear understanding of a complete solution or even see how a solution will eventually fall into place.

This situation is like taking a step in a completely dark room without having any idea of where we will end up. This natural fear of the unknown is sometimes so strong that we freeze like a deer in the headlights. However, we must learn not to let this understandable fear paralyze us intellectually; we must take a step. It is only by stumbling through many small intellectual steps that we later are able to make intellectual leaps.

For example, let's think like a soccer player with the ball at midfield. In this position we can't possibly know how a goal will be achieved, and we can't stop to envision the entire progression of the future before kicking the ball. Instead, we take small steps with the understanding that the specific goal strategy will become clear as opportunities arise.

Just try out ideas with these stories—loosen up, try to kick the ball, and don't worry if you miss. Remember:

*Truth comes out of error more easily than out of confusion.*
—Francis Bacon

After you have given considerable thought to a story, move to the corresponding part in section 1.2, Nudges, where leading questions and suggestions provide a gentle push in the right direction, in case you need a hint. There we also identify some strategies for tackling both mathematical questions, and, more importantly, questions that will arise in your life.

Section 1.3, The Punch Lines, provides solutions and commentary about how the questions and their resolutions fit into the mathematical landscape. As you think about these stories, you will discover some profound ideas that capture the essence of some deep and beautiful mathematical issues.

Remember the rules and strategies given on page 4, especially rule 6.

## Story 1. That's a Meanie Genie

On an archeological dig near the highlands of Tibet, Alley discovered an ancient oil lamp. Just for laughs she rubbed the lamp. She quickly stopped laughing when a huge puff of magenta smoke spouted from the lamp, and an ornery genie named Murray appeared. Murray, looking at the stunned Alley, exclaimed, "Well, what are you staring at? Okay, okay, you've found me; you get your three wishes. So, what will they be?" Alley, although in shock, realized what an incredible opportunity she had. Thinking quickly, she said, "I'd like to find the Rama Nujan, the jewel that was first discovered by Hardy the High Lama." "You got it," replied Murray, and instantly nine identical-looking stones appeared. Alley looked at the stones and was unable to differentiate any one from the others.

Finally she said to Murray, "So where is the Rama Nujan?" Murray explained, "It is embedded in one of these stones. You said you wished to find it. So now you have to find it. Oh, by the way, you may take only one

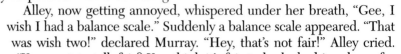

of the stones with you, so you had best be careful how you choose!" "But they look identical to me. How will I know which one has the Rama Nujan in it?" Alley questioned. "Well, eight of the stones weigh the same, but the stone containing the jewel weighs slightly more than the others," Murray responded with a devilish grin.

Alley, now getting annoyed, whispered under her breath, "Gee, I wish I had a balance scale." Suddenly a balance scale appeared. "That was wish two!" declared Murray. "Hey, that's not fair!" Alley cried. "You want to talk fair? You think it's fair to be locked in a lamp for 1,729 years? You know you can't get cable TV in there, and there's no room for a satellite dish! So don't talk to me about fair," Murray exclaimed. Realizing he had gone a bit overboard, Murray proclaimed, "Hey, I want to help you out, so let me give you a tip: That balance scale may be used only once." "What? Only once?" she said, thinking out loud. "I wish I had another balance scale." ZAP! Another scale appeared. "Okay, kiddo, that was wish three." Murray snickered. "Hey, just one minute," Alley said, now regretting not having asked for one million dollars or something more standard. "Well at least this new scale works correctly, right?" "Sure, just like the other one. You may use it only once." "Why?" Alley inquired. "Because it is a 'wished' balance scale. That means that you can use it only once since it was only one wish. It's just like you cannot wish for a hundred more wishes." "You are a very obnoxious genie." "Hey, I don't make up the rules, lady, I just follow them."

So, Alley may use each of the two balance scales exactly once. Is it possible for Alley to select the slightly heavier stone containing the Rama Nujan from among the nine identical-looking stones? Please explain why or why not.

## Story 2. Damsel in Distress

Long ago, knights in shining armor battled dragons and rescued damsels in distress on a daily basis. Although it is not often stressed in many of the surviving stories of chivalry, frequently the rescue involved logical thinking and creative problem solving, and often the damsel provided the solution. Here then is a typical knightly encounter.

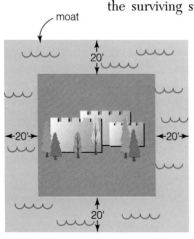

Once upon a time, a damsel was captured by a notorious knight and imprisoned in a castle surrounded by a square moat. The moat was infested with extraordinarily hungry alligators for whom the prospect of a luncheon damsel brought enormous smiles to their green faces. The moat was 20 feet across, and no drawbridge existed because the evil knight took it with him (giving his horse one major hernia).

After a time, a good knight and his squire rode up and said, "Hail sweet damsel, for I am here; and thou art there; now what are we going to do?"

The knight, though good, was not too bright and consequently paced back and forth along the moat looking anxiously at the alligators and trying feebly to think of a plan. Then, on the shore the knight found two sturdy beams of wood suitable for walking across but lacking sufficient length. Alas, the moat was 20 feet across, and the beams were each only 19 feet long and 8 inches wide. He tried to stretch them and tried to think. Neither effort proved successful. He had no nails, screws, saws, Superglue, or any other method of joining the two beams to extend their length.

What to do? What to do? Fortunately, the damsel, after a suitable time to allow the good knight to attempt to solve the puzzle himself, was able to give the knight a few hints that enabled him to rescue her and carry her home to her own castle. How did the maiden advise the knight to accomplish the rescue?

This story from medieval times foreshadows our journey into the geometric and the visual.

## Story 3. The Fountain of Knowledge

During an incredibly elaborate hazing stunt during pledge week, Trey Sheik suddenly found himself alone in the Sahara desert. His desire to become a fraternity brother was now overshadowed by his desire to find something to drink (these desires, of course, are not unrelated). As he wandered aimlessly through the desert sands, he began to regret his involvement in the whole frat scene. Both hours and miles had passed and Trey was near dehydration. Only now did Trey appreciate the advantages of sobriety. Suddenly, as though it were a mirage, Trey came upon an oasis.

There, sitting in a shaded kiosk beside a small pool of mango nectar, was an old man named Al Donte. Big Al not only ran the mango bar but was also a travel agent and could book Trey on a two-humped camel back to Michigan. At the moment, however, Trey desired nothing but a large drink of that beautifully translucent and refreshing mangoade. Al informed Trey that the juice was sold only in 8-ounce servings and that the cost for one serving was $3.50. Trey frantically searched all his pockets and found some change and much sand. Trey counted and discovered that he had exactly $3.50.

Trey's jubilation at the thought of liquid coating his dried and chapped throat was quickly shattered when Al casually announced that there were no 8-ounce glasses available. Al had only a 6-ounce glass and a 10-ounce glass—neither of which had any markings on them. Al, being a man of his word, would not hear of selling any more or any less than an 8-ounce serving of his libation. Trey, in desperation, wondered whether it was possible to use the two glasses to produce exactly 8 ounces of mango juice in the 10-ounce glass. Trey thought and thought. Do you think it is possible to use only the unmarked 6- and 10-ounce glasses to produce exactly 8 ounces in the 10-ounce glass? If so, explain how, and if not, explain why not. This pledge week prank does whet our appetites for a world of numbers.

6-oz glass

10-oz glass

## Story 4. Dropping Trou

Before reading on, remember that truth is sometimes stranger than fiction. The highlight of Professor Burger's April, 1993 talk to more than 300 Williams College students and their parents occurred when he tied his feet together with a stout rope, leaped onto the table, dramatically removed his belt, unzipped his zipper, and dropped his pants. The purple cows (Williams mascots) that were mooing about on his baggy boxer shorts completed an image not soon forgotten in the annals of mathematical talks. The more conservative members of the audience were contemplating a transfer of their sons and daughters to a less progressive school.

Edward Burger: exposed on April 24, 1993.

But then, at that moment of maximum bewilderment and absolute disbelief, Professor Burger did the seemingly impossible feat of rehabilitating his fast-sinking reputation. He turned his pants inside out without removing the rope attached to his feet and pulled his trousers back to their accustomed position (though inside out). Thus he simultaneously restored his modesty and his credibility by demonstrating the mathematical triumph of reversing his pants without removing the rope that was tying his feet together.

Please attempt to duplicate Professor Burger's amazing feat—in the privacy of your room, of course. You will need a rope or cord about 5 feet long. One end of the rope should be tied snugly around one ankle and the other end tied equally snugly about the other ankle. Now, without removing the rope, try to take your pants off, turn them inside out, and put them back on so that you, the rope, and your pants are all exactly as they were at the start, with the exception of your pants being inside out. While some may find this experiment intriguing, others may find it in poor taste. Everyone will agree, however, that surprising outcomes can arise if we are allowed to bend and contort objects and space.

## Story 5. Dodge Ball

Dodge Ball is a game for two players—Player One and Player Two (although any two people can play it, even if they are not named "Player One" and "Player Two"). Each player has his or her own special board and is given six turns. Here is a copy of each player's board.

Player One begins by filling in the first horizontal row of his table with a run of X's and O's. That is, on the first line of his board, he will write six letters—one in each box—each letter being either an X or an O. Then Player Two places either one X or one O in the first box of her board. So at this point, Player One has filled in the first row of his board with six letters, and Player Two has filled in the first box of her board with one letter.

Player One's game board:

|   |   |   |   |   |   |
|---|---|---|---|---|---|
| 1 |   |   |   |   |   |
| 2 |   |   |   |   |   |
| 3 |   |   |   |   |   |
| 4 |   |   |   |   |   |
| 5 |   |   |   |   |   |
| 6 |   |   |   |   |   |

Player Two's game board:

| 1 | 2 | 3 | 4 | 5 | 6 |
|---|---|---|---|---|---|
|   |   |   |   |   |   |

The game continues with Player One writing down a run of six letters (X's and O's), one in each box of the second horizontal row of his board, followed by Player Two writing one letter (an X or an O) in the second box of her board. This game proceeds in this fashion until all Player One's boxes are filled with X's and O's; thus, Player One has produced six rows of six marks each, and Player Two has produced one row of six marks. All marks are visible to both players at all times. Player One wins if any horizontal row he wrote down is identical to the row that Player Two created (Player One matches Player Two). Player Two wins if Player Two's string is not one of the six strings made by Player One (Player Two dodges Player One).

Would you rather be Player One or Player Two? Who has the advantage? Can you devise a strategy for either side that will always result in victory? This little game holds within it the key to understanding the sizes of infinity.

## Story 6. A Tight Weave

Sir Pinsky, a famous name in carpets, has a worldwide reputation for pushing the limits of the art of floor coverings. The fashion world stands agog at the clean lines and uncanny coherence of his purple and gold creations. Some call him square because his designs so richly employ that quaint quadrilateral. But squares in the hands of a master can create textures beyond the weavers' world, although not beyond human imagination.

One day Sir Pinsky began a creation with, as always, a perfect, purple square. However, one square was too plain, so in the exact center of it he added a gold square. He saw that the central square implicitly defined eight purple squares surrounding it. As he pondered, he realized that those eight purple squares were identical to his original large square except for two things: (1) Each was reduced to 1/3 the size of the whole square; and (2) the eight squares did not have gold squares in their centers.

He wondered whether he could further modify his design so that each of the eight reduced copies would be identical to the entire design except for being reduced by 1/3. After much thought, he solved this puzzle and created a design with which his name is associated. Can you sketch and describe his design? Create this design in stages, adding more gold squares at each stage.

Suppose the entire rug is 1 yard by 1 yard. How much gold material is needed for the second stage? How much for the third stage? Continue computing the area of the gold squares at various stages of the process, and then guess how much gold material will be needed to create the final floor covering. The answer is surprising.

Though our carpet designer is thoroughly modern in all ways, this type of design is ancient. We will see, in Chapter 6: Chaos and Fractals, an example of this style in the 14th century Buddhist tapestry, *vajradhatu mandala*.

## Story 7. Let's Make a Deal

"Let's Make a Deal!" Monty Hall enthuses to the gentleman dressed as a giant singing raisin. The gleeful raisin, whose name is Warren Piece, is ready to wheel and deal as Monty Hall explains the game. "Behind one of these three doors is the Cadillac of your dreams. It is as long as a train and comes complete with a Jacuzzi. Of course, if you spend too much time in the Jacuzzi, your skin will wrinkle, but hey, you're a raisin, your skin's already wrinkled." Monty Hall continues by warning that, "Behind the other doors, however, are two other modes of transportation: two old pack mules. They don't come with Jacuzzis, although given their exotic odor, you may want to give them a bath." Of course, the crowd is laughing and applauding just as the studio sign instructs.

Monty sums it up, "So, to sum it up, there are three closed doors. Behind one is a luxurious car, and behind the other two are mules. Now

comes the moment of truth. What door do you pick?" We now hear the traditional shouts of: "Take Door Number One, take Door Number One!!" from hundreds of frenzied fans. "Door Number Two, Door Number Two!!" scream hundreds more. "Door Number Three's the one. Choose Three," yell a competing contingent as poor Warren Piece looks around at the crowd, confused and nervous. He considers Door Number One, then Two, then Three. Finally Monty prompts, "Okay Warren, which do you want?"

The raisin-clad Warren shouts, "Okay, okay, I'll take Door Number Three, Door Number Three." As Monty Hall quiets the overly excited audience, he tells Warren, "I'll tell you what I'm going to do. I'm going to show you what's behind one of the doors you didn't pick. Let's take a look at what's behind Door Number Two." With that, Monty Hall turns to the Vanna White of the 1960s, "Please show us what is behind Door Number Two." The door dramatically swings open, the audience erupts, and Warren breathes once more—behind Door Number Two is a mule! Monty, knowing where the mules are, always opens one of the mule doors.

Monty continues, "We now see that the Cadillac is *not* behind Door Number Two. You guessed Door Number Three. I'll tell you what I'm going to do. If you want, I'll let you change your mind and choose Door Number One instead. It's up to you. Do you want to stick to your original choice, or do you want to switch?" The audience helpfully erupts again. "Stick, stick," yell half. "Switch, switch," advise the others. What to do, what to do?

We now invite you to add your voice to the cacophony—although you need not shout. What should Warren Piece do? Should he switch choices, stick to his original guess, or does it not matter? Here a classic TV game show raises the question: How can we accurately measure the uncertain?

## *Story 8. Dot of Fortune*

One day three college students were selected at random from the studio audience to play the ever-popular TV game show, "Dot of Fortune." One of the students already had discovered the power and beauty of mathematical thinking, while the other two were not nearly so fortunate. The stage had no mirrors, reflecting surfaces, or television monitors. The three students were led blindfolded to their places around a small round table. As the rules of the game were explained by Pat, Vanna affixed to each of the three youthful foreheads a conspicuous but small colored dot.

"So, contestants," Pat explained, "at the bell your blindfolds will be removed. You will see your two companions sitting quietly at the table, each with a dot on his or her forehead. Each dot is either red or white. You cannot, of course, see the dot on your own forehead. After you have observed the dots on your companions' foreheads, you will raise your hand if you see at least one red dot. If you do not see a red dot, you will keep your hands on the table. The object of the game is to deduce the color of your own dot. As soon as you know the color of your dot, you are to hit the buzzer in front of you. Do you understand the rules of the game?" All the students understood the rules, although the math fan understood them better.

"Are you ready?" asked Vanna after affixing three red dots to the foreheads of the three students. After the three contestants nodded, Vanna instructed them to simultaneously remove their blindfolds as the studio audience quivered with anticipation. The three students looked at one another's dots, and all raised their hands. After some time, the math fan hit her buzzer knowing what color dot she had. Please explain how she knew this. Why did the other students not know? This game show requires creative logical reasoning—a powerful means to make discoveries whether they are in math, in life, or even (although rarely) on prime-time TV.

**YELLOW CAUTION!**
Proceed to the next section
only after you have given
considerable thought to
each of the stories.

# 1.2 Nudges:
## Leading Questions and Hints for Resolving the Stories

*When we cannot use the compass of mathematics or the torch of experience . . . it is certain we cannot take a single step forward.*
—Voltaire

### Story 1. That's a Meanie Genie

Initially, we might think that it is impossible to find the jewel since Alley is allowed to make only two comparisons. Instead of comparing stones individually, perhaps she should compare one *collection* of stones with another *collection* of stones. Now suppose Alley compares one group with another using the first scale. What can she conclude? What should she do next?

### Story 2. Damsel in Distress

Thinking about variations on a situation helps us understand which features are essential and which are unnecessary. In this case we might consider a variation in which the damsel in distress is on the other side of a

Often we discover a solution only after we move beyond what appears to be the obvious or straightforward approach.

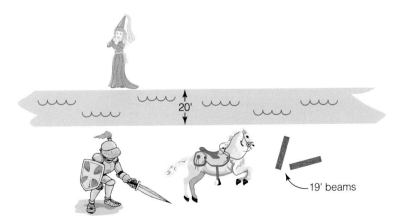

20-foot river rather than surrounded by a square moat. Unfortunately for the maiden, if she were separated from bliss by a river, she would go blissless, because the two 19-foot beams, in the absence of tools, would not enable the knight to rescue her. Somehow, the square shape of the moat must come into play in the solution.

Do not overlook small details; they often lead to tremendous discoveries.

♦ ♦ ♦ ♦ ♦ ♦

Looking at extremes is a potent technique of analysis in many situations and may be helpful here. The extremes, either geometrical ones as in this situation or conceptual extremes in other situations, frequently reveal features that are otherwise hidden.

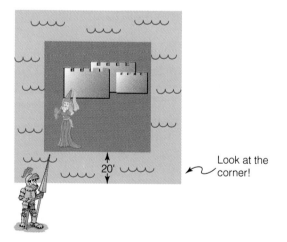

We shouldn't be afraid to experiment, especially when we are unsure where the outcomes will lead.

♦ ♦ ♦ ♦ ♦ ♦

## Story 3. The Fountain of Knowledge

Attempt this puzzle by trial and error together with careful observation. As we observe the outcomes of various attempts, we will teach ourselves what is possible. Try filling up the 10-ounce glass, and then use it to fill the 6-ounce glass. What do you now have—anything new?

*Visualization and experimentation often lead to surprising and even counterintuitive results.*

## Story 4. Dropping Trou

We hope that you physically attempt this exercise. By actually trying a task on your own, we often discover insights that otherwise may have been hidden from view (particularly in this case).

You will notice that the rope does restrict the amount of movement of your pants. Your mission is to discover means to work around such constraints. For example, try moving parts of the pants through other parts. An easier scenario may be to first try this task wearing shorts rather than long pants.

## Story 5. Dodge Ball

*Minor differences early on may lead to dramatically different outcomes.*

Play this game a few times with a friend; switch roles so that each of you has the opportunity to be Player One and Player Two. Remember, if you are Player One, your goal is to match one of your rows with your opponent's row. If you are Player Two, you want to dodge all six of your opponent's rows; that is, you want your row to differ in at least one spot from each of the six rows of your opponent. Who would you rather be: Player One or Player Two?

## Story 6. A Tight Weave

Consider a square having a smaller gold square in its center. How do each of the eight reduced surrounding squares differ from the whole picture? They are much the same except that the whole picture has a gold square in the middle, and each of the eight surrounding squares are solid purple.

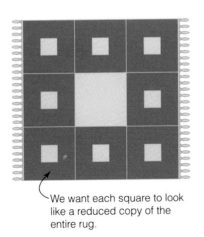

We want each square to look like a reduced copy of the entire rug.

How could you modify those eight 1/3-size surrounding squares to make them look like reduced copies of the entire picture you see here?

Now let's ask the question again: "In the picture you now have is each of the eight 1/3-size squares identical to reduced copies of the whole picture?" No. How would you modify each 1/3-size square-with-a-gold-center to make it identical to the whole new figure? Are you done?

Draw several steps of this repetitive process. At each stage, add up the areas of all the gold squares. When should you stop this process?

*Don't quit.*

*Consider various scenarios.*

## Story 7. Let's Make a Deal

Suppose the raisin had been wrong with his initial guess. What would be the result if he switched?

*Extreme or exaggerated cases often reveal essential features.*

♦ ♦ ♦ ♦ ♦ ♦

Suppose instead of three doors, there were ten doors. After Warren Piece guessed Door Three, suppose Monty Hall opened eight of the remaining doors and all had mules. Should our raisin switch in that game? Why?

## Story 8. Dot of Fortune

Sometimes no action is action enough. Put yourself in the position of one of the three contestants. You know that the dot on your forehead is either red or white. The trick to figuring out this conundrum is to suppose you have a white dot and see what would happen.

*Explore the consequences of alternatives.*

♦ ♦ ♦ ♦ ♦ ♦

Suppose you are sitting at the table looking at two red dots, and you assume that you have a white dot on your forehead. What would each of the two others at the table see? What could they deduce? What would they do? What did they do? What can you conclude?

**RED CAUTION!**
Do not enter the next section until you have thought about the stories, read the previous section, and tried to come up with answers. No peeking!

# 1.3  The Punch Lines:
## Solutions and Further Commentary

### Story 1. That's a Meanie Genie

Alley gets the jewel with no problem since she has read *The Heart of Mathematics*. She groups the stones into three groups of three and places one group of three on one side of the first balance scale and another group of three on the other side. What can she conclude? If both sides weigh the same, then she knows that the (heavier) jewel must be in the third group of three. If, however, one side is heavier than the other, then she knows that the jewel is one of the three that weighed more. In either case, after only one weighing, Alley is able to identify a group of only three stones among which is the Rama Nujan.

She then takes two of these three stones and places one on each side of the second scale. If one weighs more than the other, then she knows that this stone is the one containing the jewel. If they both weigh the same, then she knows that the third stone must contain the jewel. Thus, by weighing the stones only twice, Alley is able to find the jewel.

Take partial steps whenever possible. Notice that, instead of trying to find the jewel immediately, Alley first reduces the number of possibilities from nine to three. Thus she makes the problem easier and then cracks the easier problem. "Divide and Conquer" is an important and useful technique in both mathematics and in life.

*Break a hard problem into easier ones.*

### Story 2. Damsel in Distress

Focusing attention on the corner of the moat suggests using one of the beams to span the corner. Of course, we need to check that the two 19-foot beams are long enough to make the configuration in the illustration.

There are at least two ways to check that this picture is correct. One way is to construct a physical model. The figure is a physical model scaled down so that 1 foot in the story corresponds to 1 millimeter in the picture. You can now measure and ensure that this configuration is possible.

An alternative method to check that this pictorial solution works is to observe that we have some right triangles. These triangles provide us with a nice opportunity to foreshadow our look at the Pythagorean Theorem. After we examine good old Pythagoras's theorem (Chapter 4), the following paragraphs will seem soothing and comforting. If for now you find them less so, feel free to glance through them and just move on.

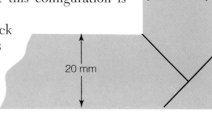

We notice that the corner of the moat forms a 20-foot-by-20-foot square. By the Pythagorean Theorem, the distance from the outer corner of the shore to the inner corner of the castle island is equal to the square root of $20^2 + 20^2$. Using a calculator, we see that the distance is 28.2842 . . . feet.

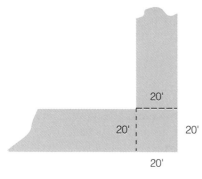

Placing the 19-foot beam diagonally across the corner of the moat as far out as it can go creates a triangle that cuts off the corner. If we draw a line from the center of the beam to the outer corner of the moat, we create two identical 45-degree right triangles, as shown. Since the length of half the beam is 9½ feet, we learn that the center of the beam is also 9½ feet from the outer corner of the moat.

Since the total diagonal distance from the outer corner of the moat to the corner of the castle island is 28.2842 . . . feet, the center of the beam is (28.2842 . . . feet − 9.5 feet) = 18.7842 . . . feet. Since that distance is just less than 19 feet, the other beam will just barely span the remaining distance, and the damsel can be rescued. In gratitude for her rescue, the damsel provided the good knight with a romantic lesson in *geometry*.

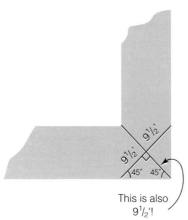

## Story 3. The Fountain of Knowledge

Suppose we fill up the 10-ounce glass and slowly pour it into the 6-ounce glass, stopping at the moment the 6-ounce glass is filled. Notice that the 10-ounce glass will contain precisely 4 ounces of mango juice. We now empty the 6-ounce glass back into the pool and refill it with the 4 ounces from the other glass. If we now submerge the 10-ounce glass into the pool and refill it, we can again slowly pour its contents into the 6-ounce glass until the 6-ounce glass is filled. We have added exactly 2 more ounces in the smaller glass and thus are left with 8 ounces in the larger glass. Happily, those 8 ounces of mango juice can then be served to Trey (on a tray). If Trey himself had found a solution, he would have made his first discovery in an area of mathematics known as *number theory*.

There is more than one solution to this puzzle. For example, we could have begun by filling the 6-ounce glass and pouring its entire contents in the 10-ounce glass. See if you can use this starting point to find an alternative solution.

## Story 4. Dropping Trou

The sequence of diagrams in the figure illustrates a solution to this knotty puzzle. Notice that by bending, contorting, and twisting our pants

Method: With pants on rope, bring one of the ends (cuff) of the right leg through the inside of the left leg; pull all the way through. When done, pants will be right side out (still) but the rope will now go through the pants. Now reach each hand into the inside of each pants leg and grab the cuffs. Simultaneously, pull the cuffs up through the pants. The pants will be inside out and the rope will no longer be around pants.

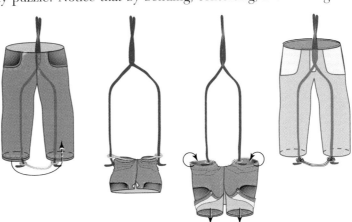

around we are able to produce different configurations. Issues involving bending, contorting, and twisting lead to interesting and surprising results. The notion of bending space is the fundamental notion in an area of mathematics called *topology*.

Often we are unable to conceive of an outcome when we sit back passively and think in the abstract. Make the issue concrete and physical whenever possible.

Many people believe mathematical issues exist outside our circle of life experience. In truth, many surprising and even counterintuitive mathematical discoveries can be made by freeing ourselves from old, unsubstantiated biases and experimenting with new objects and thoughts.

*By doing we often discover valuable insights.*

◆ ◆ ◆ ◆ ◆ ◆

## Story 5. Dodge Ball

We want to be Player Two. Here is a strategy that will guarantee victory. Player One fills in the first row of six boxes in his table. As Player Two, we look at the first letter and ignore the last five. If his first letter is an X, we write an O; if his first letter is an O, we write an X. Notice that, no matter what happens later, after this point, we are certain that the string we will create will definitely not be the same as Player One's first row. The two rows will differ in at least the first box. Player One now writes down his second string of six letters. We examine only the second letter in this new string. If that letter is an X, we write an O; if that letter is an O, we write an X. Now we are sure that no matter what follows, our string will not be the same as Player One's second string for the strings definitely differ in the second letter. If we repeat this process, we will have created a string of X's and O's that is different from the six strings created by Player One.

Creating a string that does not match any of our opponent's strings has a powerful application in the study of *infinity*. Although this modest little game has only six steps, the concept behind it has tremendous ramifications, as we shall see.

*Often simple observations can have deep consequences.*

◆ ◆ ◆ ◆ ◆ ◆

As a final note, we pose the following question: Suppose that we are Player One, but our opponent—who is trying to follow the strategy described above to win—makes just one mistake by placing the wrong letter in the first box. Can you now describe a strategy for us, as Player One, to ensure a win? Give this new challenge a try.

## Story 6. A Tight Weave

The solution is to repeat the process infinitely often. We start with a purple square. Then at the first stage, a single gold square of size $1/3 \times 1/3$ is placed in the center. At the next stage eight more gold squares

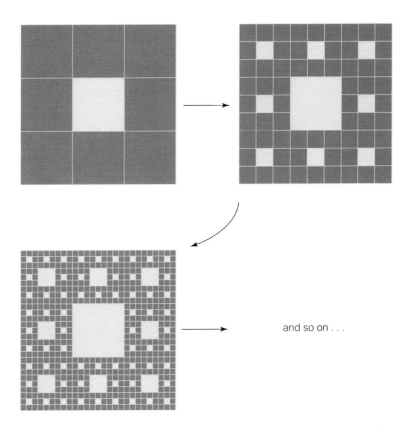

and so on . . .

of size $1/9 \times 1/9$ are placed in the centers of each of the eight surrounding squares. At the next stage, $8 \times 8$, or 64, more golden squares of size $1/27 \times 1/27$ are placed in the centers of each of the eight surrounding squares that surround each of the eight squares that surround the original square. At each stage, $8^n$ gold squares, each of size $1/3^{n+1} \times 1/3^{n+1}$, are added. So the final picture actually has infinitely many gold squares, but each of the eight squares surrounding the central square is an exact reduced copy of the whole picture. This intricate purple and gold carpet is an example of a self-similar object known as a *fractal*. In Chapter 6: Chaos and Fractals, we will examine many such infinitely intricate objects.

What is the area of all the (infinitely many) gold squares? Since all those gold squares lie within the rug that is 1 yard square, we know the area cannot be more than 1. At the first stage, we have one gold square $1/3 \times 1/3$; so its area is $1/9$. At the next stage, we add eight more gold squares, each of size $1/9 \times 1/9$, so their areas total $8 \times (1/9)^2$, making the total area of gold squares at stage two equal to $1/9 + 8 \times (1/9)^2 = .2098$ . . . . At the third stage, we add $8^2$ more squares, each of area $(1/9)^3$. Thus, the total area of gold squares at stage three equals $1/9 + 8 \times (1/9)^2 + 8^2 \times (1/9)^3 = .2976\ldots$. Repeating, we begin to see a pattern. The fourth stage, for example, would have a gold area equal to $1/9 + 8 \times (1/9)^2 + 8^2 \times (1/9)^3 + 8^3 \times (1/9)^4 = .3757\ldots$. Thus, the total area of gold squares in the final pattern would be the infinite sum:

$$\frac{1}{9} + 8 \times \left(\frac{1}{9}\right)^2 + 8^2 \times \left(\frac{1}{9}\right)^3 + 8^3 \times \left(\frac{1}{9}\right)^4 + 8^4 \times \left(\frac{1}{9}\right)^5 + 8^5 \times \left(\frac{1}{9}\right)^6 + \ldots$$

What does it equal? Even though there are infinitely many terms, we know that the whole area must be a number not greater than 1. What number is it?

> The gold area at the 5th stage is .4450 ...;
> at the 10th stage it is .6920 ...;
> at the 15th stage it is .8291 ...;
> at the 25th stage it is .9474 ...;
> at the 50th stage it is .9972 ...;
> at the 100th stage it is .999992 ....

From this pattern of numbers, it becomes clear that the gold area becomes increasingly close to 1—and that is a great guess for the area.

A clever way to calculate the total area is to add up all the infinitely many terms. We start by giving a name to the total; let's call that number SUM. Below you see the infinite sum that SUM represents. Directly under that, you see what $(8/9)$SUM equals. Notice that multiplying each term of SUM by $8/9$ just shifts that term to the right. For example, $(8/9)(1/9) = 8 \times (1/9)^2$.

*Data can help uncover surprising observations and help build intuition and understanding.*

$$\text{SUM} = \frac{1}{9} + 8 \times \left(\frac{1}{9}\right)^2 + 8^2 \times \left(\frac{1}{9}\right)^3 + 8^3 \times \left(\frac{1}{9}\right)^4 + 8^4 \times \left(\frac{1}{9}\right)^5 + 8^5 \times \left(\frac{1}{9}\right)^6 + \ldots$$

$$\left(\frac{8}{9}\right)\text{SUM} = \quad 8 \times \left(\frac{1}{9}\right)^2 + 8^2 \times \left(\frac{1}{9}\right)^3 + 8^3 \times \left(\frac{1}{9}\right)^4 + 8^4 \times \left(\frac{1}{9}\right)^5 + 8^5 \times \left(\frac{1}{9}\right)^6 + \ldots$$

Since all the terms of $(8/9)$SUM are directly under an identical term of SUM, it is easy to subtract $(8/9)$SUM from SUM, because all the terms drop out except the first term:

$$\text{SUM} - \left(\frac{8}{9}\right)\text{SUM} = \frac{1}{9} \quad \text{and so:}$$

$$\left(\frac{1}{9}\right)\text{SUM} = \frac{1}{9}$$

Since $(1/9)$SUM $= 1/9$, what is SUM? It must equal 1! In other words, the area of the gold squares is equal to the area of the entire rug. Thus, even though there are many purple threads remaining in the final pattern, as we begin to see in the illustration on page 21, the purple contributes no area to the rug. Surprise! We will see many more counter-intuitive mysteries of infinity in our studies of numbers, fractals, and, of course, in our studies of infinity itself.

## Story 7. Let's Make a Deal

Fortunately, Warren Piece enjoys mathematics as a hobby and so believes he can solve this conundrum. He thinks carefully, assesses the chances each way, and confidently proclaims (while still jumping up and down, of course), "I switch my guess to Door Number One, Monty."

Monty Hall turns and says, "OK. Let's see what deal you've made. What is behind Door Number One?" The door swings slowly open, and the crowd gasps as they see behind Door Number One the most beautiful finned chassis that General Motors ever painted pink. Bedlam reigns. Warren waves shyly, and happiness descends over the land. "How did you know?" asks Monty Hall over the din. Warren Piece explains.

"When I originally guessed Door Number Three, I had a 1/3 chance of being correct and a 2/3 chance of being wrong. Thus it's more likely I was wrong than right. When you opened Door Number Two and revealed no car, I hoped I was wrong originally—which, recall, was more likely than not. If I was wrong originally, then I knew for sure that the car must be behind the remaining door, Door Number One. So I switched, knowing that my probability of winning after switching was 2 out of 3, whereas my chance of having picked it correctly the first time was only 1 out of 3." Monty Hall was compelled to ask, "How did you figure that out?"—to which our hero raisin sagely replied, "By studying *The Heart of Mathematics: An invitation to effective thinking*." (Not a bad coast-to-coast TV plug for our book, even coming from a guy named Warren Piece in a raisin suit.)

*Keep an open mind and be willing to understand new ideas that first appear counterintuitive.*

◆ ◆ ◆ ◆ ◆ ◆

Some people might think that it doesn't matter if he switches or not. However, the chances of finding the car are indeed greater by switching. One way to demonstrate this is to list all the possible ways the Cadillac and the mules can be placed behind the doors:

|        | Door Number One | Door Number Two | Door Number Three |
|--------|-----------------|-----------------|-------------------|
| Case 1 | Cadillac        | mule            | mule              |
| Case 2 | mule            | Cadillac        | mule              |
| Case 3 | mule            | mule            | Cadillac          |

Warren Piece first picked Door Number Three, and the likelihood of finding the car there was 1 out of 3; that is, 1/3, which is not too likely. Next, Monty opened a door showing a mule. Let's see what happens if Warren were to switch in each of the three possible scenarios. In case 1, Monty opens Door Two. If Warren switches in this case, he wins the Cadillac. In case 2, Monty opens Door One. If Warren switches in this case, he would win. In case 3, Monty could open either Door One or Two. If Warren switches in this case, sadly he would be the owner of a mule. Therefore, overall, the likelihood of winning the car by switching is 2 in 3 or 2/3, which is twice as likely as the 1/3 chance of selecting the car if he sticks to his original guess. This brief encounter with probability illustrates that counterintuitive outcomes can occur when one attempts to measure the unknown.

## Story 8. Dot of Fortune

The math fan sees two red dots on the foreheads of the other two players. She knows she has either a white dot or a red dot on her own forehead. So let us suppose her dot is white and think about the consequences.

What would each of the two others at the table see? Each would see one red dot and one white dot, and each would see two arms raised. Each person would be thinking, "Do I have a red dot or a white dot on my forehead? If I have a white dot, then the red-dotted person would not have her hand up. Therefore, I must have a red dot." After that easy deduction, that person would hit the buzzer.

What did these two people actually do? Or, more to the point, what did they not do? They did not hit their buzzers! If either one of them were seeing a white dot and a red dot and two hands up, he or she would be able to deduce that his or her own dot is red. Since neither person buzzed right away, neither must have seen a white dot on the math student's forehead. Therefore, the math fan waits just long enough to know that the other two players cannot deduce their own dot colors, and then she buzzes, confident that her dot is red.

A final question of the story is, Why did the other students not know? The answer to that question is obvious: Because they had not read *The Heart of Mathematics*.

*There is great power to be found in logical and creative thinking.*

◆ ◆ ◆ ◆ ◆ ◆

# 1.4 From Play to Power:
## Discovering Strategies of Thought for Life

*Imagination is more important than knowledge.*

—Albert Einstein

Our stories illustrated strategies of thinking. Even in such a lighthearted setting, certain techniques of thought begin to emerge as powerful forces for combating the unknown—techniques applicable to any situation we may face in life. We'll encounter more "life lessons" elsewhere in *The Heart of Mathematics;* below we've summarized a few. Some may seem obvious or trivial, but we shouldn't take them lightly—they give us surprising strength for analyzing and conquering life's issues.

> **LESSONS FOR LIFE**
> 1. Just do it.
> 2. Make mistakes and fail, but never give up.
> 3. Keep an open mind.
> 4. Explore the consequences of new ideas.
> 5. Seek the essential.
> 6. Understand the issue.
> 7. Understand simple things deeply.
> 8. Break a difficult problem into easier ones.
> 9. Examine issues from several points of view.
> 10. Look for patterns and similarities.

## MINDSCAPES   Invitations to Further Thought

We now provide some additional stories for further amusement and enlightenment. We call them "Mindscapes" because they are vistas for the mind that encourage you to expand your way of thinking.

For each of the following situations, contemplate, analyze, and resolve the dilemma. Also, guess which branch of mathematics each situation represents: Logic, Number Theory, Infinity, Geometry, Topology, Chaos, or Probability. Of course, we haven't discussed any of these areas yet, but just take a guess—being wrong is fine.

Finally, we invite you to provide an aesthetic critique of each question and each of your solutions. Did you find either the question or your solution interesting? Which questions were the most challenging? Do you like one of your solutions better than the others? At the end of this section we provide some hints for some of the questions. Use them with caution.

> *"Contrariwise," continued Tweedledee, "if it was so, it might be; and if it were so, it would be; but as it isn't, it ain't. That's logic."* —Lewis Carroll

1. **Late night cash.** Suppose that David Letterman and Paul Shafer have the same amount of money in their pockets. How much must Dave give to Paul so that Paul would have $10 more than Dave?

2. **Politicians on parade.** There were 100 politicians at a certain convention. Each politician was either crooked or honest. We are given the following two facts:
   a. At least one of the politicians was honest.
   b. Given any two of the politicians, at least one of the two was crooked.
   Can it be determined from these facts how many of the politicians were honest and how many were crooked? If so, how many? If not, why not?

3. **The profit.** A dealer bought an article for $7, sold it for $8, bought it back for $9, and sold it for $10. How much profit did she make?

4. **The truth about . . .** Fifty-six biscuits are to be fed to ten pets; each pet is either a cat or a dog. Each dog is to get six biscuits, and each cat is to get five. How many dogs are there? (Try to find a solution without performing any algebra.)

5. **It's in the box.** There are two boxes: one marked A and one marked B. Each box contains either a million dollars or a deadly snake that will kill you instantly. You must open one box. On box A there is a sign that reads: "At least one of these boxes contains one million dollars." On box B there is a sign that reads: "A deadly snake that will kill you instantly is in box A." You are told that either both signs are true or both are false. Which box do you open? Be careful, the wrong answer is fatal!

6. **Lights out.** Two rooms are connected by a hallway that has a bend in it so that it is impossible to see one room while standing in the other. One of the rooms has three light switches. You are told that exactly one of the switches turns on a light in the other room, and the other two are not connected to any lights. What is the fewest number of times you would have to walk to the other room to fig-

ure out which switch turns on the light? And the follow-up question is: Why is the answer to the preceding question "one"? (Look out, this question uses properties of real lights as well as logic.)

7. **Out of sight but not out of mind.** The infamous band Slippery Even When Dry ended their concert and checked into the Fuzzy Fig Motel. The guys in the band (Spike, Slip, and Milly) decided to share a room. They were told by Chip, the night clerk who was taking a home study course on animal husbandry, that the room cost $25 for the night.

   Milly, who took care of the finances, collected $10 from each band member and gave Chip $30. Chip handed Milly the change, $5 in singles. Milly, knowing how bad Slip and Spike were at arithmetic, pocketed two of the dollars, turned to the others, and said, "Well guys, we got $3 change, so we each get a buck back." She then gave the other two members each one dollar and took the last one for herself.

   Once the band members left the office, Chip, who witnessed this little piece of deception, suddenly realized that something strange had just happened. Each of the three band members first put in $10 so there was a total of $30 at the start. Then Milly gave each guy and herself $1 back. That means that each person put in only $9, which is a total of $27 ($9 from each of the three). But Milly had skimmed off $2, so that gives a total of $29. But there was $30 to start with. Chip wondered what happened to that extra dollar and who had it. Can you please resolve and explain the issue to Chip?

8. **The cannibals and the missionaries.** In 1853 in the wilds of central Iowa, three missionaries and three cannibals were walking in a group. The missionaries were trying to convert the cannibals to their religion, while the cannibals were looking for a chance to practice their culture on the missionaries. After a time, they came to a river that they wished to cross. None of the six could swim, but all could row. Fortunately, on the river bank was a small rowboat available for use.

   Since the boat was small and the cannibals and the missionaries were all on the large side, it was clear that only two persons could cross at one time. It was late in the day and neither cannibals nor missionaries had eaten much recently, and the missionaries began to notice that the cannibals were indicating greater and greater appreciation for the missionaries' ample girths. The missionaries decided that being prudent was better than being a main course, so they decided that at no time should they allow any group of missionaries to be outnumbered by cannibals during the crossing. For their part, the cannibals did not fear being outnumbered by the missionaries since they realized that an excess of missionaries would result only in more discussion among the missionaries, thus relieving the cannibals of the burden of polite conversation.

Is it possible for the cannibals and the missionaries to all cross the river using only the one boat so that at no time do the cannibals outnumber the missionaries on either side of the river?

9. ***Whom do you trust?*** Congresswoman Smith opened the *Post* and saw that the bean-counting scandal had been leaked to the press. Outraged, Smith immediately called an emergency meeting with the five other members of the Special Congressional Scandal Committee, the busiest committee on Capitol Hill.

   Once they were all assembled in Smith's office, Smith declared, "As incredible as it sounds, I know that three of you always tell the truth. So now I'm asking all of you, who leaked the beans to the press?"

   Congressman Schlock spoke up, "It was either Wind or Pocket."

   Congressman Wind, outraged, shouted, "Neither Slie nor I leaked the scandal."

   Congressman Pocket then chimed in, "Well both of you are lying!"

   This provoked Congressman Greede to say, "Actually, I know that one of them is lying and that the other is telling the truth."

   Finally, Congressman Slie, with steadfast eyes, stated, "No, Greede, that is not true."

   Assuming that Congresswoman Smith's first declaration is true, can you determine who spilled the beans?

10. ***A commuter fly.*** A passenger train traveling at a steady 50 miles per hour left Austin, Texas at 12:00 noon bound for Dallas, exactly 210 miles away. At the same instant, a freight train traveling at 20 miles per hour left Dallas headed for Austin on the same track. At this same high noon, a fly leaped from the nose of the passenger train and flew along the track at 100 miles per hour. When the fly touched the nose of the oncoming freight train, she turned and flew back along the track at 100 miles per hour toward the passenger train. When she reached the nose of the passenger train, she instantly turned and flew back toward the freight train. She continued turning and flying until the expected tragedy occurred—she was squashed as the trains collided head on.

    Figure out how far the fly had flown before her untimely demise.

11. ***A fair fare.*** Three strangers, Bob, Mary, and Ivan, meet at a taxi stand and decide to share a cab to cut down the cost. They have different destinations, but all the destinations are right on the highway leading from the airport, so no circuitous driving is required. Bob's destination is 10 miles away, Mary's is 20 miles, and Ivan's is 30 miles. The taxi costs $1.50 per mile including the tip regardless of the number of passengers. How much should each person pay? (*Caution:* There is more than one way of looking at this situation.)

12. ***Getting a pole on a bus.*** For his 13th birthday, Adam was allowed to travel down to Sarah's Sporting Goods store to purchase a brand new fishing pole. With great excitement and anticipation, Adam boarded the bus on his own and arrived at Sarah's store.

Although the collection of fishing poles was tremendous, there was only one pole for Adam and he bought it: a 5-foot, one-piece fiberglass "Trout Troller 570" fishing pole.

When Adam's return bus arrived, the driver reported that Adam could not board the bus with the fishing pole. Objects longer than 4 feet were not allowed on the bus. Adam remained at the bus stop holding his beautiful 5-foot Trout Troller. Sarah, who had observed the whole ordeal, rushed out and said, "We'll get your fishing pole on the bus!" Sure enough, when the same bus and the same driver returned, Adam boarded the bus with his fishing pole, and the driver welcomed him aboard with a smile. How was Sarah able to have Adam board the bus with his 5-foot fishing pole without breaking or bending the bus-line rules or the pole?

13. **Tea time.** Carmilla Snobnosey lifted the delicate Spode tea pot and poured exactly 3 ounces of the aromatic brew into the flowered, shell china teacup. She placed the cream pitcher, also containing exactly 3 ounces, on the Revere silver tray and carried the offering to Podmarsh Hogslopper.

    "Would you like some tea and cream, Mr. Hogslopper?" she asked.

    "Yup. Thanks. Ow doggie, sure looks hot. I'd better cool it down with this here milk," he responded politely and carefully poured exactly 1 ounce of cream into his steaming tea and stirred. "That oughta do it," he said when the steam stopped rising from the tea. "Here, I'll just give you back that there cream." Whereupon he carefully spooned exactly 1 ounce from his teacup back into the creamer. Podmarsh blushed as he looked at a tea leaf or two floating in the cream and realized his faux pas. Caught at an awkward pass, he decided to smooth things over with an intriguing puzzle.

    "Ya know, Mrs. Snobnosey, I wonder if the tea is more diluted than the cream, or whether the cream is more diluted than the tea?"

    Resolve the dilution problem.

14. **A shaky story.** Stacy and Sam Smyth were known for throwing a heck of a good party. At one of their wild gatherings, five couples were present (this included the Smyths, of course). The attendees were cordial, and some even shook hands with other guests. Although we have no idea who shook hands with whom, we do know that no one shook hands with themselves and no one shook hands with his or her own spouse. Given these facts, a guest might not shake anyone's hand or might shake as many as eight other people's hands. At midnight, Sam Smyth gathered the crowd and asked the nine other people how many hands each of them had shaken. Much to Sam's amazement, each person gave a different answer. That is, someone didn't shake any hands, someone else shook one hand, someone else shook two hands, someone else shook three hands, and so forth, down to the last person, who shook eight hands. Given this outcome, determine the exact number of hands that Stacy Smyth shook.

**15. *Murray's brother.*** On another archeological dig, Alley discovered another ancient oil lamp. Again she rubbed the lamp, and a different genie named Curray appeared. After Alley explained her run-in with Murray, Curray responded, "Well, since you know my brother Murray, it's like we're almost family. I'm going to give you four wishes instead of three. What do you say?" Since things had worked out so well the last time, she said, "I already found the Rama Nujan, so now I'd like to find the Dormant Diamond." "You got it," replied Curray. And instantly twelve identical-looking stones and three balance scales appeared. Each scale was clearly labeled, "One Use Only." Alley looked at the stones and was unable to differentiate any one from the others. Curray explained, "The diamond is embedded in one of the stones. Eleven of the stones weigh the same, but the stone containing the jewel weighs either slightly more or slightly less than the others. I am not telling you which—you must find the right stone and tell me whether it is heavier or lighter."

Alley could use each of the three balance scales exactly once. She was able to select the stone containing the Dormant Diamond from among the twelve identical-looking stones and determine whether it was heavier or lighter than each of the eleven other stones. This puzzle is a challenge. Try to figure out how Alley might have accomplished this feat.

## HINTS AND COMMENTARY FOR THESE MINDSCAPES

In this section we present hints and identify some problem-solving techniques suggested by the preceding Mindscapes. Hints and solutions to selected Mindscapes in later chapters appear at the end of the book.

1. ***Late night cash.*** Try it! After you act it out, explain what happened.
2. ***Politicians on parade.*** What if more than one politician is honest? Read fact b carefully.

*Carefully understand and analyze the facts at hand.*

◆ ◆ ◆ ◆ ◆ ◆ ◆ ◆ ◆ ◆ ◆ ◆ ◆ ◆ ◆ ◆ ◆

3. ***The profit.*** Different people will get different answers, and each person will argue that his or hers is correct. Act out the transactions and see what happens. After you try this, go back and figure out why other answers are incorrect.

*Experimentation is an effective means to resolve difficult issues.*

◆ ◆ ◆ ◆ ◆ ◆ ◆ ◆ ◆ ◆ ◆ ◆ ◆ ◆ ◆ ◆ ◆

4. ***The truth about . . .*** What if all the animals were cats? How many extra biscuits would you have? Now try to turn some of those cats into dogs. This transformation leads to an algebra-free solution.

    Often a clever idea can be more potent than conventional wisdom.

    ◆ ◆ ◆ ◆ ◆ ◆ ◆ ◆ ◆ ◆ ◆ ◆ ◆ ◆ ◆ ◆ ◆ ◆ ◆

5. ***It's in the box.*** Consider the two possibilities carefully. You don't want to slip up on this one.

    Carefully consider the outcomes of various scenarios.

    ◆ ◆ ◆ ◆ ◆ ◆ ◆ ◆ ◆ ◆ ◆ ◆ ◆ ◆ ◆ ◆ ◆ ◆ ◆

6. ***Lights out.*** Suppose you turn on a switch, wait a half hour, and then turn the switch off. If you walk into the other room now, could you tell if the light had been on for half an hour? Think about this question, and use it to resolve the original issue.

    Don't overlook or dismiss facts that seem insignificant or irrelevant.

    ◆ ◆ ◆ ◆ ◆ ◆ ◆ ◆ ◆ ◆ ◆ ◆ ◆ ◆ ◆ ◆ ◆ ◆ ◆

7. ***Out of sight but not out of mind.*** Don't be fooled by all the numbers. Force yourself to figure out what was paid out and what was given back.

    Don't believe claims that are unsubstantiated, even if they sound scientific. Until you understand the issue for yourself, be skeptical!

    ◆ ◆ ◆ ◆ ◆ ◆ ◆ ◆ ◆ ◆ ◆ ◆ ◆ ◆ ◆ ◆ ◆ ◆ ◆

    If that doesn't help, get 30 one-dollar bills and act out the entire episode. Once you discover the truth, go back and find out where the problem is in the story.

    Experimentation is a powerful means for
    discovering patterns and developing insights.

    ◆ ◆ ◆ ◆ ◆ ◆ ◆ ◆ ◆ ◆ ◆ ◆ ◆ ◆ ◆ ◆ ◆ ◆ ◆

    See how many different ways you can devise to understand and explain what truly happened.

Once you find an argument that resolves an issue, it is a great challenge to find a different argument. However, often in attempting to find other arguments, we gain further insight into and understanding of the situation. Also, often the first argument we come up with is not the best one.

◆ ◆ ◆ ◆ ◆ ◆ ◆ ◆ ◆ ◆ ◆ ◆ ◆ ◆ ◆ ◆ ◆ ◆ ◆

8. ***The cannibals and the missionaries.*** Professor Starbird shares his grandmother's solution.

>When my grandmother was 92, I gave this problem to her along with three nickels to represent the cannibals and three Life Savers candies to represent the missionaries. We set up a line on her table to represent the river so that she could slide the cannibals and the missionaries (the nickels and the Life Savers) back and forth singly or in pairs, thereby solving the problem. When I arrived the next day for my visit, she was delighted to tell me that she had solved the problem.
>"How did you do it?" I asked.
>She replied triumphantly, "I ate the missionaries."

We give her half credit. A good aspect of her method was to model the problem using a concrete representation. Making a written table with two columns would also be a good way to represent the setting. One column would be one bank of the river, and the other column would be the other bank. Each row would represent the situation after a crossing. So the first row would have three C's, three M's, and a B for the boat in the left-hand box and nothing in the right. The next row might have two C's and two M's in the left box and one C, one M, and the B in the right box. Going from row to row must be obtainable by moving one or two C's or M's along with the B to the other column.

Once you have an effective representation of this question, a little experimentation will lead to an answer.

*A good representation of a problem is frequently the biggest step toward a solution.*

◆ ◆ ◆ ◆ ◆ ◆ ◆ ◆ ◆ ◆ ◆ ◆ ◆ ◆ ◆ ◆ ◆ ◆

9. ***Whom do you trust?*** This question asks you to find the person who leaked the story. But to do that you must determine who is telling the truth. Ask yourself if you can determine the truthful or deceitful character of any one person.

If Pocket is telling the truth, then Schlock and Wind are liars, and the remaining three—Pocket, Greede, and Slie—are telling the truth. Could those three all be telling the truth? If not, then you know for certain that Pocket is lying.

Since Slie contradicts Greede, you know that one of them is lying. Which one?

*A rock of certainty can be the foundation of a tower of truth.*

◆ ◆ ◆ ◆ ◆ ◆ ◆ ◆ ◆ ◆ ◆ ◆ ◆ ◆ ◆ ◆ ◆ ◆

10. ***A commuter fly.*** On close inspection, notice that the fly changes directions an infinite number of times during her travels. It is possible to compute how far the fly has flown before she encounters the freight train for the first time. It would then be possible to compute the distance traveled before the fly encountered the passenger train on the return trip. You could compute those distances and find a pattern and then solve the problem by adding up the infinite list of distances. However, there is a much easier way.

How long will the trains travel before they collide? How far will the fly fly in that length of time? Case closed.

This story is not complete without telling an anecdote about the famous mathematician John von Neumann. Von Neumann was notorious for being extremely fast and accurate at calculating numbers in his head—oddly enough, not a skill that all mathematicians possess. One day he was walking with a friend who asked him the question of the fly between the trains. Instantly, von Neumann stated the answer. The questioner said, "Oh, you saw the trick." To which von Neumann replied, "Yes, it was an easy infinite series."

If you are not von Neumann, the fly between the trains story provides a good life lesson.

Look at problems from different perspectives.

Go out of your way to think about how you could see a problem in a different light. In this case, if you know how long the fly flies, you can compute the distance the fly travels. You have now reduced the original problem to a related, but different problem. In this case, the different problem is much simpler than the original one.

Look at related situations.

11. ***A fair fare.*** This question does not have one definitely correct answer. However, looking at a related problem may persuade you that one possibility is best. What if, instead of staying in one taxi the whole time, the three travelers went the first 10 miles together and then all got out and paid the first cabby. The first traveler then left, and the remaining two got another cab, rode 10 miles, again got out and paid the second cabby. Then the last traveler took a cab alone for the remaining 10 miles. This situation is similar to the first, but the rephrasing makes a division of payment seem more obvious.

12. ***Getting a pole on a bus.*** It seems impossible to get the 5-foot pole on the bus, given that the largest length of an item allowed on the bus is 4 feet. Sarah gave Adam a large box to put the pole in.

Now give the dimensions of the box and explain why it does the trick.

*Often an inventive solution arises from looking at a situation in an unusual way.*

❖ ❖ ❖ ❖ ❖ ❖ ❖ ❖ ❖ ❖ ❖ ❖ ❖ ❖ ❖ ❖ ❖ ❖ ❖

13. **Tea time.** This question is mostly composed of unnecessary and distracting information. Obviously, the description of the dinnerware and the names of the people are not pertinent to the puzzle. But it may not be quite so obvious that the number of ounces in the teacup and the creamer and the amounts poured and spooned are also irrelevant.

Don't be distracted by extraneous information. Suppose the problem did not contain those facts at all, and instead were stated as follows.

A creamer and a teacup have exactly the same amount of cream and tea, respectively. An undisclosed amount of mixing of the cream and tea goes on, but after the mixing, each of the two containers still contains the same amount of liquid as the other. Is the tea more diluted than the cream, or is the cream more diluted than the tea?

You could still solve this puzzle, and having less information might force you to look at the situation differently. In fact, looking at it differently will probably also make it easier to solve and understand.

*Look at problems from different perspectives.*

❖ ❖ ❖ ❖ ❖ ❖ ❖ ❖ ❖ ❖ ❖ ❖ ❖ ❖ ❖ ❖ ❖ ❖ ❖

14. **A shaky story.** Exactly one person at the party said to Sam that he or she shook eight hands. Note the obvious fact that each person with whom that person shook hands must have shaken hands with at least one person. Now determine how many hands that person's spouse shook. See if this approach leads to any conclusions. If it still does not, consider an easier problem: Suppose that there were just three couples, or even two couples. Search for a pattern.

*If you have a hard problem, first work on a simpler, related problem to develop insight.*

❖ ❖ ❖ ❖ ❖ ❖ ❖ ❖ ❖ ❖ ❖ ❖ ❖ ❖ ❖ ❖ ❖ ❖ ❖

**15. *Murray's brother.*** This genie has posed a difficult challenge. It can be fun to work on, but do not work on it too long if you get frustrated. In this puzzle, we must squeeze every ounce (or even gram) of information from every weighing.

Don't ignore information.

❖ ❖ ❖ ❖ ❖ ❖ ❖ ❖ ❖ ❖ ❖ ❖ ❖ ❖ ❖ ❖ ❖ ❖ ❖

Each weighing must be designed to give us maximum information. After a weighing, we learn many things. Let's begin by putting four stones on each side of a scale and record what we observe. If the scale balances, we know that all eight stones weigh the same, and the diamond is not among those eight. So the mystery stone is among the remaining four, but we do not know whether it is heavier or lighter than the others. Can you now find the Dormant Diamond and determine whether it is heavy or light?

Suppose, however, that the four-against-four weighing does not balance. Then we know many things. We know that the unweighed four stones all weigh the same. We know that each of the four stones on the light side of the scale are potentially light, but none of them is potentially heavier than the eleven other stones. We know similar things about the four stones on the other side of the scale. We will have to keep track of the stones and consider putting potentially light stones with potentially heavy ones to help sort things out. For example, suppose we weigh a potentially light stone with a potentially heavy stone on one side of the scale and two stones that are known to be normal on the other side. Then, depending on which way the scale tips, we can conclude which of the two stones is the Dormant Diamond.

You might think about the last step to help you aim for an intermediate target. That is, you might specify what collections of stones and knowledge would allow you to find the diamond in one more weighing. For example, suppose you figure out that the diamond is among three stones that are potentially heavier than the others. Could you find the diamond in one more weighing? Or suppose you had narrowed the field to three stones, one potentially heavier than normal and two potentially lighter than normal. Could you find the diamond in one more weighing? This technique of working backward is often useful.

This balance-scale conundrum is tricky and difficult for anyone. Everyone, including experienced mathematicians, would have to think hard to solve it. It can be fun to work on if you enjoy this type of puzzle. Play with it; think carefully about what you know; carefully keep track of all the information you gather. But if you don't enjoy it, then just move on.

CHAPTER TWO

# Number Contemplation

*Arithmetic has a very great and elevating effect, compelling the soul to reason about abstract number...*

—Plato

**2.1** Counting

**2.2** Numerical Patterns in Nature

**2.3** Prime Cuts of Numbers

**2.4** Crazy Clocks and Checking Out Bars

Life is full of numbers. The moment we were born, our parents probably noted the time, our weight, our length, possibly our width, and most important, counted our toes. Numbers accompany us throughout life. We use numbers to measure our age, keep track of how much we owe on our charge cards, measure our wealth. (Notice how negative numbers may sneak in on that last one if we're in debt.) In fact, does any aspect of daily life not involve counting or measurement? Numbers are a part of human life.

In this chapter we explore the notion of number. Just as numbers play a fundamental role in our daily lives, they also play a fundamental role within the realm of mathematics. We will come to see the richness of numbers and delve into their surprising traits. Some collections of numbers fit so well together that they actually lead to notions of aesthetics and beauty, whereas other numbers are so important that they may be viewed as basic building blocks. Relationships among numbers turn out to have powerful implications in our modern world, for example, within the context of secret codes and the Internet. Exploring the numbers we know leads us to discover whole new worlds of numbers beyond our everyday understanding. Within this expanded universe of number, many simple questions are still unanswered—mystery remains.

One of the main goals of this book is to illustrate methods of investigating the unknown, regardless of where in life it occurs. In this chapter we highlight some guiding principles and strategies of inquiry by using them to develop ideas about numbers.

Remember that an intellectual journey does not start with clear definitions and a list of facts—it ends there. A journey of the mind requires us to stumble about, experiment, and search for patterns. Our goal, therefore, is to use our investigations into the world of number as an illustration of a powerful mode of thought and analysis, which has profound consequences in our daily lives.

❖ ❖ ❖ ❖ ❖ ❖ ❖ ❖ ❖ ❖ ❖

**2.5**

*Secret Codes and How to Become a Spy*

**2.6**

*The Irrational Side of Numbers*

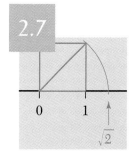

**2.7**

*Get Real*

# 2.1 Counting:

## How the Pigeonhole Principle Leads to Precision Through Estimation

*The simple modes of number are of all other the most distinct; even the least variation, which is a unit, making each combination as clearly different from that which approacheth nearest to it, as the most remote; two being as distinct from one, as two hundred; and the idea of two as distinct from the idea of three, as the magnitude of the whole earth is from that of a mite.*

—John Locke

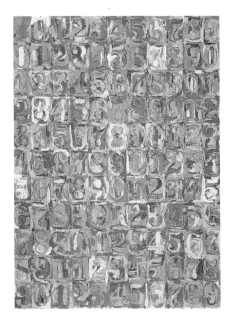

*Numbers in Color* (1958–1959) by Jasper Johns

We begin with the numbers we first learned as children: 1, 2, 3, 4, . . . (the ". . ." indicates that there are more, but we don't have enough room to list them). These numbers are so natural to us they are actually called *natural numbers*. These numbers are familiar, but often familiar ideas lead to surprising outcomes, as we will soon see.

The most basic use of numbers is counting, and we will begin by just counting approximately. That is, we'll consider the power and the limitations of making rough estimates. In a way, this is the weakest possible use we can make of numbers, and yet we will still find some interesting outcomes. So let's just have some fun with plain old counting.

## *Quantitative Estimation*

Make it quantitative.

One powerful technique for increasing our understanding of the world is to move from the qualitative to the quantitative whenever possible. Some people still count: "1, 2, 3, many." Counting in that fashion is effective for a simple existence but does not cut it in a world of trillion-dollar debts and gigabytes of hard-drive storage. In our modern world there are practical differences between thousands, millions, billions, and trillions. Some collections are easy to count exactly because there are so few things in them: the schools in the Big Ten Conference, the collection of letters you've written home in the past month, and the clean underwear in your dorm room. Other collections are more difficult to count exactly—such as the grains of sand in the Sahara Desert, the stars in the sky, and the hairs on your roommate's body. Let's look more closely at this last example.

It would be difficult, awkward, and frankly just plain weird to count the number of hairs on your roommate's body. Without undertaking that perverse task, we nevertheless pose a related question:

*Do there exist two nonbald people on the planet who have exactly the same number of hairs on their bodies?*

It appears that we cannot answer this question since we don't know (and don't intend to find out) the body-hair counts for anyone. But can we estimate body-hair counts well enough to get some idea of what that number might be? In particular, can we at least figure out a number that we could state with confidence is larger than the number of hairs on the body of any person on earth?

## *Big Hair, but How Big?*

Let's take the direct approach to this body-hair business. One of the authors counted the number of hairs on a 1/4" × 1/4" square area on his scalp and counted about 100 hairs—that's roughly 1,600 hairs per square inch. From this modest follicle count, we can confidently say that no person on earth has as many as 16,000 hairs in any square inch anywhere on his or her body. The author is about 72 inches tall and 32 inches around. If the author were a perfect cylinder, he would have 72 × 32 or about 2,300 square inches of skin on the sides and about another 200 square inches for the top of his head and soles of his feet, for a total of 2,500 square inches of skin. Since the author is not actually a perfect cylinder (he has, for example, a neck), 2,500 square inches is an overestimate of his skin area. There are people who are taller and bigger than this author, but certainly there is no one on this planet who has 10 times as much skin as this author. Therefore, no body on earth will have more than 25,000 square inches of skin. We already agreed that each square inch can have no more than 16,000 hairs on it. Thus we deduce that no person on this planet can have more than 400 million (400,000,000) hairs on his or her body.

## We Are the World

An almanac or a Web site tells us that there are about 5.8 billion (5,800,000,000) people on this planet. Given this information, can we answer our question: Do there exist two nonbald people on the planet who have *exactly the same* number of hairs on their bodies? We urge you to think about this question and try to answer it before reading on.

## The Reason That Many People Are Equally Hairy

There are more than 5 billion people on Earth, but each person has many fewer than 400 million hairs on his or her body. Could it be that no two people have the same number of body hairs? What would that mean? It would mean that each of the 5 billion people would have a different number of body hairs. But the number of body hairs is less than 400 million. So, there are less than 400 million different possible body-hair numbers. Therefore, we know that not all 5 billion people can have different body-hair counts.

*Looking at an issue from a new point of view often enables us to understand it more clearly.*

Suppose we have 400 million rooms—each numbered in order. Suppose each person did know his or her body-hair count, and we asked each person in the world to go into the room whose number is equal to his or her body-hair number. Could everyone go into a different room? Of course not! We have 5 billion people and only 400 million room choices—some room or rooms must have more than one person. In other words, there definitely exist two people, in fact many people, who have the same number of body hairs.

By using some simple estimates, we have been able to answer a question that first appeared unanswerable. The surprising twist is that in this case a rough estimate led to a conclusion about an exact equality. However, there are limitations to our analysis. For example, we are unable to name two *specific* people who have the same body-hair counts even though we know they are out there.

## The Power of Reasoning

In spite of the silliness of our hair-raising question we see the power of reasoned analysis. We were faced with a question that on first inspection appeared unanswerable, but through creative thought we were able to crack it. When we are first faced with a new question or problem, the ultimate path of logical reasoning is often hidden from sight. When we try, think, fail, think some more, and try some more, we finally discover a path.

We solved the hairy-body question, but that question in itself is not of great value. However, once we have succeeded in resolving an issue, it is worthwhile to isolate the approach we used, because the method of thought may turn out to be far more important than the problem it solved. In this case, the key to answering our question was the realization that there are more people on the planet than there are body hairs on any individual's body. This type of reasoning is known as the *Pigeonhole principle.* If we have an antique desk with slots for envelopes (known as *pigeonholes*), and we have more envelopes than slots, then certainly some slot must contain at least two envelopes. This Pigeonhole principle is a simple idea, but it is a useful tool for drawing conclusions when the size of a collection exceeds the number of possible variations of some distinguishing trait.

Once we have identified this principle, we see it in a large number of situations. For example, in a large swim meet, some pairs of swimmers will get exactly the same times to the tenths of a second. Some days more than 100 people will die in car wrecks. With each breath, we breathe an atom that Einstein breathed before us. Each person will arrive at work during the exact same minute many times during his or her life. Many trees have the same number of leaves. Many people get the same SAT score. Once we understand the Pigeonhole principle, we become conscious of something that has always been around us—we see it everywhere.

*Often after we learn a principle of logical reasoning, we see many instances where it applies.*

◆ ◆ ◆ ◆ ◆ ◆

## *Beyond Counting*

The natural numbers 1, 2, 3, . . . , besides being useful in counting, have captured the imagination of people around the world from different cultures and different eras. The study of natural numbers began several thousand years ago and continues to this day. Mathematicians who are intrigued by numbers come to know them individually. In the eye of the mathematician, individual numbers have their own personalities—unique characteristics and distinctions from other numbers. In subsequent sections of this chapter, we will discover some intriguing properties of numbers and uncover their nuances. For now, however, we wish to share a story that captures the human side of mathematicians. Of course, mathematicians, like people in other professions, display a large range of personalities, but this true story of Ramanujan and Hardy depicts almost a caricature of the "pure" mathematician. It illustrates part of the mythology of mathematics and provides insight into the personality of an extraordinary mathematician.

This interaction of two mathematicians on such an abstract plane even during serious illness is poignant. They clearly thought each number was worthy of special consideration. To affirm their special

Srinivasa Ramanujan

# RAMANUJAN AND HARDY

One of the most romantic tales in the history of the human exploration of numbers involves the life and work of the Indian mathematician Srinivasa Ramanujan. Practically isolated from the world of academics, libraries, and mathematicians, Ramanujan made amazing discoveries about natural numbers.

In 1913, Ramanujan wrote to the great English mathematician G. H. Hardy at Cambridge University describing his work. Hardy immediately recognized that Ramanujan was a unique jewel in the world of mathematics, because Ramanujan had not been taught the standard ways to think about numbers and thus was not biased by the rigid structure of a traditional education; yet he was clearly a mathematical genius. Since the pure nature of mathematics transcends languages, customs, and even formal training, Ramanujan's imaginative explorations have since given mathematicians everywhere an exciting and truly unique perspective on numbers.

Ramanujan loved numbers as his friends, and found each to be a distinct wonder. A famous illustration of Ramanujan's deep connection with numbers is the story of Hardy's visit to Ramanujan in a hospital. Hardy later recounted the incident: "I remember once going to see him when he was lying ill at Putney. I had ridden in taxi cab number 1729 and remarked that the number seemed to me rather a dull one and that I hoped it was not an unfavorable omen. 'No,' he replied, 'it is a very interesting number; it is the smallest number expressible as the sum of two cubes in two different ways.'" Notice that, indeed, $1729 = 12^3 + 1^3$, and also $1729 = 10^3 + 9^3$.

G.H. Hardy

regard for each number, we now demonstrate conclusively that every natural number is interesting by means of a whimsical, though ironclad, proof.

**THE INTRIGUE OF NUMBERS.** *Every natural number is interesting.*

## Proof

Let's first consider the number 1. Certainly 1 is interesting, because it is the first natural number and it is the only number with this property: if we pick any number and then multiply it by 1, the answer is the original number we picked. So, we agree that the first natural number is interesting.

Let us now consider the number 2. Well, 2 is the first even number, and that is certainly interesting—and, if that weren't enough, remember that 2 is the smallest number of people required to make a baby. Thus, we know that 2 is genuinely interesting.

We now consider the number 3. Is 3 interesting? Well, there are only two possibilities: Either 3 *is* interesting, or 3 *is not* interesting. Let us suppose that 3 *is not* interesting. Then notice that 3 has a spectacular property: it is the *smallest* natural number that is not interesting—which is certainly an interesting property! Thus we see that 3 is, after all, quite interesting.

Knowing now that 1, 2, and 3 are all interesting, we can make an analogous argument for 4 or any other number. In fact, suppose now that $k$ is a certain natural number with the property that the first $k$ natural numbers are all interesting. That is, $1, 2, 3, \ldots, k$ are all interesting. We know this fact is true if $k$ is 1, and, in fact, it is true for larger values of $k$ as well (2, 3, and 4, for example).

We now consider the very next natural number: $k + 1$. Is $k + 1$ interesting? Suppose it were not interesting. Then it would be the smallest natural number that is not interesting (all the smaller natural numbers would be known to be interesting). Well, that's certainly interesting! Thus, $k + 1$ must be interesting, too. Since we have shown that there can be no smallest uninteresting number, we must conclude that every natural number is interesting.

**A LOOK BACK**

Natural numbers are the natural place to begin our journey. This deceptively simple collection of numbers plays a significant role in our lives. We can understand our world more deeply by moving from qualitative to quantitative understanding. Counting and estimating, together with the Pigeonhole principle, lead to surprising insights about everyday events. Natural numbers help us understand our world, but they also constitute a world of their own. The whimsical assertion that every natural number is interesting foreshadows our quest to discover the variety and individual essence of these numbers.

Our strategy for understanding the richness of numbers was to start with the most basic and familiar use for numbers—counting. Looking carefully at the simple and the familiar is a powerful technique for creating and discovering new ideas.

*To speak algebraically, Mr. M. is execrable, but Mr. C. is $(x + 1)$-ecrable* —Edgar Allan Poe

Understand simple things deeply.

◆ ◆ ◆ ◆ ◆ ◆ ◆ ◆ ◆ ◆ ◆ ◆ ◆ ◆ ◆ ◆ ◆

# MINDSCAPES  Invitations to Further Thought*

### I. SOLIDIFYING IDEAS

1. *Treasure chest* **(H).** Someone offers to give you a million dollars in one-dollar bills. To receive the money, you must lie down; the million one-dollar bills will be placed on your stomach. If you keep them on your stomach for 10 minutes, the money is yours! Do you

---

*In the Mindscapes section, exercises marked (H) have hints for solutions at the back of the book. Exercises marked (S) have solutions.

accept the offer? Carefully explain your answer using quantitative reasoning.

2. ***Order please.*** Order the following numbers from smallest to largest: number of telephones on the planet; number of honest congressmen; number of people; number of grains of sand; number of states in the United States; number of cars.

3. ***Penny for your thoughts*** (H). Two thousand years ago, a noble Arabian king wished to reward his minister of science. Although the modest minister resisted any reward from the king, the king finally forced him to state a desired reward. Impishly the minister said that he would be content with the following token: "Let us take a checkerboard. On the first square I would be most grateful if you would place one piece of gold. Then on the next square twice as much as before, thus placing two pieces, and on each subsequent square, placing twice as many pieces of gold as in the previous square. I would be most content with all the gold that is on the board once your majesty has finished." This sounded extremely reasonable, and the king agreed. Given that there are 64 squares on a checkerboard, roughly how many pieces of gold did the king have to give to our "modest" minister of science? Why did the king have him executed?

4. ***29 is fine.*** Find the most interesting property you can, unrelated to size, that the number 29 has and that 27 does not have.

5. ***Perfect numbers.*** The only natural numbers that divide evenly into 6, other than 6 itself, are 1, 2, and 3. Notice that the sum of all those numbers equals the original number 6 (1 + 2 + 3 = 6). What is the next number that has the property of equaling the sum of all the natural numbers, other than itself, that divide evenly into it? Such numbers are called *perfect numbers*. No one knows whether or not there are infinitely many perfect numbers. In fact, no one knows whether there are *any* odd perfect numbers. These two unsolved mysteries are examples of long-standing open questions in the theory of numbers.

6. ***Many fold*** (S). Suppose you were able to take a large piece of paper of ordinary thickness and fold it in half 50 times. What is the height of the folded paper? Is it less than a foot? About one yard? As long as a street block? As tall as the Empire State Building? Taller than Mount Everest?

7. ***Only one cake.*** Suppose we had a room filled with 370 people. Must there be two people who celebrate their birthdays on the same day?

8. ***For the birds.*** Years ago, before overnight delivery services and e-mail, people would send messages by carrier pigeon and would keep an ample supply of pigeons in pigeonholes on their rooftops.

Suppose you have a certain number of pigeons, let's say *P* of them, but you have only *P* − 1 pigeonholes. If every pigeon must be kept in a hole, what can you conclude? How does the principle we discussed in this section relate to this question?

9. **Sock hop.** You have 10 pairs of socks, 5 black and 5 blue, but they are not paired up. Instead, they are all mixed up in a drawer. It's early in the morning, and you don't want to turn on the lights in your dark room. How many socks must you pull out to guarantee that you have a pair of one color? How many must you pull out to have two matched pairs (each pair is the same color)? How many must you pull out to be certain you have a pair of black socks?

10. **The last one.** Here is a game to be played with natural numbers. You start with any number. If the number is even, then you divide it by 2. If the number is odd, you triple it (multiply it by 3), and then add 1. Now you repeat the process with this new number. Keep going. You win (and stop) if you get to 1. Here is an example. If we start with 17, then we would have:

17, 52, 26, 13, 40, 20, 10, 5, 16, 8, 4, 2, 1 → we see a 1, so we win!

Play this game with the starting numbers 19, 11, 22, and 30. Do you think you will always win no matter what number you start with? No one knows the answer!

## II. CREATING NEW IDEAS

1. **See the three.** What proportion of the first 1,000 natural numbers have a 3 somewhere in them? For example, 135, 403, and 339 all contain a 3, whereas 402, 677, and 8 do not.

2. **See the three II (H).** What proportion of the first 10,000 natural numbers contain a 3?

3. **See the three III.** Explain why almost all million-digit numbers contain a 3.

4. **Commuting.** One hundred people in your neighborhood always drive to work between 7:30 and 8:00 A.M. and arrive 30 minutes later. Why must two people always arrive at work at the same time, within a minute?

5. **RIP (S).** The Earth has 5.8 billion people and almost no one lives 100 years. Suppose this longevity fact remains true. How do you know that some year soon, more than 50 million people will die?

## III. FURTHER CHALLENGES

1. ***Say the sequence.*** The following are the first few terms in a sequence. Can you figure out the next few terms and describe how to find all the terms in the sequence?

          1
         11
         21
        1211
       111221
      312211
       . . .

2. ***Lemonade.*** You want to buy a new car, and you know the model you want. The model has three options, each one of which you can either take or not take, and you have a choice of four colors. So far 100,000 cars of this model have been sold. How many of these cars must have the identical color and the same options?

## IV. IN YOUR OWN WORDS

1. ***With a group of folks.*** In a small group, discuss and work through the reasoning for why there are two people on Earth having the same number of hairs on their bodies. After your discussion, write a brief narrative describing your method.

## 2.2 Numerical Patterns in Nature:
### Discovering Nature's Beauty and the Fibonacci Numbers

*There is no inquiry which is not finally reducible to a question of Numbers; for there is none which may not be conceived of as consisting in the determination of quantities by each other, according to certain relations.*

—Auguste Comte

Often when we see beauty in nature, we are subconsciously sensing hidden order—order that itself has an independent richness. Thus we stop and smell the roses—or, more accurately, count the daisies. In the previous section, we contented ourselves with estimation, whereas, here we move to exact counting. The example of counting daisies is an illustration of discovering numerical patterns in nature through direct observation. The pattern we find in the daisy appears elsewhere in nature and also gives rise to issues of aesthetics that touch such diverse fields as architecture and painting. We begin our investigation, however, firmly rooted in nature.

Look for patterns.

❖ ❖ ❖ ❖ ❖

Have you ever examined a daisy? Sure, you've picked off the white petals one at a time while thinking: "Loves me . . . loves me not," but have you ever taken a good hard look at what's left once you've finished plucking? A close inspection of the yellow in the middle of the daisy reveals unexpected structure and

intrigue. Specifically, the yellow area contains clusters of spirals coiling out from the center. If we examine the flower closely, we see that there are, in fact, two sets of spirals—a clockwise set and a counterclockwise set. These two sets of spirals interlock to produce a hypnotic interplay of helical form.

Interlocking spirals abound in nature. The cone flower and the sunflower both display nature's signature of dual, locking spirals. Flowers are not the only place in nature where spirals occur. A pinecone's exterior is composed of two sets of interlocking spirals. The rough and prickly facade of a pineapple also contains two collections of spirals.

## Be Specific

In our observations we should not be content with general impressions. Instead, we move toward the specific. In this case we ponder the quantitative quandary: How many spirals are there? An approximate count is: lots. Is the number of clockwise spirals the same as the number of counterclockwise spirals? You can physically verify that the pinecone has 5 spirals in one direction and 8 in the other. The pineapple has 8 and 13. The daisy and cone flower each has 21 and 34. The sunflower has a staggering 55 and 89. In each case, we observe that the number of spirals in one direction is nearly twice as great as the number of spirals in the opposite direction. Listing all those numbers in order we see

5, 8, 13, 21, 34, 55, 89.

Is there any pattern or structure to these numbers?

Suppose we were given just the first two numbers, 5 and 8, on that list of spiral counts. How could we use these two numbers to build the next number? How can we always generate the next number on our list?

We note that 13 is simply 5 plus 8, whereas 21, in turn, is 8 plus 13. Notice that this pattern continues. What number would come after 89? Given this pattern, what number should come before 5? How about before that? How about before that? And before that?

## Leonardo's Sequence

The rule for generating successive numbers in the sequence is to add up the previous two terms. So the next number on the list would be $55 + 89 = 144$.

Through spiral counts, nature appears to be generating a sequence of numbers with a definite pattern that begins

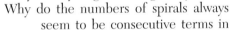

1   1   2   3   5   8   13   21   34   55   89   144 . . . .

This sequence is called the *Fibonacci sequence,* named after the mathematician Leonardo of Pisa (better known as Fibonacci—a shortened form of Filius Bonacci, *son of Bonacci*), who studied it in the 13th century. After seeing this surprising pattern, we hope you feel compelled to count for yourself the spirals in the previous pictures of flowers. In fact, you may now be compelled to count the spirals on a pineapple every time you go to the grocery store.

Why do the numbers of spirals always seem to be consecutive terms in this list of numbers? The answer involves issues of growth and packing. The yellow florets in the daisy begin as small buds in the center of the plant. As the plant grows, the young buds move away from the center toward a location where they have the most room to grow—that is, in the direction that is least populated by older buds. If one simulates this tendency of the buds to find the largest open area as a model of growth on a computer, then the spiral counts in the geometrical pattern so constructed will appear in our list of numbers. The Fibonacci numbers are an illustration of surprising and beautiful patterns in nature. The fact that nature and number patterns reflect each other is indeed a fascinating concept.

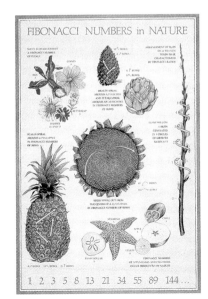

A powerful method for finding new patterns in nature is to take the abstract patterns that we directly observe and look at them by themselves. That is, let's move beyond the vegetable origins of the Fibonacci numbers and just think about the Fibonacci sequence as an interesting entity in its own right. We conduct this investigation with the hope and expectation that interesting relationships that may arise among Fibonacci numbers may also be represented in our lives.

## *Fibonacci Neighbors*

We observed that flowers, pinecones, and pineapples all display consecutive pairs of Fibonacci numbers. These observations point to some natural bond between adjacent Fibonacci numbers. In each case, the number of spirals in one direction was not quite twice as great as the number of spirals in the other direction. Perhaps we can find richer structure and develop a deeper understanding of the Fibonacci numbers by moving from an estimate ("not quite twice") to a precise value. So, let's measure the relative size of each Fibonacci number in comparison to the next one. We measure the relative size of one number in comparison to another by considering their ratio—that is, by dividing one of the numbers into the other. Here we list the quotients of adjacent Fibonacci numbers. Use a calculator to compute the last three terms in this chart:

> Unexpected patterns are often a sign of hidden, underlying structure.
> ◆ ◆ ◆ ◆ ◆ ◆

| Fraction of adjacent Fibonacci numbers | | Decimal Equivalent |
|---|---|---|
| $\frac{1}{1}$ | = | 1.0 |
| $\frac{2}{1}$ | = | 2.0 |
| $\frac{3}{2}$ | = | 1.5 |
| $\frac{5}{3}$ | = | 1.666 . . . |
| $\frac{8}{5}$ | = | 1.6 |
| $\frac{13}{8}$ | = | 1.625 |
| $\frac{21}{13}$ | = | 1.6153 . . . |
| $\frac{34}{21}$ | = | 1.6190 . . . |
| $\frac{55}{34}$ | = | 1.6176 . . . |
| $\frac{89}{55}$ | = | _____ |
| $\frac{144}{89}$ | = | _____ |
| $\frac{233}{144}$ | = | _____ |

What do we notice about these answers? In the display to the left, notice that the pairs of Fibonacci numbers are getting larger and larger in size. But what about their *relative* sizes?

## Converging Quotients

The relative sizes—that is, the quotients of consecutive Fibonacci numbers in the right-hand column—seem to oscillate. They get bigger, then smaller, then bigger, then smaller, but they are apparently becoming increasingly close to one another and are converging toward some intermediate value. What is the exact value for the target number toward which these ratios are heading?

To find it, let's look again at those quotients of Fibonacci numbers, but this time let's write those fractions in a different way. Looking at the same information from a different vantage point often leads to insight. If we're careful with the arithmetic and remember the rule for building the Fibonacci numbers, we will uncover an unusual pattern of 1's. Notice how we use the pattern of 1's from one quotient to produce a pattern of 1's for the next quotient. Each step below involves the facts that $a/b = 1/(b/a)$ and that each Fibonacci number can be written as the sum of the previous two.

$$\frac{1}{1} = 1$$

$$\frac{2}{1} = \frac{1+1}{1} = 1 + \frac{1}{1}$$

$$\frac{3}{2} = \frac{2+1}{2} = \frac{2}{2} + \frac{1}{2} = 1 + \frac{1}{\frac{2}{1}} = 1 + \frac{1}{1 + \frac{1}{1}}$$

$$\frac{5}{3} = \frac{3+2}{3} = \frac{3}{3} + \frac{2}{3} = 1 + \frac{1}{\frac{3}{2}} = 1 + \frac{1}{1 + \frac{1}{1 + \frac{1}{1}}}$$

$$\frac{8}{5} = \frac{5+3}{5} = \frac{5}{5} + \frac{3}{5} = 1 + \frac{1}{\frac{5}{3}} = 1 + \frac{1}{1 + \frac{1}{1 + \frac{1}{1 + \frac{1}{1}}}}$$

Let's look at what we're doing. Replacing the Fibonacci number in the numerator of our fraction by the sum of the previous two Fibonacci numbers allows us to see a pattern. For example,

$$\frac{233}{144} = \frac{144 + 89}{144} = 1 + \frac{89}{144} = 1 + \frac{1}{\frac{144}{89}}.$$

Now notice that 144/89 would be the previous fraction on our list. So 144/89 would have already been written as a long fraction of 1's.

If we continue this process we see that the ratio of any two adjacent Fibonacci numbers is a number that looks like this:

$$1 + \cfrac{1}{1 + \cfrac{1}{1 + \cfrac{1}{1 + \cfrac{\cdot}{\cdot \cdot} + 1}}}$$

## Unending 1's

As we compute the quotient of ever larger Fibonacci numbers, the ratios head toward the strange expression: $1 + 1/(1 + 1/(1 + \ldots ))$ in which we mean by "$\ldots$" that this fraction never ends, the quotient goes on forever. Only in mathematics can we create something that is truly unending. Let's give this unending number a name. We call this number $\varphi$. $\varphi$ is the lowercase Greek letter *phi*, and in our journey through geometry a few chapters from now, we'll find out why it is called $\varphi$—stay tuned. So we see that

$$\varphi = 1 + \cfrac{1}{1 + \cfrac{1}{1 + \cfrac{1}{1 + \cdot \cdot \cdot}}}$$

where "$\ldots$" means this process goes on forever. Remember, our goal is to figure out what number the quotients of consecutive Fibonacci numbers approach, and we now see that the number we approach is $\varphi$. But what exactly does $\varphi$ equal? Since it's described in such a strange form—as an infinitely long fraction—it seems impossible to know the precise value of $\varphi$, or is it possible? Right now, the answer is not clear. So let's look for some pattern within that exotic expression for $\varphi$.

Before attempting to answer the preceding question, we first ask a warm-up question: We are going to write $\varphi$ out again; however, this time, notice that we placed a frame around part of $\varphi$. Here is our question: What does the number inside the frame equal?

*Often new perspectives reveal new insights.*

$$\varphi = 1 + \cfrac{1}{\boxed{1 + \cfrac{1}{1 + \cfrac{1}{1 + \cdot \cdot \cdot}}}}$$

The answer is: The number in the frame is $\varphi$ again. Why? Well, suppose we were just shown the number inside the frame without any of that other stuff around it. We'd look at that new number and realize that the

1's go on forever, and thus that number is just φ. Stay with this picture until you see the idea behind it. Therefore, we just discovered that

$$\varphi = 1 + \frac{1}{\varphi}.$$

## Solving for φ

Now we have an equation involving just φ, and this will allow us to solve for the exact value of φ. First, we can subtract 1 from both sides to get

$$\varphi - 1 = \frac{1}{\varphi}.$$

Multiplying through by φ we get

$$\varphi^2 - \varphi = 1$$

or just

$$\varphi^2 - \varphi - 1 = 0.$$

This quadratic equation can be solved using the quadratic formula, which states that

$$\varphi = \frac{1 \pm \sqrt{5}}{2}.$$

But since φ is bigger than 1, we must have

$$\varphi = \frac{1 + \sqrt{5}}{2}.$$

Using a calculator, express $(1 + \sqrt{5})/2$ as a decimal and compare it with the data from our previous calculator experimentation on the quotients of consecutive Fibonacci numbers. Well, there we have it—our goal was to find the exact value of φ, and through a process of observation and thought we succeeded.

## The Golden Ratio

At the moment, we have no reason to consider the number φ to be especially interesting; however, it is somewhat curious that the quotients of consecutive Fibonacci numbers do seem to approach this fixed value. We started with simple observations of flowers and pinecones. We saw a numerical pattern among our observations. The pattern led us to the number $(1 + \sqrt{5})/2$.

The number $\varphi = (1 + \sqrt{5})/2$ is called the *Golden Ratio* and, besides its connection with nature's spirals, it captures the proportions of some especially pleasing shapes in art, architecture, and geometry. Just to foreshadow what is to come when we revisit the Golden Ratio in the geometry chapter, here is a question: What are the proportions of the most attractive rectangle? In other words, when someone says "rectangle" to you, and you think of a shape, what is it? Light some scented candles, put on some Sinatra CDs, close your eyes, and dream about the most attractive and pleasing rectangle you can imagine. Once that image is etched in your mind, open your eyes, put out the candles, and pick from the four choices below the rectangle that you think is most representative of that magical rectangle dancing in your mind.

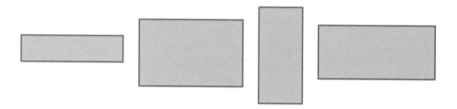

Many people think that the second rectangle from the left is the most aesthetically pleasing—the one that captures the notion of "rectangleness." That rectangle is called the *Golden Rectangle,* and we will examine it in detail in Chapter 4. The ratio of the dimensions of the sides of the Golden Rectangle is a number rich with intrigue. If we divide the length of the longer side by the length of the shorter side, we get $\varphi$: the Golden Ratio. A 3" × 5" index card is close to being a Golden Rectangle. Notice that its dimensions, 3 and 5, are consecutive Fibonacci numbers. In the geometry chapter we will consider the aesthetic issues involving $\varphi$ and make some interesting connections between the Fibonacci numbers and the Golden Rectangle in art.

## *Failure to Be Fibonacci*

After finding Fibonacci numbers hidden in the spirals of nature, it saddens us to realize that not all numbers are Fibonacci. We do not wish to end on a sad note, so we are delighted to announce that, although not all numbers are Fibonacci, every natural number is a neat sum of Fibonacci numbers. In particular, *every* natural number is either a Fibonacci number, or it is expressible uniquely as a sum of Fibonacci numbers whereby no two are adjacent Fibonacci numbers. Here is one way to do it:

1. Write down a natural number.
2. Find the largest Fibonacci number that does not exceed your number. That Fibonacci number is the first term in your sum.
3. Now subtract that Fibonacci number from your number and look at this new number.

4. Next find the largest Fibonacci number that does not exceed this new number. That Fibonacci number is the second number in your sum.

5. Continue this process.

For example, consider the number 38. The largest Fibonacci number not exceeding 38 is 34. So consider $38 - 34 = 4$. The largest Fibonacci number not exceeding 4 is 3, and $4 - 3 = 1$, which is a Fibonacci number. Therefore, $38 = 34 + 3 + 1$. Similarly, we can build any natural number by just adding Fibonacci numbers in this manner. In one sense, Fibonacci numbers are building blocks for the natural numbers through addition.

## *Fun with Fibonacci*

Fibonacci numbers not only appear in nature. If we wager on a sure thing, Fibonacci numbers can also be used to accumulate wealth. (Moral: Math pays.) The game is called *Fibonacci nim*, and it is played with two people: Person One and Person Two. All that is needed is a pile of sticks (toothpicks, straws, or even pennies will do). Person One moves first by taking any number of sticks (at least one but not all) away from the pile. After Person One moves it is Person Two's move, and the moves continue to alternate between them. Each person (after the first move) may take away as many sticks as he or she wishes with the restrictions that he or she must take at least one stick but no more than two times the number of sticks the previous person took. The player who takes the last stick wins the game.

## *A Sample Round of Play*

Suppose we start with 10 sticks, and Person One removes 3 sticks thus leaving 7. Now Person Two may take any number of the remaining sticks from 1 to 6 (6 is two times the number Person One took). Suppose Person Two removes 5, leaving 2 in the pile. Now Person One may take any number of sticks from 1 to 10 ($10 = 2 \times 5$). Because there are only 2 sticks left, Person One takes the 2 sticks and wins. Play Fibonacci nim with various friends and with different numbers of starting sticks. Get a feel for the game and its rules—but don't wager quite yet.

If we are careful and use the Fibonacci numbers, we can always win. Here is how. First, we make sure that the initial number of sticks we start with is **not** a Fibonacci number. Now we must be Person One, and we find some poor soul to be Person Two. If we play it just right, we will always win. The secret is to write the number of sticks in the pile as a sum of nonconsecutive Fibonacci numbers. Figure out the *smallest* Fibonacci number occurring in the sum, and remove that many sticks from the pile. Now it is your luckless opponent's turn. No matter what he or she does, we will repeat the preceding procedure. That is, once he or she is done, we count the number of sticks in the pile, express the number as a sum of nonconsecutive Fibonacci numbers, and then remove the number of sticks that equals the smallest Fibonacci number in the sum. It is a fact that, no matter what our poor opponent does, we will always be able to remove that number of sticks without breaking the rules. Experiment with this game and try it. Wager at will—or not.

## A LOOK BACK

The Fibonacci numbers are defined successively by starting with 1, 1, and then adding the previous two terms to get the next term. These numbers are rich with structure and appear in nature. The numbers of clockwise and counterclockwise spirals of flowers and other plants are equal to adjacent Fibonacci numbers. The ratio of adjacent Fibonacci numbers approaches the Golden Ratio, which is involved in the search for the most aesthetically pleasing rectangle. While not all numbers are Fibonacci, every natural number can be expressed as the sum of distinct, nonadjacent Fibonacci numbers.

The story of Fibonacci numbers is a story of pattern. As we look at the world, we can often see order, structure, and pattern. The order that we see provides a mental concept that we can then explore on its own. As we discover relationships in the pattern, we frequently find that those same relationships refer back to the world in some intriguing way.

*Understand simple things deeply.*

◆ ◆ ◆ ◆ ◆ ◆ ◆ ◆ ◆ ◆ ◆ ◆ ◆ ◆ ◆ ◆ ◆

## MINDSCAPES  Invitations to Further Thought*

### I. SOLIDIFYING IDEAS

1. **Baby bunnies.** This question gave the Fibonacci sequence its name. It was posed and answered by Leonardo of Pisa, better known as Fibonacci.

    Suppose we have a pair of baby rabbits; one male and one female. Let us also assume that rabbits cannot reproduce until they are 1 month old and that they have a 1-month gestation period. Once they start reproducing, they produce a pair of bunnies each month (1 of each sex). Assuming that no pair ever dies, how many pairs of rabbits will exist in a particular month?

    During the first month, the bunnies grow into rabbits. After 2 months, they are the proud parents of a pair of bunnies. There will

---

*In the Mindscapes section, exercises marked (H) have hints for solutions at the back of the book. Exercises marked (S) have solutions.

now be two pairs of rabbits: the original, mature pair and a new pair of bunnies. The next month, the original pair produces another pair of bunnies, but the new pair of bunnies is unable to reproduce until the following month. Thus we have:

| Time in Months | Number of Pairs |
|---|---|
| start | 1 |
| 1 | 1 |
| 2 | 2 |
| 3 | ___ |
| 4 | ___ |
| 5 | ___ |
| 6 | ___ |
| 7 | ___ |

Continue to fill in this chart and search for a pattern. Here is a suggestion: Draw a family tree to keep track of all the offspring.

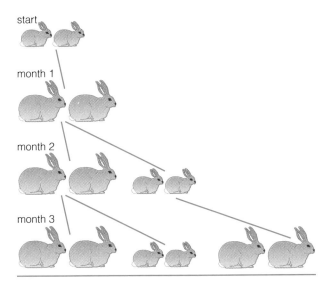

We'll use the symbol $F_1$ to stand for the first Fibonacci number, $F_2$ for the second Fibonacci number, $F_3$ for the third Fibonacci number, and so forth. So $F_1 = 1$, and $F_2 = 1$, and, therefore, $F_3 = F_2 + F_1 = 2$, and $F_4 = F_3 + F_2 = 3$, and so on. In other words, we write $F_n$ for the $n$th Fibonacci number where $n$ represents any natural number. So, the rule for generating the next

Fibonacci number by adding up the previous two can now be stated symbolically as:

$$F_n = F_{n-1} + F_{n-2}.$$

2. **Discovering Fibonacci relationships (S).** By experimenting with numerous examples in search of a pattern, determine a simple formula for $(F_{n+1})^2 + (F_n)^2$—that is, a formula for the sum of the squares of two consecutive Fibonacci numbers.

3. **Discovering more Fibonacci relationships.** By experimenting with numerous examples in search of a pattern, determine a simple formula for $(F_{n+1})^2 - (F_{n-1})^2$—that is, a formula for the difference of the squares of two Fibonacci numbers.

4. **Late bloomers (H).** Suppose we start with one pair of baby rabbits, and again they create a new pair every month, but this time let's suppose that it takes 2 months before a pair of bunnies is mature enough to reproduce. Make a table for the first 10 months indicating how many pairs there would be at the end of each month. Do you see a pattern? Describe a general formula for generating the sequence of rabbit-pair counts.

5. **A new start.** Suppose we build a sequence of numbers using the method of adding the previous two numbers to build the next one. This time, however, suppose our first two numbers are 2 and 1. Generate the first 15 terms. This sequence is called the *Lucas sequence* and is written as $L_1, L_2, L_3, \ldots$. Compute the quotients of consecutive terms of the Lucas sequence as we did with the Fibonacci numbers. What number do these quotients approach? What role do the initial values play in determining what number the quotients approach? Try two other first terms and generate a sequence. What do the quotients approach?

6. **Discovering Lucas relationships.** By experimenting with numerous examples in search of a pattern, determine a formula for $L_{n-1} + L_{n+1}$—that is, a formula for the sum of a Lucas number and the Lucas number that comes after the next one. (*Hint:* The answer will not be a Lucas number. See question 4.)

7. **Still more Fibonacci relationships.** By experimenting with numerous examples in search of a pattern, determine a formula for $F_{n-1} + F_{n+1}$— that is, a formula for the sum of a Fibonacci number and the Fibonacci number that comes after the next one. (*Hint:* The answer will not be a Fibonacci number. Try question 5 first.)

8. **Even more Fibonacci relationships.** By experimenting with numerous examples in search of a pattern, determine a formula for $F_{n+2} - F_{n-2}$—that is, a formula for the difference between a Fibonacci number and the fourth Fibonacci number before it. (*Hint:* The answer will not be a Fibonacci number. Try question 5 first.)

9. ***Discovering Fibonacci and Lucas relationships.*** By experimenting with numerous examples in search of a pattern, determine a formula for $F_n + L_n$—that is, a formula for the sum of a Fibonacci number and the corresponding Lucas number.

10. ***The enlarging area paradox* (S).** The square below has sides equal to 8 (a Fibonacci number) and thus has area equal to $8 \times 8 = 64$. The square can be cut up into four pieces whereby the short sides have lengths 3 and 5, as illustrated. You can find these pieces in your kit. Now use those pieces to construct a rectangle having base 13 and height 5. The area of this rectangle is $13 \times 5 = 65$. Therefore, by moving around the four pieces, we were able to increase the area by 1 unit! Using the puzzle pieces from your kit, do the experiment and then explain this impossible feat. Show this puzzle to your friends and record their reactions.

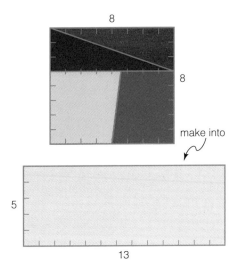

11. ***Sum of Fibonacci* (H).** For each natural number given, express it as a sum of distinct, nonconsecutive Fibonacci numbers: 52, 143, 13, 88.

12. ***Some more sums.*** For each natural number given, express it as a sum of distinct, nonconsecutive Fibonacci numbers: 43, 90, 2000, 609.

13. ***Fibonacci nim: The first move.*** Suppose you are about to begin a game of Fibonacci nim. You start with 50 sticks. What is your first move?

14. ***Fibonacci nim: The first move II.*** Suppose you are about to begin a game of Fibonacci nim. You start with 100 sticks. What is your first move?

15. ***Fibonacci nim: The first move III.*** Suppose you are about to begin a game of Fibonacci nim. You start with 500 sticks. What is your first move?

16. ***Fibonacci nim: The next move.*** Suppose you are playing a round of Fibonacci nim with a friend. You start with 15 sticks. You start by removing 2 sticks; your friend then takes 4. How many sticks should you take next to win?

17. ***Fibonacci nim: The next move II.*** Suppose you are playing a round of Fibonacci nim with a friend. You start with 50 sticks. You start by removing 3 sticks; your friend then takes 5; you then take 8; your friend then takes 10. How many sticks should you take next to win?

18. ***Fibonacci nim: The next move III.*** Suppose you are playing a round of Fibonacci nim with a friend. You start with 90 sticks. You start by removing 1 stick; your friend then takes 2; you take 3; your friend takes 6; you take 2; your friend takes 1; you take 2; your friend takes 4; you take 1, and then your friend takes 2. How many sticks should you take next to win?

19. ***Beat your friend.*** Play Fibonacci nim with a friend: Explain the rules, but do not reveal the secret to winning. Use 20 sticks to start. Play carefully, and beat your friend. Play again with another (non-Fibonacci) number of sticks to start. Finally, reveal the secret. Record the number of sticks removed at each stage of each game and your friend's reaction to the game, as well as the secret of winning.

20. ***Beat another friend.*** Play Fibonacci nim with another friend: Explain the rules, but do not reveal the secret to winning. Use 30 sticks to start. Play carefully and beat your friend. Play again with another (non-Fibonacci) number of sticks to start. Finally, reveal the secret. Record the number of sticks removed at each stage of each game and your friend's reaction to the game, as well as the secret of winning.

## II. CREATING NEW IDEAS

1. ***Discovering still more Fibonacci relationships.*** By experimenting with numerous examples in search of a pattern, determine a formula for $F_{n+1} \times F_{n-1} - (F_n)^2$—that is, a formula for the product of a Fibonacci number and the Fibonacci number that comes after the next one, minus the square of the Fibonacci number in between them. (*Hint:* The answer will be different depending on whether $n$ is even or odd. Consider examples of different cases separately.)

2. ***Finding factors*** (S). By experimenting with numerous examples, find a way to factor $F_{2n}$ into the product of two natural numbers that are from famous sequences. That is, consider every other Fibonacci number starting with the second 1, and factor each in an interesting way. Discover a pattern. (*Hint:* Question 5 from the previous collection of questions may be relevant.)

3. ***The rabbits rest.*** Suppose we have a pair of baby rabbits; one male and one female. As before, the rabbits cannot reproduce until they are 1 month old. Once they start reproducing, they produce a pair of bunnies (one of each sex) each month. Now, however, let us assume that each pair dies after 3 months, immediately after giving birth. Create a chart showing how many pairs we have after each month from the start through month nine.

4. ***Digging up Fibonacci roots.*** Using the square root key on a calculator, evaluate each number in the left column and record the answer in the right column.

| Number | Computed Value |
|---|---|
| $\sqrt{(F_3/F_1)} =$ | _____ |
| $\sqrt{(F_4/F_2)} =$ | _____ |
| $\sqrt{(F_5/F_3)} =$ | _____ |
| $\sqrt{(F_6/F_4)} =$ | _____ |
| $\sqrt{(F_7/F_5)} =$ | _____ |
| $\sqrt{(F_8/F_6)} =$ | _____ |
| $\sqrt{(F_9/F_7)} =$ | _____ |

Looking at the chart, make a guess as to what special number $\sqrt{F_{n+2}/F_n}$ approaches as $n$ gets larger and larger.

5. ***Tribonacci.*** Let's start with the numbers 0, 0, 1, and generate future numbers in our sequence by adding up the previous three numbers. Write out the first 15 terms in this sequence, starting with the first 1. Use a calculator to evaluate the value of the quotients of consecutive terms (dividing the smaller term into the larger one). Do the quotients seem to be approaching a fixed number?

6. ***Fibonacci follies.*** Suppose you are playing a round of Fibonacci nim with a friend. You start with 15 sticks. You start by removing 2 sticks; your friend then takes 1; you take 2; your friend takes 1.

What should your next move be? Can you make it without breaking the rules of the game? Did you make a mistake at some point? If so, where?

7. ***Fibonacci follies II.*** Suppose you are playing a round of Fibonacci nim with a friend. You start with 35 sticks. You start by removing 1 stick; your friend then takes 2; you take 3; your friend takes 6; you take 3, your friend takes 2. What should your next move be? Can you make it without breaking the rules of the game? Did you make a mistake at some point? If so, where?

8. ***Fibonacci follies III.*** Suppose you are playing a round of Fibonacci nim with a friend. You start with 21 sticks. You start by removing 1 stick; your friend then takes 2. What should your next move be? Can you make it without breaking the rules of the game? What went wrong?

9. ***A big fib.*** Suppose we have a natural number that is not a Fibonacci number—let's call it $N$. Suppose that $F$ is the largest Fibonacci number that does not exceed $N$. Show that the number $N - F$ must be smaller than the Fibonacci number that comes right before $F$.

10. ***Decomposing naturals*** (**H**). Use the result of the previous exercise, together with the notion of systematically reducing a problem to a smaller problem, to show that every natural number can be expressed as a sum of distinct, nonconsecutive Fibonacci numbers.

### III. FURTHER CHALLENGES

1. ***How big is it?*** Is it possible for a Fibonacci number greater than 2 to be exactly twice as big as the Fibonacci number immediately before it? Explain why or why not. What would your answer be if we removed the phrase "greater than 2"?

2. ***Too small.*** Suppose we have a natural number that is not a Fibonacci number—let's call it $N$. Let's write $F$ for the largest Fibonacci number that does not exceed $N$. Show that it is impossible to have a sum of two distinct Fibonacci numbers each less than $F$ add up to $N$.

3. ***Beyond Fibonacci.*** Suppose we create a new sequence of natural numbers starting with 0 and 1. Only this time, instead of adding the two previous terms to get the next one, let's generate the next term by adding 2 *times* the previous term to the term before it. In other words: $F_{n+1} = 2F_n + F_{n-1}$. Such a sequence is called a *generalized Fibonacci sequence*. Write out the first 15 terms in this generalized Fibonacci sequence. Adapt the methods used in this section to figure out that the quotient of consecutive Fibonacci numbers approaches $(1+\sqrt{5})/2$ to discover the exact number that $F_{n+1}/F_n$ approaches as $n$ gets large.

4. **Generalized sums.** Let $F_n$ be the generalized Fibonacci sequence defined in the previous exercise. Can every natural number be expressed as the sum of distinct, nonconsecutive generalized Fibonacci numbers? Show why, or give several counterexamples. What if we are allowed to use consecutive generalized Fibonacci numbers? Do you think you can do it then? Illustrate your hunch with four or five specific examples.

5. **It's hip to be square (H).** Adapt the methods of this section to prove that the numbers $\sqrt{F_{n+2}/F_n}$ approach $\varphi$ as $n$ gets larger and larger. (Here, $F_n$ stands for the usual Fibonacci number. See question III.4.)

## IV. IN YOUR OWN WORDS

1. **Personal perspectives.** Write a short essay describing the most interesting or surprising discovery you made in exploring this section's material. If any material seemed puzzling or even unbelievable, address that as well. Explain why you chose the topics you did. Finally, comment on the aesthetics of the mathematics and ideas in this section.

2. **With a group of folks.** In a small group, discuss and work through the reasoning for how the quotients of consecutive Fibonacci numbers approach the Golden Ratio. After your discussion, write a brief narrative describing the rationale in your own words.

3. **Creative writing.** Write an imaginative story (it can be comical, dramatic, whatever you like) that involves or evokes the ideas of this section.

4. **Power beyond the mathematics.** Describe several real-life situations—ideally, from your own experience—for which some of the strategies of thought presented in this section would provide effective methods for approaching and resolving them.

# 2.3 Prime Cuts of Numbers:

## How the Prime Numbers Are the Building Blocks of All Natural Numbers

*... number is merely the product of our mind.*

—Karl Friedrich Gauss

The natural numbers, 1, 2, 3, . . . , help us describe and understand our world. However, they in turn form their own invisible world filled with relationships among themselves. Some basic connections can be distilled from simple addition and multiplication. These basic operations lead to subtle insights about our familiar numbers.

Our strategy for uncovering the structure of the natural numbers is to break down complex objects into simple building blocks. This basic yet powerful technique recurs frequently throughout this book and throughout our lives. As we become accustomed to this idea, we will see that complicated situations are frequently best analyzed by first understanding each fundamental component and then understanding the methods by which the fundamental pieces are combined to create a more complex whole. The natural numbers are a good model for observing this principle in action.

*I Saw the Figure 5 in Gold* (1928) by Charles Demuth

### Building Blocks

What are the fundamental components of the natural numbers? How can we follow the suggestion of breaking down numbers into smaller components?

Many ways exist to deconstruct large natural numbers into smaller ones. We might first think of addition: Every natural number can be

constructed by just adding

$$1 + 1 + 1 + 1 + \cdots + 1$$

enough times. This method demonstrates perhaps the most fundamental feature of natural numbers: They are simply a sequence of counting tools, each number one bigger than its predecessor. However, this feature provides only a narrow way of distinguishing one natural number from another.

## *Multiplicative Building Blocks*

How can one natural number be expressed as the product of smaller natural numbers? This innocent-sounding question leads to a vast field of interconnections among the natural numbers that mathematicians have been exploring for literally thousands of years. The adventure begins by recalling the arithmetic from our youth and looking at it afresh.

One method of writing a natural number as a product of smaller ones is to divide and see if there is a remainder. We were introduced to division a long time ago in third or fourth grade—we weren't impressed. Somehow it paled in comparison to, say, recess. The basic reality of long division is that either it comes out even or there is a remainder. If the division comes out even, we have factored our number, and the smaller number divides evenly into the larger number. For example, 12 divided by 3 is 4, so $3 \times 4 = 12$ and 3 and 4 are factors of 12. More generally, suppose that $n$ and $m$ are any two natural numbers. We say that $n$ divides evenly into $m$ if there is an integer $q$ such that

$$nq = m.$$

The integers $n$ and $q$ are called *factors* of $m$.

If the division does not come out even, the remainder is less than the number we tried to divide. For example, 16 divided by 5 is 3 with a remainder of 1. This whole collection of elementary school flashbacks can be summarized in a statement that sounds far more impressive than long division, namely, the *Division Algorithm*.

> **THE DIVISION ALGORITHM.** *Suppose n and m are natural numbers. Then there exist unique integers q (for quotient) and r (for remainder) such that*
>
> $m = nq + r$    *and*
>
> $0 \leq r \leq n - 1$ *(r is greater than or equal to 0 but less than or equal to $n - 1$).*

## Primes

Factoring a big number into smaller ones gives us some insights into the larger number. This quest for breaking up natural numbers into their basic components leads us to the important notion of prime numbers.

There certainly are natural numbers that cannot be factored at all as the product of two smaller natural numbers. For example, 7 cannot be factored into two smaller natural numbers. There are others:

2, 3, 5, 11, 13, 17

are the first seven such numbers (let's ignore the number 1). These unfactorable numbers are called *prime numbers*. A natural number greater than 1 is a prime number if it cannot be expressed as the product of two smaller natural numbers.

The prime numbers form the multiplicative building blocks for all natural numbers greater than 1. That is, every natural number greater than 1 is either prime or it can be expressed as a product of prime numbers.

> **THE PRIME FACTORIZATION OF NATURAL NUMBERS.**
> *Every natural number greater than 1 is either a prime number or it can be expressed as a product of prime numbers.*

*Understanding a specific case is often a major step toward discovering a general principle.*

Let's first look at a specific example and then see why the Prime Factorization of Natural Numbers is true for all natural numbers greater than 1.

## Divide and Conquer

Is 1,386 prime? No, 1,386 can be factored as

$$1{,}386 = 2 \times 693.$$

Is 693 prime? No. It can be factored as

$$693 = 3 \times 231.$$

The number 3 is prime, but

$$231 = 3 \times 77.$$

So far we have

$$1{,}386 = 2 \times 3 \times 3 \times 77.$$

Is 77 prime? No,

$$77 = 7 \times 11,$$

but 7 and 11 are both primes. Thus we have

$$1{,}386 = 2 \times 3 \times 3 \times 7 \times 11.$$

So, we have factored 1,386 into a product of primes. This simple "divide and conquer" technique works for any number!

## Proof of the Prime Factorization of Natural Numbers

Let $n$ be any natural number greater than 1 that is not a prime. Then there must be a factorization of $n$ into two smaller natural numbers both greater than 1 (why?), say $n = a \times b$. We now look at $a$ and $b$. If both are primes, we end our proof, since $n$ is equal to a product of two primes. If either $a$ or $b$ is not prime, then it can be factored into two smaller natural numbers, each greater than 1. If we continue in this manner with each of the factors, since the factors are getting smaller, we will find in a finite number of steps a factorization of $n$ where all the factors are prime numbers. So, $n$ is a product of primes, and our proof is complete.

## We Can, but We Can't

Even though we know that every natural number can theoretically be factored, as a practical matter, factoring really large numbers is impossible. For example, suppose someone asks us to factor this long number:

2203019871351092560192573410934710923871024835618972658273
6458764058761984569183246501923847019237418726501826354109
8327401923847019234601982561092384760198237401923867450872
3640982367401982360192386092364019283470129834701293487102
9386102398461039871023984710923871029387102398719238710 93.

We might have some difficulty. In fact, the fastest computers in the world working full time would take centuries to factor this number. Our inability to factor really big numbers is the key to devising public key codes for sending private information over the Web and using e-mail and ATMs. We consider this modern application of prime numbers in section 2.5.

## What's the Largest Prime?

The Prime Factorization of Natural Numbers reveals that the prime numbers are actually the building blocks for the natural numbers: To build any natural number, all we need to do is take the necessary prime numbers and multiply them. Of course, whenever we are presented with building blocks, whether they are Lincoln Logs, Legos, elementary particles, or primes, we wonder: How many blocks do we have? In this case, we ask how many prime numbers are there? Specifically, are there infinitely many primes?

Since there are infinitely many natural numbers, it seems reasonable to think that we must have bigger and bigger primes to be able to build up the bigger and bigger natural numbers. Therefore, we might conclude that there are infinitely many primes. This argument, however, isn't valid. Do you see why? The argument is invalid because we can build natural numbers as large as we wish by just multiplying a couple of small prime numbers. For example, using just 2 and 3, we are able to make huge numbers: Consider

$$45{,}349{,}632 = 2^8 \, 3^{11}.$$

The point is that, since we are allowed to use tons of 2's and 3's in our product, we can construct numbers as large as we wish. Therefore, at this moment it is plausible that there may be only finitely many prime numbers, even though every large (and small) natural number is a product of primes.

## *Infinitely Many Primes*

It turns out that there are infinitely many prime numbers, and thus there is no largest prime number. Unfortunately, the argument we gave at the beginning of the previous paragraph remains flawed. We have to prove this result another way. The proof we are about to give is a famous one discovered by Euclid more than 2,000 years ago. It is beautiful in that the idea isn't very complicated, just extremely clever. In fact, on first inspection the proof may appear as though there is some sleight of hand going on: as if some fast-talking salesman in a polyester plaid sports jacket is trying to sell us a used car—before we know it, we're the not-so-proud owner of a 1973 Gremlin. Think along with us as we develop this argument. The ideas fit together beautifully, and, if you stay with us, the argument will suddenly "click." Let's now examine one of the great triumphs of human reasoning.

> **THE INFINITUDE OF PRIMES.** *There are infinitely many prime numbers.*

## *Proof*

Our strategy for proving that there are infinitely many prime numbers is to show that, for each and every given natural number, we can always find a prime number that is larger than that given natural number. Since we can consider larger and larger numbers as our given natural number, this claim would imply that there are larger and larger prime numbers. Thus we would show that there must be infinitely many primes—because we could find primes as large as we want without bound.

## *Seeing the Strategy of the Proof Through an Example*

Before moving forward with the general idea of the proof, let's illustrate the key ingredient with a specific example. Suppose we wanted to find a prime number that is greater than 4. How could we proceed? Well, we could just say "5" and be done with it, but that is not in the spirit of what we are trying to do. Our goal is to discover a method that will lend itself to be generalized and that can be used to find a prime number that exceeds an *arbitrary* natural number, not just the pathetically small number 4. So we seek a systematic means of finding a prime that exceeds, in this case, 4. What should we do? Our challenge is to:

1. find a number bigger than 4 that is not evenly divisible by 2;
2. find a number bigger than 4 that is not evenly divisible by 3; and
3. find a number bigger than 4 that is not evenly divisible by 4.

Each of these tasks individually is easy. To satisfy (1), we just pick an odd number. To satisfy (2), we just pick a number that has a remainder when we divide by 3. To satisfy (3), we just pick a number that has a remainder when we divide by 4. We now build a number that meets all those conditions simultaneously. Let's call this new number $N$ for *New*. Here is an $N$ that meets all three conditions simultaneously:

$$N = (1 \times 2 \times 3 \times 4) + 1.$$

So, $N$ is really just the number 25, but, since we are trying to discover a general strategy, let's not think of $N$ as merely 25 but instead think of $N$ as the more impressive $(1 \times 2 \times 3 \times 4) + 1$.

We notice that $N$ is definitely larger than 4. By the Prime Factorization of Natural Numbers, we know that $N$ is either a prime number or a product of prime numbers. In the first possibility, if $N$ is a prime number, then we have just found a prime number that is larger than 4—which was our goal. We now must consider the only other possibility: if $N$ is not prime. If $N$ is not prime, then $N$ is a product of prime numbers. Let's call one of those prime factors of $N$, OUR-PRIME. So OUR-PRIME is a prime that divides evenly into $N$.

Now what can we say about OUR-PRIME? Is OUR-PRIME equal to 2? Well, if we divide 2 into $N$ we see that, since

$$N = 2 \times (1 \times 3 \times 4) + 1,$$

we have a remainder of 1 when 2 is divided into $N$. Therefore, 2 does not divide evenly into N, and so OUR-PRIME is not 2. Is OUR-PRIME equal to 3? No, for the same reason:

$$N = 3 \times (1 \times 2 \times 4) + 1,$$

so we get a remainder of 1 when 3 is divided into N; therefore, 3 does not divide evenly into N and hence is not a factor of N. Likewise, we will get a remainder of 1 when 4 is divided into N, and, therefore, 4 is not a factor of N. So 2, 3, and 4 are not factors of N. Hence we conclude that all the factors of N must be larger than 4 (there are no factors of N that are 4 or smaller). But that means that, since OUR-PRIME is a prime that is a factor of N, OUR-PRIME must be a prime number greater than 4. Therefore, we have just found a prime number greater than 4. Mission accomplished!

Following this same strategy, show that there must be a prime greater than 5. Can you use the preceding method to show there must be a prime greater than 10,000,000? Try it now.

## *The General Proof*

Now let's use the method we developed in the specific example to prove our theorem in general. Remember that we wish to demonstrate that, for any particular natural number, there is a prime number that exceeds that particular number. Let $m$ represent an arbitrary natural number. Our goal now is to show that there is a prime number that exceeds $m$. To accomplish this lofty quest, we will, just as before, construct a new number using all the numbers from 1 to $m$. We'll call this new number N (for *New* number) and define it to be 1 plus the product of all the natural numbers from 1 to $m$, in other words (symbols):

$$N = (1 \times 2 \times 3 \times 4 \times \ldots \times m) + 1.$$

It is fairly easy to see that N is larger than $m$. By the Prime Factorization of Natural Numbers we know that there are only two possibilities for N: Either N is prime or N is a product of primes. If the first possibility is true, then N is a prime number, and, since it is larger than $m$, we found what we wanted. We now must consider the more challenging possibility: if N is a product of prime numbers.

If N is a product of primes, then we can choose one of those prime factors and call it BIG-PRIME. So, BIG-PRIME is a prime factor of N. Let's now try to pin down the value of BIG-PRIME. We'll start off small.

Does BIG-PRIME equal 2? Well, if we divide 2 into N we see that, since

$$N = 2 \times (1 \times 3 \times 4 \times \ldots \times m) + 1,$$

we have a remainder of 1 when 2 is divided into N. Therefore, 2 does not divide evenly into N, and so BIG-PRIME cannot equal 2.

Does BIG-PRIME equal 3? No, for the same reason:

$$N = 3 \times (1 \times 2 \times 4 \times \ldots \times m) + 1,$$

so we get a remainder of 1 when 3 is divided into $N$. In fact, what is the remainder when any number from 2 to $m$ is divided into $N$? The remainder is always 1 by the same reasoning that we used for 2 and 3.

Okay, so we see that none of the numbers from 2 through $m$ divides evenly into $N$. That fact means that none of the numbers from 2 through $m$ is a factor of $N$. Therefore, what can be said about the size of the factor *BIG-PRIME*? Answer: It must be BIG since we know that $N$ has no factors between 2 and $m$. Hence any factor of $N$ must be larger than $m$. Therefore, *BIG-PRIME* is a prime number that is larger than $m$.

Well, we did it! We just showed that there is a prime that exceeds $m$. Since this procedure works for any value of $m$, this argument shows that there are arbitrarily large prime numbers. Therefore, we must have infinitely many prime numbers, and we have completed the proof.

## *The Clever Part of the Proof*

In the proof, each step by itself wasn't too hard, but the entire argument, taken as a whole, was subtle. What was the most clever part of the proof? In other words, where was the most imagination required? What step in the argument would have been hardest to think up on your own?

We believe that the most ingenious part of the proof was the idea of constructing the auxiliary number $N$ (one more than the product of all the numbers from 1 to $m$). Once we have the idea of considering that $N$, we can finish the proof. But it took creativity and contemplation to arrive at that choice of $N$. We might well say to ourselves, "Gee, I wouldn't have thought of making up that $N$." Generally, these proofs were arrived at only after many attempts and false starts—just as Euclid no doubt experienced before he thought of the idea. Very few people can understand arguments of this type on first inspection, but once we can hold the whole proof in our minds, we will regard it as straightforward and persuasive and appreciate its aesthetic beauty. Ingenuity is at the heart of creative mathematical reasoning, and therein lies the power of mathematical thought.

## *How Do the Primes Fit into the Pack?*

Now that we know for sure that there are infinitely many prime numbers, we wonder how the primes are distributed among the natural numbers. Is there some pattern to their distribution? There are infinitely many primes, but how rare are they among the numbers? What proportion of the natural numbers are prime numbers? Half? A third? To explore these questions, the best way to start is to look at the natural numbers and the primes among them. Here then are the first few natural numbers with the primes printed in bold:

$$1, \mathbf{2}, \mathbf{3}, 4, \mathbf{5}, 6, \mathbf{7}, 8, 9, 10, \mathbf{11}, 12, \mathbf{13}, 14, 15, 16, \mathbf{17}, 18, \mathbf{19}, 20, 21, 22, \mathbf{23}, 24, \ldots$$

Out of the first 24 natural numbers, nine are primes. We see that 9/24 = .375 of the first 24 natural numbers are primes—that's just a little over one third. Would we guess that just over one third of all natural numbers are prime numbers? We could try an experiment, namely, we could continue to list the natural numbers and find the proportions of primes and see whether that proportion remains about one third of the total number. If we do this experiment, we will learn an important lesson in life: Don't be too hasty to generalize based on a small amount of evidence.

## *Experiments*

Before high-speed computers were available, calculating (or just estimating) the proportion of prime numbers in the natural numbers was a difficult task. In fact, years ago "computers" were people who did computations—such people were amazingly accurate, but they required a great deal of time and dedication to accomplish what today's electronic computers can do in seconds. An 18th-century Austrian arithmetician, J. P. Kulik, spent 20 years of his life creating, by hand, a table of the first 100 million primes. His table was never published, and sadly the volume containing the primes between 12,642,600 and 22,852,800 has disappeared.

Today, software computes the number of primes less than $n$ for increasingly large values of $n$, and the computer prints out the proportion: (number of primes less than $n$)/$n$. Computers have no difficulty in producing such a table for values of $n$ up to the billions, trillions, and beyond. If we examine the results, we notice that the proportion of primes slowly goes downward. That is, the percentage of numbers less than a million that are prime is smaller than the percentage of numbers less than a thousand that are prime. The primes, in some sense, get sparser and sparser among the bigger numbers.

## *A Conjecture*

In the early 1800s, Karl Friedrich Gauss (on left), one of the greatest mathematicians ever—known by many as the Prince of Mathematics, and

A.-M. Legendre (on right), another world-class mathematician, made an insightful observation about primes. They noticed that, even though primes do not appear to occur in any predictable pattern, the proportion of primes is related to the natural logarithm.

Years ago, one needed to interpolate huge tables to find the logarithm. Today, we have scientific calculators that compute logarithms instantly and painlessly. Get out a scientific calculator and look for the LN key. Type "3" and then hit LN. You should see 1.09861.... We encourage you to try some natural-logarithm experiments on your calculator. How does the size of the natural logarithm of a number compare with the size of the number itself?

Gauss and Legendre conjectured that the proportion of the number of primes among the first $n$ natural numbers is approximately $1/\text{Ln}(n)$. The following chart, constructed with the aid of a computer (over which Gauss and Legendre would drool), shows the number of primes up to $n$, the proportions of primes, and a comparison with $1/\text{Ln}(n)$.

| $n$ | Number of primes up to $n$ | Proportion of primes up to $n$ (number of primes $\leq n$)/$n$ | $1/\text{Ln}(n)$ | Proportion $-$ $1/\text{Ln}(n)$ |
|---|---|---|---|---|
| 10 | 4 | .4 | .43429... | $-$.03429... |
| 100 | 25 | .25 | .21714... | 0.03285... |
| 1000 | 168 | .168 | .14476... | 0.02323... |
| 10000 | 1229 | .1229 | .10857... | 0.01432... |
| 100000 | 9592 | .09592 | .08685... | 0.00906... |
| 1000000 | 78498 | .078498 | .07238... | 0.00611... |
| 10000000 | 664579 | .0664579 | .06204... | 0.00441... |
| 100000000 | 5761455 | .05761455 | .05428... | 0.00332... |
| 1000000000 | 50847534 | .050847534 | .04825... | 0.00259... |

Notice how the last column seems to be getting closer and closer to zero—that is, the proportion of primes in the first $n$ natural numbers is approximately $1/\text{Ln}(n)$, and the fraction (number of primes less than $n$)/$n$ is becoming increasingly closer to $1/\text{Ln}(n)$ the bigger $n$ gets. Our observations show that

(number of primes up to $n$)/$n$ is roughly $1/\text{Ln}(n)$.

Multiplying both of these quantities by $n$, it appears that

(number of primes up to $n$) is roughly $n/\text{Ln}(n)$.

These observations culminate in what is called the *Prime Number Theorem*, which gives the approximate number of primes less than or equal to $n$ as $n$ gets really big.

> **THE PRIME NUMBER THEOREM.** *As $n$ gets larger and larger, the number of prime numbers less than or equal to $n$ approaches $n/\text{Ln}(n)$.*

## The Proof Had to Wait

Having sufficient insight into the nature of numbers to make such a conjecture is a tremendous testament to the intuition of Gauss and Legendre. But, even though they conjectured that the Prime Number

Theorem was true, they were unable to prove it. This famous conjecture remained unproved for nearly a century until, in 1896, two mathematicians, Hadamard and de la Vallee Poussin, simultaneously but independently gave proofs of the Prime Number Theorem. That is, working on their own, they each found, in the same year, a proof of this extremely important and difficult result. The proof of this theorem is extremely long and uses the machinery of many branches of mathematics (including, of all things, imaginary numbers). This mathematical episode illustrates a hopeful and powerful fact about human thought: The human mind is capable of solving extremely difficult problems and answering truly hard questions. This observation should give us all hope for finding solutions to some of the many problems we face in our world today.

Some problems are solved in 10 years; others take longer. In some cases long-standing mysteries concerning numbers are resolved only after centuries of vain attempts. When such stubborn questions are finally answered, the solutions represent a triumph for humanity. Recently, with the diligence and drive of several generations of mathematicians, a 350-year-old problem was finally solved. Such a solution is a celebration of the human spirit and justly made the front page of *The New York Times* and the *Wall Street Journal*. This is the story of Fermat's Last Theorem.

*Truly difficult problems can be solved. We must have tenacity and patience.*

◆ ◆ ◆ ◆ ◆ ◆

## *The Story of Fermat's Last Theorem ( . . . Mostly True)*

It was a dark and stormy night. Rain pelted the wavy glass. Wind seeped through cracks in the window frame, and the candle flickered uncertainly in the cold breeze. The year was 1637.

Pierre de Fermat sat in the great chair of his library poring over the

leather-bound, Latin translation of the ancient tome of Diophantus's *Arithmetica*. He concentrated intensely on the ideas that Diophantus had written 2,000 years before. During those 2,000 years whole civilizations had risen and fallen, but the study of numbers bridged Fermat and Diophantus, spanning centuries and even transcending death itself.

The night wore on, and inchoate ideas, at first jumbled and ill-defined, began to find one another and weave a pattern in Fermat's mind. He gasped at the emergence of insight and understanding. He saw how the parts fit together, how the numbers played on one another. Finally, it was clear to him.

On that fateful night in 1637, he wrote in the margin of Diophantus's *Arithmetica* the statement that would make him famous forever. He wrote in Latin:

*It is impossible to write a cube as a sum of two cubes, a fourth power as a sum of two fourth powers, and, in general, any power beyond the second as a sum of two similar powers. For this, I have discovered a truly wondrous proof, but the margin is too small to contain it.*

It was clear to him. He saw that, for any two natural numbers, if we take the first number raised to the third power, then add it to the second number raised to the third power, that sum will *never* equal a natural number raised to the third power. For example, suppose we consider the natural numbers 2 and 14. We notice that

$$2^3 = 2 \times 2 \times 2 = 8, \quad 14^3 = 14 \times 14 \times 14 = 2{,}744,$$

so,

$$2^3 + 14^3 = 2{,}752,$$

and there is no natural number that, when raised to the third power, equals 2,752 ($14^3 = 2{,}744$ is too small, and $15^3 = 3{,}375$ is too large).

In fact, Fermat saw then that for *any* natural number greater than or equal to 3, let's call that number $n$, it is impossible to find three natural numbers, let's call them $x$, $y$, and $z$, such that $x$ raised to the $n$ power ($x$ multiplied by itself $n$ times) added to $y$ raised to the $n$ power equals $z$ raised to the $n$ power. In other words, in the language of mathematical symbols, for any natural number $n$ greater than or equal to 3, there do not exist any natural numbers $x$, $y$, and $z$ that satisfy the equation

$$x^n + y^n = z^n.$$

Fermat claimed that he saw a "wondrous proof," but did he? Did he discover a correct proof, or had he deceived himself? No one knows for certain, because Fermat never wrote his "proof" down. In fact, Fermat was notorious for making such statements with usually little or no justification or proof. By the 1800s, all of Fermat's statements had been resolved; all but the preceding one—his "last" one. Thus Fermat's unproved assertion became known as *Fermat's Last Theorem*.

We will never know whether Fermat had discovered a correct proof of his Last Theorem, but we do know one thing. He did not discover the proof that Andrew Wiles of Princeton University produced in 1994, 357 years after Fermat wrote his tantalizing marginal note. If Fermat had somehow conceived of Wiles's deep and complicated proof, he would not have written, "The margin of this book is not large enough to contain it." He would have written, "The proof would require a moving van to carry it. It involves developing entirely new branches of mathematics using ideas undreamed of in the 17th century and for more than three centuries to come."

Some of the greatest minds in mathematics have worked on Fermat's Last Theorem. While working on a specific problem, one often solves other problems or develops ideas of independent importance. The statement of Fermat's Last Theorem does not strike us as intrinsically important or interesting—it just states that a certain type of equation never can be solved with natural numbers. What are undoubtedly interesting and important are all the deep mathematical ideas, many of which led to new branches of mathematics, that were discovered during attempts to prove Fermat's Last Theorem. Although Fermat's Last Theorem has not yet been used for practical purposes, the new theories developed to attack it turned out to be valuable in many practical technological advances. Within the aesthetics of mathematics, it is difficult to determine which questions are more important than others. When problems resist attempts by the best mathematical minds over many years, the problems gain prestige. Fermat's Last Theorem resisted all attacks for 357 years, but it finally succumbed.

Andrew Wiles's complete proof of Fermat's Last Theorem is over 130 pages long, and it relies on many important and difficult theorems including some new theorems from geometry (although it appears surprising that geometry should play a role in solving this problem involving natural numbers). When mathematicians expose connections between seemingly different areas of mathematics, they feel an electric excitement and pleasure. In nature and mathematics, things fit together. As we will begin to discover, deep and rich connections thread their way through the various mathematical topics and form the very fabric of truth; the very fabric of mathematics.

## The Vast Unknown

Every answer allows us to recognize and formulate new questions.

To many people, mathematics may seem like a cold, ancient body of facts, formulas, and techniques. In reality, much of the subject remains unknown, unanswered, and even unasked. Many people have the impression that what we currently know and think we will know soon is all there is to know. But throughout history such people have often been proved wrong.

> *Everything that can be invented has been invented.*
> —Charles H. Duell, Commissioner U.S. Patent Office, 1899

> *We don't know a millionth of one percent about anything.*
> —Thomas Edison

Human thought represents a never-ending universe of mystery—especially in mathematics. We know a small amount, and our knowledge allows us to recognize a small part of what we do not know. Vastly larger is our *ignorance* of what we do not know. An important shift in perspective on mathematics and other areas of human knowledge is the move from the sense that we know most of the answers to the more accurate and comforting realization that we will not run out of mysteries.

So, after celebrating, as we have, some of the great mathematical achievements that were solved only after many decades of human creativity and thought, we close this section by gazing forward to questions that remain unsolved—*open* questions. From among the thousands of questions on which mathematicians are currently working, we state here two famous questions about prime numbers that were posed hundreds of years ago and are still unsolved. Fermat's Last Theorem has been conquered; but somehow the mathematical force is not ready to let go its hold and let these two fall.

> **THE TWIN PRIME QUESTION.** *Are there infinitely many pairs of prime numbers that differ from one another by two? (11 and 13, 29 and 31, 41 and 43 are examples of some such pairs.)*

> **THE GOLDBACH QUESTION.** *Can every positive, even number greater than 2 be written as the sum of two primes? (Pick some even numbers at random, and see whether you can write them each as a sum of two primes.)*

Computer analysis allows us to investigate a tremendous number of cases, but the results of such analysis do not provide ironclad proof for all cases.

We have seen how to decompose natural numbers into their fundamental building blocks, and we have discovered further mysteries and structures in this realm. Can we use these antique, abstract results about numbers in our modern lives? The amazing and perhaps surprising answer is a resounding: "YES."

## A LOOK BACK

The prime numbers are the basic multiplicative building blocks for natural numbers, since every natural number greater than 1 can be factored into primes. We can prove that there are infinitely many primes by showing that we can always find a prime number larger than any specified number. The strategy is to take the product of all numbers up to the specified number and then add 1. This new large integer must have a prime factor greater than the original specified number.

The study of primes goes back to ancient times, and some questions remained unanswered for hundreds of years before being resolved. Many questions, however, still remain unanswered.

We discovered proofs of the Prime Factorization of Natural Numbers and the Infinitude of Primes by carefully exploring specific examples and searching for patterns. Considering specific examples while thinking about the general case guided us to new discoveries.

*Understanding a specific case well is a major step toward discovering a general principle.*

# MINDSCAPES   Invitations to Further Thought*

## I. SOLIDIFYING IDEAS

1. *A silly start.* What is the smallest number that looks prime but really isn't?

2. *Waiting for a nonprime.* What is the smallest natural number $n$, greater than 1, for which $(1 \times 2 \times 3 \times \ldots \times n) + 1$ is *not* prime?

3. *Always, sometimes, never.* Does a prime multiplied by a prime ever result in a prime? Does a nonprime multiplied by a nonprime ever result in a prime? Always? Sometimes? Never? Explain your answers.

4. *The dividing line.* Does a nonprime divided by a nonprime ever result in a prime? Does it ever result in a nonprime? Always? Sometimes? Never? Explain your answers.

5. *Prime power.* Is it possible for an extremely large prime to be expressed as a large integer raised to a very large power? Explain.

6. *Nonprimes.* Are there infinitely many natural numbers that are not prime? If so, prove it.

7. *Prime test.* Suppose you are given a number $n$ and are told that 1 and the number $n$ divide into $n$. Does that mean $n$ is prime? Explain.

8. *Twin primes.* Find the first 15 pairs of twin primes.

9. *Goldbach.* Express the first 15 even numbers greater than 2 as the sum of two prime numbers.

10. *Odd Goldbach* (H). Can every odd number greater than 3 be written as the sum of two prime numbers? If so, prove it; if not,

---

*In the Mindscapes section, exercises marked (H) have hints for solutions at the back of the book. Exercises marked (S) have solutions.

find the smallest counterexample and show that the number given is definitely not the sum of two primes.

11. **Still the 1 (S).** Consider the following sequence of natural numbers: 1111, 11111, 111111, 1111111, 11111111, . . . . Are all these numbers prime? If not, can you describe infinitely many such numbers that are definitely not prime?

12. **Zeros and ones.** Consider the following sequence of natural numbers made up of 0's and 1's: 11, 101, 1001, 10001, 100001, 1000001, 10000001, . . . . Are all these numbers prime? If not, find the first such number that is not prime and express it as a product of prime numbers.

13. **Zeros, one and three.** Consider the following sequence of natural numbers made up of 0's, 1's, and 3's: 13, 103, 1003, 10003, 100003, 1000003, 10000003, . . . . Are all these numbers prime? If not, find the first such number that is not prime and express it as a product of prime numbers.

14. **A rough count.** Using results discussed in this section, estimate the number of prime numbers that are less than $10^{10}$.

15. **Generating primes (H).** Consider the list of numbers: $n^2 + n + 17$, where $n$ first equals 1, then 2, 3, 4, 5, 6, . . . . What is the smallest value of $n$ for which $n^2 + n + 17$ is not a prime number? (*Bonus:* Try this for $n^2 - n + 41$. You'll see an amazingly long string of primes!)

16. **Generating primes II.** Consider the list of numbers: $2^n - 1$, where $n$ first equals 2, then 3, 4, 5, 6, . . . . What is the smallest value of $n$ for which $2^n - 1$ is not a prime number?

17. **Floating in factors.** What is the smallest natural number that has three different factors?

18. **Lucky 13 factor.** Suppose a certain number when divided by 13 yields a remainder of 7. What is the smallest number we would have to subtract from our original number to have a number with a factor of 13?

19. **Remainder reminder (S).** Suppose a certain number when divided by 13 yields a remainder of 7. If we add 22 to our original number, what is the remainder when this new number is divided by 13?

20. **Remainder roundup.** Suppose a certain number when divided by 91 yields a remainder of 52. If we add 103 to our original number, what is the remainder when this new number is divided by 7?

## II. CREATING NEW IDEAS

1. **Related remainders (H).** Suppose we have two numbers that both have the same remainder when divided by 57. If we subtract the

two numbers, are there any numbers that we know will definitely divide evenly into this difference? What is the largest number that we are certain will divide into the difference? Use this observation to state a general principle about two numbers that have the same remainder when divided by another number.

2. **Prime differences.** Write out the first 15 primes all on one line. On the next line, underneath each pair, write the difference between the larger number and the smaller number in the pair. Under this line, below each pair of the previous line, write the difference between the larger number and the smaller number. Continue in this manner. Your "triangular" table should begin with:

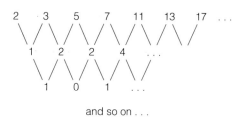

and so on . . .

Once your chart is made, imagine that all the primes were listed on that first line. What would you guess is the pattern for the sequence of numbers appearing in the first entry of each line? The actual answer is not known. It remains an open question! What do you think?

3. **Minus two.** Suppose we take a prime number greater than 3 and then subtract 2. Will this new number always be a prime? Explain. Are there infinitely many primes for which the answer to the question is yes? How does this last question relate to a famous open question?

4. **Prime neighbors.** Does there exist a number $n$ such that both $n$ and $n + 1$ are prime numbers? If so, find such an $n$; if not, show why not.

5. **Perfect squares.** A perfect square is a number that can be written as a natural number squared. The first few perfect squares are 1, 4, 9, 16, 25, 36. How many perfect squares are less than or equal to 36? How many are less than or equal to 144? In general, how many perfect squares are less than or equal to $n^2$? Using all these answers, estimate the number of perfect squares less than or equal to $N$. (*Hint:* Your estimate may involve square roots and should be the exact answer whenever $N$ is itself a perfect square.)

6. **Perfect squares versus primes.** Using a calculator or a computer, fill in the last two columns of the following chart.

| n | Number of Primes up to n | Approx. Number of Perfect Squares up to n | Number of Primes / Number of Perfect Squares |
|---|---|---|---|
| 10 | 4 | | |
| 100 | 25 | | |
| 1000 | 168 | | |
| 10000 | 1229 | | |
| 100000 | 9592 | | |
| 1000000 | 78498 | | |
| 10000000 | 664579 | | |
| 100000000 | 5761455 | | |
| 1000000000 | 50847534 | | |

Given the information found in your chart, what do you conclude about the proportion of prime numbers to perfect squares? Are prime numbers more or less common than perfect squares? Using the Prime Number Theorem, estimate the quotient of the number of primes up to $n$ divided by the number of perfect squares up to $n$. Use a computer or a graphing calculator to graph your answer. Does the graph confirm your original conjecture?

7. **Prime pairs.** Suppose that $p$ is a prime number greater than or equal to 3. Show that $p + 1$ cannot be a prime number.

8. **Remainder addition.** Let $A$ and $B$ be two natural numbers. Suppose that, when $A$ is divided by $n$, the remainder is $a$, and, when $B$ is divided by $n$, the remainder is $b$. How does the remainder when $A + B$ is divided by $n$ compare with the remainder when $a + b$ is divided by $n$? Try some specific examples first. Can you prove your answer?

9. **Remainder multiplication.** Let $A$ and $B$ be two natural numbers. Suppose that the remainder when $A$ is divided by $n$ is $a$, and the remainder when $B$ is divided by $n$ is $b$. How does the remainder when $A \times B$ is divided by $n$ compare with the remainder when $a \times b$ is divided by $n$? Try some specific examples first. Can you prove your answer?

10. **A Prime-free gap (S).** Find a run of six consecutive natural numbers, none of which is a prime number. (*Hint:* Prove that you can start with $(1 \times 2 \times 3 \times 4 \times 5 \times 6 \times 7) + 2$.)

## III. FURTHER CHALLENGES

1. ***Prime-free gaps.*** Using the previous exercise, show that, for a given number, there exists a run of that many consecutive natural numbers, none of which is a prime number.

2. ***Three primes.*** Prove that it is impossible to have three consecutive integers, all of which are prime.

3. ***Prime plus three.*** Prove that, if you take any prime number greater than 11 and add 3 to it, that sum is not prime.

4. ***A small factor.*** Prove that if a number greater than 1 is not a prime number, then it must have a prime factor that is less than or equal to its square root. (*Hint:* Suppose that all the factors are greater than its square root, and show that the product of any two of them must be larger than the original number.) This exercise shows that, to determine if a number is a prime, we have only to check divisibility of the primes up to the square root of the number.

5. ***Prime products* (H).** Suppose we make a number by taking a product of prime numbers and then adding the number 1 (for example, $(2 \times 5 \times 17) + 1$). Compute the remainder when any of the primes used are divided into the number. Show that none of the primes used can divide evenly into the number. What can you conclude about the primes that divide evenly into the number? Can you use this line of reasoning to give another proof that there are infinitely many prime numbers?

## IV. IN YOUR OWN WORDS

1. ***Personal perspectives.*** Write a short essay describing the most interesting or surprising discovery you made in exploring this section's material. If any material seemed puzzling or even unbelievable, address that as well. Explain why you chose the topics you did. Finally, comment on the aesthetics of the mathematics and ideas in this section.

2. ***With a group of folks.*** In a small group, discuss and work through the proof that there are infinitely many primes. After your discussion, write a brief narrative describing the proof in your own words.

3. ***Creative writing.*** Write an imaginative story (it can be comical, dramatic, whatever you like) that involves or evokes the concepts in this section.

4. ***Power beyond the mathematics.*** Provide several real-life situations—ideally, from your own experience—for which some of the strategies of thought presented in this section would provide effective methods for approaching and resolving them.

# 2.4 Crazy Clocks and Checking Out Bars:

## Checking Bar Codes on Products with Cyclical Clock Arithmetic

*A rule to trick th' arithmetic.*

—Rudyard Kipling

Cycles are familiar parts of life. The seasons, phases of the moon, day and night, birth and death—all are among the most powerful natural forces that define our lives, and all are cycles. Whole cultural traditions revolve around this cyclic reality of life; consider, for example, the notion of reincarnation, unless you already considered it in a previous life. We can use these cycles as models to develop analogous constructs in the realm of numbers. Such explorations create yet another kind of cycle, because the abstract mathematical insights refer back to the world, and we find applications of these abstractions in our daily lives.

Our strategy for examining cycles in the world of numbers is to find a phenomenon in nature (in this case cyclicity) and to develop a mathematical model that captures some features of the natural processes. This method of reasoning by analogy is a powerful way to develop new ideas, because we use existing ideas, events, and phenomena to guide us in creating new insights.

Most people would not believe that there is deep and powerful number theory going on when they glance at their watch or when they check out at the grocery store: "Sure, numbers are involved: The time of day is expressed in numbers, as is the price of an item—but these numbers are neither deep nor powerful anything!" Most people, however, are sadly mistaken. In reality, there's a world of exotic number theory hidden in everyday objects.

### Time

What time is it? Suppose our watch says it is 9:00 (9 o'clock), and we are to meet the love of our life in 37 hours. What time will our watch read when we fall into the arms of our soul mate? Careful—the answer is not 46 o'clock. The answer is 10:00. So we must be doing some type of strange arithmetic, since 9 plus 37 equals 10—wacky! How does one perform arithmetic in the context of telling time by a clock? Unlike the natural numbers, which get larger and larger when we add them together, a clock cycles around, and in 12 hours the clock returns to its original position (we're assuming we are using a 12-hour clock). Counting with a clock in some sense is easier than standard counting, because the numbers never get too large. For example, to add 37 to 9 we could count as follows:

9, 10, 11, 12,

1, 2, 3, 4, 5, 6, 7, 8, 9, 10, 11, 12,

1, 2, 3, 4, 5, 6, 7, 8, 9, 10, 11, 12,

1, 2, 3, 4, 5, 6, 7, 8, 9, 10.

Use your 3D glasses from your kit.

Notice how, once we get to 12, we start all over again and cycle back to 1. This procedure involves a kind of arithmetic different from standard arithmetic.

## Clock Arithmetic

For the moment, let's refer to this arithmetic—where we return to 1 after 12—as *clock arithmetic*. Let's look at a few examples of clock arithmetic. We have already seen that $9 + 37 = 10$. What is $6 + 12$? The answer is 6, since adding 12 just spins us back around to where we started. So 12 is just like 0 in this clock arithmetic. This observation allows us to perform this new arithmetic in a different way. For example, with $9 + 37$, we could notice that 37 is equal to $12 + 12 + 12 + 1$. But remember that adding 12 is just like adding 0, so really 37 is equivalent to 1. Therefore, $9 + 37 = 9 + 1 = 10$.

Let's consider a different kind of question: What does $(4 \times 7) + 20$ equal in clock arithmetic? Well, $4 \times 7 = 28$, but 28 is equivalent to 4, since $28 = 12 + 12 + 4$. Now 20 is equivalent to 8, since $20 = 12 + 8$. Therefore, $(4 \times 7) + 20 = 4 + 8 = 12$. So the answer is 12.

What would happen to our arithmetic if we had a crazy clock that looked like this:

Notice now that adding 7 spins us back to where we started, and thus adding 7 is the same as adding 0. So now $6 + 4 = 10$ is equivalent to 3, since $10 = 7 + 3$. In other words, with this crazy clock, 4 hours after 6 o'crazy clock is actually

3 o'crazy clock. Why would anyone ever bother with such a crazy clock? Well, actually we use this crazy clock, not for telling time, but for telling days in a week. Once again we see that the notion of cycles is natural and important. In fact, as we will now discover, this kind of crazy-clock arithmetic helps us find errors in grocery prices; our checking accounts; UPS package deliveries; airline tickets; driver's license numbers; and even helps us check out Shakespeare—*read on MacDuff.*

## *Equivalence*

As we look at cycles, we are developing an idea of *equivalence*. The notion of equivalence occurs in clock arithmetic, for example, when we note that 37 is equivalent to 1. As we develop the idea of cyclical arithmetic, this concept of equivalence will become a central theme.

> Identifying similarities among different objects is often the key to understanding a deeper idea.
>
>

Let's carefully define a type of arithmetic that will capture the spirit of our previous observations and generalize the notion of clock arithmetic. First we'll explore the notion of a new, hip or "mod" clock. Suppose we are given a number—for example, 9, and we have an unusual clock that has 9 hours marked on it: 0, 1, 2, 3, 4, 5, 6, 7, 8. Suppose the hour hand is on the 5. Then 9 hours later the hand returns to the 5, and thus according to this clock, adding 9 doesn't change what the clock reads. So, we can now perform arithmetic using this clock by remembering that 9 is equivalent to 0. We'll write the fact that 9 is equivalent to 0 as $9 \equiv 0$, where the symbol "$\equiv$" means

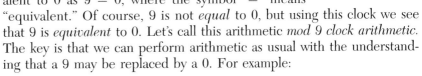

"equivalent." Of course, 9 is not *equal* to 0, but using this clock we see that 9 is *equivalent* to 0. Let's call this arithmetic *mod 9 clock arithmetic.* The key is that we can perform arithmetic as usual with the understanding that a 9 may be replaced by a 0. For example:

$$13 + 25 = (9 + 4) + (9 + 9 + 7) \equiv (0 + 4) + (0 + 0 + 7) =$$
$$4 + 7 = 11 = 9 + 2 \equiv 0 + 2 = 2 \text{ mod } 9,$$

so $13 + 25 \equiv 2 \mod 9$. We write the phrase "mod 9" at the end to remind us and indicate to others what kind of mod clock we are using to perform the arithmetic. In terms of the clock itself, we could have computed the preceding sum by placing the hour hand on the 0 and then moving the hand around 13 hours (which brings us to 4) and then moving from that 4 another 25 hours, which brings us to 2.

Once we see how mod 9 clock arithmetic works, we can abstract the idea by dispensing with the visual aid of the clock and calling it *mod 9 arithmetic.* This notation is convenient, since in mathematical jargon "mod" actually stands for "modulo" or "modular," which means just doing arithmetic using the equivalence of $9 \equiv 0 \mod 9$ (translation: "9 is equivalent to 0 modulo 9" or "9 is equivalent to 0 in mod 9 arithmetic").

*Explore ideas systematically.*

In this next example, notice how we are able to replace large numbers with smaller equivalent numbers by just writing them in terms of 9's (also notice how remainders are making an appearance in our work):

$$(3 \times 5) + (7 \times 100) = (9 + 6) + (7 \times ((9 \times 11) + 1)) \equiv$$
$$(0 + 6) + (7 \times ((0 \times 11) + 1)) = 6 + 7 = 13 = 9 + 4 \equiv$$
$$0 + 4 = 4 \bmod 9,$$

so, $(3 \times 5) + (7 \times 100) \equiv 4 \bmod 9$. Once we get the hang of it, this arithmetic is pretty easy. We just pull out multiples of 9 and replace them by 0's. Of course, we can now think about other mod clocks. We can do this modular arithmetic with any clock, as long as we know how many hours it has on it. For any particular natural number $n$, we write "mod $n$" to mean that we are thinking about arithmetic on a mod clock that has $n$ hours on it (marked $0, 1, 2, \ldots, n - 1$), and so adding $n$ to any number just brings us back to where we started—thus $n$ is equivalent to 0.

## Practice Makes Perfect

Let's really make this new arithmetic our own. Below we ask a few questions. Some present true equivalences, while others are false. Check each one, and determine which are correct and which are wrong. For the ones that are incorrect, figure out a correct answer. Notice that in each case we are using a different mod number, so, first have a look at the "mod $n$" part to see what kind of arithmetic to use. As a warm up, we'll answer the first one.

(1) Is $26 + 31^5 \equiv 0 \bmod 29$?

## Warm-Up Answer

This statement is true since we are considering a mod 29 clock, so 29 is equivalent to 0. Therefore, $26 + 31^5 = 26 + (29 + 2)^5 \equiv 26 + 2^5 = 26 + 32 = 26 + (29 + 3) \equiv 26 + 3 = 29 \equiv 0 \bmod 29$. (Notice how we did not need to figure out that $31^5 = 28{,}629{,}151$.) Now it's your turn.

(2) Is $7^2 + (5 \times 57) \equiv 40 \bmod 48$?

(3) Is $2^4 + 5^{301} + (6 \times 31) \equiv 3 \bmod 5$?

(4) Is $9^{2000} \equiv 1 \bmod 80$? (*Hint*: Write 2000 as $2 \times 1000$.)

## Cool-Down Answers

(2) is incorrect: $7^2 + (5 \times 57) \equiv 49 + (5 \times 9) \equiv 1 + 45 = \mathbf{46}$ mod 48.

(3) is also incorrect: $2^4 + 5^{301} + (6 \times 31) \equiv 16 + 0^{301} + (1 \times 1) \equiv 1 + 1 = \mathbf{2}$ mod 5.

(4) is correct: $9^{2000} = 9^{2 \times 1000} = (9^2)^{1000} = (81)^{1000} \equiv 1^{1000} = \mathbf{1}$ mod 80.

Notice how we can work with enormous powers of numbers without even breaking a sweat by just carefully reducing the numbers in clever ways to smaller equivalent numbers in the modular arithmetic. Now let's see how modular arithmetic is used in our daily lives without our even realizing it.

## *The Clairvoyant Kleenex Consultant*

You have the flu and feel awful. You moan, you groan, you sneeze, you wheeze—let's face it, you're sick. As you sit up in bed, you feel light-headed—not because you have a fever, but because you have been watching too many hours of mind-numbing daytime TV. Having no one to talk to and feeling lonely, you notice that on the bottom of your Kleenex box there is printed a toll-free number for consumer service—1-800-KLEENEX, which amuses you (because of your lightheadedness). Although you are extremely satisfied with the tissues, you decide to dial the number and talk to somebody. The perky Kleenex representative on the other end of the telephone line asks you to read the 12-digit product code appearing on the bottom of the box. You look and see those thin, fat, and medium lines making up the Universal Product Code (bar code), which is now tattooed on nearly every product. As all those lines dance in your head you read:

0  3  6  0  0  0  2  8  1  5  0  9

The chipper voice immediately responds by saying, "I think you made a mistake, could you please read them again?" You glance back, still bleary-eyed, and realize that you indeed made a mistake. In fact, the numbers appearing under the bar code are 0 3 6 0 0 0 2 8 5 1 0 9; you reversed the 5 and the 1. But how did your telephone partner immediately know you made a mistake? Perhaps Kleenex reps are clairvoyant, but they definitely use modular arithmetic.

## Check Digits

Bar codes and associated numbers make up the Universal Product Code (UPC). Each bar code is usually made up of a 12-digit number. The first six digits encode information about the manufacturer, and the next five digits encode information about the product. That leaves us with the last digit, which is called the *check digit*. The check digit provides a means of detecting if a UPC number is incorrect. Here is how the check digit is determined: We line up the first 11 digits of the UPC—let's call them: $d_1$, $d_2$, $d_3$, $d_4$, $d_5$, $d_6$, $d_7$, $d_8$, $d_9$, $d_{10}$, $d_{11}$. We now combine them in an unusual way. We multiply every other number by **3** and then add up all the numbers. That is, we compute

$$3d_1 + d_2 + 3d_3 + d_4 + 3d_5 + d_6 + 3d_7 + d_8 + 3d_9 + d_{10} + 3d_{11}.$$

We now select the check digit (we'll call it $c$) to be the number from 0 to 9 such that when we add the preceding sum to $c$, we get an answer that is equivalent to 0 mod 10. In our Kleenex example, we would want to find the number $c$ so that:

$$(\mathbf{3} \times 0) + 3 + (\mathbf{3} \times 6) + 0 + (\mathbf{3} \times 0) + 0 + (\mathbf{3} \times 2) + 8 + (\mathbf{3} \times 5) + 1 + (\mathbf{3} \times 0) + c \equiv 0 \bmod 10,$$

or, in other words (numbers),

$$51 + c \equiv 0 \bmod 10, \text{ or equivalently } (\bmod\ 10): 1 + c \equiv 0 \bmod 10.$$

Notice that $1 + 9 = 10$, which is equivalent to 0 in mod 10 arithmetic. So the check digit $c$ should equal 9. Note that the last digit of the UPC is indeed a 9. The Kleenex customer service rep was able to take the UPC you read and check it by multiplying every other number by 3, adding all those numbers (including the check digit) and determining if that sum is equivalent to 0 mod 10. Let's compute this sum for the number you originally read off:

$$0\ 3\ 6\ 0\ 0\ 0\ 2\ 8\ 1\ 5\ 0\ 9$$

$$(\mathbf{3} \times 0) + 3 + (\mathbf{3} \times 6) + 0 + (\mathbf{3} \times 0) + 0 + (\mathbf{3} \times 2) + 8 + (\mathbf{3} \times 1) + 5 + (\mathbf{3} \times 0) + 9 = 52 \equiv 2 \bmod 10.$$

Since that sum is not equivalent to 0 mod 10, the service rep knew that there must have been an error. This check digit system is actually used on all 12-digit UPCs. Find some UPCs on various products and compute the sum (multiplying every other number by 3). Your final answer should be equivalent to 0 mod 10. If a single error is made (one number was read incorrectly), the check digit will always detect the error. This knowledge of UPCs is great for impressing friends, family, and people at parties. With practice, you can have someone give you all the digits except a particular one (say the first one), and, as long as you know which digit is left out, you can predict what the number should be. We provide some practice in the Mindscapes.

We must be careful, however. This system is able to detect certain errors, but not all possible errors. For example, if the 2 and the 5 were switched in the original, correct UPC, this system would not detect that an error was made. This simple system is always able to detect an error if there is exactly one wrong digit, and it is able to find, in most cases, errors when two adjacent numbers are switched. The point, however, is that mod 10 arithmetic is the key to this error-checking code. There are many different types of error-detection systems—many of which use modular arithmetic. We'll just point out one more here and several others (including Shakespeare) in the Mindscapes.

## *Checking Checks*

Take a look at a check from any U.S. bank. In the lower left-hand corner we notice nine digits written in that funky 1960s computer font. That is the bank identification number, and the last digit is the check digit. To check the check's check digit, we compute a slightly more complicated sum. Let's call the nine digits $n_1, n_2, n_3, n_4, n_5, n_6, n_7, n_8, n_9$ (that last digit is the check digit). With this error-checking system we compute:

$$7n_1 + 3n_2 + 9n_3 + 7n_4 + 3n_5 + 9n_6 + 7n_7 + 3n_8 + 9n_9$$

and again consider this number mod 10. If the sum is not equivalent to 0 mod 10, then there was an error in reading the number. For example, suppose the nine numbers are:

$$1 \quad 2 \quad 0 \quad 0 \quad 1 \quad 0 \quad 1 \quad 4 \quad 3,$$

then we would compute:

$$(7 \times 1) + (3 \times 2) + (9 \times 0) + (7 \times 0) + (3 \times 1) + (9 \times 0) +$$
$$(7 \times 1) + (3 \times 4) + (9 \times 3) = 62 \equiv 2 \bmod 10,$$

and, since we do not get 0, we conclude that there must be an error somewhere. We know for certain those numbers are definitely not a bank identification number. Neat, huh? The Williamstown Savings Bank identification number begins with:

$$2 \quad 1 \quad 1 \quad 8 \quad 7 \quad 2 \quad 9 \quad 4 \quad \blacksquare.$$

What must its last (check) digit be?

The answer is 6. This error-detection system is able to catch any one error (when one digit is read incorrectly) and is able to catch most switches of two consecutive digits and even most switches of two digits one apart from each other (for example, if you read 5 3 9

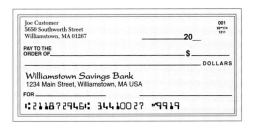

when the number was actually 9 3 5). In the Mindscapes, we provide other illustrations, including airline tickets, so the next time you are waiting around in an airport for your connecting flight, look at your boarding pass and do some modular arithmetic.

In the next section we will discover that modular arithmetic is at the core of keeping all our secrets secure on the Information Superhighway.

## A LOOK BACK

The notion of clock arithmetic or modular arithmetic mathematically models the idea of cycles and the cyclic nature of our lives. This form of arithmetic allows us not only to figure out what day of the week it will be in 210 days but also allows us to check for errors in data through various check digit methods. As we will see, modular arithmetic also leads to the notion of public key secret codes. These last two advances are extremely valuable in our technological world.

When we look at the world, we see many examples of cycles. One way to develop mathematical ideas is to look at natural phenomena, model them using mathematics, and then explore the abstract ideas contained in the model. Ideas can be developed by taking general notions and refining them. We can then explore our new world of the mind without referring back to nature. However, we often find that our thought experiments are useful in the real world.

*Create abstract ideas by modeling nature.*

*Explore ideas systematically, investigate consequences, and formulate general principles.*

◆ ◆ ◆ ◆ ◆ ◆ ◆ ◆ ◆ ◆ ◆ ◆ ◆ ◆ ◆ ◆ ◆ ◆ ◆

## MINDSCAPES ▸ Invitations to Further Thought*

### I. SOLIDIFYING IDEAS

1. **Hours and hours.** The clock now reads 10:45. What time will the clock read in 96 hours? What time will the clock read in 1,063 hours? Suppose the clock reads 7:10. What did the clock read 23 hours earlier? What did the clock read 108 hours earlier?

2. **Days and days.** Today is Saturday. What day of the week will it be in 3,724 days? What day of the week will it be in 365 days?

3. **Months and months (H).** It is now July. What month will it be in 219 months? What month will it be in 120,963 months? What month was it 89 months ago?

---

*In the Mindscapes section, exercises marked (H) have hints for solutions at the back of the book. Exercises marked (S) have solutions.

4. **Celestial seasonings (S).** Which of the following is the correct UPC for Celestial Seasonings Ginseng Plus Herb Tea? Show why the other numbers are not valid UPCs.

   0   7 1 7 3 4   0 0 0 2 1   8
   0   7 0 7 3 4   0 0 0 2 1   8
   0   7 0 7 4 3   0 0 0 2 1   8

5. **Spaghettios.** Which of the following is the correct UPC for Franco-American Spaghettios? Show why the other numbers are not valid UPCs.

   0   5 1 0 0 0   0 2 5 6 2   4
   0   5 1 0 0 0   0 2 5 2 6   4
   0   5 1 0 0 0   0 2 5 2 6   5

6. **Progresso.** Which of the following is the correct UPC for Progresso Minestrone soup? Show why the other numbers are not valid UPCs.

   0   4 1 1 9 6   0 1 0 1 2   1
   0   5 2 0 1 0   0 0 1 2 1   2
   0   0 5 0 5 5   0 0 5 0 5   3

7. **Tonic water.** Which of the following is the correct UPC for Canada Dry tonic water? Show why the other numbers are not valid UPCs.

   0   1 6 9 0 0   0 0 3 0 3   4
   0   2 4 0 0 1   1 0 6 9 1   3
   0   1 0 0 1 0   2 0 1 1 0   5

8. **Real mayo (H).** The following is the UPC for Hellmann's 8-oz. Real Mayonnaise. Find the covered digit.

   0   4 8 0 0 1   2 6 ■ 0 4   2

9. **Applesauce.** The following is the UPC for Lucky Leaf Applesauce. Find the covered digit.

   0   2 8 5 0 0   1 1 0 7 0   ■

10. **Grand Cru.** The following is the UPC for Celis Ale Grand Cru. Find the covered digit.

    ■   3 5 8 8 8   4 1 2 0 1   9

11. ***Mixed nuts.*** The following is the UPC for Planter's 6.5-oz. Mixed Nuts. Find the covered digit.

    0   2 9 ■ 0 0   0 7 3 6 7   8

12. ***Blue chips.*** The following is the UPC for Garden of Eatin' 10-oz. Blue Corn Chips. Find the covered digit.

    0   1 5 8 3 9   ■ 0 0 1   5

13. ***Lemon.*** The following is the UPC for RealLemon Lemon Juice. Find the covered digit.

    0   5 3 0 0 0   1 5 1 0 8   ■

14. ***Decoding (S).*** A friend with lousy handwriting writes down a UPC number. Unfortunately, you can't tell his 4's from his 9's or his 1's from his 7's. If the code looks like 9 0 3 0 6 8 8 2 3 5 1 7, is there any way to deal with the ambiguity? If so, what is the actual UPC number? If it is not possible to determine the correct UPC number, explain why.

15. ***Check your check.*** Look up your bank code on your check. Verify that it is a valid bank code.

16. ***Bank checks.*** Find the check digits for the following bank codes:

    3 1 0 6 1 4 8 3 ___    0 2 5 7 1 1 0 8 ___

17. ***More bank checks.*** Find the check digits for the following bank codes:

    6 2 9 1 0 0 2 7 ___    5 5 0 3 1 0 1 1 ___

18. ***UPC your friends.*** Have your friend find a product that has a 12-digit UPC. Ask your friend to carefully read the digits, skipping one digit and saying "blank" in its place. Figure out the missing digit. Do this with several different products if you wish. Explain to your friend how you did it. Record the UPCs, the blank digit, and your friend's reactions.

19. ***Whoops.*** A UPC number for a product is

    0   5 1 0 0 0   0 2 5 2 6   5.

    Explain why the errors in the following misread versions of this UPC would not be detected as errors:

    0   5 1 0 0 0   0 2 6 2 5   5;
    0   5 0 0 0 0   0 5 5 2 6   5.

20. *Whoops again.* A bank code is

   0  1  1  7  0  1  3  9  8.

   Explain why the errors in the following misread versions of this bank code would not be detected:

   7  1  1  0  0  1  3  9  8;

   0  1  1  7  0  8  3  9  1.

## II. CREATING NEW IDEAS

1. *Mod remainders* (S). Where would 129 be on a mod 13 clock (clock goes from 0 to 12)? What is the remainder when 129 is divided by 13?

2. *More mod remainders.* Where would 2,015 be on a mod 7 clock? What is the remainder when 2,015 is divided by 7? Generalize your observations and state a connection between mod clocks and remainders.

3. *Money orders.* U.S. Postal Money Orders have a 10-digit serial number and a check digit. The check digit is the number between 0 and 6 that represents what the 10-digit serial number is equivalent to using a mod 7 clock. This check digit is the same as the remainder when the serial number is divided by 7. What is the check digit for the postal money order with serial number 6830910275?

4. *Airline tickets.* An airline ticket identification number is a 14-digit number. The check digit is the number between 0 and 6 that represents what the identification number is equivalent to using a mod 7 clock. Thus, the check digit is just the remainder when the identification number is divided by 7. What is the check digit for the airline ticket identification number 1 006 1559129884?

5. *UPS.* The United Parcel Service uses the same check digit method as the U.S. Postal Money Orders and airline tickets for their pickup record numbers. What would be the check digit for UPS pickup record number 84200912?

6. *Check a code.* U.S. Postal Money Order serial numbers, airline ticket identification numbers, UPS pickup record numbers, and Avis and National rental car identification numbers all use the mod 7 check digit procedure. Find an example and check the check digit. For instance, get a copy of an airline ticket and check the identification number.

7. *ISBN.* The 10-digit book identification number called the International Standard Book Number (ISBN), has its last digit as the check digit. The check digit works on a mod 11 clock. If the ISBN has digits $d_1\ d_2\ d_3\ d_4\ d_5\ d_6\ d_7\ d_8\ d_9\ d_{10}$, then to check if this number is valid, we compute the following number

$$1d_1 + 2d_2 + 3d_3 + 4d_4 + 5d_5 + 6d_6 + 7d_7 + 8d_8 + 9d_9 + 10d_{10}.$$

If the ISBN is correct, this new calculated number should be equivalent to 0 mod 11. We use the digit X to stand for 10 on the mod 11 clock. For example, consider the ISBN 0-387-97993-X. To check this ISBN, we compute (mod 11) (Remember that X = 10):

$$(\mathbf{1} \times 0) + (\mathbf{2} \times 3) + (\mathbf{3} \times 8) + (\mathbf{4} \times 7) + (\mathbf{5} \times 9) +$$
$$(\mathbf{6} \times 7) + (\mathbf{7} \times 9) + (\mathbf{8} \times 9) + (\mathbf{9} \times 3) + (\mathbf{10} \times 10) =$$
$$6 + 24 + 28 + 45 + 42 + 63 + 72 + 27 + 100 \equiv$$
$$6 + 2 + 6 + 1 + 9 + 8 + 6 + 5 + 1 \equiv$$
$$44 \equiv 0 \bmod 11.$$

Therefore, this number is a valid ISBN. Verify this check method for the ISBN on the copyright page of this book.

8. **ISBN check (H).** Find the check digits for the following ISBNs:

   0-219-60512-____;   1-101-38216-____.

9. **ISBN error.** The ISBN 3-540-06395-6 is incorrect. Two adjacent digits have been transposed. The check digit is not part of the pair of reversed digits. What is the correct ISBN?

10. **Brush up your Shakespeare.** Find a play by Shakespeare and check its ISBN.

### III. FURTHER CHALLENGES

1. **Mods and remainders.** Use the Division Algorithm (see section 2.3) to show that the remainder when a number $n$ is divided by $m$ is equal to the position $n$ would be on a mod $m$ clock (a mod $m$ clock goes from 0 to $m - 1$).

2. **Catching errors (H).** Give some examples in which the UPC check digit does not detect an error of two switched adjacent digits. Try to determine a general condition whereby a switching error in those digits would not be detected. (*Hint:* Consider the difference of the digits.)

3. **Why three?** In the UPC, why is 3 the number every other digit is multiplied by rather than 6? (*Hint:* Multiply every digit from 0 to 9 by 3 and look at the answers mod 10. Do the same with 6 and compare your results.) Are there other numbers besides 3 that would function effectively? What number might you try?

4. **A mod surprise.** For each number $n$ from 1 to 4, compute $n^2$ mod 5. Then for each $n$, compute $n^3$ mod 5 and finally $n^4$ mod 5. Do you notice anything surprising?

**5. *A prime magic trick.*** Pick a prime number and call it $p$. Now pick any natural number smaller than $p$ and call it $a$. Compute $a^{p-1}$ mod $p$. What do you notice? You can use this observation as the basis for a magic trick. Have a friend think of a natural number less than $p$ (but don't have the person say the number). Tell that person that you will predict and write what the remainder will be when $a^{p-1}$ is divided by $p$. Write your answer and seal it in an envelope, and then ask what the person's number was. Now compute the remainder when $a^{p-1}$ is divided by $p$ and reveal the hidden prediction to the amazement of your friend. Record the reactions of your friends. The next section uses this observation in a powerful way. Check it out.

### IV. IN YOUR OWN WORDS

1. ***Personal perspectives.*** Write a short essay describing the most interesting or surprising discovery you made in this section's material. If any material seemed puzzling or even unbelievable, address that as well. Explain why you chose the topics you did. Finally, comment on the aesthetics of the mathematics and ideas in this section.

2. ***With a group of folks.*** In a small group, discuss and work through the four mod questions posed in the section and the Clairvoyant Kleenex Consultant story. After your discussion, write a brief narrative describing the methods in your own words.

3. ***Creative writing.*** Write an imaginative story (it can be comical, dramatic, whatever you like) that involves or evokes the ideas of this section.

4. ***Power beyond the mathematics.*** Provide several real-life situations—ideally, from your own experience—for which some of the strategies of thought presented in this section would provide effective methods for approaching and resolving them.

# 2.5 Secret Codes and How to Become a Spy:
## How Modular Arithmetic and Primes Lead to Secret Public Codes

*No more fiction: we calculate; but that we may calculate, we had to make fiction first.*

—Friedrich Nietzsche

With the thawing of the Cold War, have spies fallen on hard times? Might we expect to see people in trench coats standing on street corners with signs that read, "Ex-spy. Will break codes for food?" Actually, no, because although the spy business is a bit slack now, in this age of the Information Superhighway, secret codes have become indispensable. Whether you're withdrawing money from an ATM, sending your VISA card number over the Web to make a purchase, or making a stock transaction, stolen data could mean stolen money. Information on the Internet needs to be secured through the use of codes. Modular arithmetic is at the heart of modern coding, and thus coding is possibly the most powerful example of the unforeseen applicability of abstract mathematical ideas—in this case to the digital world. Who would have thought that *cryptography*—the study of secret codes—would become an important part of daily life and that the exploration of numbers would be central to coding?

This section is difficult. To master every part of the mathematics involved requires a significant effort. Luckily for us, the idea of public key cryptography is interesting even without fully delving into the mathematical details that

*Mona Lisa with Keys* (1930) by Fernand Léger

make it work. This challenging section is evidence that as the world changes, marginal ideas of today may become central tomorrow. Good luck.

## *Top Secret Codes*

How can we code and decode messages? One possibility is to replace one letter by another letter. For example, suppose the authors created the following coding scheme:

| Message: | A | B | C | D | E | F | G | H | I | J | K | L | M | N | O | P | Q | R | S | T | U | V | W | X | Y | Z |
|---|---|---|---|---|---|---|---|---|---|---|---|---|---|---|---|---|---|---|---|---|---|---|---|---|---|---|
| Coded as: | T | H | E | Q | U | I | C | K | B | R | O | W | N | F | X | J | M | P | D | V | L | A | Z | Y | G | S |

If we wanted to send the message:

   YOUR JOKES ARE LAME,

then we could send the coded message:

   GXLP RXOUD TPU WTNU.

The major problem with this code is that it is easy to break without knowing the key. That is, any enemy who captured a sufficiently long encrypted message could figure out the original message. More elaborate coding methods are harder to break but still can be deciphered if the codes are shared. That is, suppose you are receiving messages from both Bill and Hillary, and each encodes his or her message using the same scheme. If Hillary captures Bill's encrypted message, couldn't she decode it by simply reversing the encoding procedure? It seems that we must trust our friends—a grave drawback to any shared coding scheme. Ideally, we would prefer a code by which people are able to encode messages to us but are at the same time unable to decode other messages that have been encoded by the same process. Is such a coding scheme possible? If someone could encode a message, then all he or she would have to do to decode messages in the same type of coding scheme is reverse the coding procedure. However, this plausible statement turns out to be false, and therein lies the core of modern coding methods.

In this section we will look at a coding technique invented during the last few decades that uses a 350-year-old theorem about modular arithmetic to encrypt and decode secret messages.

*Attractive ideas in one realm often have unexpected uses elsewhere.*

## *Public Key Codes*

The method uses an encrypting and decoding scheme that is fundamentally new in the coding business. The new wrinkle is the invention of the public key code. *Public key codes* are codes that allow us to encode any message but prevent us from decoding other messages encrypted by the same technique. Such codes are called *public key codes* since we can tell the entire world how to encode messages to us. We can even tell our

## THE BASIC THEME OF THE "PUBLIC" ASPECT OF THE RSA CODING SCHEME IN TEN SENTENCES

◆ ◆ ◆ ◆ ◆ ◆ ◆ ◆ ◆ ◆ ◆ ◆ ◆ ◆ ◆ ◆ ◆ ◆ ◆ ◆ ◆ ◆ ◆ ◆ ◆

Let's select two enormous prime numbers—and we mean enormous—say each having about 300 digits—and multiply those numbers together. How can we multiply them together? Computers are whizzes at *multiplying* natural numbers—even obscenely long ones. *Factoring* large numbers, however, is hard—even for computers. Computers are smart but not infinitely smart—there are limits to the size of natural numbers that they can factor. In fact, our product is much too large for even the best computers to factor. So if we announce that huge product to the world, even though it can be factored in theory; in practice it cannot. Thus we are able to announce the gigantic number to everyone, and yet no one but we would know its two factors. This huge product is the *public* part of the RSA public key code. Somehow, the fact that only we know how to factor that number allows us to decode messages while others cannot. It's not obvious why this factoring fact is helpful in making secret codes, but we'll see that it's really at the heart of the matter.

Rivest and Adleman wondering where Shamir is.

enemies. We can take out ads in the papers telling everyone how to encode messages to us; it's no secret; it's public. The key is that we and only we can *decode* an encrypted message. Isn't this notion counterintuitive? How can such a coding scheme work? We'll take a look at the RSA public key code. RSA stands for the names of the three mathematicians who discovered the system: Ronald Rivest, Adi Shamir, and Leonard Adleman.

Before jumping into the technical details of this coding scheme, let's try to make the basic idea of the encoding and decoding aspects of the RSA code plausible. For this purpose, we journey to Carson City.

### *The Carson City Kid and the Perfect Shuffle Code*

The Carson City Kid was the master of cards (actually he was no kid, but it sounds better than "The Carson City Yuppie"). His hands were quicker than the eye, and his morals were just as fast. One thing the Kid could do without fail was what is known in the trade as a *perfect shuffle*. That is, he would cut the deck of 52 cards precisely in half and then shuffle them perfectly—one card from the top half, then one from the bottom half, and so on, intermixing the cards exactly—one from one side, one from the other. For the larcenous among our readers, the advantage of a perfect shuffle is that, contrary to typical random shuffles, perfect shuffles only *appear* to bring disorder to the deck. The original ordering is restored after exactly 52 perfect shuffles.

Kid Carson was an enterprising soul who did not want to spend his life in casinos, rolling in money and surrounded by glamorous and attractive people. Instead, he decided to go into the secret message biz and be surrounded by glabrous and atrocious people. His method was simple. He knew that most people could not execute perfect shuffles. They could do only five or six shuffles before messing up. The Kid's method was straightforward: The code sender would take a deck of 52 blank playing cards and write the message using one letter per card. Then the sender spy would carefully do five perfect shuffles—leaving the deck of cards in an apparently random order. The spy receiving the shuffled deck would then hand the coded message (the shuffled deck) to Kid Carson.

Kid Carson knew exactly what to do with the shuffled deck. He quickly shuffled the deck with 47 more perfect shuffles and voilà! The cards had rearranged themselves exactly into their original order, and so the message could be read.

Of course, Kid Carson's technique is too simple to use in practice. With determination, a person who captured the five-shuffled deck could do the reverse of those perfect shuffles. However, the Kid's technique demonstrates a mathematical fact that revolutionized the coding business.

## *RSA*

In 1977, Ronald Rivest, Adi Shamir, and Leonard Adleman discovered a public key coding scheme that uses modular arithmetic. This public key coding method is referred to as the *RSA Coding Scheme* and is now used millions of times each day. Kid Carson's 47 perfect shuffles that return the deck to its original order captures the spirit of this RSA public key coding scheme. A shuffling procedure encodes a message, and only the receiver knows how to continue to shuffle the message further in a way that unshuffles the message—no one besides the receiver can perform that additional shuffling. So, now there are two basic questions we hope you are wondering: (1) What are we shuffling? and (2) How do we keep shuffling to get back to where we started?

## *Shuffling Numbers*

We will shuffle numbers. That is, we will first convert our message to numbers and then shuffle those numbers. How do we shuffle the numbers? Let's take a prime number, say 5. Pick any number that does not have 5 as a factor, for example, 8. To shuffle 8, let's just raise 8 to higher powers and look at the remainders of those powers of 8 when we divide by 5. In other words, raise 8 to higher powers and look at those numbers mod 5.

| Powers of 8 | | Powers of 8 mod 5 |
|---|---|---|
| $8^1 =$ | 8 | 3 |
| $8^2 =$ | 64 | 4 |
| $8^3 =$ | 512 | 2 |
| $8^4 =$ | 4096 | 1 |
| $8^5 =$ | 32768 | 3 |

**Look for patterns.**

◆ ◆ ◆ ◆ ◆ ◆

The second column represents a type of shuffling of a four-card deck where the "shuffling" is accomplished by multiplying by 8. Notice that after 5 shuffles of multiplication by 8 mod 5 we get back to 3.

Let's try this again with a different-size deck. Suppose we pick the prime 7 and choose a number that does not have 7 as a factor—say 10. Let's shuffle 10 by raising it to higher powers and considering those powers mod 7.

| Powers of 10 | | Powers of 10 mod 7 |
|---|---|---|
| $10^1 =$ | 10 | 3 |
| $10^2 =$ | 100 | 2 |
| $10^3 =$ | 1000 | 6 |
| $10^4 =$ | 10000 | 4 |
| $10^5 =$ | 100000 | 5 |
| $10^6 =$ | 1000000 | 1 |
| $10^7 =$ | 10000000 | 3 |

Here we notice that after seven shuffles of powers mod 7 we get back to 3. Now it's your turn. Try this shuffling yourself. Let's set the prime number to be 5. Now pick some numbers that have no factor of 5 and shuffle them by raising them to powers mod 5. Try this shuffling with at least two different numbers. What do you notice? How many shuffles get us back to where we started mod 5? Let's look for patterns.

By experimenting, we discover that, if we shuffle 5 times mod 5, we get back to where we started. We also notice that, after we shuffle 4 times mod 5, we always get 1 as the answer. This observation turns out to be a mathematical fact—known as *Fermat's Little Theorem*.

> **FERMAT'S LITTLE THEOREM.** *If $p$ is a prime number and $n$ is any integer that does not have $p$ as a factor, then $n^{p-1}$ is equal to 1 mod $p$. In other words, $n^{p-1}$ will always have a remainder of 1 when divided by $p$.*

It is Fermat's Little Theorem, proved more than 350 years ago, that is the basis of our shuffling procedure. Now let's tackle the RSA public key code scheme.

## An Illustrative, Cryptic Example

We introduce the RSA public key code method by considering a specific example. Using a diabolically clever idea that will be explained later, we construct and publicize a pair of numbers to the world—in this example (7, 143). In real life the numbers would be much larger, perhaps having several hundred digits each. At the same time we construct the public numbers, we construct and keep a secret decoding number, in this case 103. The public part of the public key code does not contain the key to unlock the code; instead, the key is the secret decoding number that is kept only by the receiver of encrypted messages. It never needs to be transmitted to anyone else. We'll explain later how all these numbers were created.

*Ground your understanding in the specific.*

We publicize not only the numbers 7 and 143, but we also explain exactly how to use those numbers to encrypt a message. Here are the instructions, which could be published in the newspaper.

## Instructions for Encrypting a Secret Number Message (Less Than 143) Using the Public Numbers (7, 143)

Suppose that $W$ is a secret Swiss bank account number (less than 143) that the sender wants to encrypt and send to someone. The sender computes $W^7$ (remember that 7 is the first number announced) and then computes the remainder when $W^7$ is divided by 143 (remember that 143 was the second number announced). That is,

$$W^7 = 143q + C$$

where the remainder $C$ is an integer between 0 and 142. Or, expressed in modular arithmetic,

$$W^7 \equiv C \bmod 143.$$

The number $C$ is now the coded version of $W$.

## Decoding Messages

The receiver receives the coded message $C$ and now must decode it. This decoding process requires the receiver to compute $C^{103}$ (remember that 103 is the secret number that no one but the receiver knows) and then to compute the remainder when $C^{103}$ is divided by 143. That is,

$$C^{103} = 143q + D$$

where the remainder $D$ is an integer between 0 and 142. Or, expressed in modular arithmetic,

$$C^{103} \equiv D \bmod 143.$$

The amazing fact is that $D$ (the decoded message) will always be identical to $W$ (the original, uncoded message). Thus the receiver decoded the coded message $C$ to produce the original message $W$.

Suppose someone sets up the public key code described above, announces the public numbers (7, 143) and the coding method, and keeps the number 103 secret. Now let's further suppose that a friend wishes to secretly send her Swiss bank account number, 71. Here is a table showing the sequence of events to code the message 71.

| Receiver | Sender |
|---|---|
| Announces to the world the numbers (7, 143) and coding instructions. Tells no one secret decoding number 103. | |
| | Wants to send "71" to her friend. Knows the coding numbers (7, 143). |
| | Sender computes $71^7$ mod 143, which is 124. 124 is the coded version of 71. Sender sends 124 to receiver. |
| Receiver receives the coded message "124." Using the secret number 103, receiver computes $124^{103}$ mod 143, which is 71—the original message! | |

To encode the number 71, the sender computes $71^7$, which equals 9,095,120,158,391, and then computes the remainder when this number is divided by 143. The remainder turns out to be 124. So 124 is the encoded version of 71, and that is what the sender sends to the receiver. Now the receiver has to decode 124.

Remember that 103 is the secret decoding key known only to the receiver. To decode 124, the receiver first computes 124 to the 103rd power and then finds the remainder when that number is divided by 143. $124^{103}$ equals

41921187047849896446113000569294530888483668997732045634627122565220914671133939555703940592675185212029512808239919702590414929088043093696556512787027350058759384015077439569484127475589434834019120344958849410 6624, which is a pretty big number; however, the remainder when it is divided by 143 is (drum roll please) 71.

Of course, if we wanted to send or receive words (rather than just numbers) we could first convert the letters into numbers in some straightforward way (for example, A = 01, B = 02, ... Z = 26, space = 27) and then encode the number, just as we did with 71. Raising the encrypted number to the 103rd power and taking the remainder when that result is divided by 143 results in the original message. The incredible fact is that this method of retrieving the original number always works.

Take any number $n$ from 0 to 142, compute the remainder when the number $n^7$ is divided by 143, and call that remainder $C$. Then compute the remainder when $C^{103}$ is divided by 143, and we will miraculously have the original number $n$ with which we started! Of course, there's a reason behind such seemingly amazing coincidences.

## New Questions

The preceding example leads to many questions. Here are a few:

1. We said that, in real life, the numbers we use would have perhaps several hundred digits each. Is it practical, even for a computer, to raise numbers with hundreds of digits to powers of several hundred digits?
2. Where did the numbers 7, 143, and 103 come from?
3. Why in the world does this process work? Is this coding and decoding process just an amazing fluke?

## Practical Ways to Raise Huge Numbers to Large Powers

As we illustrated, with the aid of a computer, we can raise 124 to the 103rd power and write out the whole 200+ digit answer.

*Question:* If you take a 100-digit number and raise it to a 100-digit power, approximately how many digits would the answer have? Would it be physically possible to write out such a number?

*Answer:* No. A 100-digit number raised to a 100-digit power would have about **$10^{102}$ digits.** If we wanted to write out such a number, we would have to write pretty small, since physicists' best current estimate for the number of atoms in the universe is less than $10^{80}$, a tiny fraction of $10^{102}$. Thus a huge number of digits would have to be written on each atom in the universe. In other words, no computer could possibly do that calculation.

Well, then, aren't we stuck? We have just seen that no computer can actually raise a 100-digit number to a 100-digit power. We must be clear about what we really want.

Actually, we are not stuck. The trick is to realize that all we are really after is the *remainder* that we get after dividing that huge power by some number. Finding this remainder can be accomplished by modular arithmetic without ever computing the big power. Such a procedure does not strain even a laptop computer. We and our computers are able to compute remainders efficiently without needing more atoms than our universe possesses.

## Where Those Numbers Come From

The numbers 143, 7, and 103 come from an ingenious combination of just a few ideas from number theory. Here is how we choose the numbers. We first give the method in flowchart form and then describe the procedure in prose.

**Receiver**

Selects two different prime numbers, in this case 11 and 13, but tells no one what they are.

Multiplies them together: $11 \times 13 = 143$. This is one of the public numbers. The public will know the product but will not be able to factor it since the number, in practice, would be too large.

Subtracts 1 from each of the two primes, $11 - 1 = 10$; $13 - 1 = 12$, and then selects a number at random that has no common factor with these two numbers—in this case, 10 and 12. Here the receiver selects 7, so this is the other number publicly announced.

Using the numbers 120 and 7, the receiver finds integers $d$ and $y$ so that they satisfy the equation: $7d - 120y = 1$. One such solution is $(7 \times 103) - (120 \times 6) = 1$. The value of $d$—in this case 103—is the secret decoding number that only the receiver knows or can figure out, since figuring out a solution to the equation required the factorization of 143—which no one knows but the receiver.

## The Receiver's Preparation in Words

The process begins by picking two different prime numbers. In this case, we pick $p = 11$ and $q = 13$. Their product is 143, and that is one of the public numbers. The first public number in this coding scheme is always the product of two different primes. In real life, it is the product of two primes, each having a couple hundred digits.

To select the numbers 7 and 103, we follow some intermediate steps. The next step is to compute $(p - 1)(q - 1)$ or, in this case, $10 \times 12 = 120$. Now we select any number $e$ that shares no common factors with 120 ($e$ stands for *encoding*). It can be any such number. In our example, we chose $e$ to be 7. That number is the other public number.

Okay, in all honesty, what we are about to do next appears to be coming out of thin air, but we'll soon see its value. We now find integers $d$ and $y$ such that $7d - 120y = 1$. In this case, we find that we can take $d$ to equal 103 and $y$ to be 6, since those values satisfy the equation

$$(7 \times 103) - (120 \times 6) = 1.$$

The number $d$ ($d$ for *decoding*), which in this case equals 103, is the secret decoding number that we keep to ourselves. That's it! Whenever we select the numbers $p$, $q$, $e$, and $d$ in the manner described above, the coding scheme will always work.

## Why Those Numbers Work

Let's see why this coding and decoding scheme always works. Before we give an overview of why the RSA coding scheme works, we have to confess that what follows is difficult. What makes it difficult? The answer is that there are many steps—any one of which is no great intellectual feat. However, when we string them together, one after another, the logic and modular arithmetic can get out of hand. These details are more interesting to some than to others. So readers who decide to invest the energy to learn what follows must expect to struggle and to reread the information several times. Other readers may decide to limit their investment in this topic and move on—remember Vietnam.

We begin our explanation by a quick recap: First we picked two primes:

$$p = 11 \quad \text{and} \quad q = 13.$$

Their product is 143, and that is one of the public numbers. We next computed

$$(p - 1)(q - 1)$$

or, in this case,

$$10 \times 12 = 120$$

and selected a number $e$ that shares no common factors with 120—we chose $e$ to be 7. That number is the other public number. We then found integers 103 and 6 that satisfy

$$(7 \times 103) - (120 \times 6) = 1.$$

The number 103 is the secret decoding number, and we announce the pair (7, 143) to the entire world. We're now ready to see why coding and decoding works with the message 71. Remember that, to encode the message, we computed the remainder when $71^7$ is divided by 143 and got 124. So, in modular arithmetic terms,

$$71^7 \equiv 124 \bmod 143.$$

Then, to decode, we raised 124 to the 103rd power and found the remainder when we divided the answer by 143 (that is, we figured out $124^{103}$ mod 143). We now want to determine why this remainder must always equal the original message (in this case 71) that our sender sent. Specifically, we wish to understand why

$$124^{103} \equiv 71 \bmod 143.$$

## The Numerical Details

Recall that 143 is $11 \times 13$. We consider the two primes 11 and 13 separately and start with 11. Recall that 124 is the remainder when $71^7$ is divided by 143. Therefore, by the Division Algorithm,

$$71^7 = 143k + 124,$$

for some integer $k$. We now write 143 as $11 \times 13$ and notice that

$$71^7 = (11 \times 13 \times q) + 124 \equiv 0 + 124 \bmod 11.$$

So we can conclude that

$$(71^7)^{103} \equiv (124)^{103} \bmod 11.$$

Remember now how we selected 103:

$$(7 \times 103) = 1 + (120 \times 6).$$

Using this fact we see that

$$(71^7)^{103} = 71^{(7 \times 103)} = 71^{1+(120 \times 6)} = 71 \times (71)^{(120 \times 6)}$$
$$= 71 \times (71^{72})^{10}.$$

Since the prime 11 is not a factor of 71, we know by Fermat's Little Theorem—our shuffling fact—that

$$(71^{72})^{10} \equiv 1 \bmod 11 \text{ (notice that the power 10 is just } 11 - 1).$$

So, putting all these observations together we see that

$$124^{103} \equiv (71^7)^{103} = 71 \times (71^{72})^{10} \equiv 71 \times 1 = 71 \bmod 11.$$

Therefore, by subtracting 71 from both sides, we see that

$$124^{103} - 71 \equiv 0 \bmod 11.$$

In other words, we now see that 11 divides evenly into $124^{103} - 71$.

If we repeat with the prime 13 every step we did with 11, we see that 13 also divides evenly into $124^{103} - 71$.

We have just shown that both 11 and 13 divide evenly into $124^{103} - 71$. Since 11 and 13 are different primes, we know that the product of 11 and 13, namely 143, must divide evenly into $124^{103} - 71$. In other words, when $124^{103} - 71$ is divided by 143 the remainder is 0. Given this fact, what is the remainder when just $124^{103}$ alone is divided by 143? It has to equal 71, since once we subtract off the 71 we have seen its remainder is zero. Thus the remainder when $124^{103}$ is divided by 143 is 71. We decoded and got the original message back. This numerical illustration includes all the essential ideas why this procedure works in general.

## *The RSA Code in General*

To build an RSA public key code, the receiver first selects two different prime numbers—let's call them $p$ and $q$. In practice the receiver chooses primes that have perhaps 200 digits each. Now we define the integers $n$ and $m$ as follows: $n = pq$, and $m = (p-1)(q-1)$.

Next the receiver selects any positive integer $e$ (*encoding* power) such that $m$ and $e$ share no common factors. Since $e$ and $m$ share no common factors, it turns out we can always find positive integers $d$ (*decoding* power) and $y$ such that $ed = 1 + my$.

The receiver announces to the entire world the pair of numbers ($e$, $n$). The important thing, however, is to tell no one what $d$ is. All the other numbers are no longer needed. The receiver probably should destroy any documents with the numbers $p$ or $q$ on them. Although we know that $n$ can be factored into the product of $p$ and $q$, the number $n$ is so large that nobody besides the receiver can find those factors. The receiver no longer needs the factors $p$ and $q$, and, if they get into the wrong hands, the security of our code could be violated.

If a sender wishes to send the receiver the message $W$, the sender computes the remainder when $W^e$ is divided by $n$—let's call the remainder $C$ (for *coded* message). The integer $C$ is the coded version of the word $W$. Since the message $W$ has now been encoded to the message $C$, the sender could take out an ad in the paper and tell everyone the message $C$. Only the receiver, however, will be able to decode it by finding the remainder when $C^d$ is divided by $n$—let's call it $D$ (for *decoded* message). It turns out that $D$ will be exactly $W$. That is, $D$ is the original word, so the receiver just decoded the message. The proof of why it always works involves the exact same logic used in our previous example.

The RSA scheme is both interesting and extremely practical. In fact RSA is becoming widely accepted commercially. Think we're just kidding? RSA Data Security, the company formed to promote and sell RSA solutions, was recently sold for about $400 million to Security Dynamics. This astronomical figure gives some indication of how seriously this coding scheme is being taken, and it all comes from divisibility, remainders, and crazy clocks. More proof that math pays! Innovative ideas are valuable in our society, regardless of their source.

### Breaking the Code

Before closing this section, we address a natural question, namely, "How could someone break this code?" Remember that the world knows both $n$ and $e$. If someone could factor $n$ into its prime factorization $p \times q$, then knowing $p$, $q$, and $e$, he or she could figure out $m$ $(= (p - 1)(q - 1))$ and then figure out $d$ (do you see how?). Once $d$ is known, the code can be broken. So how could we avoid having someone break the code? Well, certainly we don't want anyone to be able to factor our number $n$. How could we arrange that? We take a look at current computer technology. We determine what the largest integer is that can be factored by the best computers. We then select our primes $p$ and $q$ so that each is larger than the largest integer that can be factored by machine.

Might some clever insight into how numbers work allow computers to factor even extremely large numbers and thus factor our $n$? If someone came up with such a scheme, the RSA public key code scheme would fail. Is there another way of breaking the code without factoring $n$? This question is an unsolved problem in number theory. In other words, no one knows if the only way to break the code is to actually factor $n$. Perhaps there is a devilishly sneaky way of breaking the code without ever factoring $n$. This possibility is a big mystery. Since no one now knows how to break the code, the RSA system provides a safe coding scheme for now. It's interesting to note where the borders of knowledge are. This problem is on a border being investigated currently by research mathematicians.

**A LOOK BACK**

A secret number can be encoded by raising it to a power mod $n$ where $n$ is the product of two large primes. It can be decoded and the original secret number retrieved by raising the result to yet another power mod $n$. The secrecy of this scheme is dependent on our being unable to factor numbers that are the products of two primes, each having hundreds of digits. It seems plausible that, if we can encrypt a message using some process, then we could reverse the process and decode other messages that were encrypted by that method. However, such is not the case. Public key encryption exists in the realm between the plausible and the true.

Understanding the public key encryption and decoding scheme is best attempted using a specific example. All the ideas for the general scheme are clearly present in the special case. Often we learn generally applicable principles by concentrating on specific cases.

*Ground your understanding in the specific.*

❖ ❖ ❖ ❖ ❖ ❖ ❖ ❖ ❖ ❖ ❖ ❖ ❖ ❖ ❖ ❖ ❖ ❖ ❖

# MINDSCAPES  Invitations to Further Thought*

### I. SOLIDIFYING IDEAS

1. *Petit Fermat 5.* Compute $2^4$ (mod 5). Compute $4^4$ (mod 5). Compute $3^4$ (mod 5). Oh, what the heck, compute $n^4$ (mod 5) for all numbers $n$ from 1 through 4.

2. *Petit Fermat 7.* Compute $4^6$ (mod 7). Compute $5^6$ (mod 7). Compute $2^6$ (mod 7). Oh, why not, compute $n^6$ (mod 7) for all numbers $n$ from 1 through 6.

3. *Top secret.* In our discussion, the two public numbers 7 and 143 were given. How would you encode the word "2"? The secret decoding number is 103. Without performing the calculation (unless you have a computer that can do modular arithmetic for you), how would you decode the encrypted message you just made if you are the receiver?

4. *Middle secret* (**H**). In our discussion, the two public numbers 7 and 143 were given. How would you encode the word "3"? The secret decoding number is 103. Without performing the calculation (unless you have a computer that can do modular arithmetic for you), how would you decode the encrypted message you just made if you are the receiver?

5. *Bottom secret.* In our discussion, the two public numbers 7 and 143 were given. How would you encode the word "10"? The secret decoding number is 103. Without performing the calculation (unless you have a computer that can do modular arithmetic for you), how would you decode the encrypted message you just made if you are the receiver?

6. *Creating your code* (**S**). Suppose you wish to devise an RSA coding scheme for yourself. You select $p = 3$ and $q = 5$. Compute $m$, and then find (by trial and error if necessary) possible values for $e$ and $d$.

---

*In the Mindscapes section, exercises marked (H) have hints for solutions at the back of the book. Exercises marked (S) have solutions.

7. **Using your code.** Given the coding scheme you devised in Creating your code, show how a friend would encode HI (as in our discussion, use 01 for A, 02 for B, ..., 26 for Z to convert the letters to numbers). Now decode the coded message. Did you return to your original HI?

---

The following list of random information may be useful in the subsequent three questions:

$73^7 \equiv 83 \bmod 143$  $\quad 83^{143} \equiv 58 \bmod 103 \quad$  $8^{103} \equiv 83 \bmod 143$
$74^7 \equiv 35 \bmod 143$  $\quad 74^{143} \equiv 51 \bmod 103 \quad$  $74^{103} \equiv 61 \bmod 143$
$61^7 \equiv 74 \bmod 143$  $\quad 38^{143} \equiv 29 \bmod 103 \quad$  $83^{103} \equiv 73 \bmod 143$
$83^7 \equiv 8 \bmod 143$  $\quad 35^{143} \equiv 5 \bmod 103 \quad$  $38^{103} \equiv 103 \bmod 143$
$38^7 \equiv 25 \bmod 143$  $\quad 8^{143} \equiv 72 \bmod 103 \quad$  $35^{103} \equiv 74 \bmod 143$

8. **Public secrecy.** Using the preceding list, with the public numbers 7 and 143, how would you encode "83"? How would you decode the message using the decoding number 103? What numbers in the list must you refer to for the encoding and decoding operations?

9. **Going public.** Using the preceding list, with the public numbers 7 and 143, how would you encode "61"? How would you decode the message using the decoding number 103? What numbers in the list must you refer to for the encoding and decoding operations?

10. **Secret says (H).** Using the preceding list, with the public numbers 7 and 143, and decoding number 103, how would you decode "38"? What numbers in the list must you refer to for this decoding operation?

## II. CREATING NEW IDEAS

1. **Big Fermat (S).** Compute $5^{600}$ (mod 7). (*Hint:* Remember your answer to I. 2, Petit Fermat 7.) Compute $8^{1,000,000}$ (mod 11).

2. **Big and powerful Fermat (H).** Recall how exponents work, for example, $7^{15} = 7^{(12+3)} = 7^{12} \times 7^3$. Now, using exponent antics, compute $5^{668}$ (mod 7).

3. **The value of information.** How large should you select the primes $p$ and $q$ in the RSA coding scheme? Of course, if you pick two ridiculously large primes, then the product of the two would be impossible to factor from a practical point of view. But do you really need the primes to be that large? What if you just want to send a little love message to a special friend? Do you think the CIA will want to break your code? How would you determine the size of the primes you need?

4. **Something in common.** Suppose that $p$ is a prime number and $n$ is a number that has $p$ as a factor. What is $n^{p-1} \bmod p$ in this case?

5. **Faux pas Fermat.** Compute $1^5 \bmod 6$, $2^5 \bmod 6$, $3^5 \bmod 6$, $4^5 \bmod 6$, and $5^5 \bmod 6$. What if you raise the numbers to the power

2? Compute $n^6$ mod 9 for numbers $n$ from 1 to 8. What is the answer when $n$ and 9 have no common factors? Do you think there is a way to extend Fermat's Little Theorem when the mod number is not prime? Make a guess (yes or no).

## III. FURTHER CHALLENGES

1. ***Breaking the code.*** If you could factor the large public number into its two prime factors, how could you break the code? Outline a procedure.

2. ***Signing your name.*** Suppose you get a message that claims to be coming from your friend Joseph Shlock. It says that you should invest all your savings in pork kidneys. How do you know if the message really came from your friend Shlock and not your arch-enemy Irving Satan? Ideally, it would be great if each message were "signed" by the sender in such a manner that no one could "forge" the signature. Using the RSA scheme (in reverse), devise a method by which the origins of the messages can be verified. (*Hint:* Could the encoding procedure be used to reveal the signature?)

## IV. IN YOUR OWN WORDS

1. ***Personal perspectives.*** Write a short essay describing the most interesting or surprising discovery you made in exploring this section's material. If any material seemed puzzling or even unbelievable, address that as well. Explain why you chose the topics you did. Finally, comment on the aesthetics of the mathematics and ideas in this section.

2. ***With a group of folks.*** In a small group, discuss and work through the RSA public key coding scheme. After your discussion, write a brief narrative describing the method in your own words.

3. ***Creative writing.*** Write an imaginative story (it can be comical, dramatic, whatever you like) that involves or evokes the ideas of this section.

4. ***Power beyond the mathematics.*** Provide several real-life situations—ideally, from your own experience—for which some of the strategies of thought presented in this section would provide effective methods for approaching and resolving them.

# 2.6 The Irrational Side of Numbers:
## Are There Numbers Beyond Fractions?

*God made integers, all else is the work of humankind.*

—Leopold Kronecker

The natural numbers are the first and most natural measures of quantity; however, suppose we have more than one of something but less than two? Clearly we need fractional quantities to make such measurements. Fractions let us measure any quantity to any desired degree of precision. In principle, we could measure a length to within one millionth of an inch. But are there quantities that even the most precise fraction cannot measure exactly? Specifically, is every number a fraction? That question challenged the ancient Greeks and eventually opened their minds to a totally new and surprising realization about the notion of "number." This discovery of the Greeks, which we will soon discover for ourselves, is another powerful illustration of our major theme: By asking clear questions and examining the familiar in a careful and logical manner, we uncover hidden richness.

Is every number a fraction? To answer this question we make an assumption and follow its consequences. Letting logic lead, we suppose that every length could be measured exactly as a fraction, and then we see what other results we would be compelled, by logic, to accept. Exploring the logical consequences of an assumption is a valuable way to determine whether the assumption is reasonable.

### A Rational Mindset

The ancient Greeks, and probably people before them, devised a reasonable method of measuring parts of things. If we take an object and break it into 10 equal pieces and take 9 of them, then we have measured 9 tenths (9/10). Corner stores that sell gasoline have learned this lesson well—

every gas price ends in 9/10 of a cent. This clever ploy allows the neighborhood convenience store to milk us for a smidgen extra on each gallon without our really noticing.

Theoretically we could take an object, break it into a million pieces, and take 375,687 of them and get 375,687 millionths (375,687/1,000,000). So by taking a large enough number of pieces, we can measure parts of things to any degree of accuracy we want; unfortunately, even then we may not be correct. But we are getting a bit ahead of ourselves. The Greeks thought that the natural numbers were natural gifts from the gods. Ratios of those number essences together with their negatives and zero produce the rational numbers. A *rational number*, therefore, is a number that can be written as a fraction $a/b$ or $-a/b$ where $a$ and $b$ are natural numbers or where $a = 0$. Some examples of rational numbers are 1/2, 22/7, 109/51, −35/219, 15 (15/1), and 0.

rational thought

To get accustomed to this idea, find a rational number between 1 and 2. Now find a rational number between 1001/1003 and 1002/1003. Why is there always a rational number between any two other rational numbers?

Notice that, even if two rational numbers are very close together, we can always find many (in fact, infinitely many) rational numbers between them. Since we can cut things up into as many equal-size pieces as we wish, it seems reasonable to conjecture that every number is rational, which the early Greeks believed. Given common observations and life experiences this idea seems both natural and rational (excuse the pun). In fact, the atomic theory of matter and quantum mechanics suggests that matter has a limit to its divisibility, and, hence, for physical objects, there may literally be a specific number of indivisible units that make up the object. So if we break an object in two and wish to measure how big each piece is, we could count the number of particles in one piece and divide that number by the total number of particles in the original object to see what fraction of the object we have.

Of course, mathematics is not constrained by mere physical reality. Physical reality is just the starting point for mathematics.

## *The Pythagoreans' Secret Society and the Square Root of 2*

Explore the consequences of assumptions.

Let's now assume, as the Greeks did, the reasonable hypothesis that all numbers are rational numbers and see where that assumption leads us. Unfortunately, it leads us into some deep trouble. Along the way, however, we will learn that an effective method for discovering new ideas and truths is to explore the consequences of assumptions.

Pythagoras (580 B.C.–500 B.C.) and his followers formed a school devoted to discovering great ideas, many of which were mathematical. The Pythagorean School was a secret society. They developed important mathematical concepts and kept them to themselves.

The Pythagorean theorem, which we will physically see and feel in Chapter 4, Geometric Gems, tells us that in a right triangle the square of the hypotenuse equals the sum of the squares of the lengths of the two shorter sides; that is, $a^2 + b^2 = c^2$ where

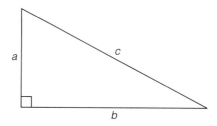

After the Pythagoreans discovered this relationship, they considered the triangle with $a = b = 1$:

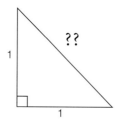

and wondered what the length of the hypotenuse was. If they called the length $H$, they knew that $1^2 + 1^2 = H^2$. So $H^2 = 2$, which is to say that $H$ is the square root of 2, denoted as $H = \sqrt{2}$.

## Assume It's Rational

The Pythagoreans believed that all numbers were rational, so the number $H$, the square root of 2, would have to be equal to some fraction, say $a/b$, where both $a$ and $b$ are natural numbers. Let's see where that assumption led the Pythagoreans and where it leads us.

The first thing we can do with our assumption that $H$ equals $a/b$ is to cancel any common factors in $a$ and $b$ until we are left with an equivalent rational number with the top number and the bottom number having no common factors. For example (notice how we factor every number into primes and then cancel common primes),

$$\frac{60}{45} = \frac{2^2 \times 3 \times 5}{3^2 \times 5} = \frac{2^2}{3} = \frac{4}{3}.$$

So the rational 60/45 is equal to 4/3. Notice that 3 and 4 have no common factors.

## No Common Factors

Let's return to $H$, the square root of 2. We are assuming that $H$ is a rational number $a/b$, and by following the cancellation process, we find that we can write $H$ as $c/d$, where $c$ and $d$ share no common factor other than 1. In particular (and this observation is important) if 2 divides evenly into either of the numbers $c$ or $d$, then 2 will not divide evenly into the other one. That is, $H = c/d$ and

**not both** $c$ and $d$ are even numbers

(since otherwise they would have a common factor of 2).

On the one hand, $H = \sqrt{2}$, and on the other hand, $H = c/d$. So $\sqrt{2} = c/d$. To simplify that equation, let's square both sides of $\sqrt{2} = c/d$. Doing so would produce

$$2 = \frac{c^2}{d^2}.$$

Since natural numbers are easier to visualize than fractions, let's multiply each side by $d^2$:

$$2d^2 = c^2.$$

## Here's Looking at c

Now let's see where this equation leads. What kind of number is $c^2$? Well, we see it equals $2d^2$ so 2 is a factor of $c^2$ and therefore $c^2$ must be even. But if $c^2$ is even, $c$ itself is even. (Why? Well, we know 2 is a prime number, and if 2 divides evenly into $c \times c$ then 2 must divide evenly into just $c$ alone.)

So $c$ is an EVEN number.

Since $c$ is an even number, then it must equal 2 times some other number, say $c = 2n$ (since that's what it means to be even, after all). If we substitute $2n$ for $c$ in the equality above, we see that

$$2d^2 = c^2 = (2n)^2 = (2n)(2n) = 4n^2.$$

Looking at this equation, we have this unstoppable desire to divide both sides by 2, which leads to $d^2 = 2n^2$, which, in turn, leads to trouble.

## A Troubling d Tour

What kind of number is $d^2$? It must be even.

Thus $d$ itself must also be EVEN.

Remember that we started by assuming that the square root of 2 was equal to a rational number *a/b*. We made some legal deductions from that assumption, namely, after canceling, *a/b* = *c/d* where *c* and *d* are not both even. Then we deduced that *c* and *d* must both be even—directly contradicting ourselves. So this situation is impossible. Therefore, whatever we assumed must be false.

But what did we assume? If you reread our argument, we assumed only one thing, namely that $\sqrt{2}$ was a rational number. Since our assumption led us to a contradiction, our assumption must be false. So, $\sqrt{2}$ is not a rational number; $\sqrt{2}$ is not equal to the ratio of two integers. A number that is not rational is called *irrational*. The observation about the square root of 2 is so important that we want to display it.

> **SQUARE ROOT OF 2 IS IRRATIONAL.** *The square root of 2 is an irrational number.*

## Make It Your Own

You can make this argument your own by working through the ideas. Explain it to a friend. It truly is an example of a beautiful and elegant proof, because a powerful and surprising result arises from some basic ideas creatively strung together. This proof appears in Euclid's book entitled *Elements* and is usually attributed to Euclid (ca 300 B.C.). However, evidence exists that Aristotle (384 B.C.–322 B.C.) also knew about this argument. Of course, we'll never know for sure who was the first to discover this counterintuitive revelation together with this elegant line of reasoning. Among most mathematicians, the proofs that there are infinitely many prime numbers and that the square root of 2 is irrational are considered to be among the most beautiful arguments in the field.

## Accepting Reality

We may not like the idea that an entirely new kind of number must exist—numbers that are not rational; however, there is no use fighting it. No matter how much our previous world view led us to believe that all numbers are expressible as the ratio of two integers, it just isn't so. We must accept the proven truth, accept a new world view, and explore the reality of numbers as they are. Once we prove something, we must add it to our list of truths and move on.

Of course, we are modern people, and we are unlikely to find the existence of irrational numbers a challenge to our philosophical biases about how the world is organized. But for the ancient Greeks, the ideas of proportion and ratios of whole numbers played a more central role in their understanding of reality. For them, accepting irrational numbers was a significant challenge to their ideas about reality.

The Pythagoreans reacted strongly to the disturbing discovery of irrational numbers, and they kept the idea secret. When one of their members was caught telling the secret of irrational numbers, they took him out in a boat and threw him overboard. Think we're joking? Proclus around the 5th century A.D. gave a brief account: "It is well known that the man who first made public the theory of irrationals perished in a shipwreck in order that the inexpressible and unimaginable should ever remain veiled. And so the guilty man, who fortuitously touched on and revealed this aspect of living things, was taken to the place where he began and there is forever beaten by the waves." In contrast, students today who divulge the mysteries of the irrational numbers and other scientific phenomena find themselves scooped up by major corporations offering impressive salaries.

## Beyond the Square Root of 2

Once we discover an important idea, we should use it to deduce new or more general consequences.

Mathematicians are extremely waste conscious when it comes to ideas: Once they have one, they try to recycle and reuse it. By pushing an idea to its limits, we often uncover more than we first expected.

To illustrate this lesson, let's see if we can adapt the ideas used in the proof that the square root of 2 is irrational to show that the square root of 3 is irrational. If you try to work through the ideas on your own before reading on, you will gain a much deeper understanding of the ideas at work.

## Square Root of 3

Let's assume that $\sqrt{3}$ is a rational number $a/b$, where $a$ and $b$ are natural numbers. By following the cancellation process, we find that we can write $\sqrt{3}$ as $c/d$, where $c$ and $d$ share no common factor other than 1. In particular (and this observation is important) if 3 divides evenly into either of the numbers $c$ or $d$, then 3 will not divide evenly into the other one. That is, $\sqrt{3} = c/d$ and **not both** $c$ and $d$ have the factor of 3 (since otherwise we could cancel). If we now square both sides of the equation $\sqrt{3} = c/d$, we see that

$$3 = c^2/d^2.$$

If we multiply each side by $d^2$ to get integers on both sides of the equation, we have

$$3d^2 = c^2.$$

Now let's see what this equation means. We see that $c^2$ equals $3d^2$, so 3 is a factor of $c^2$. But if $c^2$ has a factor of 3, then $c$ itself must have a factor of 3. (Why? Well, we know 3 is a prime number, and if 3 divides evenly into $c \times c$ then 3 must divide evenly into just $c$ alone.) Since $c$ has a factor of 3, it must equal 3 times some other integer, say $c = 3n$. If we substitute $3n$ for $c$ in the preceding equality, we see that

$$3d^2 = (3n)^2 = (3n)(3n) = 9n^2.$$

We can now divide both sides by 3 and get $d^2 = 3n^2$. This last equation shows that $d^2$ must have a factor of 3 in it, and therefore $d$ must have a factor of 3 as well. Thus we see that $c$ and $d$ share a common factor of 3. But we selected $c$ and $d$ so that they have no common factor greater than 1. This is a contradiction, and, therefore, this situation is impossible—hence our assumption must be false. So, $\sqrt{3}$ is not a rational number—it is irrational. Notice how this argument parallels our first one.

## *Other Irrationals*

In fact we can use this method to show that the $\sqrt{6}$ and many other examples are irrational. We invite you to try these in the Mindscapes.

With a bit of care, we can extend this idea even further and show that $\sqrt{2} + \sqrt{3}$ is irrational. Again, let's assume that $\sqrt{2} + \sqrt{3}$ is actually a rational number, say $a/b$. If we square both sides, $(\sqrt{2} + \sqrt{3})^2 = (a/b)^2$, we have to be a bit careful to expand the left side. We do it here:

$$(\sqrt{2} + \sqrt{3})^2 = (\sqrt{2} + \sqrt{3})(\sqrt{2} + \sqrt{3}) = (\sqrt{2} + \sqrt{3})\sqrt{2} + (\sqrt{2} + \sqrt{3})\sqrt{3}$$
$$= \sqrt{2}\sqrt{2} + \sqrt{3}\sqrt{2} + \sqrt{2}\sqrt{3} + \sqrt{3}\sqrt{3} = 2 + \sqrt{6} + \sqrt{6} + 3 = 5 + 2\sqrt{6}.$$

So we see that $5 + 2\sqrt{6} = a^2/b^2$, which means that $\sqrt{6} = (a^2 - 5b^2)/2b^2$. But the number on the right side is a fraction, so that means that $\sqrt{6}$ is a rational number. However, you will show in the Mindscapes that $\sqrt{6}$ is irrational. Therefore, we have reached another contradiction. Our first assumption must have been false, and we conclude that $\sqrt{2} + \sqrt{3}$ is irrational.

## *Irrational Power*

If we whittle the idea we are using down to its core, we can use it to prove that other more exotic numbers are irrational. Suppose $3^A = 9$. What would $A$ equal? If $3^C = 27$, what would $C$ equal? These questions are not too hard: $A = 2$, and $C = 3$. But suppose $B$ is the number such that $3^B = 10$. What is $B$? We know from the previous two questions, that

$B$ is bigger than 2 and smaller than 3, but there is no use trying to figure out exactly what decimal number it equals, we can't: It's an irrational number. Why? Well, suppose that $B$ were a rational number, say $u/v$, where both $u$ and $v$ are natural numbers. Then $3^{u/v} = 10$. If we raise both sides to the $v$th power, then the $v$'s cancel out in the power on the left side:

$$(3^{u/v})^v = 3^u = 10^v.$$

Since $u$ and $v$ are each at least 1, then 3 must divide evenly into $10^v$, which is absurd: The only prime numbers that divide evenly into $10^v$ are 2's and 5's. This contradiction means that our assumption was false, so $B$ must be an irrational number.

The number $B$ is called a *logarithm*. If $3^B = 10$, then we would say that $B$ is the logarithm of 10 in base 3. Using a calculator we can estimate $B$ and see that $B = 2.095903274289384 60 \ldots $.

## π

Our method to show certain numbers are irrational does not work for more exotic numbers. The circumference of a circle having diameter 1 is equal to the famous number $\pi$ (pi). Although the rational number 22/7 is almost equal to $\pi$, it is not exactly equal to $\pi$: $22/7 = 3.142857142857\ldots$, while $\pi = 3.141592\ldots$

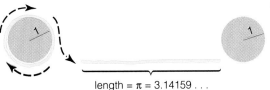

length = π = 3.14159 . . .

The Greeks and subsequent mathematicians studied $\pi$ intensely: In 1650 B.C., Egyptians estimated that $\pi \approx 256/81$, and roughly 500 years later, mathematicians in India approximated $\pi$ as 62832/20000, which is incredibly close to $\pi$. It was not until 1761, however, that someone proved that, in fact, $\pi$ is irrational. The first person to prove this important fact was Johann Lambert, and he used techniques from calculus. As an amusing postscript to Lambert's result and to the earlier works of Greek, Egyptian, and Indian mathematicians, we note that some ground was lost in 1897: The Indiana State Legislature considered a bill to declare $\pi$ equal to 4, which was "offered as a contribution to education to be used only by the State of Indiana free of cost. . . ." Fortunately, the legislature did not pass the bill.

## *Still Unknown*

In general, it is difficult to determine if numbers are rational or irrational. As a modest illustration, nobody knows if any of the numbers on the following list are irrational—it is possible (but not likely) that some are actually rational: $2^\pi$, $\pi^\pi$, $\pi^{\sqrt{2}}$. Don't they all "look" irrational? Yes, but no one knows how to prove it for sure.

We now see that numbers come in two flavors: rational and irrational. The collection of all these numbers—rational and irrational—form the real numbers, which leads us to our final journey through the notion of number.

**A LOOK BACK**

Beyond the world of natural numbers are the rational numbers, fractions. But some numbers are irrational—not rational. We can show that the $\sqrt{2}$ is irrational by assuming the contrary. If the $\sqrt{2}$ were rational, then it would be equal to a fraction written in lowest terms. That assumption implies that both the numerator and the denominator would have to have a factor of 2, which would contradict the fact that the fraction was in lowest terms. Thus $\sqrt{2}$ is not a fraction. This strategy can be used to demonstrate that other numbers are irrational. This reasoning allowed us to move from the comfortable world of natural numbers and their ratios to the real world of irrationality.

An effective strategy for analyzing life is to make an assumption and see what consequences follow logically. If a logical consequence is a contradiction, then the assumption must be wrong.

Follow assumptions to their logical conclusions.

◆ ◆ ◆ ◆ ◆ ◆ ◆ ◆ ◆ ◆ ◆ ◆ ◆ ◆ ◆ ◆ ◆ ◆ ◆ ◆ ◆

# MINDSCAPES  Invitations to Further Thought*

### I. SOLIDIFYING IDEAS

1. **Irrational rationalization.** We know that $\sqrt{2}$ is irrational. Therefore $3\sqrt{2}/5\sqrt{2}$ must also be irrational. Is this conclusion correct? Why or why not?

2. **Rational rationalization.** We know 2/5 and 7/3 are rational. Therefore (2/5)/(7/3) is also rational. Is this conclusion correct? Why or why not?

3. **Rational or not.** For each of the following numbers, determine if the number is rational or irrational. Give brief reasons justifying your answers.

   $\dfrac{4}{9}$, 1.75, $\dfrac{\sqrt{20}}{3\sqrt{5}}$, $\dfrac{\sqrt{2}}{14}$, 3.14159

4. **Irrational or not.** Determine if each of the following numbers is rational or irrational. Give brief reasons justifying your answers.

---

*In the Mindscapes section, exercises marked (H) have hints for solutions at the back of the book. Exercises marked (S) have solutions.

$$\sqrt{\frac{16}{20}}, \quad \sqrt{\frac{12}{7.5}}, \quad -147, \quad 0, \quad \frac{\sqrt{3}}{3}$$

In questions 5–11, show that the value given is irrational.

5. $\sqrt{5}$ (H)
6. $\sqrt{6}$
7. $\sqrt{7}$
8. $\sqrt{3} + \sqrt{5}$
9. $\sqrt{2} + \sqrt{7}$
10. $\sqrt{10}$ (S)
11. $1 + \sqrt{10}$
12. **An irrational exponent** (H). Suppose that $E$ is the number such that $12^E = 7$. Show that $E$ is an irrational number.
13. **Another irrational exponent.** Suppose that $E$ is the number such that $13^E = 8$. Show that $E$ is an irrational number.
14. **Still another exponent.** Suppose that $E$ is the number such that $14^E = 9$. Show that $E$ is an irrational number.
15. **Rational exponent.** Suppose that $E$ is the number such that $8^E = 4$. Show that $E$ is a rational number. In the previous two Mindscapes, you developed an argument that showed that an exponent was irrational. Where does that argument break down in this case?
16. **Rational exponent.** Suppose that $E$ is the number such that $(\sqrt{2})^E = 2\sqrt{2}$. Show that $E$ is a rational number. In Mindscapes 13 and 14, you developed an argument that showed that an exponent was irrational. Where does that argument break down in this case?
17. **Rational sums.** Show that the sum of any two rational numbers is another rational number. (*Hint*: Let $a/b$ be one rational number and $c/d$ be another rational. Now show that $a/b + c/d$ is another rational number.)
18. **Rational products.** Show that the product of any two rational numbers is another rational number. (*Hint*: Adapt the previous hint.)
19. **Root of a rational.** Show that $\sqrt{(1/2)}$ is irrational.
20. **Root of a rational** (S). Show that $\sqrt{(2/3)}$ is irrational.

## II. CREATING NEW IDEAS

1. **$\pi$.** Using the fact that $\pi$ is irrational, show that $\pi + 3$ is also irrational.
2. **$2\pi$.** Using the fact that $\pi$ is irrational, show that $2\pi$ is also irrational.

3. **$\pi^2$.** Suppose that we know only that $\pi^2$ is irrational. Use that fact to show that $\pi$ is irrational.

4. **A rational in disguise.** Show that the number $(\sqrt{2}^{\sqrt{2}})^{\sqrt{2}}$ is a rational number even though it might look irrational. What familiar number does it equal?

5. **Cube roots (H).** The cube root of 2, denoted as $\sqrt[3]{2}$, is the number such that if it were cubed (raised to the third power), it would equal 2. That is, $(\sqrt[3]{2})^3 = 2$. Show that the $\sqrt[3]{2}$ is irrational.

6. **More cube roots.** Show that $\sqrt[3]{3}$ is irrational.

7. **One fourth root.** Show that the fourth root of 5, $\sqrt[4]{5}$, is irrational.

8. **Irrational sums (S).** Does an irrational number plus an irrational number equal an irrational number? If so, show why. If not, give some counterexamples.

9. **Irrational products.** Does an irrational number multiplied by an irrational number equal an irrational number? If so, show why. If not, give some counterexamples.

10. **Irrational plus rational.** Does an irrational number plus a rational number equal an irrational number? If so, show why. If not, give some counterexamples.

### III. FURTHER CHALLENGES

1. **$\sqrt{p}$.** Show that for any prime number $p$, $\sqrt{p}$ is an irrational number.

2. **$\sqrt{pq}$.** Show that, for any two different prime numbers $p$ and $q$, $\sqrt{pq}$ is an irrational number.

3. **$\sqrt{p} + \sqrt{q}$.** Show that, for any prime numbers $p$ and $q$, $\sqrt{p} + \sqrt{q}$ is an irrational number.

4. **$\sqrt{4}$.** The square root of 4 is equal to 2, which is a rational number. Carefully modify the argument for showing that $\sqrt{2}$ is irrational to try to show that $\sqrt{4}$ is irrational. Where and why does the argument break down?

5. **Sum or difference (H).** Let $a$ and $b$ be any two irrational numbers. Show that either $a + b$ or $a - b$ must be irrational.

### IV. IN YOUR OWN WORDS

1. **Personal perspectives.** Write a short essay describing the most interesting or surprising discovery you made in exploring this section's material. If any material seemed puzzling or even unbelievable, address that as well. Explain why you chose the topics you did. Finally, comment on the aesthetics of the mathematics and ideas in this section.

2. ***With a group of folks.*** In a small group, discuss and work through the arguments that the square root of 2 and the square root of 3 are irrational. After your discussion, write a brief narrative describing the arguments in your own words.

3. ***Creative writing.*** Write an imaginative story (it can be comical, dramatic, whatever you like) that involves or evokes the ideas of this section.

4. ***Power beyond the mathematics.*** Provide several real-life situations—ideally, from your own experience—for which some of the strategies of thought presented in this section would provide effective methods for approaching and resolving them.

## 2.7 Get Real:
### The Point of Decimals and Pinpointing Numbers on the Real Line

*Why are wise few, fools*
*numerous in the excesse?*
*'Cause, wanting number,*
*they are numberlesse.*

—Augusta Lovelace

Our development of the notion of "number" took us from the familiar natural numbers and rationals to the more mysterious realm of irrational numbers. While these collections of numbers are distinctive, they all fit together in a basic way: Given any two different numbers, one is bigger than the other. The numbers are all ordered. That orderly hierarchy of numbers by size allows us to represent all numbers on one line. In this final section on number, we explore the connections between the number line and the notion of number.

The guiding principle for this part of the exploration of number is to bring global coherence to separate ideas. By examining the totality of numbers as one entity, we will discover new surprises and develop a better understanding of both the rational and irrational. Initially some of these discoveries may contradict our intuition. Our exploration involves looking at the types of numbers we know and deducing how those numbers must be interconnected on the number line. This point of view leads to the representation of numbers in decimal form. We must be open-minded and accept logical consequences that we deduce. Once we accept correct conclusions, we will understand the collection of all numbers on the number line—the real numbers—as a coherent idea aptly called the *continuum*.

### Lining up

The real number line has appeared in elementary school textbooks as long as school cafeterias have been serving students sloppy joes. Here we start from scratch but soon make unexpected discoveries—just as we did with our sloppy joes—about the familiar idea of the number line.

We begin with the number line itself:

Now only the integer points are labeled, but we would like to be able to label or describe every point on this line. To make progress in this direction, let's consider the points halfway between each consecutive pair of integers. For example, the number 5/2 is the point that sits exactly midway between 2 and 3. In fact, any rational number corresponds to a specific point on this line. For example, we can locate the point to which 37/23 corresponds by dividing each interval between consecutive integers into 23 equal pieces. Then we start at 0 and jump from mark to mark: 1/23, then 2/23, then 3/23, and so on. When we get to 23/23, we see that is the point that is also labeled 1. Jumping 14 more times gets us to 37/23. A similar procedure allows us to find a point on the line corresponding to any rational number.

## Rationals Everywhere

The points associated with rational numbers are all over the line: No matter where we are standing on the line, we can always find a rational number point as close as we wish. Suppose we want to find a rational number point that is within a distance of 1/10,000 of where we are standing. We just divide each segment between every two consecutive integers into 10,000 equal pieces and make those 10,000 marks, then mark off all the points that correspond to rational numbers having 10,000 in their denominators (5,876/10,000, for example). Therefore, no matter where we are, we will be within 1/10,000 of one of these rational points.

Now that we see that the rationals are essentially everywhere, we may ask: Are there any unlabeled or undescribed points left on our line? The previous section provides us with the answer to this question. Let's construct a point on the number line that definitely does not correspond to a rational number.

## An Irrational Point

Here is a way of finding a number on the number line that is not a rational number: Build a square whose base is the interval from 0 to 1. Next draw the diagonal from 0 to the upper-right corner of the square. Using a compass, copy the length of that diagonal onto the number line and make a mark there. What number did we just mark? The square root of 2.

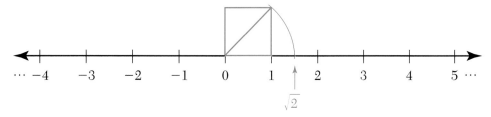

Look for new ways of expressing an idea.

✦ ✦ ✦ ✦ ✦ ✦ ✦

In the previous section we showed that the square root of 2 is irrational. Thus there are points on the line that cannot be labeled with rational numbers, and we are faced with the question: Is there a uniform method to label every single point on the line—rational and irrational?

## *The Decimal Point*

Let's label each point by describing ever more precisely where it sits on our line. This process is familiar, because it is the idea that generates the decimal expansion of numbers. The *decimal expansion* of a number provides us with a road map that allows us to hone in and locate the number on our line. For example, let's consider the decimal expansion of the square root of 2:

$$\sqrt{2} = 1.414213562\ldots.$$

The number to the left of the decimal point tells us that our number will be somewhere between 1 and 2. Where between? We cut the interval from 1 to 2 into 10 equal pieces (10 since we are looking at the deci[10]mal expansion). The next digit, in this case 4, tells us in which small interval our number is located. We then take that small interval and cut it up into 10 (very small) equal pieces. The next digit (in our example, 1) tells us in which very small interval our number resides. Notice that, as we continue this process, we whittle away and create smaller and smaller intervals. This process allows us to hone in on the number we seek, in this case $\sqrt{2}$.

We are getting closer and closer to the $\sqrt{2}$. Each little interval, once it is cut up into ten equal pieces, looks like the larger parent interval that it came from. For $\sqrt{2}$, this honing process never ends. We keep localizing

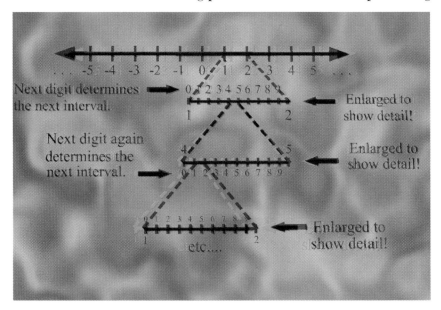

Stare deeply into the line with your 3D glasses.

our point in smaller and smaller nested intervals, and we get closer and closer, but we have to do this process infinitely many times to actually hit the $\sqrt{2}$ point exactly.

## *A Home Base*

To name the point $\sqrt{2}$ on the real line, we divided intervals into 10 equal parts, because we are accustomed to using 10 as the basic number for counting. However, we could well have located the number $\sqrt{2}$ by dividing each interval into any other number of equal parts. For example, we could always divide each subinterval into two parts. In that case, since the point $\sqrt{2}$ lies in the first half of the interval, we put a 0 after the point. Dividing that half interval into half again, we would note that $\sqrt{2}$ is in the second half, so we would put a 1 as the next digit. Dividing into halves at each time gives a way to locate $\sqrt{2}$ using just 0's and 1's. This representation is known as the *base 2 expansion*. In base 2, $\sqrt{2}$ would have a representation of the form:

$$\sqrt{2} = 1.01101010000010011110\ldots{}_2.$$

The key to remember is that each preceding digit just tells us which of the two subintervals—the left = 0, the right = 1—we fall into as we head toward the square root of 2.

Base 2 (sometimes referred to as *binary*) representation of numbers is quite useful since each point on the number line can be located using only 0's and 1's. This economy of symbols is convenient for computers which store information by sequences of ons and offs, represented by 1's and 0's. Other bases are useful for other purposes, but the strategy for finding a representation in any base remains the same.

## *Rational vs. Irrational Decimals*

We have seen that some of the points on the real line, like $\sqrt{2}$, have decimal expansions that require infinitely many digits before they are completely specified. The number 1/2, however, has a decimal expansion of 0.5. This distinction between the decimal representation of the irrational $\sqrt{2}$ and the rational 0.5 leads us to consider the relationship between the decimal expansions of rational numbers compared to the decimal expansions of irrational numbers.

*Question: Given a decimal expansion of a number, can we tell if it is a rational or an irrational number?*

A reasonable guess is that a decimal expansion represents a rational number precisely when its decimal expansion terminates (like 0.5 or 12.76), and it is an irrational number precisely when it has a decimal expansion that goes on forever (like the decimal expansion of $\sqrt{2}$). What do you think about this guess? Is it correct? Partly correct? Completely incorrect? None of the above?

Our guess is *partly correct*. If the decimal expansion of a number terminates, then the number must be rational. To see why, we notice that, if the decimal expansion terminates, then we can just shift the decimal point all the way over to the right and then divide by 10 raised to the power equal to the number of places we moved the decimal point. For example,

$$12.76 = \frac{1276}{100}, \quad 6.3709 = \frac{63709}{10000}, \quad 14.35670381 = \frac{1435670381}{10^8}.$$

However, just because a decimal expansion goes on forever does not imply that the number is irrational. Consider the rational number

$$\frac{1}{3} = 0.3333333333\ldots.$$

In this case we just see 3 repeating forever. We call a decimal expansion *periodic* if from some point on and then forever onward the pattern of digits repeats. For example,

$$3.5959595959\ldots \quad \text{and} \quad 9.345276994994994994\ldots$$

are examples of periodic decimal expansions. Periodic decimal expansions go on forever and do not terminate. The interesting fact is that they are all rational numbers.

## *Periodic Decimals*

To see why periodic decimal numbers are rational, consider the number $3.5959595959\ldots$. We now illustrate a method that will allow us to figure out what rational number this decimal represents. The key idea is to multiply the decimal by a power of 10 (that just shifts the decimal point to the right) so that we can align the periodic part with the periodic part of another copy of the decimal expansion. Then if we subtract these numbers, the infinitely long periodic part cancels away. In our example (we'll call the number $W$ for "What number is it?"), we multiply the decimal expansion by 100 and thus shift the decimal two places to the right. We then subtract the original decimal expansion. Notice how the periodic parts line up perfectly and how they all drop out when we subtract.

$$\begin{aligned} 100W &= 359.59595959\ldots \\ -\quad W &= \phantom{000}3.5959595959\ldots \\ \hline 99W &= 356.0000000000\ldots \end{aligned}$$

So, we see that $99W = 356$, so, $W = 356/99$, a rational number. Using a calculator we can check that $356/99$ has a decimal expansion of $3.5959595959\ldots$. This method always works. We just have to make sure that the periodic parts line up. So numbers whose decimal representation is periodic are always rational.

## *Divide and Conquer (Repeat)*

To find the decimal expansion for a rational number, we use long division. We notice that, as we perform the long division, the intermediate differences we get (through the subtraction) are always between 0 and the number by which we are dividing. Since there are only finitely many natural numbers between 0 and the number we are dividing by, at some point we must see a difference that we already saw before. Notice that this is just another example of the Pigeonhole principle. Once this repetition happens, the decimal expansion must repeat forever. The following example illustrates this observation more clearly.

Let's find the decimal expansion of 1141/990 using long division:

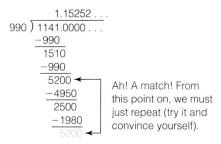

We just discovered exactly when a decimal expansion is a rational number. We record this insight formally as:

> **DECIMAL EXPANSION OF RATIONALS AND IRRATIONALS.**
> *A decimal expansion represents a rational number precisely when either the decimal expansion terminates or the decimal expansion is eventually periodic. Thus, a decimal expansion is an irrational number precisely when it does not terminate and is not periodic.*

We can now use this theorem to prove some interesting results. For example:

> **0.123 ... IS IRRATIONAL.** *The real number*
>
> 0.12345678910111213141516171819202122232425 26 ...
>
> *(formed by writing the natural numbers in order, in juxtaposition) is an irrational number.*

## *Proof*

There are only two possibilities: The number in the preceding statement is either a rational number or an irrational number. Let us assume that the preceding number is a rational number. If we could show that this possibility is in fact impossible, that would prove that the other possibility (that the number is irrational) must be true.

If this number were a rational number, its decimal expansion would have to be eventually periodic. In other words, there would have to be a finite string of numbers in the decimal expansion that, from some point on, repeats forever. What would be the length of this finite string that repeats? We don't know. All we are sure of is that the length is finite—let's call that finite length of the repeating pattern $L$. For example, in the number 45.32198119811981198 11 . . . the repeating pattern "1981" has length 4, so here $L$ would be 4.

Recall that our number is created by writing down the natural numbers in order, in juxtaposition. Thus we will eventually write down all numbers that are made just from 1's. In other words, at some point we will see 11 and later 111 and later still 1111 and even later still 11111, etc. Thus we see that there will be long runs of 1's occurring in the decimal expansion. In fact, we will have arbitrarily long runs of 1's. For example, the natural number consisting of a billion 1's in a row will eventually appear. In particular, we can find infinitely many runs of 1's, each of which has more than 10 times $L$ number of 1's in a row (remember that $L$ is the length of the alleged repeating pattern). So at some point the repeating pattern must march through a sea of 1's. The only way that could happen is if the repeating pattern itself were made up exclusively of 1's. If the repeating pattern were all 1's, then from some point onward, all we would see in the decimal expansion of this number would be 1's, contrary to how the number was constructed. This contradiction tells us that our assumption that the number is rational must be false. Therefore, the decimal number .1234 . . . is not rational and must be irrational, which completes our proof.

## *No Holes, No Neighbors*

Characterizing the decimal expansions of rational numbers allows us to journey deep into the dense jungle of the real number line. *Gaze closely at the number line. Look really closely, put your nose right up to the page. Stare deeply into the hypnotic line.* Perhaps you are getting sleepy . . . well, snap out of it and wake up! Notice that there are no holes in the number line—instead the line flows smoothly and produces a continuous and unending stream of real numbers—*the continuum*. This unholey image leads to a question that has a strange answer.

Suppose we are a particular real number—to ground our thinking, let's suppose we are 0. Now who are our immediate neighbors? In particular, what is the next real number after 0? Can we name it? Suppose someone guessed 1/2. We could easily show that 1/2 is not the next real number after 0; after all, 1/4 is closer to 0 than 1/2. Suppose that someone else guessed 3/702(= 0.004273 . . .). We could take half of that number and find the number 3/1404(= 0.002136 . . .), which is even closer to 0. In fact, if anyone gave us any number greater than 0, we could just divide that number in half and find another even smaller number that is also greater than 0. What can we conclude? The answer is that there is no next real number immediately following 0. The moment we specify a number bigger than 0, we could find another number (in fact infinitely many) that is between 0 and the specified number. The real number line flows continuously without breaks and between any two points on the line; we can always find lots of points in between them. Hence, there is no next real number after 0.

Our reasoning could be used to show that there is no next real number after 1 or even after $\sqrt{2}$. We have therefore verified the following.

**NO NEXT REAL NUMBER.** *Given any particular real number, there is no next real number immediately following it.*

## *Redundancy in Representation of Reals*

We now have a sense of the connections among points on the number line, their decimal expansions, and the notions of rational and irrational. Every point on the real number line can be represented as a decimal number, but we have not considered the following question:

*Is there only one way to write a number in its decimal expansion?*

Although it would be convenient for each real number to have only one decimal expansion, unfortunately there are some real numbers that have more than one decimal expansion. We illustrate this fact with an example.

What rational number is 0.999999 . . . ? We'll call this number $N$ (for Nines). Let's answer this question using our method of multiplying by a power of 10, lining up the repeating period, and then subtracting.

$$
\begin{array}{rl}
10N = & 9.9999 \ldots \\
- \quad N = & 0.9999 \ldots \\
\hline
9N = & 9.0000 \ldots
\end{array}
$$

So $9N = 9$; and what must $N$ be? $N = 1$. We just proved that $1 = 0.99999 \ldots$ . Does this equation look strange? Sometimes mathematical results, even when proven rigorously, are so counterintuitive that we remain skeptical of their validity. The fact that $1 = 0.9999 \ldots$ is a great example of this phenomenon. Even though we have given a rigorous mathematical proof of this amazing fact, we now give an intuitive argument that may be more convincing.

Don't let personal biases get in the way of new discoveries.

Suppose we believe that 1 is not equal to 0.99999 . . . (remember that those 9's go on forever without ever stopping). Then one of these numbers would be bigger than the other. Which one would be larger? Certainly 1 would be larger than 0.9999 . . . . If these numbers were not equal, then there must be some numbers between 0.99999 . . . and 1 on the number line. For example, the average of those two numbers would have to be between them. That average must be a number that is larger than 0.99999 . . . and at the same time smaller than 1. What could that number be? It would have to start off with a 0 (otherwise it would not be less than 1). What would the next digit be? It must be a 9 since anything else would make the number smaller than 0.9999 . . . . What about the next digit? It would have to be 9 as well. If we continue in this manner we see that we are building 0.99999 . . . . But that is not bigger than 0.99999 . . . . So there are no numbers between 0.9999 . . . and 1, and hence they must be equal. Did you like this argument? It's amusing to think about.

Here is another way of looking at all those 9's. For some strange reason, people feel comfortable with the fact that 1/3 = 0.33333 . . . and 2/3 = .6666 . . . . Well, if we add those together we see 1 = 0.9999 . . . ! We have proved the following:

**0.9999 . . . = 1.**  $1 = 0.999999 \ldots$

What is another decimal expansion for the number 0.499999 . . . ? Use the ideas given above and give it a try.

## Random Reals

Finally, before closing this section and this chapter we pose an intriguing question.

> If we randomly pick a real number—that is, we take a pin, close our eyes, and place the pin on some point on the real number line—what is the likelihood that the number we picked is a rational number? Is it a 50-50 chance?

A reasonable answer would be 50-50 since we have already shown that every pair of irrational numbers is separated by a rational, and every pair of rational numbers is separated by an irrational. Unfortunately, this reasonable sounding answer is far from correct. Although we will not be able to give a rigorous answer to this question until we journey through the world of the infinite in the next chapter, we are able to give a plausible argument that answers the question.

How could we randomly pick a real number besides closing our eyes and dropping a pin on a number line? Well, one way is to randomly choose digits among 0, 1, 2, 3, . . . , 8, 9 and write them down to build a decimal number. We could get a 10-sided die with the sides numbered from 0 to 9. Let's suppose that we always start with 0 so our random number will be between 0 and 1. Now we roll a 10-sided die or have a random number generator spit out digits and we record them:

0.79356284565748388365300483628304726118394573781 . . . .

We don't stop! We do this forever and thus create a real number. What is the likelihood that this random number is a rational number? Well, for it to be rational, from some point on, the number must have a pattern that repeats forever. But what does that mean? It means that from some point on, we keep repeating the identical pattern without ever deviating. How likely is it that we will repeat a finite pattern forever given that we are generating the digits randomly? The answer is not likely at all—in fact it should "never" happen. There would have to be an amazing and even unheard of conspiracy in the random digits to have them all, from some point on, follow a periodic pattern. Thus the probability that we randomly generate a rational number is actually zero. So if we just randomly pick a real number, it is "certain" to be an irrational number.

## *Irrationals Abound*

This huge preponderance of irrational numbers might be a tough pill to swallow since we are so accustomed and comfortable with rational numbers and since we noticed earlier that the rationals seem to be everywhere on the number line. But, mathematically, rational numbers are actually hard to find. We will see exactly what "probability zero" and "certain" mean in the probability chapter. For now, if we accept the preceding informal analysis, we are faced with an extremely interesting question: If a real number selected at random is "never" a rational number and "always" an irrational number (whatever the notions of "never" and "always" mean), then does that mean that there are, somehow, more irrational numbers than rational numbers? Certainly there are infinitely many rational numbers and infinitely many irrational numbers. Could one of these infinite sets actually be greater than the other? Perhaps what first appears familiar and natural (the rational numbers) will in fact be the exotic and strange, whereas what appeared to be foreign and strange (irrational numbers) will actually turn out to be more the norm! These curious questions set the stage for our next adventure: the world of the infinite.

**A LOOK BACK**   The rational and irrational numbers taken together form the real numbers—the collection of points on a line. We are able to use the decimal expansion of a real number to locate any real number on the number line. The decimal expansion also allows us to distinguish rational numbers from irrational numbers. A number is rational precisely if its decimal expansion eventually repeats; otherwise it is irrational. Using these ideas we are able to devise means of converting repeating decimals to fractions and also to prove that certain real numbers, such as 0.123456789101112 . . . are irrational, while surprisingly, 0.9999 . . . = 1. The real line presents a picture of numbers orderly arranged. No number has an immediate neighbor, a number just above it or just below it.

Our strategy for exposing this view of the real numbers was to seek a unified view of all the types of numbers we had developed before. We looked for a relationship that encompassed all the ideas we had generated, in this case, the ideas of rational and irrational numbers. The simple ordering of numbers suggested that we could effectively represent all numbers as points on a line and that we could name each point on the line or number, rational or irrational, using a decimal representation. Some discoveries required us to give up biases and accept logical conclusions. Being open-minded about new ideas is a difficult and important lesson in every arena of life.

> . . . *an irrational number . . . lies hidden in a kind of cloud of infinity.* —Michael Stifel

Seek unifying ideas.

◆ ◆ ◆ ◆ ◆ ◆ ◆ ◆ ◆ ◆ ◆ ◆ ◆ ◆ ◆ ◆ ◆

# MINDSCAPES  Invitations to Further Thought*

### I. SOLIDIFYING IDEAS

1. **Always, sometimes, never.** A number with an unending decimal expansion is (choose one: always, sometimes, never) irrational. Explain and illustrate your answer with examples.

2. **Square root of 5.** The $\sqrt{5}$ has an unending decimal expansion, but it might eventually repeat. Is this statement true or false? Explain.

3. **A rational search.** Find a rational number that is bigger than 12.0345691 but smaller than 12.0345692.

4. **Another rational search.** Find a rational number that is bigger than 3.14159 but smaller than 3.14159001.

5. **An irrational search (H).** Describe an irrational number that is bigger than 5.7 but smaller than 5.72.

6. **Another irrational search.** Describe an irrational number that is bigger than 0.0001 but smaller than 0.00010001.

7. **Your neighborhood.** Suppose we tell you that we are thinking of a

---

*In the Mindscapes section, exercises marked (H) have hints for solutions at the back of the book. Exercises marked (S) have solutions.

7. **Your neighborhood.** Suppose we tell you that we are thinking of a number that begins with 10.0398XXXXX, where "XXXXX" are digits that we have hidden from view. What is the smallest our number could possibly be? What is the largest our number could possibly be?

8. **Another neighborhood.** Suppose we tell you that we are thinking of a number that begins with 5.5501XXXXX . . . , where "XXXXX . . . " are digits that we have hidden from view. What is the smallest our number could possibly be? What is the largest our number could possibly be?

In questions 9–11, express each fraction in its decimal expansion.

9. **6/7** (S)
10. **17/20**
11. **21.5/15**

In questions 12–19, express each number as a fraction.

12. **1.28901**
13. **20.4545**
14. **12.999**
15. **2.222222 . . .**
16. **43.12121212 . . .** (S)
17. **5.6312121212 . . .**
18. **0.0101010101 . . .**
19. **71.239999999 . . .**
20. **Just not rational** (H). Show that the number

    0.01001000100001000001000000100000000100 . . .

    is irrational.

## II. CREATING NEW IDEAS

1. **Farey fractions.** Let $F_n$ be the collection of all rational numbers between 0 and 1 (we write 0 as 0/1 and 1 as 1/1) whose numerators and denominators do not exceed $n$. So, for example,

$$F_1 = \left\{\frac{0}{1}, \frac{1}{1}\right\}, \quad F_2 = \left\{\frac{0}{1}, \frac{1}{2}, \frac{1}{1}\right\}, \quad F_3 = \left\{\frac{0}{1}, \frac{1}{3}, \frac{1}{2}, \frac{2}{3}, \frac{1}{1}\right\}.$$

$F_n$ is called the *nth Farey fractions*. List $F_4$, $F_5$, $F_6$, $F_7$, and $F_8$. Make a large number line segment between 0 and 1 and write in the Farey fractions. How can you generate $F_8$ using $F_7$? Generalize your observations and describe how to generate $F_n$. (*Hint:* Try adding fractions a wrong way.)

2. **Even irrational.** Show that the number

   0.2468101214161820222426283032343638 40 . . .

   is irrational.

3. **Odd irrational (H).** Show that the number

   1.357911131517192123252729313335373941 . . .

   is irrational.

4. **A proof for $\pi$.** Suppose someone gave us the first one billion decimal digits of $\pi$. If those digits did not repeat, would that prove that $\pi$ is irrational? Why or why not? Explain your answer. What if we had the first trillion digits?

5. **Irrationals and zero.** Is there an irrational number that is closer to zero than any other irrational? If so, describe it. If not, give a sequence of irrational numbers that get closer and closer to zero. (*Hint:* Start by considering $\sqrt{2}/2$ and $\sqrt{2}/3$.)

6. **Irrational with 1's and 2's (S).** Is it possible to build an irrational number whose decimal digits are just 1's and 2's? If so, describe such a number and show why it's irrational. If not, explain why.

7. **Irrational with 1's and some 2's.** Is it possible to build an irrational number whose decimal digits are just 1's and 2's and only finitely many 2's appear? If so, describe such a number and show why it's irrational. If not, explain why.

8. **Half steps.** Suppose you are just a point and are standing on the number line at 1 but are dreaming of 0. You take a step to the point 1/2, the midpoint between 0 and 1. You proceed to move closer to 0 by taking a step that is half of the previous one. You continue this process again and again. Will you ever land on 0? Explain. Is this observation hard to accept?

9. **Half steps again.** Suppose now that you are a very, very, very short line segment (your length is less than 1/100000000000). You are standing on the number line so that your center is right on 1, but, again, you are dreaming of 0. You shift your segment so that your center is at 1/2, midway between 0 and 1. You proceed to move closer to 0 by taking a step that is half of the previous one. You continue this process again and again. Will your segment ever contain 0? Explain. Is this observation less puzzling than the previous exercise? Why?

10. **Cutting $\pi$.** Is it possible to cut up the interval between 3 and 4 into pieces of exactly the same size such that one of the pieces has the point $\pi$ on its right edge? Why or why not?

### III. FURTHER CHALLENGES

1. ***From infinite to finite.*** Find a real number that has an unending and nonrepeating decimal expansion, with the property that if you square the number, the decimal expansion of the squared number terminates.

2. ***Rationals*** (H). Show that, between any two different real numbers, there is always a rational number between them.

3. ***Irrationals.*** Show that, between any two different real numbers, there is always an irrational number between them.

4. ***Terminator.*** If a rational number has a decimal expansion that terminates (or alternatively, has a tail of zeros that goes on forever), then the rational number can be written as a fraction where the only prime numbers dividing the denominator are 2 and 5.

5. ***Terminator II.*** If the denominator of a fraction has only factors of 2 and 5, then the decimal expansion for that number must terminate in a tail of zeros.

### IV. IN YOUR OWN WORDS

1. ***Personal perspectives.*** Write a short essay describing the most interesting or surprising discovery you made in exploring this section's material. If any material seemed puzzling or even unbelievable, address that as well. Explain why you chose the topics you did. Finally, comment on the aesthetics of the mathematics and ideas in this section.

2. ***With a group of folks.*** In a small group, discuss and work through the arguments that the number 0.12345678910 . . . is irrational and that 0.99999 . . . = 1. After your discussion, write a brief narrative describing the arguments in your own words.

3. ***Creative writing.*** Write an imaginative story (it can be comical, dramatic, whatever you like) that involves or evokes the ideas of this section.

4. ***Power beyond the mathematics.*** Provide several real-life situations—ideally, from your own experience—for which some of the strategies of thought presented in this section would provide effective methods for approaching and resolving them.

# CHAPTER THREE

# *Infinity*

> *To see the world in a grain of sand,*
> *And a heaven in a wildflower:*
> *Hold infinity in the palm of your hand,*
> *And eternity in an hour.*
> —William Blake

### 3.1

**Beyond Numbers**

### 3.2

**Comparing the Infinite**

### 3.3

**The Missing Member**

### 3.4

**Travels Toward the Stratosphere of Infinities**

Infinity—where all superlatives meet . . . bigger than the biggest, more than all . . . infinitely beyond what could be conceived by the mind . . . greater than any quantity . . . more vast than can be counted . . . beyond all numbers. The notion of infinity triggers a sense of mystical wonder and boundless incomprehensibility. Among all ideas of human thought, infinity is the most mysterious and romantic. It often resides within the realm of the spiritual where it captures the essence of ultimate grandeur and vastness without end.

Using the power of mathematical reasoning, we will conquer this seemingly incomprehensible idea of infinity and claim it for our own. Infinity will continue to evoke images of vastness and power in the realm of thought; however, after we understand it mathematically, our initial sense of mystery will be replaced by an understanding of a whole new world of richness. We will discover that there is not just one superlatively vast quantity called *infinity*; instead, even infinity itself has peers and superiors. Instead of one infinity, we will see a constellation of infinities, each different from the others, each with unique features and qualities that can be described and investigated, compared and contrasted. Our single image of one luminous but ill-defined infinity will burst into infinitely many clear infinities—richly related and intertwined.

We approach infinity by keeping our feet firmly on the ground and recalling ideas we have known since childhood. As children we shared M&M's with our best friend like this: "One for me, one for you; one for me, one for you . . . ." We take this familiar, everyday idea, specify its meaning with great precision, explore its consequences, and discover that it is the key to understanding infinity. In our everyday lives, we are often confronted with vagueness and mystery. Our journey toward the infinite will illustrate a powerful method of moving from the fuzzy to the focused. The heart of the strategy is consciously to make familiar ideas precise. Such thinking lets us journey from the familiar to the mysterious and empowers us to discover within the mysterious, the familiar.

❖ ❖ ❖ ❖ ❖ ❖ ❖ ❖ ❖ ❖ ❖ ❖

## 3.5

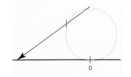

**Straightening Up the Circle**

# 3.1 Beyond Numbers:
## What Does Infinity Mean?

*The known is finite, the unknown infinite; intellectually we stand on an island in the midst of an illimitable ocean of inexplicability. Our business in every generation is to reclaim a little more land.*

—Thomas H. Huxley

We seek to put the study of infinity on a firm and logical footing. Since infinity is such a vast and forbidding topic, we prefer to begin by closely looking at the familiar ideas of numbers and counting. We want to count infinite collections, but we don't know how to count that high. Thus we seek different ways to count collections of ordinary objects in the hope that one method might work for infinite collections as well.

Where would we begin to search for the infinite? It's well beyond 1, 2, and 3; certainly past

4135972454512471511561120963071048.

Infinity is even beyond

9142452345245001282106162966380902777871210598012670019872665219873306111854210982762999117665327653327907384729589283572983994283103421671612374023720340174871023810847930471092371092471098097109230192740192380198231092309182741092730239845235029374019823019231724091823047109375019340198401973401982409173041097401924019840197340918340172368176346512349178648179486143174211907761265432109772127456928734010988824536666122888990182745311342622220984356278837676161.

In fact, as large and incomprehensible as this number is, it is sobering to realize that almost all numbers are far larger still. On the road to the infinite, we would pass this enormous number almost instantly and soon view it as a tiny jot. Even though this number appears early in the infinite list of all numbers, to our minds it seems vast. In reality, we do not even have an intuitive feel for such a large number. The number has no sensible name—no magnitudes like millions, billions, or trillions even make a dent in it. Just reading the digits aloud without error would be a challenge, and finding any of its prime divisors might be nearly impossible.

Even though we are comfortable with large numbers in the abstract, in truth we have little real understanding of such enormous quantities. Given our inability to grasp even these—relatively speaking—modestly sized numbers, is it possible for our minds to fathom the notion of the infinite? Let's first ponder a basic question.

## What Does "2" Mean?

Where should we begin our voyage toward the infinite? Our journey to infinity begins with 2. We are all familiar with 2, and we will use that intuitive intimacy to develop a concept of size that will take us well beyond what we know.

Understand simple
things deeply.

✦ ✦ ✦ ✦ ✦ ✦

If we have two apples and two hands, we could put one apple in each hand, and no hand would be un-appled, or any apple un-handed. If we have two socks and we put one sock on each hand, then our socks and our hands correspond exactly. This observation also implies that we can put one apple in each sock and demonstrate that the socks and apples also correspond; however, we do not recommend this last experiment unless the socks are clean.

Grappling with infinity requires us to imagine physical scenarios that are not quite possible but can be clearly conceived in the mind. Suppose we have a huge barrel full of Volvos and a barrel full of soccer balls, and we want to know if there are more Volvos than soccer balls, more soccer balls than Volvos, or the same number of each. How would we decide?

Probably we would just count the number of automobiles and pieces of sports paraphernalia in each of the barrels and compare the two numbers. Certainly that method works, but an alternative method allows us to grapple with magnitudes beyond what we can count. Let's devise a method for comparing the barrels without counting Volvos or soccer balls.

We could take one Volvo from the first barrel and one soccer ball from the second, pair them up and put them aside (probably putting the ball in the car's trunk). Then we could pair up another Volvo from the first barrel with another soccer ball from the second. If we continue pairing in

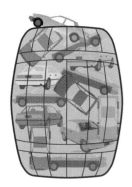

this fashion, we could tell whether we have the same number of Volvos and soccer balls without ever knowing how many we actually have.

This simple idea is important. We have just described a method for determining when two collections have the same number of objects in them without actually naming that number. Two collections whose objects can be paired together evenly—one from one collection with one from the other collection—are said to have a *one-to-one correspondence*.

## Correspondences

As we explore the notion of one-to-one correspondences, we must do something that is extremely difficult: We must forget. We must forget that 8 means something, we must forget that 37 means something. We must strip from our minds the names of numbers, leaving behind only the idea that *two collections of objects are equally numerous, precisely if there is a one-to-one correspondence between the elements of the two collections*. In taking this step, we are moving away from thinking of a number as an attribute of a collection ("I have five fingers on my right hand") to thinking instead of an idea of comparison ("I have just as many fingers on my right hand as on my left, because I can touch all the fingertips on one hand with the fingertips of the other hand").

This focus on comparison, on one-to-one correspondence, allows us to examine the infinite and turn our vague sense of awe into concrete understanding.

**A LOOK BACK**

The fundamental idea in the study of infinity is that two collections have the same size if there is a one-to-one correspondence between the members of one collection and the members of the other collection. This compelling concept of comparing sizes via one-to-one correspondence is the rock on which the whole study of infinity is built.

We distilled this important idea of one-to-one correspondence by thinking hard about something we know well—counting. Generally, the most reliable guide to the unknown is a deep understanding of the simple and familiar.

*Understand simple things deeply.*

# MINDSCAPES  Invitations to Further Thought*

## I. SOLIDIFYING IDEAS

1. ***The same, but unsure how much*** (**H**). We have used a method of checking whether two sets of objects have the same number of things by pairing up and removing one object from each set until we run out. If we run out of objects from both sets at the same time, then we know that the sets contain the same number of things. Otherwise, we know that one set is larger than the other. Describe several events whereby we are able to compare the size of two collections without computing the individual sizes—for example, people filling all seats in an auditorium.

2. ***Taking stock*** (**S**). It turns out that there is a one-to-one correspondence between the New York Stock Exchange symbols for companies and the companies themselves (for example, PE is Philadelphia Electric Company). Explain why this correspondence must be one-to-one. What would happen if it were not? Describe potential problems.

3. ***Don't count on it.*** The following are two collections of the symbols @ and ©.

   @ @ @ @ @ @ @ @ @ @ @ @ @ @ @ @ @ @ @ @ @ @ @ @

   Are there more @'s than ©'s? Describe how you can quickly answer the question without counting, and explain the connection with the notion of a one-to-one correspondence.

4. ***Here's looking @ ®.*** Below are two collections of the symbols @ and ®.

   @ @ @ @ @ @ @ @ @ @ @ @ @ @ @ @ @ @ @ @ @ @ @

   Are there more @'s than ®'s? Describe how you can quickly answer the question without actually counting, and explain the connection with the notion of a one-to-one correspondence.

5. ***Enough underwear.*** When Deb packs for a trip, she doesn't count how many days she is going to be away and then count out that many pairs of underwear. Instead she places underwear into her suitcase, one at a time, and says the name of each day she will be away: "Monday" (and places one in), "Tuesday" (and places another in), "Wednesday", etc. Using this method, does Deb know the

---

*In the Mindscapes section, exercises marked (H) have hints for solutions at the back of the book. Exercises marked (S) have solutions.

number of underwear she placed in her bag? Does she have enough underwear for her trip? Discuss the connection this true story has with the notion of a one-to-one correspondence.

6. ***791ZWV.*** Suppose a stranger tells you that the license plate number on his car is 791ZWV. If you had a listing of all automobiles in the United States together with their license plate numbers, would you be able to precisely identify the stranger's vehicle? If so, explain why. If not, explain why, and identify what additional information you would require to identify it exactly. Discuss the connection between this situation and the notion of a one-to-one correspondence.

7. ***245-2345.*** Suppose a stranger tells you that her telephone number is 245-2345. Would you be able to dial her number and be certain that you reach her? If so, explain why. If not, explain why, and identify what additional information you would require to be certain to reach her home. Discuss the connection between this situation and the notion of a one-to-one correspondence.

8. ***Social Security*** (H). Is there a one-to-one correspondence between U.S. residents and their social security numbers? Explain why or why not.

9. ***Testing one two three.*** A professor wishes to distribute one examination to each student in the class. What is the most efficient way for her to determine whether she has more students than exams: Pass the exams out or count?

10. ***Laundry day.*** Suppose you are given a bag of quarters. The laundry machine requires $1.75 worth of quarters. One way to count how many washes you can do is to take out one quarter and say 25¢, then take out another and say 50¢, and so on. In practice, however, you might well use a different method that uses a notion of one-to-one correspondence. Explain such a method.

## II. CREATING NEW IDEAS

1. ***Hair counts.*** Do there exist two nonbald people on Earth having the property that there is a one-to-one correspondence between the collection of hairs on one person's body and the collection of hairs on the other person's body? Feel free to use facts from previous chapters, but explain how they provide an answer to the one-to-one correspondence question.

2. ***Social number*** (S). A social security number is a nine-digit number. Suppose that all nine-digit numbers are allowable social security numbers. Is there a one-to-one correspondence between allowable social security numbers and U.S. residents? You may assume that the U.S. population is about 250 million. Explain your answer.

3. ***Musical chairs.*** Musical chairs is a fun game in which a group of people march around a row of chairs while music is played. There is

one more person than there are chairs. The moment the music stops, everyone scrambles for a chair. The person left chairless loses and moves off to the sidelines. Then everyone in a chair gets up, one chair is removed, and the music and marching begin again. At what points in this game do we have a one-to-one correspondence between chairs and people and at what points do we not have such a correspondence? Explain the correspondences as the game is played until there is a winner.

4. *Dining hall blues.* One day during dinner in Ralph P. Uke Dining Hall the students in line for food discovered that the dining hall had run out of forks (no clean or dirty ones were available). While there was much jubilation in the line, what can you conclude about the set of all students in the dining hall and the set of all forks? Do we know how many forks there are? How many students? Discuss how the issue of one-to-one correspondence is relevant in addressing these questions.

5. *Dorm life* (H). Every student at a certain college is assigned to one and only one dorm room. Does this imply that there is a one-to-one correspondence between dorm rooms and students? Explain your answer.

### III. FURTHER CHALLENGES

1. *Pigeonhole principle.* Recall the Pigeonhole principle from the first section in Chapter 2. Restate this principle in terms of a correspondence. Suppose you try a method of assigning pigeons to holes and, after filling all the holes, you have some pigeons left over. If you take the pigeons out and try again, is there any hope of placing each pigeon in an individual hole the second time? Suppose you have an infinite number of pigeons and pigeonholes. Is it possible that a first attempt to give each pigeon his or her own individual hole failed while a second attempt succeeded?

2. *Mother and child.* Every child has one and only one birth mother. Does this imply that a one-to-one correspondence exists between the set of all children and the set of all birth mothers? Explain your answer.

### IV. IN YOUR OWN WORDS

*With a group of folks.* In a small group, discuss the notion of a one-to-one correspondence and why two collections that have a one-to-one correspondence should be considered the same size. After your discussion, write a brief narrative describing your conclusions.

# 3.2 Comparing the Infinite:
## Pairing Up Collections via a One-to-One Correspondence

*I saw . . . a quantity passing through infinity and changing its sign from plus to minus. I saw exactly how it happened . . . but it was after dinner and I let it go.*

—Sir Winston Churchill

We now enter the world of infinity armed with one idea—a criterion for comparison: the one-to-one correspondence. We will test the consequences of this idea by comparing some infinite collections of familiar objects. As usual, our most productive strategy is to examine the familiar before journeying toward the unknown.

Since numbers are really the only infinite collections we know, it is to numbers that we turn for our first examples. We consider various col-

An early one-to-one correspondence

lections of numbers that will help us become accustomed to the idea of one-to-one correspondence. We start with the most basic collection of numbers we can think of and then consider related but different collections. Our goal always is to determine whether or not various collections can be put into one-to-one correspondence, since one-to-one correspondence is the fundamental principle on which our investigation of infinity is built.

## *Familiar, but Infinite*

What is familiar and concrete to one person may be foreign and abstract to another, but as far as numbers are concerned we probably all agree on which are the most familiar. In 1886 Leopold Kronecker, a number theorist, made a statement about what is basic in the world of mathematics: "God created the positive integers; all the rest is human creation."

The set of natural numbers (They're all there, but we didn't have time to write them all in.)

One, two, three, . . . these are the positive integers. For every positive integer there is a next bigger one. Although we may think of these positive integers successively, we may also think of all of them at once— that is, think about the collection of all positive integers. The collection (also referred to as the *set*) of positive integers is so basic and natural to our way of thinking that it is called the set of *natural numbers*.

The set of all natural numbers is our first infinite set, and it has a comfortable feel about it. Among infinite sets, the natural numbers seem the most natural. By examining this and related collections of numbers, we will begin to develop a better and more precise idea of infinity.

## *Natural Numbers with 1 Removed*

Suppose now that we are given another copy of the set of natural numbers—in a different font:

*1, 2, 3, . . . .*

Unfortunately, being a bit absentminded, we lose the number **1**. Thus this new set is the collection of natural numbers with the number **1** removed. Specifically, our set consists of

*2, 3, 4, . . . .*

Thus we have a brand new infinite set. But this new infinite set has one less element than the set of natural numbers . . . or has it?

On the one hand, we can observe through life experience that, if we have a barrel of Volvos and one is removed, we then have fewer Volvos left. It seems reasonable to conclude that, if we make a new set by removing the number *1* from the set of natural numbers, we have a set with fewer things in it. On the other hand, we may think, "Hey, infinity is infinity is infinity so the new set doesn't have fewer elements."

We find our intuition being pulled in two directions. One direction is the "infinity is infinity" camp, the other is the "take one away, you have one less" school. We will soon discover that both these arguments will lead us astray. What's wrong with our intuition? Nothing, except that our insights and life experiences involving collections of everyday objects will not always apply to infinite collections.

If our intuition leads to two opposite conclusions and both sound reasonable, we are impelled to investigate further until clarity emerges. If we believe that the set of natural numbers with the number 1 removed is the same size as the set of natural numbers, then we need a rigorous and logical reason. Vague thoughts of "infinity is infinity" will not suffice in a quantitative court of law. However, we must remember to avoid little distractions such as our entire life history that tell us that, when we remove an object, we are left with fewer things. Instead, we must remember that our criterion for determining the equivalence of two collections is not a vague, undefined feeling developed through years of experience but is instead a clear criterion stated crisply and explicitly as the existence of a one-to-one correspondence. Since we have formulated an explicit definition, let's rely on it in preference to general impressions. Infinity is a large, wild beast, but, if we keep focused on our principle of comparison, we will have infinity tamely eating out of the palm of our hand.

## A Search for a One-to-One Correspondence

If we wish to answer the question of whether there are as many natural numbers as there are integers starting with **2, 3, 4, . . .** , we are asking neither more nor less than the question: Is there a one-to-one correspondence between the elements of the set

**2, 3, 4, . . .**

and the set of natural numbers

1, 2, 3, . . . ?

If we are from the school of "take one away, we have one less," it may appear that there cannot be a one-to-one correspondence between our two sets, since, if we paired the numbers in the two sets, we'd see the following.

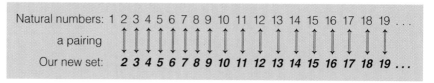

*Just because a specific attempt failed does not mean that the task at hand is impossible.*

❖ ❖ ❖ ❖ ❖ ❖

Notice how the "1" in the natural numbers is alone and is not paired with any number in our new set. Hence this pairing is not a one-to-one correspondence. Does this failure imply that no one-to-one correspondence exists? The answer is a resounding NO! A one-to-one correspondence may still exist. We merely conclude that the *particular* pairing we just created isn't a one-to-one correspondence.

## The Naturals Minus 1 Equals the Naturals

It turns out that, in fact, a one-to-one correspondence between these two sets does exist, which will not surprise members of the "infinity is infinity" camp. We provide such a one-to-one correspondence below, but members of the "infinity is infinity" camp should not be too smug just yet:

Natural numbers: 1 2 3 4 5 6 7 8 9 10 11 12 13 14 15 16 17 18 19 ...
a new pairing
Our new set:   **2 3 4 5 6 7 8 9 10 11 12 13 14 15 16 17 18 19 20** ...

*Be open to new ways of looking at a situation.*

❖ ❖ ❖ ❖ ❖ ❖

Suppose we dump all the natural numbers into the Natural Number Barrel and the natural numbers with **1** removed in the New Set Barrel. In the new pairing, we grab 1 from the Natural Number Barrel and **2** from the New Set Barrel, pair them, and toss them. Next we pair 2 with **3**, then 3 with **4**, 4 with **5**, and so on.

Notice that every element from the Natural Number Barrel is paired with exactly one of the elements from our New Set Barrel, and each number from our New Set Barrel is associated with exactly one natural number. The moment we mention a particular number from one list, we know who it pairs with from the other list. After completing this infinite process, no one is left out; both barrels are left empty.

We can use symbols to express this one-to-one correspondence to illustrate who gets paired with whom. The one-to-one correspondence could be represented in a compact manner as $n \leftrightarrow n + 1$, where this symbolic expression just indicates a generic pairing of numbers: the number $n$ from the Natural Number Barrel is paired with the number $n + 1$ from the New Set Barrel. Notice that, once we know what number $n$ is, we immediately know who its mate is, namely $n + 1$. So,

for example, if *n* is 4, then we see its mate is **4 + 1**, which is **5**. So 4 from the Natural Number Barrel is paired with **5** from the New Set Barrel. Notice that this pairing is exactly the one we described in the preceding picture.

## Cardinality

The existence of this one-to-one correspondence means that these two sets have the same number of elements. Now we must be a little bit careful here. We really should not say that these sets have the same "number" of elements, since infinity is not actually a number. How can we get around this thorn? We just create new terminology. We use the phrase *cardinality of a set* to mean the "number" of things in the set, with the understanding that the set may contain infinitely many things. If a set contains only finitely many things, then the set has finite cardinality, and we may remove those quotes from the word "number" in the previous sentence.

Given two sets, we say they have the *same cardinality* if there is a one-to-one correspondence between the elements of one set and the elements of the other set—for example, the set of stars on the U.S. flag and the set of states in the United States. In this case, both sets have finite cardinality, and that cardinality equals 50.

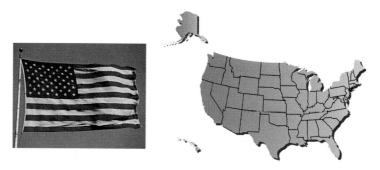

How would you show a one-to-one correspondence between the set of stars and the set of states?

As a further example, we just saw that the set of natural numbers and the set of natural numbers greater than 1 have the same cardinality.

## The Ping-Pong Ball Conundrum

Let's now consider the following tale. This story frees us from the confines of our physical reality and allows us to explore a world of our own imagination. After the story, we pose a question and encourage you to take a moment to guess an answer.

The scene opens with a very large barrel. Immediately to the right of the barrel, lined up like little round soldiers, we see an unending stream

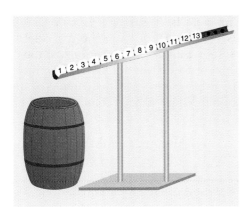

of white, numbered Ping-Pong balls ordered by number starting with 1, 2, 3, 4, and so on. There is one Ping-Pong ball for each natural number.

Now we are going to embark on a mental experiment. This experiment will last for exactly 1 minute. Sixty seconds after we start, our experiment is over and we stop. We note that our stopwatch is completely oblivious to our activities during that minute. Therefore, no matter what we do within that time period, 60 seconds later, the stopwatch's alarm will beep and we will stop—period.

## Task 1

We begin with 60 seconds on the clock. In half that time, 30 seconds, we must accomplish the following feat. We pour the first 10 Ping-Pong balls (numbered 1 through 10) into the large barrel, and, after the dumping, we reach into the barrel, find the Ping-Pong ball labeled 1, and remove it from the barrel (just throw it away). So, during the first stage of this experiment, we have 30 seconds to drop the first 10 balls in and then fish out the first ball (numbered 1). We are left with 9 balls in the barrel and 1 thrown away—neither a big nor exciting proposition.

## Task 2

Now we have 30 seconds remaining, and we begin to pick up the pace. In half the time remaining, 15 seconds, we must dump into the barrel the next 10 Ping-Pong balls (numbered 11 through 20), and then reach in and scoop out the ball numbered 2. So, in the second stage, we quickly (in 15 seconds) drop in another 10 balls and fish out the second ball (numbered 2). We are now left with 18 balls in the barrel and 2 thrown away—but we don't stop yet.

## Task 3

The third task must be accomplished in half the time remaining, 7.5 seconds. We must quickly drop in the next 10 balls (numbered 21 through 30) and then find and remove ball number 3. We are then left with 27 balls in the barrel and a few beads of perspiration on our foreheads.

We're still a long way from finished, but perhaps the pattern is emerging. We continue to cut the time remaining in half, and, in each individual half period, drop in the next 10 balls and fish out the ball in the barrel with the lowest number.

## Task 6

So, at the sixth stage we drop in balls numbered 51 through 60 and remove ball number 6, leaving a total of 54 balls in the barrel. We have to work pretty fast, because this entire stage must be completed in .9375 of a second (that is, half of half of half of half of half of half of 60 seconds).

## Remaining Tasks

Clearly, we really have to pick up the pace and work fast. In fact, we see that soon we will have to move faster than the speed of sound and even faster than the speed of light—now that's fast. Physically impossible? Why yes, of course, but it's also physically impossible to have infinitely many Ping-Pong balls—if we did, they would take up all the space in our universe, and their weight would squash us like small bugs. Happily, this is an exercise of the mind, and thus we use our power of imagination to see ourselves safe from these potentially dangerous balls and also to see ourselves dumping in and then pulling out balls at incredible speeds.

Our experiment is hurried and frantic; however, mercifully it lasts only 1 minute. Sixty seconds after we begin, the stopwatch beeps and we stop and attempt to catch our breath. As we catch our breath after the minute is over, we glance into the barrel. What do we see? Is it empty? Does it contain some Ping-Pong balls? Does it contain infinitely many Ping-Pong balls? These are the questions we invite you now to consider.

## On the Ball

Do Ping-Pong balls remain in the barrel after the minute has expired? If so, then we have an annoying request: Name one! Any ball left has a number stamped on its surface. Just tell us what that number is. Recall that the balls are numbered 1, 2, 3, . . . forever—there is no last ball. What is the number of a ball in the barrel? Could it be 5? No, because we know exactly when ball 5 was removed—namely, at the fifth stage (remember, we dumped in the 10 balls 41–50 and removed number 5). Could ball 45,671,803 remain? Well no, because we know exactly when we took that ball out of the barrel—specifically, at the 45,671,803rd stage of the experiment. Thus what balls are left? The answer is amazing and surprising: "NONE"! Since all the balls were numbered and we removed them systematically, we can state with exact precision the time when any one particular ball was removed. Thus at the end of the minute, the barrel is empty. This counterintuitive and perhaps unbelievable answer is puzzling—especially in view of the fact that the number of Ping-Pong balls left in the barrel increased by 9 at each stage.

This story, the question, and particularly the answer require serious thought, but we can convince ourselves that the barrel is empty. This Ping-Pong ball conundrum is a dynamic illustration of the dramatic difference between the finite and the infinite. In actuality, this activity produced a one-to-one pairing between the intervals of half times (stages) and the numbered balls. We were able to pair them evenly. Again, our intuition, based on finite collections, is not always accurate when applied to infinity. Nevertheless, through reason, the experience we are gaining will help us develop an understanding of infinity. To help develop this understanding, let's explore other infinite collections.

## *Looking for Giants*

When we saw that removing the number *1* from the set of natural numbers did not decrease the size or the cardinality of the set, we surprised those whose intuition previously dictated that removing an element should make a set smaller. We will also decimate the intuition of those who are still in the "infinity is infinity" camp by discovering that some infinite sets are even bigger or more infinite than the natural numbers! How can we even begin this seemingly impossible quest for giant sets?

We need to think of some infinite sets that are likely to be even larger than the set of natural numbers. A possibility that we might consider is the set of *all* integers: positive, negative, and zero. This set has infinitely many more numbers in it than are in the natural numbers—namely, all the negative integers and zero—so it appears to be a good candidate. In some sense, there appear to be *twice* as many integers as natural numbers.

## *Integers Equal Naturals*

Unfortunately, merely adding infinitely many negative numbers and zero is not enough to increase the size of the set of natural numbers. To prove this statement, as always, we need to exhibit an explicit one-to-one correspondence between the set of natural numbers and the set of all integers. You can find that correspondence yourself. To describe such a correspondence, draw a table that has the natural numbers, 1, 2, 3, . . . down the left side of the paper. Next to each natural number write one integer (positive, negative, or zero) in such a systematic fashion that all integers will eventually appear on the list. Wait to look at our answer until you've given it a try.

Notice that in this pairing, every natural number appears exactly once in the left-hand column, and every integer (positive, negative, or zero) appears exactly once at some time in the right-hand column. Thus we do get an even pairing, a one-to-one correspondence. What number does −9 in the right column get paired with? What number does 9 in the left

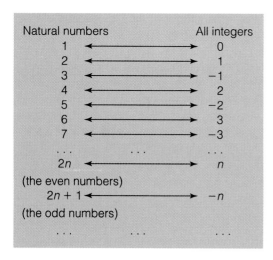

A one-to-one correspondence between the natural numbers (that is, the positive integers) and all integers (including negative integers and zero).

column get paired with? We can choose random elements from either set and figure out under this pairing with whom those random elements get paired—this experimentation is the best way to get a feel for the correspondence. Is any number not paired? No.

So we have proven that the set of natural numbers has the same cardinality as the set of all integers. Notice how the two symbolic expressions toward the bottom of our list capture the spirit of the pairing; namely, the even natural numbers are paired with the positive integers, while the odd natural numbers are paired with the negative integers or zero. For example, the natural number 32 can be written as $2 \times 16$, so the symbolic expression $2n \leftrightarrow n$ shows us that 32 is paired with 16. However, 33 can be written as $(2 \times 16) + 1$, and thus the symbolic expression $2n + 1 \leftrightarrow -n$ shows us that 33 is paired with $-16$. So, if we looked down our list, at some point we would see

## Rational Numbers

We might feel more or less stymied in our quest for an infinite set bigger than the set of natural numbers. But let's not forget that lots of numbers are not integers. For example, how about the rational numbers? Recall that the set of rational numbers is the set of all ratios of integers (fractions). Between any two integers there are infinitely many rationals. Within the set of rational numbers, we actually have infinitely many different infinite sets. The set of rational numbers must be huge.

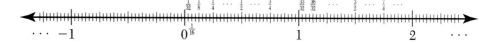

Unfortunately, in our search for different sizes of infinity, the rationals are still not numerous enough. We rely once again on the definition of same cardinality; namely, two sets have the same cardinality if there is a one-to-one correspondence between the elements of one set and the elements of the other. The trick here is to write down the rational numbers in a convenient and systematic way so that we know we have listed them all. The idea is to put all the rational numbers with numerator 1 in one column, all those with numerator 2 in another column, and so on as shown in the following diagram. Notice that all the rationals in the same row have the same denominator. For example, to find 37/112 in the diagram, we just go 37 spaces to the right of 0 and 112 spaces up. So, we can see that all the rational numbers appear somewhere in the pattern. In fact, each rational number appears many times, because the fractions are not all reduced to lowest terms. For example, 1/2 appears and so does 2/4, 3/6, and so on, but that redundancy is okay, because at this point we merely want a systematic method of writing down every rational number without leaving any out. Notice that all the positive rational numbers appear in the upper-right part of the diagram, all the negative rationals appear in the lower left, and 0 is right in the middle. So far, then, we have described a way of writing down all the rational numbers.

|     |     |     |     |     |     |
| --- | --- | --- | --- | --- | --- |
| ⋮   | ⋮   | ⋮   | ⋮   | ⋮   |     |
| 1/5 | 2/5 | 3/5 | 4/5 | 5/5 | ··· |
| 1/4 | 2/4 | 3/4 | 4/4 | 5/4 | ··· |
| 1/3 | 2/3 | 3/3 | 4/3 | 5/3 | ··· |
| 1/2 | 2/2 | 3/2 | 4/2 | 5/2 | ··· |
| 1/1 | 2/1 | 3/1 | 4/1 | 5/1 | ··· |

0

|     |      |      |      |      |      |
| --- | ---- | ---- | ---- | ---- | ---- |
| ··· | −5/1 | −4/1 | −3/1 | −2/1 | −1/1 |
| ··· | −5/2 | −4/2 | −3/2 | −2/2 | −1/2 |
| ··· | −5/3 | −4/3 | −3/3 | −2/3 | −1/3 |
| ··· | −5/4 | −4/4 | −3/4 | −2/4 | −1/4 |
| ··· | −5/5 | −4/5 | −3/5 | −2/5 | −1/5 |
|     | ⋮    | ⋮    | ⋮    | ⋮    | ⋮    |

## *Rationals Equal Naturals*

To show the one-to-one correspondence between the rational numbers and the natural numbers, we will thread a single rectangular spiral through all the rationals, starting in the middle at 0 and moving counterclockwise outward. To see the one-to-one correspondence with the natural numbers, we will just count the rational numbers as we encounter them along the spiral and make them bold-face to remind us that we have paired that rational with some natural number. We start with the rational **0** corresponding to the natural number 1; then, moving to the right and up, the rational **1/1** = 1 corresponds to the natural number 2, the rational **−1/1** = −1 corresponds to 3, the rational **2/1** = 2 corresponds to 4. We next come to 2/2, which has already been counted, so we skip it and move to **1/2,** which corresponds to 5, then **−2/1** = −2 corresponds to 6. We skip −2/2 since that equals −1, which already corresponds to 3, and move to **−1/2,** which corresponds to 7, and so on. Notice that every rational number will eventually be reached and put in correspondence with some natural number. This one-to-one correspondence shows that the set of all rational numbers has the same cardinality as the set of the natural numbers.

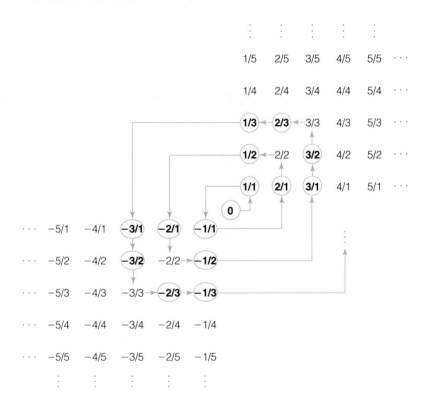

We may now straighten out the spiral and produce the following list showing the one-to-one correspondence. Notice that, if someone examined just the list, he or she would have difficulty detecting the pattern. The pattern arises from the spiral above. Our visual method generated this one-to-one correspondence:

| Natural numbers | Rational numbers |
|---|---|
| 1 | 0 |
| 2 | 1/1 |
| 3 | −1/1 |
| 4 | 2/1 |
| 5 | 1/2 |
| 6 | −2/1 |
| 7 | −1/2 |
| 8 | 3/1 |
| 9 | 3/2 |
| 10 | 2/3 |
| 11 | 1/3 |
| 12 | −3/1 |
| ... | ... |

Threading the spiral and counting along it provides an important insight into sets with the same cardinality as the set of natural numbers. If we can write a set out as an infinite list, we can make a one-to-one correspondence with the natural numbers.

We now see that the rational numbers did not provide us with an infinity larger than that of the natural numbers. Our quest for an even grander infinity has thus far failed. But perhaps when we are chasing sets larger than an infinite set, we should expect to have to go a long, long way.

## A LOOK BACK

Two sets have the same cardinality if there is a one-to-one correspondence between the contents of one and the contents of the other.

The set of natural numbers, 1, 2, 3, . . . , is a natural first infinite set to investigate. The set of natural numbers has the same cardinality as the set of natural numbers with the number *1* removed; the same cardinality as the set of all integers; and even the same cardinality as the set of all rationals. One might naturally, but mistakenly, guess that all infinite sets have the same cardinality as the natural numbers.

Our strategy for understanding a difficult topic is to explore the simplest, most familiar example we can find and do some experiments. In this case, we compared our simplest example (the natural numbers) with variations of it (the natural numbers minus 1, the integers, the rationals). As we make discoveries that are counterintuitive, we gain experience that will retrain our intuition. The discovery of counterintuitive truths liberates us to think of notions grander still.

*Experiments with the familiar help us to understand the unknown.*

# MINDSCAPES  Invitations to Further Thought*

## I. SOLIDIFYING IDEAS

1. **Even odds.** Let **E** stand for the set of all even natural numbers (so **E** = {2, 4, 6, 8, . . . .}) and **O** stand for the set of all odd natural numbers (so **O** = {1, 3, 5, 7, . . . .}). Show that the sets **E** and **O** have the same cardinality by describing an explicit one-to-one correspondence between the two sets.

2. **Naturally even.** Let **E** stand for the set of all even natural numbers (so **E** = {2, 4, 6, 8, . . . .}). Show that the set **E** and the set of all natural numbers have the same cardinality by describing an explicit one-to-one correspondence between the two sets.

3. **5's take over.** Let **EIF** be the set of all natural numbers ending in 5 (**EIF** stands for "ends in five"). That is,

   **EIF** = {5, 15, 25, 35, 45, 55, 65, 75, . . .}.

   Describe a one-to-one correspondence between the set of natural numbers and the set **EIF**.

4. **6 times as much.** If we let $\mathbb{N}$ stand for the set of all natural numbers, then we write $6\mathbb{N}$ for the set of natural numbers all multiplied by 6 (so $6\mathbb{N}$ = {6, 12, 18, 24, . . . }). Show that the sets $\mathbb{N}$ and $6\mathbb{N}$ have the same cardinality by describing an explicit one-to-one correspondence between the two sets.

5. **Any times as much.** If we let $\mathbb{N}$ stand for the set of all natural numbers, and $a$ stand for any particular natural number, then we write $a\mathbb{N}$ for the set of natural numbers all multiplied by $a$. Do the sets $\mathbb{N}$ and $a\mathbb{N}$ have the same cardinality? If so, describe an explicit one-to-one correspondence between the two sets.

6. **Missing 3 (H).** Let **TIM** be the set of all natural numbers except the number 3 (**TIM** stands for "three is missing"), so **TIM** = {1, 2, 4, 5, 6, 7, 8, 9, . . . }. Show that the set **TIM** and the set of all natural numbers have the same cardinality by describing an explicit one-to-one correspondence between the two sets.

7. **One weird set.** Let **OWS** (you figure it out) be the set defined by

   **OWS** = {1, 3, 5, 7, 8, 10, 11, 12, 13, 14, 15, 16, . . .};

   that is, after 8, the set contains all the natural numbers from 10 on. Show that the set **OWS** and the set of all natural numbers have the same cardinality by describing an explicit one-to-one correspondence between the two sets.

---

*In the Mindscapes section, exercises marked (H) have hints for solutions at the back of the book. Exercises marked (S) have solutions.

8. **Squaring off.** Let **S** stand for the set of all natural numbers that are perfect squares, so **S** = {1, 4, 9, 16, 25, 36, 49, 64, ... }. Show that the set **S** and the set of all natural numbers have the same cardinality by describing an explicit one-to-one correspondence between the two sets.

9. **Counting cubes (formerly crows).** Let **C** stand for the set of all natural numbers that are perfect cubes, so **C** = {1, 8, 27, 64, 125, 216, 343, 512, ... }. Show that the set **C** and the set of all natural numbers have the same cardinality by describing an explicit one-to-one correspondence between the two sets.

10. **Reciprocals.** Suppose **R** is the set defined by

    **R** = {1/1, 1/2, 1/3, 1/4, 1/5, 1/6, ... } .

    Describe the set **R** in words. Show that it has the same cardinality as the set of natural numbers.

11. **Hotel Cardinality (formerly California) (H).** It is the stranded traveler's fantasy. The Hotel Cardinality is a full-service luxury hotel with bar and restaurant, having as many rooms as there are natural numbers. The room numbers are 1, 2, 3, 4, 5, .... You can see why stranded travelers love the Hotel Cardinality. There appears to be no need for the sad sign: No Vacancy. Suppose, however, that every room is completely filled. Now it appears that the night manager must flash the No Vacancy sign. What if a weary traveler arrives late in the night looking for a place to stay? Could the night manager figure out a way to give the traveler her own private room (no sharing) without having anyone leave the hotel? The answer is yes. Describe how this accommodation can be made; of course, some guests will have to move to other rooms.

12. **Hotel Cardinality continued.** Given the scenario in the previous question, suppose now that two more travelers arrive, each wanting his or her own private room. Is it possible for the night manager to make room for these folks without pushing anyone onto the streets?

13. **More Hotel C.** Given the scenario in the previous question—that is, the hotel starts full—suppose now that the Infinite Life Insurance Company, which has lots of employees—in fact, there are as many employees as there are natural numbers—decides to give each of its employees a private room. Is it possible for the night manager to give each of the infinitely many employees their own rooms without kicking any of the other guests onto the streets? By the way, as you may have guessed, after this busy evening, the night manager quit and got a job as the night manager of a nearby Motel 6 (they only had 56 rooms).

14. **So much sand.** Prove that there cannot be an infinite number of grains of sand on the Earth.

15. ***Half way.*** Suppose you take the line below,

cut it in half, and then take the left piece and cut it in half, and then take the leftmost piece and cut it in half, and so on, without ever stopping. How many different pieces of the line would you have? Does the set of all pieces have the same cardinality as the set of all natural numbers? Justify your answer.

16. ***Pruning sets.*** Suppose you have a set. If you remove some of the things from the set, will the collection of remaining things always have fewer things in it than the original set? If so, demonstrate why. If not, illustrate with an example.

17. ***A natural prune.*** Describe a collection of numbers you could remove from the set of natural numbers so that the set of remaining numbers contains fewer numbers than the set of all natural numbers. (*Bonus:* How many times did the word *numbers* appear in the previous sentence?)

18. ***Prune growth.*** Is it possible to remove things from a set so that the collection of remaining things is larger than the original set? Explain why or why not. Illustrate with an example.

19. ***Same cardinality?*** Suppose we have two sets and we are able to pair elements from one set with elements from the other set in such a way that all the elements from the first set are paired with elements from the second set, but there are elements from the second set that were never paired. Does this pairing imply that the two sets do not have the same cardinality? Justify your answer.

20. ***Still the same?*** (**S**). Suppose we have two sets, and every pairing of elements from one set with elements from the other set results in having elements from the second set never paired. Do these pairings imply that the two sets do not have the same cardinality? Justify your answer.

## II. CREATING NEW IDEAS

1. ***Modest rationals*** (**H**). Devise and then describe a method to systematically list all rational numbers between 0 and 1.

2. ***A window of rationals.*** Using your answer to the previous exercise, show that the set of rationals between 0 and 1 has the same cardinality as the set of natural numbers.

3. ***Bowling ball barrel.*** Suppose you have infinitely many bowling balls and two huge barrels. You take the bowling balls and put each ball in one of the two barrels. What can you conclude about the cardinality of at least one of the barrels? Prove your answer.

4. **Not a total loss.** Take the set of natural numbers and remove a finite number of numbers from the set. Prove that this new set has the same cardinality as the set of natural numbers.

5. **Mounds of mounds.** In a Peter-Paul Mounds package are two delicious chocolate-covered coconut candy bars. Suppose you had infinitely many Mounds packages: one for each natural number. Does the collection of individual candy bars have the same cardinality as the set of natural numbers? If not, explain why. If so, provide a one-to-one correspondence.

6. **Piles of peanuts.** You have infinitely many piles of peanuts. In the first pile you have one peanut; in the second pile you have two; in the third you have three; and so on. How many nuts do you have? Does the set of all these nuts have the same cardinality as the set of natural numbers? If not, explain why. If so, provide a one-to-one correspondence.

7. **The big city (S).** Not Finite City (also known as The Really Big Apple) is made up of infinitely many avenues running north and south (one avenue for each natural number) and infinitely many streets running east and west (one street for each natural number). A traffic light is placed at every intersection of a street with an avenue. How many traffic lights are there? Does the set of all traffic lights have the same cardinality as the set of natural numbers? If not, explain why. If so, provide a one-to-one correspondence.

8. **Don't lose your marbles.** Suppose you have infinitely many large boxes (one for each natural number). In each box, you have infinitely many marbles (one for each natural number)—so the boxes are really big! Does the set of all marbles have the same cardinality as the set of natural numbers? Just make a guess and explain why you guessed what you guessed (you need not justify your answer rigorously).

9. **Make a guess.** Guess an infinite set that does not have the same cardinality as the set of natural numbers.

10. **Coloring.** Consider the infinite collection of circles below:

Suppose you have two markers, one red and one blue, and you color each circle one of the two colors. How many different ways can you color the circles? Do you think that the set of all possible circle colorings has the same cardinality as the set of all natural numbers? This question is tricky; just make a guess and explain why you guessed what you guessed (you need not justify your answer rigorously). You may first want to answer the question for just these four circles as a warm-up. We'll return to this question later.

## III. FURTHER CHALLENGES

1. *Ping-Pong balls on parade* (**H**). This question is based on the experiment described in the section on putting in and removing infinitely many Ping-Pong balls from a barrel. This time suppose you dump in the 10 Ping-Pong balls numbered 1–10 as before and remove number 1. But next you put in 100 Ping-Pong balls, numbers 11–110, and remove number 2. Then you put in 1,000 Ping-Pong balls, numbers 111–1110 and remove number 3, and so on. The question is, How many Ping-Pong balls remain in the barrel after the timer goes off? Infinitely many? Finitely many? Can you name one?

2. *Naked Ping-Pong balls.* This question is based on the experiment described in the section on putting in and removing infinitely many Ping-Pong balls from a barrel. But this time the Ping-Pong balls are not numbered! You play the same game as in the story; however, now at each stage you just reach in and remove one Ping-Pong ball (you cannot fish around for a particular one since they now all look the same). How many Ping-Pong balls might remain in the barrel now? This question is an interesting conundrum.

3. *Primes.* Show that the set of all prime numbers has the same cardinality as the set of all natural numbers.

4. *A grand union.* Suppose you have two sets, each having the same cardinality as the set of natural numbers. Take the elements of both sets and put them together to make one huge set. Prove that this new huge set has the same cardinality as the set of natural numbers.

5. *Unnoticeable pruning.* Suppose you have any infinite set. Is it always possible to remove some things from that set such that the collection of remaining things has the same cardinality as the original set? Explain why or why not, and illustrate your answer with an example.

## IV. IN YOUR OWN WORDS

1. *Personal perspectives.* Write a short essay describing the most interesting or surprising discovery you made in exploring this section's material. If any material seemed puzzling or even unbelievable, address that as well. Explain why you chose the topics you did. Finally, comment on the aesthetics of the mathematics and ideas in this section.

2. *With a group of folks.* In a small group, discuss and work through the argument showing that the set of rational numbers has the same cardinality as the set of natural numbers. After your discussion, write a brief narrative describing the argument in your own words.

3. ***Creative writing.*** Write an imaginative story (it can be comical, dramatic, whatever you like) that involves, or evokes, the ideas of this section.

4. ***Power beyond the mathematics.*** Provide several real-life situations—ideally, from your own experience—for which some of the strategies of thought presented in this section would provide effective methods for approaching and resolving them.

# 3.3 The Missing Member:
## Georg Cantor Answers: Are Some Infinities Larger Than Others?

*From the paradise created for us by Cantor, no one will drive us out.*

—David Hilbert

Around 1872 the German mathematician Georg Cantor shook the foundations of infinity when he showed that the set of real numbers has more elements than the set of natural numbers. In other words, he proved that infinity is not one size but that some infinities are more infinite than others. At first such a notion seems almost nonsensical. Once we have reached infinity, surely we cannot climb farther. But Cantor showed that there were yet higher mountains to scale.

To show that the real numbers are more numerous than the natural numbers, Cantor focused intently on what it would mean for the real numbers and the natural numbers to have the same cardinality. It would mean that the real and natural numbers could be put in one-to-one correspondence. Writing down what such a correspondence might look like gives us a visual clue how to demonstrate conclusively that any attempted correspondence between the natural numbers and the real numbers could not include every real; some real is missing—*the missing member.*

### From Bizarre to Intuitive

When Cantor conceived his idea near the end of the last century, many mathematicians resisted it strongly and attacked Cantor in a personal, abusive manner. These attacks and other psychological woes contributed to a bleak existence for Cantor, who spent much of the last part of his life in an insane asylum. People do not easily give up their intuition or beliefs, and the resistance of mathematicians to Cantor's "taming" of infinity was neither the first nor the last time that people resisted ideas that contradicted their preconceptions.

In 1637, Galileo was imprisoned for saying that the earth moves around the sun. The idea that the earth moves is extremely counterintuitive. We accept the idea of a moving earth principally because everyone else believes it. When we were children, we were told that the earth moves, and so a moving earth does not present a challenge to our beliefs. In fact, anyone who would today assert that the earth is stationary and the sun revolves around the earth would be regarded as a kook—and for good reason. However, historically speaking, a moving earth was not easy to prove. Galileo was imprisoned because authorities of the time were not able to see how they could preserve their belief in the centrality of humanity and accept the radical idea that the earth was just one of several planets revolving about the sun.

*Consider the counterintuitive.*

❖ ❖ ❖ ❖ ❖ ❖

Most people do not consider infinity to be of life-threatening importance. Cantor was not imprisoned for heretical thinking. Infinity is a little too "out there" for most people to get worked up about. But, for those mathematicians who were deeply immersed in such issues, the idea of having many different infinities was a tremendous blow. Out of vague and ill-defined notions of sizes of sets, Cantor distilled the fundamental idea of one-to-one correspondence. This idea is so basic that one feels compelled to accept its logical consequences. But these consequences contradict the intuition of most people—until their intuition is adjusted, after which the concept of more than one infinity becomes perfectly natural. Thanks to Cantor, who reached out and considered the counterintuitive, no mathematician today has a problem encompassing the idea of multiple infinities.

Sometimes understanding a fact requires us to change our minds in a dramatic way. However that initially counterintuitive fact may at some future date attain the level of intuitive truth. For the Greeks, the existence of irrational numbers presented such a challenge. For a century now, mathematicians have come to understand the hierarchy of infinities. Time and again the bizarre and rare, after their discovery and assimilation, become the natural and familiar. These mental transitions are some of the great joys of thought.

## *Unequal Decimals*

We now turn to the task of demonstrating that the set of real numbers has a strictly larger cardinality than the set of natural numbers. In other words, we now show that the set of real numbers is more infinite than the set of natural numbers.

Recall that each real number can be expressed as an infinitely long decimal expansion. For example,

243.47666687544680088767287584934578844532l . . .

is a real number. Before moving forward, we must first make an easy observation about real numbers. Suppose we examine two decimal numbers, but we cover up all the digits in the numbers with ?s except for the digit that is in, say, the fifth place after the decimal point. So, we have a piece of paper with two funny looking numbers on it: ??.????2???? . . . and ??.????4???? . . . . We do not know what these numbers are because we can read only the fifth digit after the decimal point. But one thing we do know is that these two numbers are different. If they were the same, we could not have a 2 in the fifth place after the decimal point of one number and a 4 in the fifth place in the other. Likewise, if we have two numbers and one has a 2 and the other has a 4 in the 87th place after the decimal, then the two numbers must be different. This observation is not hard to understand, but it is a key to Cantor's reasoning.

## *Two Long Lists*

Cantor proved that there are more real numbers than natural numbers through a clever, yet simple idea. If the set of real numbers and the set of natural numbers had the same cardinality, then there would be a one-to-one correspondence between the set of natural numbers and the set of real numbers. So his idea was to list the natural numbers down the left-hand side of a page, list reals in the right-hand column, and then show how to construct a real number that could not appear on the list. He showed that, once we commit ourselves to a list of reals in the right-hand column, one real number corresponding to each natural number, then we can describe a real decimal number that does not appear anywhere on that infinite list. So, we could not have listed all the real numbers in the right-hand column. Thus, the natural numbers and the real numbers could not be put in one-to-one correspondence, and so there are more real numbers than natural numbers. Cantor's basic strategy was to attempt an impossible task in order to understand why it couldn't be done.

To see Cantor's idea in action, let's make a list of all the natural numbers in one column and in another column list real numbers. Imagine a barrel containing all the natural numbers and another barrel containing all the real numbers. We will now grab the natural numbers out, one at a time, in order, pairing each one with a real number that we will grab out of the other barrel. We will then record the pairing and grab another two and repeat. To illustrate this process, the pairing might begin with something like this:

| Natural numbers | | Real numbers |
|---|---|---|
| 1 | ↔ | 0.55627363495617384921348... |
| 2 | ↔ | 142.02732981638472734718734... |
| 3 | ↔ | 7.61235987364823519197234... |
| 4 | ↔ | 238.18521936478912092519027... |
| 5 | ↔ | −0.00083738265191836548713... |
| 6 | ↔ | 31.84722235675444566903346... |
| 7 | ↔ | 658.33333333333543356708632... |
| 8 | ↔ | −37.83958382139857446882891... |
| ... | ... | ... |
| 11 | ↔ | 29.99907982742111199853769... |
| ... | ... | ... |

We can view this correspondence as two infinitely long columns: On the left we have a complete list of all natural numbers, and on the right we have a list of real numbers. We are now wondering whether every single real number will appear somewhere in the right-hand list. If the set of real numbers and the set of natural numbers have the same cardinality, then it would be possible to list all the reals in some order—one for each natural number. But, in fact, we will construct a real number in decimal form that does not appear anywhere in the right column. That is, we will show that there are so many more real numbers than natural numbers that given *any* pairing between the natural numbers and the reals, a real number will always be left out—it is impossible to produce a one-to-one correspondence. Put another way, if we have the natural numbers in one barrel and the real numbers in another barrel and we take out one natural number and take out one real number, pair them, and repeat, then after we run out of natural numbers, there will be real numbers left over! In fact, most of the real numbers would still be left in the barrel.

## *A Missing Real*

We are going to write down a particular real number that we will call $M$, for "missing." We will write it in its decimal expansion. Our number $M$ will be between 0 and 1, so its decimal expansion begins with 0.??? .... Now we must decide what the digits "??? ..." are. Each digit will be one of two possibilities: a 2 or a 4. We will decide on the digits of our number $M$ one at a time, successively, so we must be patient. We now describe the criterion by which we choose each digit of our number $M$.

We start with the first digit after the decimal point. Remember that we have a table that pairs one real number with each natural number. So some real number is paired with the natural number 1. We take a look at that real number and look at its first digit after the decimal point. Although this insight will not shake the very foundations of your universe, we boldly state that there are only two possibilities for that first digit: It is either 2 or it is not 2. We will use the first digit of that first real number to decide on the first digit of our number *M*—the real number we are building. If the first digit of that first real number is 2, then we will set the first digit after the decimal point of our number *M* to be 4. If, however, the first digit of that first real number is not 2, then we will set the first digit after the decimal point of our number *M* to be 2. Observe that, no matter what digits come next in *M*, we know for sure that the number *M* will not equal the real number paired with 1. Why? Because *M* and the real number paired with 1 have different first digits after the decimal point!

How will we define the digit that is in the second place after the decimal point of our number *M*? We take a look at the real number paired with the natural number 2, see what its second place digit after the decimal point is, and ask if it equals 2. If that digit is 2, then we will set the second digit of our number *M* to be 4. If, however, that digit is not 2, we will set the second digit of our number *M* to be 2. Notice that we have defined *M* such that *M*'s second digit after the decimal point is not the same as the second digit after the decimal point of the real number corresponding to 2. In particular, *M* cannot equal the second real number in the list—the real number corresponding to 2.

We continue to define the digits of *M* in this fashion. So, for example, to determine the 11th digit of *M*, we look at the 11th digit in the real number that is paired with the natural number 11. If that digit is 2, then we define the 11th digit of our number *M* to be 4; if that digit is not 2, then we define the 11th digit of our number *M* to be 2.

### So Far, So Good?

To determine whether this process is clear, let's look at the lists of natural numbers and real numbers in the previous chart and write down what *M* would be. The answer appears in the next paragraph, so write your answer first. Why is Cantor's argument referred to as Cantor's *diagonalization proof*?

If the one-to-one correspondence is the one given in the preceding illustration, then *M* would equal 0.24442424 . . . . Incidentally, *M*'s 11th digit after the decimal point is 4. Notice that, no matter what the given correspondence is, the number *M* will have only 2's and 4's in its decimal expansion.

### Is M Really Missing?

At this point we know how to construct *M* if we are given a table with natural numbers in one column, each corresponding to a real number. Of course, *M* is some real number. Could *M* appear anywhere on this list of reals? If so, where could it appear?

Is *M* the first real number on the list—the real number that corresponds to 1? No, because we selected the first digit of *M* so that it differs from the first digit of the real number paired with the natural number 1.

Is $M$ the second number on the list? No, because we selected the second digit of $M$ so that it differs from the second digit of the number associated with the natural number 2.

Is $M$ the 1,582,987th number in the list? No, because we selected the 1,582,987th digit of $M$ to differ from the 1,582,987th digit of the real number paired with the natural number 1,582,987.

What does all this mean? $M$ cannot equal any real number in the table, and, therefore, $M$ is not on the list! If we wrote down a different list, with different reals, then we would build a different $M$. But the point is that, for any particular attempt we make to list the natural numbers on the left and correspond a real with each natural number, we can create a real decimal number $M$ that is not on that list. In other words, it is impossible for any correspondence of natural numbers and reals to contain all the real numbers. In particular, there is no one-to-one correspondence between the natural numbers and the reals!

## ~~Infinity~~ → Infinities

We have just shown that, if we are given any correspondence of the natural numbers to the real numbers, we can produce a real number that does not correspond to any one of the natural numbers. So, no matter how we try to pair numbers from the two sets, we will always have real numbers left over. Therefore, there are **different sizes of infinity!** We have found two different infinities: The cardinality of the natural numbers is **not the same** as the cardinality of the real numbers, even though both sets are infinite. There are more real numbers than there are natural numbers, or, phrased another way, the cardinality of the real numbers is larger than the cardinality of the natural numbers, even though the natural numbers are already infinite. Incredible!

> **CANTOR'S THEOREM.** *There are more real numbers than natural numbers.*

This whole notion of infinities being larger than other infinities is not an easy one to digest. We must think about it, work through the preceding argument many times over, and explain it to someone else. It is a great challenge to try to understand something that appears to contradict our personal hunches and intuition, especially for those members of the now defunct "infinity is infinity" camp. We must master the logical arguments so solidly that they become irrefutable. Only then will our preexisting biases make way for a new idea of the infinite. This process requires much effort but is well worth the investment.

## A LOOK BACK

Cantor's diagonalization argument shows that, for any given correspondence from the natural numbers to the reals, we can always construct a new real number that does not appear on the list—that is, that does not correspond to any one of the natural numbers. Consequently, the cardinality of the real numbers is not the same as the cardinality of the natural numbers. There are more real numbers than natural numbers.

What strategy allowed us to discover a proof that there are more real numbers than natural numbers? We began by looking carefully at what must be true if the real numbers and the natural numbers had the same cardinality. We wrote down a possible correspondence—naturals on the left, reals on the right—and thought about the question: Are any real numbers missing? We noticed that real numbers have infinitely many digits past the decimal point, and for two real numbers to be different, they need to differ only in one of those places. Keeping this in mind while looking at the lists led us to the diagonalization argument.

When we feel that something is impossible or if we just don't know whether it is true or false, a good way to find the truth is to explore carefully what would happen if the impossible or unknown were true. Systematically design what-if scenarios and play them to their conclusions.

*I can see it, but don't believe it!* —Georg Cantor

Explore consequences of possibilities.

## MINDSCAPES  Invitations to Further Thought*

### I. SOLIDIFYING IDEAS

1. **Dodge Ball.** Revisit the game of Dodge Ball from Chapter 1, Fun and Games. Using the game pieces included in the kit, play it several times with several people. Get the strategy down, and then explain to your opponents the underlying principle. Record the results of the games.

2. **Don't dodge the connection (S).** Explain the connection between the Dodge Ball game and Cantor's proof that the cardinality of the reals is greater than the cardinality of the natural numbers.

3. **Cantor with 3's and 7's.** Rework Cantor's proof from the beginning but this time, if the digit under consideration is 3, then make the corresponding digit of $M$ a 7, and if the digit is not 3, make the associated digit of $M$ a 3.

---

*In the Mindscapes section, exercises marked (H) have hints for solutions at the back of the book. Exercises marked (S) have solutions.

4. **Cantor with 4's and 8's.** Rework Cantor's proof from the beginning but this time, if the digit under consideration is 4, then make the corresponding digit of $M$ an 8, and if the digit is not 4, make the associated digit of $M$ a 4.

5. **Think positive.** Prove that the cardinality of the positive real numbers is the same as the cardinality of the negative real numbers. (*Caution:* You need to describe a one-to-one correspondence; however, remember that you cannot list the elements in a table.)

6. **Diagonalization.** Cantor's proof is often referred to as "Cantor's diagonalization argument." Explain why this is a reasonable name.

7. **Digging through diagonals.** First consider the following infinite collection of real numbers. Describe in your own words how these numbers are constructed (that is, describe the procedure for generating this list of numbers). Then, using Cantor's diagonalization argument, find a number not on the list. Justify your answer.

    0.12345678910111213141516 1718 . . .

    0.24681012141618202224262830 32 . . .

    0.36912151821242730333639 4245 . . .

    0.4812162024283236404448 525660 . . .

    0.510152025303540455055 606570 . . .

    . . .

    . . .

8. **Coloring revisited (H).** In the Mindscapes of the previous section we considered the following infinite collection of circles

    and all the different ways of coloring the circles with either red or blue markers. Show that the set of all possible circle colorings has a greater cardinality than the set of all natural numbers.

9. **A penny for their thoughts.** Suppose you had infinitely many people, each one wearing a uniquely numbered button: 1, 2, 3, 4, 5, . . . (you can use all the people in the Hotel Cardinality if you don't know enough people yourself). You also have lots of pennies (infinitely many, so you're *really* rich; but don't try to carry them all around at once). Now you give each person a penny and ask everyone to flip their pennies at the same time. You then ask them to shout out in order what they flipped (H for heads and T for tails). So you might hear: HHTHHTTTHTTHTHTHHHTH . . . or you might hear THTTTHTHHTTHTHTTTTHHHTHTHTH . . . ,

and so forth. Consider the set of all possible outcomes of their flipping (all possible sequences of H's and T's). Does the set of possible outcomes have the same cardinality as the natural numbers or not? Justify your answer.

10. ***The first digit*** **(H).** Suppose that, in constructing the number $M$ in the Cantor diagonalization argument, we declare that the first digit to the right of the decimal point of $M$ will be 7, and then the other digits are selected as before (if the second digit of the second real number has a 2, we make the second digit of $M$ a 4; otherwise, we make the second digit a 2, and so on). Show by example that the number $M$ may in fact be a real number on our list.

## II. CREATING NEW IDEAS

1. ***1's and 2's*** **(H).** Show that the set of all real numbers between 0 and 1 just having 1's and 2's after the decimal point in their decimal expansions has a greater cardinality than the set of natural numbers. (So, the number 0.1121111222121222211112 . . . is a number in this set, but 0.1161221212122122 . . . is not since it contains digits other than just 1's and 2's.)

2. ***Pairs*** **(S).** In Cantor's argument, is it possible to consider pairs of digits rather than just single digits? That is, suppose we look at the first two digits of the first real number on our list, and, if they are not 22, then we make the first two digits of $M$ be 22. If the first two digits are 22, then we make the first two digits of $M$ be 44. Similarly, let the next two digits of the next real number on our list determine the next two digits of $M$, and so on. If this procedure would still produce a number $M$ not on our list, then provide the details for such a method. If this procedure does not work, explain or illustrate why not.

3. ***Three missing.*** Given a list of real numbers, as in the Cantor argument, explain how to construct three different real numbers that are not on the list.

4. ***No Vacancy.*** Recall the Hotel Cardinality, described in the Mindscapes of the previous section. Create a collection of people so that it would be impossible for the (new) night manager to give each of them a room. Thus, for a really big group of people, a No Vacancy sign (or actually a Not Enough Room sign) might actually be necessary. Explain why it is not possible to give each person from your group a room.

5. ***Just guess.*** This is just a "guessing question." Do you think there are sets whose cardinality is actually larger than that of the set of real numbers? Or do you think the infinity of reals is the largest infinity? Just make a guess and informally explain it.

## III. FURTHER CHALLENGES

1. *9's.* Would Cantor's argument work if instead of 2 and 4 as the digits we used 2 and 9? That is, for each digit we ask whether the digit is a 2. If it is 2, we make the analogous digit of $M$ a 9, and otherwise we make the digit a 2. Using this method, we are *not* guaranteed that the number $M$ we construct is not on our list. Provide a scenario in which the constructed number is on the list! (*Hint:* Remember from Chapter 2 that $.1999999\ldots = .2$. The number 9 is key here.)

2. *Missing irrational.* Could you modify the diagonalization procedure so that the missing real you produce is a rational number? How could you modify the diagonalization argument so that the missing real number you produce is an irrational number? (*Hint:* Using the construction in this section, each digit of $M$ is a 2 or a 4. Modify the construction so the 10th place, the 100th place, the 1,000th place, and so on are either a 3 or a 5, while the rest are 2's and 4's. Why will such a number be irrational?)

## IV. IN YOUR OWN WORDS

1. *Personal perspectives.* Write a short essay describing the most interesting or surprising discovery you made in exploring this section's material. If any material seemed puzzling or even unbelievable, address that as well. Explain why you chose the topics you did. Finally, comment on the aesthetics of the mathematics and ideas in this section.

2. *With a group of folks.* In a small group, discuss and work through Cantor's argument showing that the set of real numbers has a greater cardinality than the set of natural numbers. After your discussion, write a brief narrative describing the argument in your own words.

# 3.4 Travels Toward the Stratosphere of Infinities:
## The Power Set and the Question of an Infinite Galaxy of Infinities

*There is no smallest among the small and no largest among the large; but always something still smaller and something still larger.*

—Anaxagoras

The theory of infinity described in the previous sections is truly beautiful. Why? Because we were able to analyze something logically that initially appeared unanalyzable, and our intellectual journey revealed new and surprising structure to ideas that were previously ill-formed and vague. Logical reasoning, interwoven with well-defined objects and a clever idea, completely disproved an intuitive feeling and replaced it with the solid foundation of a new understanding of the infinite. That's the power of mathematics, or perhaps more accurately, the power of mathematical thinking—the ability to identify and describe objects at hand, create a logical framework wherein these objects can exist, understand them deeply, analyze them carefully, and arrive at a valid conclusion. Of course, this way of thinking is not restricted to mathematical issues. We can use these techniques of logical reasoning every day to help resolve life issues and enrich our lives.

So, we've just discovered that there is an infinity larger than another infinity. Specifically, we have seen that the cardinality of the set of real numbers is greater than the cardinality of the set of natural numbers. This surprising fact now leads to a vast number of questions—here are just a few.

1. Is there an infinity that is greater than the infinity of the set of natural numbers yet is less than the larger infinity of the set of real numbers? (Is the cardinality of the reals the next bigger infinity after the cardinality of the natural numbers?)

People have always tried to measure the infinite.

*The Ancient of Days* (1794) by William Blake.

2. Is there an infinity greater than the infinity of the set of real numbers?
3. Are there infinitely many different sizes of infinity?
4. Is there a largest infinity—one that encompasses all others?

We will be able to answer only three of these questions. The remaining question is much more complicated and has a surprising answer of sorts. Why not think about these questions right now and try to guess which question has the unusual resolution. It's a great intellectual challenge in life to determine which questions are harder than others, even if we are unable to answer them.

By the end of this section, we will have settled three of these questions and discussed the unexpected status of the remaining question. Go ahead and guess which question is the odd ball.

The strategy we employ here is a potent method for making progress in any situation. In other words, once we have discovered an idea, let's milk it for all it's worth. Recall that the diagonalization argument was discovered first in the Dodge Ball game and then was used to show that there are more real numbers than natural numbers. Can we use the same idea to see whether we can find infinities even greater than the reals? If an idea provides an answer to one question, it is often powerful enough to answer many related questions as well.

By attempting to order the difficulty of issues we are taking the first steps toward resolving them.

◆ ◆ ◆ ◆ ◆ ◆

## Endless Infinities

We shall soon see that no infinity is without superiors. An infinity's very existence implies another infinity vaster still. In fact, the collection of all infinities has no bound. There are infinitely many different sizes of infinity—so vast a collection that no infinity can even count the number of different sizes of infinity!

We all know that there is no integer larger than all the others. Each integer is bettered by its successor. With infinite sets, a similar pattern occurs. As we have done before, we use the phrase *cardinality of a set* to mean the size of the set. For any infinite set we are given, we will show how to construct a related but larger infinite set with cardinality strictly bigger than the cardinality of the set we were given. When we have absorbed and understood this construction, we will see that Cantor's diagonalization proof, which showed that there are more real numbers than natural numbers, provides a model for how to generate ever larger infinities. In fact, both arguments have at their root the exact same idea used in Dodge Ball. Sometimes an answer to one question immediately provides the answers to more general questions.

## Sets Within Sets

We want to start with a set and then create a larger set, but we begin by creating smaller sets. One method of generating a new set is to look at a few elements in the starting set. Let's solidify our thinking by looking at a specific example.

Suppose that **Suits** is the set containing ♣, ♦, ♥, ♠. We could write this set as **Suits** = {♣, ♦, ♥, ♠} (the fancy brackets indicate that **Suits** is a set containing all the symbols inside). We can make new sets by selecting just some of the elements from **Suits**. For example, we could define a set **Reds** to be the set containing the two elements ♥, ♦ and a set **Blacks** to be the set containing ♣, ♠. In other words, **Reds** = {♦, ♥} and **Blacks** = {♣, ♠}.

So, the elements of both **Reds** and **Blacks** are all from the original set **Suits**. Thus, the sets **Reds** and **Blacks** are each an example of a *subset* of **Suits**. In particular, given a set $S$, we say that another set $T$ is a *subset* of $S$ if every element of $T$ is also in $S$. So, a subset of a set is just a collection of some of the elements from the original set. How many different subsets of the set of all card suits **Suits** are there? Here is the complete list of all **Suits'** subsets (there are a lot):

{ },
{♣}, {♦}, {♥}, {♠},
{♣, ♦}, {♣, ♥}, {♣, ♠}, {♦, ♥}, {♦, ♠}, {♥, ♠},
{♣, ♦, ♥}, {♣, ♦, ♠}, {♣, ♥, ♠}, {♦, ♥, ♠},
{♣, ♦, ♥, ♠}.

We see all the collections of single elements, all possible pairs, and all triples. But looking at the preceding list, we may wonder about the first and the last subsets. The first looks strange: It has nothing in it. It is called the *empty set*—the set that contains nothing. The empty set is a subset of every set. The other subset that may look strange is the last one; it consists of the entire set itself. Well, according to our definition of *subset*, that last set is a subset of **Suits.** Those are all the subsets of **Suits.** By counting, we see that there are 16 subsets of **Suits.**

## *How Many Subsets?*

Now we want you to experiment. Write down all the subsets of the set containing two coins: a penny and a dime. How many subsets are there?

Write down all the subsets of the set of musicians in a trio. How many subsets are there?

Do you see a pattern between the number of elements in a set and the number of subsets the set has?

Make guesses, even if they are incorrect.

◆ ◆ ◆ ◆ ◆ ◆

Guess (or better yet "conjecture") how many subsets there are of a set containing seven elements. Did you guess 14? Did you guess 49? Did you guess 32? Did you look at your previous examples and try to find the pattern? If you guessed 14, 49, 32, or many other plausible answers, then you were doing the right thing. Making concrete, reasonable guesses is an effective strategy for figuring something out. Even if it's wrong, a good guess gives us ideas. We can analyze why it failed and then make appropriate adjustments. The repeated process of guessing, analyzing reasons for failure, and making adjustments often leads to insight and the correct answers.

Let's return to the question: How many subsets does a set of seven things have? If you guessed 128, you probably have discerned the pattern.

How many subsets does a set containing 583 elements have? Did you guess

3165829138855738035974432269051484032449681268495511550900007117989084481363607899780049933583910975866850194253006583543697472439126934150785304232543256668350334894408?

If so, then you really need to get out a bit more. Okay, seriously, notice that, in each case, the number of subsets of a set is a power of 2. That is, $2^4 = 2 \times 2 \times 2 \times 2 = 16$, $2^7 = 2 \times 2 \times 2 \times 2 \times 2 \times 2 \times 2 = 128$, and $2^{583} =$ 31,658,291, blah, blah, blah, 489,408. Why is the number of subsets always a power of 2?

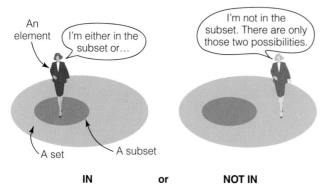

## The Power of 2

Let's think about that last question for a moment. Suppose we were an element of a set. For any particular subset we think of, we know that it was made by picking out some of the elements from the original set. Therefore, there are exactly two possibilities: Either we are in that subset (we got picked) or we are not in that subset (we didn't get picked). This basic observation is the key to unlocking the pattern.

Let's start with a set containing one element and work our way up to larger sets. Suppose we have a set consisting of one element—say the set is {♣}. Then there are exactly two subsets corresponding to the two possibilities—either ♣ is not in the subset or ♣ is in the subset. So the subsets of {♣} are

{ },
{♣}.

Now let's suppose we consider the two-element set {♣, ♦}. Its subsets consist of all the subsets that do not contain ♦ together with all the subsets that do contain ♦ (remember that, for any subset, either ♦ is in that subset or it is not in that subset). Every subset of {♣, ♦} that does not contain ♦ must be a subset of just {♣}, so those subsets are { } and {♣}. We could create the subsets that do contain ♦ by throwing ♦ into each of those previous subsets to get the four subsets of {♣, ♦}:

{ }, {♦} (notice this is just { } where we throw in ♦),
{♣}, {♣, ♦} (notice this is just {♣} where we throw in ♦).

Now let's suppose we consider the three-element set {♣, ♦, ♥}. Its subsets consist of all the subsets that do not contain ♥ together with all the subsets that do contain ♥. But, as before, every subset that does not contain ♥ is a subset of {♣, ♦}. So, we have already listed all the subsets that do not contain ♥. We could create the remaining subsets by including ♥ in each of those subsets to get the eight subsets of {♣, ♦, ♥}.

{ },     {♥},
{♣},    {♣, ♥},
{♦},    {♦, ♥},
{♣, ♦}, {♣, ♦, ♥}.

Following this pattern, we see that adding a new element doubles the number of subsets. So, we see that, if our set has one element, it has

178  ♦  INFINITY

$2^1 = 2$ subsets. If we have a set with two elements, it has twice as many subsets—that is, $2^2 = 4$ subsets. A three-element set has $2^3 = 8$ subsets, and so on. We can record these observations as a theorem.

> **THE SUBSET COUNT.** *A set containing n elements has $2^n$ subsets.*

## The Power Set

Okay, suppose we are given a set (let's call the set S). We are now about to build a new set using our idea of subsets. This new set consists of all the subsets of S. That is, this new set is a set whose elements are actually subsets of S. This set seems strange at first, but let's absorb the idea. A set contains things that we call *elements;* however, we never said what the elements are. They could be numbers, or suits of playing cards, or musicians. The elements could even be sets themselves. So, for example, if **Suits** = {♣, ♦, ♥, ♠}, then the set of all subsets of **Suits** is the set consisting of the following 16 things:

{ },
{♣}, {♦}, {♥}, {♠},
{♣, ♦}, {♣, ♥}, {♣, ♠}, {♦, ♥}, {♦, ♠}, {♥, ♠},
{♣, ♦, ♥}, {♣, ♦, ♠}, {♣, ♥, ♠}, {♦, ♥, ♠},
{♣, ♦, ♥, ♠}.

Given a set S, we define the *power set of S*, written as $\mathcal{P}(S)$, to be the set consisting of all possible subsets of S. In other words, the power set of a set contains all the different possible subcollections we can create from the original set. So, the power set of {♣, ♦, ♥, ♠} is the set consisting of the 16 subsets of {♣, ♦, ♥, ♠} that are listed above. Notice that, in the preceding example, {♦, ♥} is an element of $\mathcal{P}(\textbf{Suits})$; {♠} is another element of $\mathcal{P}(\textbf{Suits})$; { } is an element of $\mathcal{P}(\textbf{Suits})$. Remember that the elements of $\mathcal{P}(S)$ are themselves sets. So $\mathcal{P}(\textbf{Suits})$ would be written within big brackets, and each of its elements would also have brackets. Here are some sets and their associated power sets.

| Set S | Power set $\mathcal{P}(S)$ |
|---|---|
| {1, 2}, | { { }, {1}, {2}, {1, 2} } |
| {a, b, c}, | { { }, {a}, {b}, {a, b}, {c}, {a, c}, {b, c}, {a, b, c} } |
| {♣, ♦, ♥, ♠} | { { }, {♣}, {♦}, {♥}, {♠}, {♣, ♦}, {♣, ♥}, {♣, ♠}, {♦, ♥}, {♦, ♠}, {♥, ♠}, {♣, ♦, ♥}, {♣, ♦, ♠}, {♣, ♥, ♠}, {♦, ♥, ♠}, {♣, ♦, ♥, ♠} }. |

Notice that we could rephrase the Subset Count theorem by saying that, if a set has $n$ elements, its power set has $2^n$ elements.

## *Power Sets Are Big*

In all the examples we have seen, the number of elements in the power set of a set is much greater than the number of elements in the original set. For example, there are four elements in the set of all suits in a deck of cards; however, there are $2^4 = 16$ elements in the power set of this set. Let's face it, 16 is much bigger than 4. We know this observation is pretty obvious, but it's important—just keep reading.

It turns out that power sets are *always* larger than the original set *even if the original set is already infinite!* This theorem was one of Georg Cantor's important achievements.

> **CANTOR'S POWER SET THEOREM.** *Let S be a set (finite or infinite). Then the cardinality of the power set of S, $\mathcal{P}(S)$, is strictly greater than the cardinality of S.*

Let's think about what this theorem means and its consequences. If we take an infinite set $S$ and look at the set of all its subsets [that is, look at its power set $\mathcal{P}(S)$], the cardinality of the set $\mathcal{P}(S)$ will be greater than the cardinality of $S$. How can we use this fact to make more and more increasingly larger infinities? This is a challenging question! Once again, in mathematics we see our intuition turned inside out. Please think about this question and attempt to answer it before reading on.

## *An Infinity of Infinities*

Suppose we take the set of natural numbers and call this set $\mathbb{N}$. Of course, $\mathbb{N}$ is an infinite set. Let's look at the power set $\mathcal{P}(\mathbb{N})$—that is, the set of all subsets of $\mathbb{N}$. Here are some examples of elements of $\mathcal{P}(\mathbb{N})$ [remember that each element of $\mathcal{P}(\mathbb{N})$ is a subset of $\mathbb{N}$]: $\{1\}, \{2\}, \{3\}, \ldots, \{1, 2\}, \{1, 3\}, \{2, 3\}, \ldots$ each triple of numbers, each set of four numbers, each finite set of numbers. But also $\{2, 4, 6, 8, \ldots\}$, the subset consisting of all even numbers, is one element of $\mathcal{P}(\mathbb{N})$. The collection of prime numbers forms one element of $\mathcal{P}(\mathbb{N})$. The collection of Fibonacci numbers is one element of $\mathcal{P}(\mathbb{N})$. Each subset of $\mathbb{N}$ is an element of $\mathcal{P}(\mathbb{N})$.

Cantor's Power Set Theorem says that $\mathcal{P}(\mathbb{N})$ is a set that has more elements than $\mathbb{N}$ itself! Now $\mathcal{P}(\mathbb{N})$ is itself a set. So, suppose we consider the power set of it! That is, suppose we consider the set of all subsets of the set of all subsets of $\mathbb{N}$. We would call this new set $\mathcal{P}(\mathcal{P}(\mathbb{N}))$. It would require considerable care just to name any element of the set $\mathcal{P}(\mathcal{P}(\mathbb{N}))$.

But what do we know about the size of $\mathcal{P}(\mathcal{P}(\mathbb{N}))$? By Cantor's Power Set Theorem, we know that the size of $\mathcal{P}(\mathcal{P}(\mathbb{N}))$ is greater than the size of $\mathcal{P}(\mathbb{N})$. So, we have just found a third infinity that is greater than the previous two!

Of course, we could repeat this process indefinitely. For example, we could consider the set of all subsets of the set of all subsets of the set of all subsets of $\mathbb{N}$, denoted by $\mathcal{P}(\mathcal{P}(\mathcal{P}(\mathbb{N})))$. The size of this set is greater than the size of $\mathcal{P}(\mathcal{P}(\mathbb{N}))$. So, we now see a way of producing infinitely many sets each with a larger cardinality than the last. (Just for fun, try describing in words what the set $\mathcal{P}(\mathcal{P}(\mathcal{P}(\mathcal{P}(\mathbb{N}))))$ is. It's a kind of mathematical tongue twister.)

## *Why the Power Set Theorem Is True*

Cantor's Power Set Theorem opens the door to an infinity of infinities. Let's see why Cantor's Power Set Theorem is true by proving it. First we'll just state the theorem again.

> **CANTOR'S POWER SET THEOREM.** *Let S be a set (finite or infinite). Then the cardinality of the power set of S, $\mathcal{P}(S)$, is strictly greater than the cardinality of S.*

## *A Plan of Attack*

Remember that cardinality is determined by one-to-one correspondences. Suppose we look at a correspondence that has elements of $S$ on one side and elements of the power set of $S$ on the other. For example, let's look at the set $S = \{a, b, c, d, e, f, \ldots\}$ (we are just letting the letters a, b, c, d, and so on stand for some of the elements of the set $S$). A pairing between the set $S$ and its power set $\mathcal{P}(S)$ might look like the following (remember, elements of $\mathcal{P}(S)$ are subsets of $S$):

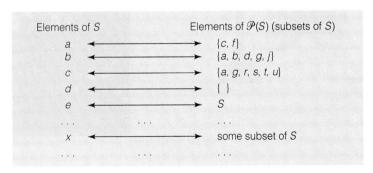

A sample correspondence between elements of $S$ and elements of $\mathcal{P}(S)$

To see that the cardinality of $\mathcal{P}(S)$ is greater than the cardinality of $S$, we will show how to construct some subset of $S$ (in other words, some element of $\mathcal{P}(S)$) that does not appear anywhere in the right-hand column. That is, given *any* pairing between the elements of $S$ and the elements of $\mathcal{P}(S)$, we will show that there will always be elements of $\mathcal{P}(S)$ that are not paired with a partner from $S$. If this were a prom, all the elements of $S$ would have dance partners from $\mathcal{P}(S)$, but there would be people from $\mathcal{P}(S)$ who would have no one from $S$ with whom to dance. It's sad but true—there is no one-to-one pairing; we will always run out of people from $S$ before running out of people from $\mathcal{P}(S)$ (even if there are infinitely many people in each set—one heck of a big prom!).

Since this method of constructing an unpartnered element of $\mathcal{P}(S)$ will work for any pairing we make, we will know that it is impossible to have a one-to-one correspondence between $S$ and all of $\mathcal{P}(S)$. We will see that this method is really a recycled version of the idea that Cantor used to show that there are more real numbers than natural numbers. Thereby, Cantor not only proved two of the great theorems of mathematics, but he also started the recycling craze a hundred years before its time. Terrific ideas deserve to be reused, expanded, exploited, modified, generalized, and developed. This strategy of mining ideas for all they are worth is a powerful take-home lesson from Cantor.

> Whenever we have an idea, we should push it as far as possible beyond its original context.
>
> ◆ ◆ ◆ ◆ ◆ ◆

### *An Upset Subset—Building a Missing Subset*

Remember that we're assuming we have a fixed correspondence that associates with each element of $S$ a subset of $S$—that is, an element of $\mathcal{P}(S)$. Our goal is to construct an element of $\mathcal{P}(S)$ that is not paired with any element of $S$. To construct this missing subset of $S$, we consider each element of $S$ and make a decision about whether that element will be in this mystery subset or not. Recall that in Cantor's proof there are more real numbers than natural numbers; our choice of the fifth digit, for example, of the number we were constructing guaranteed us that the number we were constructing was not the one corresponding to the natural number 5. We did the same for each of the natural numbers and thereby concluded that the newly constructed number did not correspond to any natural number. Here we proceed in a similar manner.

Remember that we are given a pairing that associates each element of $S$ with a subset of $S$ and that we now wish to construct a subset of $S$ that doesn't correspond to any element of $S$ in this pairing. Let's call this mystery subset **Mystery** and start constructing it by considering the element $a$ from set $S$. We need to decide whether $a$ should be in **Mystery** or not. To answer this question, let's look at the set to which $a$ corresponds in our given correspondence. As we see in the preceding table, $a$ is paired with the subset $\{c, f\}$. Since $a$ is not in $\{c, f\}$, we will put $a$ in **Mystery.** Once we have made the decision to put $a$ in **Mystery,** then we know for certain that whatever else **Mystery** may or may not contain, **Mystery** is not equal to $\{c, f\}$, because $a$ is in **Mystery,** and $a$ is not in $\{c, f\}$. Notice how similar this logic is to Cantor's previous argument.

Let's now move on to element $b$. How will we decide whether or not to put $b$ in **Mystery**? Well, again let's look at the subset of $S$ to which $b$ corresponds. In the table we see that $b$ corresponds to $\{a, b, d, g, j\}$. So should we put $b$ in **Mystery**? No. Why? Because by making the decision to omit $b$ from **Mystery,** then whatever else may or may not be in **Mystery,** we know for sure that **Mystery** is not $\{a, b, d, g, j\}$, since we have decided that $b$ is not in **Mystery,** whereas $b$ is in $\{a, b, d, g, j\}$.

We're on a roll, so let's do a couple more. Should $c$ be in **Mystery**? First we consult the table and see that $c$ corresponds to $\{a, g, r, s, t, u\}$. So we will place $c$ in **Mystery,** since $c$ is not in the subset to which $c$ corresponds. Once we have made that decision to put $c$ in **Mystery,** we know for certain that whatever **Mystery** turns out to be, it is not equal to the subset corresponding to $c$ since $c$ is in **Mystery** and not in the subset corresponding to $c$.

Should $d$ be in **Mystery**? Yes, because $d$ is not in $\{\ \}$ (the empty set). Notice that we're just using the same logic in each case.

Should $e$ be in **Mystery**? No, because $e$ is in $S$. Once again, the logic is the same.

Should $x$ be in **Mystery**? Since we did not tell you the specific subset to which $x$ corresponds, let's just write down our decision process. We will put $x$ in **Mystery** if $x$ is not in the subset corresponding to $x$, and we will not put $x$ in **Mystery** if $x$ is in the subset corresponding to $x$.

Finally, for the big finish, could **Mystery** appear anywhere in the right-hand column of our table? No, because if **Mystery** did appear, it would correspond to some specific element. That element would either be in **Mystery** or not be in **Mystery**. However, neither of those possibilities can be true. If that element were in **Mystery,** then it would be an element of the set it corresponds with and consequently *would not* be in **Mystery.** However, if that element were not in **Mystery,** then it would not be included in the set it corresponds to and hence *would* be in **Mystery.** Since that element can be neither in **Mystery** nor out of **Mystery,** it simply cannot exist. Thus no element of $S$ can correspond to **Mystery,** and, therefore, **Mystery** is a subset that cannot appear in the right-hand column of our table.

Since we have shown that no one-to-one correspondence can pair the elements of a set with its subsets, the cardinality of the power set cannot equal the cardinality of a set, and the proof of Cantor's theorem is complete.

## *The Whole Proof Shorter*

In fact, the previous three paragraphs are the entire proof of Cantor's Power Set Theorem. If we had any set $S$ at all, and any correspondence taking each element of $S$ to a subset of $S$, then we could construct a

**Mystery** subset that does not correspond to any element of S. How? For each element x of S, we will put x in **Mystery** if x is not in the subset corresponding to x, and we will not put x in **Mystery** if x is in the subset corresponding to x. By using that criterion, we know that **Mystery** cannot equal the subset corresponding to any element of S. (Notice that, in constructing the **Mystery** subset, we did not care whether S was finite or infinite.)

Since **Mystery** cannot correspond to any element of S, **Mystery** is not on the right side of our table and the pairing cannot be a one-to-one correspondence. Therefore, there is no one-to-one correspondence, and we have proved that the power set of S must have a greater cardinality than S. So, we have proved Cantor's Power Set Theorem!

### *The Whole Proof Really Short*

Just for fun, we will now write down the exact same proof using the extremely abbreviated, cryptic notation that makes mathematics succinct but difficult to read. To unravel and understand this one-line proof, a reader would have to produce the preceding explanation. We write it only for your amusement, so we will not bother to clearly define the terms or explain the notation. Feel free to use it to impress family and friends.

### *Complete Proof of Cantor's Power Set Theorem*

Consider $f: S \to \mathcal{P}(S)$. Then $\{x \in S \mid x \notin f(x)\} \notin f(S)$. Q.E.D.

### *Literal Translation*

Consider a pairing from S to $\mathcal{P}(S)$. Then the particular subset consisting of all elements x from S, such that x is not an element of the set corresponding to x with the given pairing, is not paired with any element of S. *Quod erat demonstrandum* ("which was to be proved").

Just think: If we had thought of that first, we could have gotten a Ph.D. and become famous with less than a line of work. The importance and value of a new idea should not be measured by how much space it takes to write it down. Length is not a good measure of depth.

### *The Answer Round*

Remember that our entire discussion about power sets arose from our desire to answer some questions about infinity. Now we can conclude that there is indeed a set having greater cardinality than the cardinality of the set of real numbers—namely, the power set of the real numbers. The set of all subsets of the real numbers is a collection even larger than the set of reals. By repeating this process (successively building the power set of the previous set), we see that there are infinitely many different sizes of infinity. These observations immediately (well, we put in a huge amount of work, so it wasn't so immediate) answer questions 2 and 3 from the beginning of this section. What about questions 1 and 4?

## *Cardinality Comparisons*

We saw that the cardinality of the set of natural numbers is smaller than the cardinality of the set of real numbers. The cardinality of the set of natural numbers is also smaller than the cardinality of the power set of the natural numbers. How do the cardinality of the real numbers and the cardinality of the power set of the natural numbers compare?

It turns out that the cardinality of the power set of the natural numbers, $\mathcal{P}(\mathbb{N})$, is exactly the same as the cardinality of the set of real numbers. Although the proof is not too hard, we suppress our desire to describe the details here. We mention that the cardinality of the reals is the same as the cardinality of $\mathcal{P}(\mathbb{N})$ because it allows us to state a famous conjecture about the number of points on the real line.

## *The Continuum Hypothesis*

Using Cantor's Power Set Theorem we have seen how to create larger infinities. But how big are these infinities? Are we skipping a whole bunch of infinities in between the ones we are generating?

*Question: Is the cardinality of the set of real numbers the **next** larger infinity bigger than the cardinality of $\mathbb{N}$?*

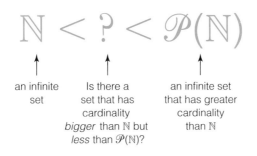

an infinite set

Is there a set that has cardinality *bigger* than $\mathbb{N}$ but *less* than $\mathcal{P}(\mathbb{N})$?

an infinite set that has greater cardinality than $\mathbb{N}$

We know that the cardinality of the real numbers is larger than the cardinality of $\mathbb{N}$, but perhaps there is an infinity between them. In other words, is there a collection of real numbers whose cardinality is strictly greater than the cardinality of the natural numbers yet strictly less than the cardinality of the real numbers? Is that possible? Is that impossible? The famous *Continuum Hypothesis* was a conjecture about this question.

> **CONTINUUM HYPOTHESIS.** *There is no cardinality between the cardinality of the set of natural numbers and the cardinality of the set of real numbers (sometimes referred to as the* **continuum***).*

If you thought that the study of infinity was pretty wild so far, hold on to your hat. If you don't have a hat, get one quick and hold on. The amazing fact is that the Continuum Hypothesis is neither true nor false!

Please do not throw our book out the window; but if you do, please buy another copy. We have not taken leave of our senses. You read correctly, the Continuum Hypothesis is neither true nor false, and it can be *proven* that it is neither true nor false. Thus, question 1 was the one with the wild answer.

The Continuum Hypothesis pushes us, in some sense, right to the edge of mathematical truth. Kurt Gödel in 1940 proved that it is impossible to disprove the Continuum Hypothesis; then in 1963 Paul Cohen proved that it is impossible to prove the Continuum Hypothesis. In other words, the Continuum Hypothesis can be neither proved nor disproved using the standard mathematical axioms and machinery! The Continuum Hypothesis is independent of our entire mathematical structure. This result is truly amazing. It brings us to the very edge of knowledge—the cliff of mathematics. Perhaps we should step back before we fall off.

## *Russell's Paradox*

Before closing this section, given all the strange phenomena we have discussed, we should discuss a grand idea bigger than everything. Suppose we take every set of every size and simply imagine the set comprising every set—in other words, the ultimately largest set, since it would contain every set as an element, including itself. This set would be the grandmother of infinities, the all-embracing set of everything. But this idea, satisfyingly grand, is inherently contradictory.

Cantor's Power Set Theorem implies that there is no greatest of all infinities—so the answer to question 4 is no. By Cantor's Power Set Theorem, for any set, the set of all its subsets is strictly bigger. If we consider this hypothesized biggest set of all sets, we run into a contradiction, because it would be impossible for all its subsets to be elements of it—there are too many of them. So it cannot contain every set. The only way around this contradiction is the realization that this huge set is not a set! This statement is an insight of Bertrand Russell and is known as *Russell's Paradox*.

What went wrong? Why isn't the set of all sets a set? The answer is fundamental. If you look back through this chapter, you may notice that we never defined a particular term: *set*. Russell's Paradox shows us that even basic-sounding ideas have inherent limitations. The intuitive idea that a set is just a "bunch of things" turns out to be inadequate for truly enormous collections. Russell's Paradox forces us to confront the strange fact that some collections are just too unwieldy to call sets. Mathematicians were thus forced to move beyond their intuition, create an appropriate definition for the word *set*, and then carefully investigate its nuances and consequences. These studies led to an entire branch of mathematics: *Set*

*Theory*. Instead of delving into the arcane essence of "setness," we thought it would be best to move on and explore infinity within the geometrical world. All set?

## A LOOK BACK

The power set of a set $S$, denoted $\mathcal{P}(S)$, is the collection of all its subsets. A finite set with $n$ elements has a power set with $2^n$ elements.

Cantor's Power Set Theorem shows that the cardinality of the power set of a set is always greater than the cardinality of the original set, even for infinite sets. The proof of this theorem uses the same diagonalization idea that Cantor had used to show that the real numbers are more numerous than the natural numbers.

Using Cantor's Power Set Theorem, we can start with the natural numbers $\mathbb{N}$ and build an infinite collection of sets each with a greater cardinality than its predecessor—namely, $\mathbb{N}$, $\mathcal{P}(\mathbb{N})$, $\mathcal{P}(\mathcal{P}(\mathbb{N}))$, $\mathcal{P}(\mathcal{P}(\mathcal{P}(\mathbb{N})))$, and so forth. There is no largest infinite set, because we can always take the set of all its subsets to create a larger set.

This theory illustrates the power and value of exploiting the same idea as much as possible. A new idea is a new tool. Go beyond its original intended use and modify it to create yet more tools.

*Apply ideas widely.*

◆ ◆ ◆ ◆ ◆ ◆ ◆ ◆ ◆ ◆ ◆ ◆ ◆ ◆ ◆ ◆ ◆ ◆ ◆

## MINDSCAPES   Invitations to Further Thought*

### I. SOLIDIFYING IDEAS

1. **All in the family.** There is a family of four that tries to eat dinner together as much as possible. On Tuesday, some of them sit down for dinner. How many different possible groups of diners are there? (Their position at the table does not matter.) List all the possible groupings.

2. **Making an agenda** (H). There are eight members on the board of directors of the Windbag Corporation, not including the Chairman of the Board. Each member of the board e-mails to the Chair one agenda item that he or she would like to discuss at an upcoming meeting. All the agenda items sent to the Chair are different. The Chair chooses the agenda from the list received; however, the Chair does not necessarily have to use all the items. How many different possible agendas (not counting order) could there be for their next meeting?

---

*In the Mindscapes section, exercises marked (H) have hints for solutions at the back of the book. Exercises marked (S) have solutions.

3. ***The power of sets*** **(S).** Let $S = \{!, @, \#, \$, \%, \&\}$. Below are two columns: on the left are names of certain sets and on the right are confusing looking thingies—each thingie is an element of one of the sets in the left-hand column. Your mission, if your instructor chooses to assign it, is for each thingie in the right-hand column, determine which set from the left-hand column contains that thing as a member. As examples we do the first two for you: $\{!, @\}$ is a subset of $S$, so it is an element of the power set $\mathcal{P}(S)$, and @ is a member of $S$ itself. Okay, now you're on your own! Good luck.

| Set | Thingie |
| --- | --- |
| $S$ | $\{!, @\}$ |
| $\mathcal{P}(S)$ | @ |
| $\mathcal{P}(\mathcal{P}(S))$ | $\{!, \#, \%\}$ |
| $\mathcal{P}(\mathcal{P}(\mathcal{P}(S)))$ | $\{\, \{\{@, !\}\}, \{\{\$\}\} \,\}$ |
| $\mathcal{P}(\mathcal{P}(\mathcal{P}(\mathcal{P}(S))))$ | $\{\, \{!\}, \{@\}, \{\#\}, \{\%\}, \{\&\} \,\}$ |
| $\mathcal{P}(\mathcal{P}(\mathcal{P}(\mathcal{P}(\mathcal{P}(S)))))$ | $\{\, \{\{@\}\}, \{\{\#, \$\}\}, \{\{!\}, \{\%, \&\}\} \,\}$ |
|  | $\{\{\{!\}\}\}$ |
|  | $\{\#\}$ |
|  | $\{\, \{@\}, \{\$,!\} \,\}$ |
|  | ! |

4. ***Powerful words.*** Suppose that $S$ is a set. Describe the following sets in words as clearly as possible: $\mathcal{P}(\mathcal{P}(S))$; $\mathcal{P}(\mathcal{P}(\mathcal{P}(S)))$; $\mathcal{P}(\mathcal{P}(\mathcal{P}(\mathcal{P}(S))))$. (For example, $\mathcal{P}(S)$ might be described as *the set of all subsets of the set $S$*.)

5. ***Identifying the power.*** Let $S$ be the set given by $S = \{m, a, t, h, f, u, n\}$. Which of the following are elements of $\mathcal{P}(S)$? $\{m, a, t, h\}$; a; $\{\ \}$; $\{\{m, a, t, h\}\}$; $\{\{m\}, \{a\}, \{t\}, \{h\}\}$; $\{m\}$.

6. ***Two Cantor.*** Suppose $S$ is the set defined by $S = \{1, 2\}$. Consider the following pairing of elements of $S$ with elements of $\mathcal{P}(S)$:

| Elements of $S$ | Elements of $\mathcal{P}(S)$ |
| --- | --- |
| 1 | $\{2\}$ |
| 2 | $\{1, 2\}$ |

Using the idea of Cantor's proof, describe a particular subset of $S$ that is not on this list.

7. ***Another two.*** Suppose $S$ is the set defined by $S = \{1, 2\}$. Consider the following pairing of elements of $S$ with elements of $\mathcal{P}(S)$:

| Elements of $S$ | Elements of $\mathcal{P}(S)$ |
| --- | --- |
| 1 | $\{2\}$ |
| 2 | $\{1\}$ |

Using the idea of Cantor's proof, describe a particular subset of $S$ that is not on this list.

8. **Cantor code.** Suppose that *Words* is the set defined by *Words* = {all, you, infinity, found, them, search, the, it}. Consider the following pairing of elements of *Words* with elements of $\mathcal{P}(Words)$:

| Elements of *Words* | Elements of $\mathcal{P}(Words)$ |
|---|---|
| all | {all, infinity, found} |
| you | {it, search, them} |
| infinity | {all, them, infinity} |
| found | {you, the, it} |
| them | {found, them} |
| search | {all, infinity, search} |
| the | {the, search} |
| it | {infinity, you, all} |

Using the idea of Cantor's proof, describe a particular subset of *Words* that is not on this list.

9. **Finite Cantor (H).** Suppose that $S$ is the set defined by $S$ = {@, &, %, $, #, !}. Consider the following pairing of elements of $S$ with elements of $\mathcal{P}(S)$:

| Elements of $S$ | Elements of $\mathcal{P}(S)$ |
|---|---|
| @ | {@, !, $} |
| & | { } |
| % | {&, $} |
| $ | {$} |
| # | {@, %} |
| ! | {@, #, !} |

Using Cantor's proof, describe a particular subset of $S$ that is not on this list.

10. **One real big set.** Describe (in words) a set whose cardinality is greater than the cardinality of the set of real numbers.

## II. CREATING NEW IDEAS

1. **The Grand Real Hotel.** Suppose there was a really huge hotel called the Grand Real Hotel. This hotel is so large that there is an individual room associated with each individual real number (imagine the room numbers: "I'm staying in room number 63.7269711294 ..."). Does there exist a set that would cause the Grand Real Hotel to put up a No Vacancy (or Not Enough Room) sign? If so, describe such a set. If not, explain why not.

2. **The Ultra Grand Hotel (S).** Could there be an infinite hotel so large that there is never a need to put up the Not Enough Room sign? Explain your answer.

3. **Powerful counting.** Let the set $S$ be given by $S$ = {@, &, %, $, #, !}. How many elements does the set $\mathcal{P}(\mathcal{P}(\mathcal{P}(\mathcal{P}(\mathcal{P}(\mathcal{P}(S))))))$ contain?

4. ***Russell's barber's puzzle*** **(H).** In a certain village there is one male barber who shaves all those men, and only those men, who do not shave themselves. Does the barber shave himself? Show that the answer cannot be yes or no and that this question is a paradox. This barber paradox is related to Russell's insight that the set of all sets cannot be a set. In particular, consider the set **NoWay,** whose elements are special sets—namely, all sets that do not contain themselves as elements. Is **NoWay** an element of itself?

5. ***The number name paradox.*** Let S be the set of all natural numbers that are describable in English words using no more than 50 characters (so, 240 is in S since we can describe it as "two hundred forty," which requires fewer than 50 characters). Assuming that we are allowed to use only the 27 standard characters (the 26 letters of the alphabet and the space character), show that there are only finitely many numbers contained in S. (In fact, perhaps you can show that there can be no more than $27^{50}$ elements in S.) Now let the set T be all those natural numbers not in S. Show that there are infinitely many elements in T. Next, since T is a collection of natural numbers, show that it must contain a smallest number. Finally consider the smallest number contained in T. Prove that this number must simultaneously be an element of S and not an element of S—a paradox!

## III. FURTHER CHALLENGES

1. ***Adding another.*** Suppose that you have any infinite set (it could be really huge), and you wish to add one new element to the set. Prove that this new set (the old set with the new element thrown in) has the same cardinality as the original set. (*Hint:* The infinite set has a subset that is in one-to-one correspondence with the natural numbers. Use previous ideas to add the new element to this subset, and let the rest of the set correspond with itself.)

2. ***1's and 2's.*** Describe a one-to-one correspondence between the set of all decimal numbers between 0 and 1 that are made with only 1's and 2's (like 0.1211212212212 . . .) and the set of all decimal numbers between 0 and 1 that are made with only 3's, 4's, 5's, and 6's (like 0.36543546354554 . . .). (*Hint:* Decimals made of 1's and 2's only can have their first two places .11 . . . , .12 . . . , .21 . . . , .22 . . .).

## IV. IN YOUR OWN WORDS

1. ***Personal perspectives.*** Write a short essay describing the most interesting or surprising discovery you made in exploring this section's material. If any material seemed puzzling or even unbelievable, address that as well. Explain why you chose the topics you did.

Finally, comment on the aesthetics of the mathematics and ideas in this section.

2. ***With a group of folks.*** In a small group, discuss and work through Cantor's proof that the cardinality of a set is always smaller than the cardinality of its power set. After your discussion, write a brief narrative describing the proof in your own words.

3. ***Creative writing.*** Write an imaginative story (it can be comical, dramatic, whatever you like) that involves or evokes the ideas of this section.

4. ***Power beyond the mathematics.*** Provide several real-life situations—ideally, from your own experience—for which some of the strategies of thought presented in this section would provide effective methods for approaching and resolving them.

# 3.5 Straightening Up the Circle:
## Exploring the Infinite Within Geometrical Objects

*I could be bounded in a nutshell, and count myself a king of infinite space.*

—William Shakespeare

In the previous introduction to infinity, all the sets and correspondences have been rather abstract. Here we consider somewhat more concrete occurrences of infinity by looking at objects that we can draw. We will still look for one-to-one correspondences, but we can sometimes make connections by just drawing lines. So let's turn now to exploring infinity in the visual field.

### Short Versus Long Lines

Consider the following two line segments $S$ and $B$. (We just cut them out of a number line; that's why there are numbers written under them.)

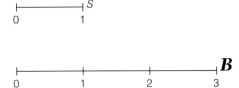

Remember that a line segment is just a bunch of points all lined up like soldiers but incredibly close together. Are there more points in line segment $B$ than in line segment $S$? Think about this question and come up with a guess and a justification. We'll wait right here.

There are many reasons why one could argue that there are more points in line $B$ than $S$. First is that segment $B$ is three times longer than segment $S$. The other is that the letter $B$ to the right of the second line segment is much bigger than the letter $S$ to the right of the first line segment. Let's think about the first reason a bit more.

## Wrong Reasoning

The following is a reasonable-sounding, though fallacious, argument:

> I've already learned that two sets have the same cardinality if there is a one-to-one correspondence between them. In other words, is there a way of pairing up the elements of the two sets so that each element of the first set is paired with exactly one element of the second set and each element of the second set is paired with exactly one element of the first set? Here is why the cardinality of the set of points in line $B$ is greater than the cardinality of the set of points in line $S$. Consider the following pairing:

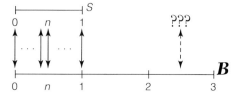

> This pairing is easy to describe. Take the point on $S$ located at the number $n$ and pair it up with the point on $B$ located at $n$. So 0 in $S$ pairs with 0 in $B$, ½ in $S$ pairs with ½ in $B$, and 1 in $S$ pairs with 1 in $B$. Now this pairing is a one-to-one correspondence between the points of $S$ and the points of $B$ FROM 0 TO 1. But now notice that the point 2½ in $B$ is not paired with anyone from $S$. In fact, no point in $B$ after 1 is paired with any point in $S$. So, since we have a one-to-one correspondence between all the points of $S$ and just some and not all of the points in $B$, the cardinality of points in $B$ must not equal the cardinality of points in $S$. This must show that the cardinality of the set of points in $B$ must be greater than the cardinality of the set of points in $S$.

*Examine arguments critically—don't just accept them.*

❖ ❖ ❖ ❖ ❖ ❖

Let's think about the preceding argument. It's pretty clever and sounds reasonable. The big question is, what's the problem? Let's examine this argument critically and objectively. We must not believe things just because they sound reasonable.

Okay, the argument is not correct, but can we figure out where it went wrong? One sentence is false, and, from that point on, everything else is wrong. Can you find it? Give it a try before reading on.

Before analyzing the situation, remember that an argument, even an incorrect one, is a good way to focus our thoughts on important issues. In this case, someone thought about an issue and put ideas together into a coherent explanation. This process is truly great. Every time someone does this, it is a time to celebrate.

*Making mistakes is a sign of creativity and strength; it is not a sign of weakness.*

❖ ❖ ❖ ❖ ❖ ❖

Okay, enough celebrating—after all, the argument is wrong. Now let's think about the issue a bit more and see if we can discover the truth.

## The Meaning of Not

Recall that, for two sets to have the same cardinality, there must exist a one-to-one correspondence between the two sets. So what does it mean for two sets *not* to have the same cardinality? It means it is impossible to find a one-to-one correspondence between the two sets. The point is that just finding a particular pairing that is not a one-to-one correspondence is not enough. Finding such a pairing just shows that that particular pairing didn't work. It does not show that a one-to-one correspondence doesn't exist. Maybe some other pairing is a one-to-one correspondence, even though the first one was not. This possibility is very important in the study of infinite sets, as the line segments $S$ and $B$ will illustrate.

## It's a Tie

It turns out that the set of points of line $S$ has the same cardinality as the set of points of line $B$. How can we prove this? All we need to do is construct a one-to-one correspondence between the two sets, and we can use geometry to help us by connecting the two segments as follows.

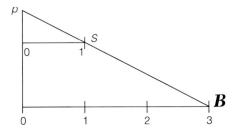

All we did is draw a line that passes through both of the 0's and a line that passes through the 1 on $S$ and the 3 on $B$. Since $S$ is shorter than $B$, these two lines must intersect. We call the point where they meet $p$ (for "point," clever, huh?). We now describe the one-to-one correspondence. Suppose we have a point on the line $S$, let's call it $w$ (for "wonder who this point will be paired with"). We will now pair up $w$ with exactly one point from $B$. Which one? Draw a line from the point $p$ through the point $w$ and see where it intersects $B$. Let's call that point of intersection $h$ (for "hit"). So, here's what it looks like:

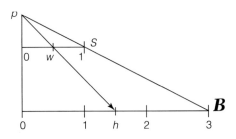

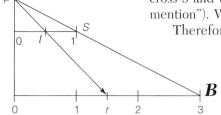

Notice that the line through $p$ and $w$ hits $B$ and hits $B$ at *exactly* one point! Thus, to every point of $S$ we have paired it with exactly one point of $B$. Now suppose we have a point from $B$—let's call it $r$ (for "running out of clever names for points"). To see which point of $S$ we will pair $r$ with, consider the line that goes from $p$ to $r$. Notice that it will have to cross $S$ and will cross $S$ at exactly one point—say $l$ (for "last point we will mention"). We will pair $r$ up with $l$.

Therefore, in this manner, every point of $B$ is paired with exactly one point of $S$. Thus, we have constructed a one-to-one correspondence between the set of points of $S$ and the set of points of $B$, and so the sets have the same cardinality. In fact, this exact idea proves the following fact.

> **ALL LINE SEGMENTS ARE CREATED EQUAL.** *For any two line segments, the cardinality of points of one segment equals the cardinality of points of the other.*

Is this theorem surprising? It may seem a bit strange, especially since one line segment could be very long and the other could be very short. But now we see how to find a one-to-one correspondence. In some sense, there are so many points in a line segment that we can stretch the segment out like a rubber band and make a longer line segment without adding more points. This is the spirit of the one-to-one correspondence we created previously—a pairing through stretching.

## Pick on Someone Your Own Size

Now we are experts about cardinalities of line segments. But what about an entire line that goes on forever in each direction? Let $S$ (for "small segment") be the set of all points on a line segment strictly between 0 and 1 (so we will not include 0 or 1—only the points between them). And let's call the entire number line $\mathbb{R}$ (for "real numbers"). Here is a picture of the players:

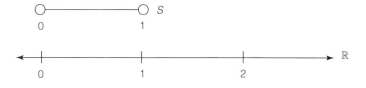

Notice the circles at the ends of $S$. Those circles are to remind us that neither 0 nor 1 is part of $S$. $S$ is just the set of points in between 0 and 1.

Is the cardinality of the set of points in $\mathbb{R}$ greater than the cardinality of the set of points in $S$? We've seen that all line segments have the same cardinality, but now we are asking what happens if we consider a line that extends out infinitely in both directions. What do you think? Make a guess before reading on.

We are asking whether there is a one-to-one correspondence between the set of points in $S$ and the set of real numbers, $\mathbb{R}$. In fact, a one-to-one correspondence does exist. That is, the cardinality of the set $S$ equals the cardinality of the set $\mathbb{R}$. This claim seems incredible. How could there be just as many points between 0 and 1 as there are points on the entire real number line? After all, the number line goes on forever.

## Round-Up

The one-to-one correspondence between these two sets turns out to be useful in several different settings besides this one. It is clever and tricky, so we'll proceed with caution. First, we notice that we will not change the cardinality of the line segment $S$ if we bend it a little bit. We would be changing its shape, but not its cardinality.

Here are some different ways of bending $S$ around.

That last smiley face semicircle interests us the most. In fact, if we continued to bring 0 and 1 around, we would end up with

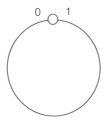

In this shape, the circle at 0 is exactly on top of the circle at 1. We just took the segment and brought the ends together so that we created a big circle with one point missing: the north pole. Why is it missing? Because the segment $S$ consists of all points between 0 and 1. So 0 and 1 are not part of the segment. Thus after we place the missing point 0 on top of the missing point 1, we are left with just one hole.

Now we bring the number line $\mathbb{R}$ into the picture—actually, right below the picture—as follows:

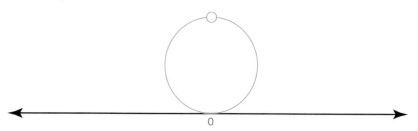

## *Rolling Up the Real Line*

The circle sits on the line at its south pole and touches the line at exactly one point on $\mathbb{R}$, namely, 0. We again draw lines to show the one-to-one correspondence. Let $a$ (for "arbitrary point") be a point on the circle minus the north pole (so $a$ cannot be the north pole). To see who gets paired with $a$ on the number line, we draw a line from the north pole hole through the point $a$ and continue until we hit the number line.

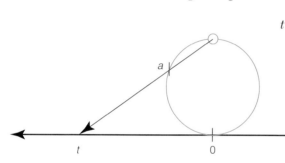

The line hits the number line at one point, say, $t$ (for "that point"). We will pair the point $a$ on the circle with the point $t$ on the number line. Similarly, if we had a point on the number line, we could find the point on the circle that it is paired with by connecting the point on the number line with the north pole hole. That line will intersect the circle at exactly one other point, and that is the point paired with the point on the number line. This process gives us a one-to-one correspondence between the points on the real line and the points on the circle minus the north pole. Hence the cardinality of the points on the circle minus the north pole equals the cardinality of the points on the real number line.

Recall that we already saw that the cardinality of the set of points of the circle minus the north pole is equal to the cardinality of the set of points of the interval $S$.

Putting all this together we conclude that the cardinality of the points of $S$ equals the cardinality of the set of all points on the number line. We just proved a remarkable result:

> **THE NUMBER LINE IS LIKE A LINE SEGMENT.** *The cardinality of points between any two real numbers on a number line is equal to the cardinality of the set of all points on the real number line.*

## Stereographic Projection

This cool one-to-one correspondence between the line segment curled up into a circle minus the north pole and the number line is so important that it has a fancy name: *stereographic projection*. The idea is that we are projecting all the points of the circle minus the north pole onto the number line. This pairing rule may be thought of in another way: In some sense we are wrapping the number line around the circle minus the north pole. Notice that, as we travel along the circle and approach the north pole from the left, those points are paired with points on the number line that are heading off to the left horizon. Similarly, as we approach the north pole from the right, those are the points paired with points on the number line that go off to the right horizon.

## The Solid Square

We have just discovered that not only do different length line segments have the same cardinalities, but, even if we stretch out a line segment forever (making the real number line), we don't up the cardinality. In our quest to make geometric objects with really big cardinality, let's try instead to up the dimension and consider solid objects in the plane. Specifically, consider these objects:

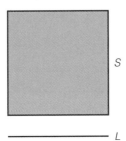

S is a square and L is a line segment. Actually it will be easier to view the line segment and the sides of the square as little pieces of the real number line so we can identify points in the square using coordinates in the usual way:

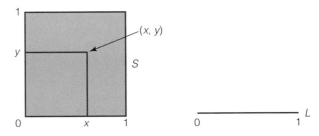

Recall that we can specify a particular point in the square by giving its horizontal distance away from 0 (called $x$) and its vertical distance away from 0 (called $y$) and denote the point as the pair $(x, y)$. Is the cardinality of points inside the square $S$ greater than the cardinality of the points of the line $L$?

Let's give this puzzler some thought. First of all, notice that the base of the square is exactly the same as line $L$. If we draw any horizontal line across the square, we would have another line that is exactly equal to $L$. Thus we have lots of copies of the line $L$ inside the square $S$. In fact, if we were to imagine inking up the line $L$ (so it's all inky) and then dragging it up the paper exactly one unit, we would make a square ink blot exactly in the same shape as $S$. So we have a different copy of $L$ in the square $S$ for every single point on the vertical edge of $S$:

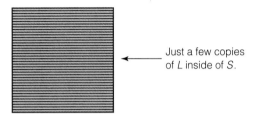

Just a few copies of $L$ inside of $S$.

So the cardinality of $L$ is clearly not bigger than the cardinality of $S$. Therefore, the cardinality of $L$ must be less than or equal to the cardinality of $S$. In fact, our observation appears to indicate that in some sense $S$ is much bigger than $L$, because it has a whole extra dimension (up!), which allows us to stack and store infinitely many $L$'s, one for each real number in the vertical direction between 0 and 1. Thus, it is reasonable to conjecture that the cardinality of points of $S$ is greater than the cardinality of points in $L$. However, so far all the examples we have looked at in this section have had the same cardinality, so perhaps the line and the square have the same cardinality as well.

Of course, since this is the last example of the chapter (the big finish), perhaps this is the one where the cardinalities will differ. But, enough mind-game distractions. In fact, the cardinality of the square is not greater than the cardinality of the line segment:

**SQUARES EQUAL LINES.** *The cardinality of points inside a square is the same as the cardinality of points on a line segment.*

## A Correspondence Gone Bad

*Failing is a powerful positive force if we can learn from it.*

❖ ❖ ❖ ❖ ❖ ❖

Why is the preceding statement true? How would we prove it? As always, we will build a one-to-one correspondence and evenly pair points in the square with points on the line. As we will discover, the first attempt we make to craft a one-to-one correspondence will actually fail. (Remember that failing is good—it is an effective way to build new insights and discover the truth. We will then see that a clever adaptation of our failed attempt will actually lead us to success.)

To show that the cardinality of the square $S$ is the same as the cardinality of the line segment $L$, we need to show that we can pair each point of $S$ with a point of $L$ and each point of $L$ with a point of $S$. In other words, suppose that we had two barrels: In one barrel we put all the points from inside the square $S$, and in the other barrel we put all the points of the line $L$. If we could demonstrate a way of pairing elements of the $S$ barrel with elements of the $L$ barrel so that, once all the pairings are made, we have nothing left in either barrel, then that would demonstrate that the cardinality of the things in the $S$ barrel is the same as the cardinality of things in the $L$ barrel.

Okay, let's give it a try. Here's how each point of the square $S$ can be paired with exactly one point of the line $L$. To describe this pairing, we first need to be able to name precisely the points in $S$ and on $L$. We do this through decimal expansions. Recall from the last section in Chapter 2 that every real number can be expressed with an infinitely long decimal expansion:

$$\frac{7}{3} = 2.3333333333333333333333333333333333333333\ldots$$
$$\pi = 3.1415926535897932384626433832795028841971 6933\ldots$$
$$\sqrt{2} = 1.4142135623730950488016887242096980785696 7187\ldots$$
$$\frac{13}{2} = 6.5000000000000000000000000000000\ldots$$

The last one could be written differently. Remember that we proved that $1 = 0.9999999999\ldots$. Similarly, we could write

$$\frac{13}{2} = 6.4999999999999999999999999999999\ldots$$

To make our naming of points unambiguous, we will always use the decimal representation ending in all nines whenever we have a choice. In other words, if a decimal ends with an infinite tail of zeros (like 0.50000 . . .), we rewrite it so it has an infinite tail of nines (0.49999 . . .). This choice of using 9's rather than 0's is arbitrary, but we have now spoken and will accept no back talk (we can be pretty forceful when we want to be).

## *Shuffling Digits*

If we take a point in the square, say $(x, y)$, we can write both $x$ and $y$ as infinite decimals. For example, say,

$$x = 0.9578086403023050996000077873225068400658 4033\ldots$$

and

$$y = 0.4444448759909842000074300924000678635780 938393725\ldots$$

To figure out which point on the line $L$ we will pair $(x, y)$ with, we do something interesting: We shuffle the digits of the decimal expansions of $x$ and $y$ together, like a deck of cards, to create a new decimal number. That is, we create a new decimal number by taking 0. and then tacking on the first digit in $x$ and then the first digit in $y$ and then the next digit in $x$ followed by the next digit in $y$ and then the third digit in $x$ followed by the third digit in $y$ and so on. We produce one infinitely long decimal expansion. That new number is between 0 and 1, and it is the number on $L$ that we will associate with the point $(x, y)$ in the square $S$. So, for example, the preceding $(x, y)$ point in $S$ would be paired with the point

$$z = 0.9454748404846847053909203908540290906000070403700\ldots$$

We just interwove the digits of $x$ and $y$ together, as if they had been shuffled. We wrote the digits of $x$ in blue and the digits of $y$ in red, so we can visually see the interwovenness of the digits in $z$. Notice that, using this method, two different points in the square get paired with two different points on the line. So, this correspondence pairs each point of $S$ with a point in $L$ in such a way that no two points of $S$ are paired with the same point of $L$. This clever pairing seems to be a one-to-one correspondence; however, something is slightly wrong. What is wrong? Can you spot the subtle problem?

## *Whoops*

On first inspection, it appears as though we have made a one-to-one correspondence between all the points of $S$ and all the points of $L$. After all, if we took a decimal point on $L$, couldn't we unshuffle it to get a point in $S$? Surprisingly, some points of $L$ were *not* hit in our shuffling correspondence. That is, there are some points in $L$ that are never paired with points in $S$. We will not keep you in suspense for long. Let's build a point of $L$ that is not hit.

Think of the point on $L$ given by the decimal $0.3790909090909\ldots$. If we unshuffle, then the point in $S$ to which we would expect this point to correspond would be $(0.399999\ldots, 0.700000\ldots)$. The problem is that we agreed that $0.700000\ldots$ would be represented as $0.699999\ldots$. So, since $(0.39999\ldots, 0.70000\ldots$ equals $(0.39999\ldots, 0.699999\ldots)$, that point in $S$ goes to the point on $L$, $0.369999\ldots$ (which equals $0.37$). So, the point $0.3790909090909\ldots$ in $L$ is left without a partner from this shuffling pairing with $S$. In other words, with this pairing, even though we will have nothing left in the $S$ barrel, there will be numbers remaining in the $L$ barrel.

## *Success from Failure*

Success often arises from failed attempts.

♦ ♦ ♦ ♦ ♦ ♦

We can try to modify good but failed attempts and ideas in order to achieve success. Let's look carefully at what went wrong and see whether we can fix it.

The problem is that alternating zeros got us in trouble. To avoid this defect, we modify the shuffling procedure so that this annoying glitch will disappear. Instead of shuffling the numbers perfectly (a digit from $x$ followed by a digit from $y$, and so on), we shuffle groups of digits—we put down a few digits of $x$ followed by a few digits of $y$, and so forth. How do we group the digits? We group all consecutive runs of zeros together and include the first nonzero digit. If there are no zeros, then the nonzero digits form their own individual groups. For example, for the numbers considered before, we would group their digits by

$$x = 0.9\ 5\ 7\ 8\ 08\ 6\ 4\ 03\ 02\ 3\ 05\ 09\ 9\ 6\ 00007\ 7\ 8\ 7\ 3\ 2\ 2\ 5\ 06\ 8\ 4\ldots$$

and

$$y = 0.4\ 4\ 4\ 4\ 4\ 8\ 7\ 5\ 9\ 9\ 09\ 8\ 4\ 2\ 00007\ 4\ 3\ 009\ 2\ 4\ 0006\ 7\ 8\ 6\ 3\ 5\ 7\ 8\ 09\ldots$$

If we now shuffle the grouped digits together, we would pair the point $(x, y)$ in $S$ with the point in $L$ given by

$$0.94547484084644803702539059090998640000727000078473300922\ldots$$

We can check that this new shuffling technique gives rise to a one-to-one correspondence. Certainly every point in the square is paired with one and only one element of the line. It is also the case that every element of the line is paired with one and only one element of the square. For example, the previous problem point on the line was $0.379090909\ldots$. Notice that this point gets paired with the point on the square at $(0.390909090\ldots, 0.709090909\ldots)$. This shuffling provides us with a one-to-one correspondence, and so the points in a square and the points on the line have the same cardinality.

## A LOOK BACK

Many geometric objects have the same cardinality, and that equivalence can sometimes be seen through creative visual correspondences. Every two intervals, no matter how long or short, have the same cardinality. The whole real line has the same cardinality as an interval. This last correspondence is produced using the stereographic projection.

A filled-in square has the same cardinality as an interval, but this one-to-one correspondence is not produced geometrically. This surprising equality of square and line is shown by demonstrating a subtle pairing between the two sets via shuffling grouped digits together.

Understanding issues of infinity is an intellectual triumph. These ideas and theories are challenging, but they are also beautiful, rich, and interesting, because they allow us to understand and analyze concepts that at first seemed beyond human comprehension. It is incredible how our minds are capable of grasping and taming notions that at first appear out of human reach. Therein lies part of the beauty and the power of mathematics. Surprising intellectual discoveries are a great tribute to the human spirit.

One important strategy illustrated in this section was the identification of failure as a step toward success. By making a mistake and then looking carefully at where the reasoning went wrong, we were able to modify an argument to make it work.

We also saw the value of taking familiar objects and ideas and viewing them from a different perspective. We can think of variations of common notions that result in uncommon insights.

Look at ordinary things in extraordinary ways.

# MINDSCAPES  Invitations to Further Thought*

## I. SOLIDIFYING IDEAS

1. ***A circle is a circle*** (H). Prove that a small circle has the same number of points as a large circle. Stated precisely, prove that the cardinality of points on a small circle is the same as the cardinality of points on a large circle. Describe a one-to-one correspondence between these two sets.

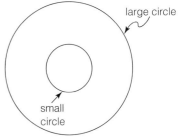

2. ***A circle is a square.*** Prove that a small circle has the same number of points as a large square. Stated precisely, prove that the cardinality of points on a small circle is the same as the cardinality of points on a large square. Describe a one-to-one correspondence between these two sets.

---

*In the Mindscapes section, exercises marked (H) have hints for solutions at the back of the book. Exercises marked (S) have solutions.

3. ***A circle is a triangle.*** Prove that a small circle has the same number of points as a large triangle. Stated precisely, prove that the cardinality of points on a small circle is the same as the cardinality of points on a large triangle. Describe a one-to-one correspondence between these two sets.

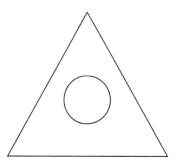

4. ***Stereo connections.*** Given the stereographic projection as a one-to-one correspondence between points on a line segment (rolled up) and points on the real line, identify which points on the line segment are paired with the points marked on the real line below.

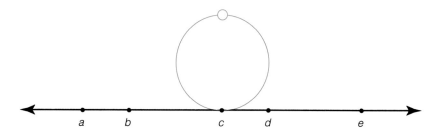

5. ***More stereo connections.*** Given the stereographic projection as a one-to-one correspondence between points on a line segment (rolled up) and points on the real line, identify which points on the real line are paired with the points marked on the line segment below.

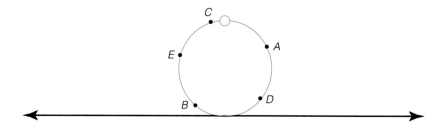

6. ***Perfect shuffle problems* (H).** Suppose we used our failed perfect shuffling of digits to mix the digits of the numbers $(x, y)$ that describe

a point on the square to get a one-to-one correspondence with the points on a line segment. What point in the square would be paired with the point

0.1200010001000100010001 . . . ?

With what point on the line does that point in the square actually get paired? Is this a problem? Explain.

7. **More perfect shuffle problems.** Suppose we used our failed perfect shuffling of digits to mix the digits of the numbers $(x, y)$ that describe a point on the square to get a one-to-one correspondence with the points on a line segment. What point on the square would be paired with the point

0.12001001001001001001 . . . ?

With what point on the line does that point on the square actually get paired? Is this a problem? Explain.

8. **Grouping digits.** Given the grouping of digits method as a way to shuffle the digits of the two numbers representing the points of a square, how would you group and then shuffle the point $(x, y)$ where

$x = 0.64340004897211098000345003703030306\ldots$

$y = 0.49867400546080009707076002033030112\ldots$ ?

9. **Where it came from.** Given the grouping of digits method as the one-to-one correspondence between the square and the line segment, with what point on the square would the point

0.990040878000404820244000501101909090909090 . . .

on the line get paired?

10. **Group fix (S).** Consider the point on the line from question 6. What point on the square would that point get paired with using the grouping-shuffling method?

## II. CREATING NEW IDEAS

1. **Is there more to a cube?** Prove that the cardinality of points in a solid cube is the same as the cardinality of points on a line segment.

2. **T and L (H).** Prove that the cardinalities of points in the following two geometrical objects are equal (these objects are made up of little line segments—so they have no thickness).

T       L

3. **Infinitely long is long.** Must it be the case that every subset of the real line, even one that goes on forever, has a one-to-one correspondence with a set of points in the interval between 0 and 1?

4. **Plugging up the North Pole.** What would happen if you filled in the north pole of the circle in the stereographic projection? Is there any point on the real line to which that extra point could correspond? Show that the cardinality of a circle is the same as the cardinality of (0, 1] (an interval where 0 is not included, but 1 is included).

5. **3-D stereo (S).** Let S' be the set of points on a the surface of a ball (a hollow sphere) minus the north pole. How does the cardinality of points on the plane compare to the cardinality of points on S? Justify your answer, and explain the title of this story.

## III. FURTHER CHALLENGES

1. **Stereo images.** Given your answer to the 3-D stereo question, suppose we had a circle drawn on the sphere that passed through the north pole (which is missing) and the south pole (longitudinal lines). What object is this circle paired with on the plane?

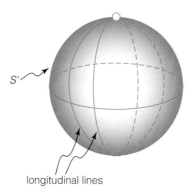
longitudinal lines

What shapes would latitudinal lines get paired with?

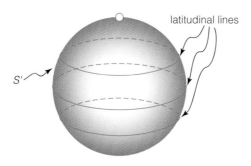
latitudinal lines

2. *Grouped shuffle.* Carefully verify that the pairing described by shuffling groups of digits is, in fact, a one-to-one correspondence.

### IV. IN YOUR OWN WORDS

1. *Personal perspectives.* Write a short essay describing the most interesting or surprising discovery you made in exploring this section's material. If any material seemed puzzling or even unbelievable, address that as well. Explain why you chose the topics you did. Finally, comment on the aesthetics of the mathematics and ideas in this section.

2. *With a group of folks.* In a small group, discuss and work through the proof that the set of points in a square has the same cardinality as the set of points on a line segment. After your discussion, write a brief narrative describing the proof in your own words.

3. *Creative writing.* Write an imaginative story (it can be comical, dramatic, whatever you like) that involves or evokes the ideas of this section.

4. *Power beyond the mathematics.* Provide several real-life situations—ideally, from your own experience—for which some of the strategies of thought presented in this section would provide effective methods for approaching and resolving them.

# CHAPTER FOUR

# Geometric Gems

*The knowledge at which geometry aims is the knowledge of the eternal.*
—Plato

**4.1**

*Pythagoras and His Hypotenuse*

**4.2**

*A View of an Art Gallery*

**4.3**

*The Sexiest Rectangle*

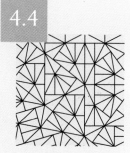

**4.4**

*Soothing Symmetry and Spinning Pinwheels*

When we look at and touch the objects in our world, we encounter the physical form and spatial relationships of matter. We see shapes—some interesting, some ordinary, and some attractive. Geometry is the study of those shapes. Geometric relationships give us a basic sense of order, coherence, and beauty. The lines, circles, and spatial patterns of the physical world give depth to our intuition about shape and enable us to develop insights and recognize patterns in things geometric. Geometry captures the structure and nuances of our physical reality.

Geometry has been studied since ancient times, and some of the foundations of geometry were developed thousands of years ago. However, advances in geometry are still being made, and many questions remain unanswered. Here we will examine both ancient geometric discoveries and modern developments. The geometry of the plane includes not only the ancient Pythagorean Theorem but also such issues as guarding museums, investigating aesthetics, creating nonrepeating floor patterns, and searching for symmetry. Of course, geometry extends beyond the flat plane and also describes objects in space. There too we see contributions ancient and modern, from the symmetric Platonic solids, which have been admired for millennia, to curved geometries that provide more modern tools for describing our universe. Finally, we journey into a world created by imagination—the fourth dimension.

Exploration of geometry allows us to experience ways of thinking that can be applied in many areas, both within mathematics and beyond it: keeping an open mind to all possibilities—even those that first appear impossible; understanding simple things deeply; exploiting our insights; searching for patterns; and breaking up difficult tasks into many easy ones. But perhaps the technique most powerfully illustrated by geometry is the idea of building understanding by actually doing and trying. Physically holding and manipulating objects or looking carefully at illustrations allows us to understand ideas more viscerally. We begin our travels through geometry by applying this strategy—physically doing and trying—to one of the most important and ancient geometrical gems, the Pythagorean Theorem.

◆ ◆ ◆ ◆ ◆ ◆ ◆ ◆ ◆ ◆ ◆ ◆ ◆

**The Platonic Solids Turn Amorous**

**The Shape of Reality?**

**The Fourth Dimension**

# 4.1 *Pythagoras and His Hypotenuse:*
## *How a Puzzle Leads to the Proof of One of the Gems of Mathematics*

*I'm very well acquainted, too, with matters mathematical
I understand equations, both the simple and quadratical
About Binomial Theorem I'm teeming with a lot of news
With many cheerful facts about the square of the hypotenuse.*
—William S. Gilbert, *The Pirates of Penzance*

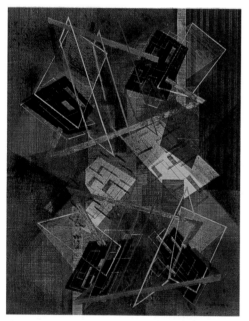

*Oblique Progression* (1948) by I. Rice Pereira

Some geometric relationships are so profound that they have changed the shape of civilization—both literally and figuratively. Perhaps the best known and most fundamental theorem in all of mathematics is the *Pythagorean Theorem*. While we have seen this basic result in high school, here we examine it more deeply.

> **THE PYTHAGOREAN THEOREM** *In a right triangle, the square of the length of the hypotenuse is equal to the sum of the squares of the lengths of the other two sides.*

This theorem and its proof form one of the classical intellectual accomplishments of humankind, comparable to a work of Shakespeare. The proof presented here, which was discovered by the Indian mathematician Bhaskara in the 12th century, exemplifies aesthetics and beauty in mathematical arguments.

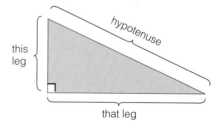

$(\text{this leg})^2 + (\text{that leg})^2 = (\text{hypotenuse})^2$

Powerful and important discoveries can come from play.

◆ ◆ ◆ ◆ ◆ ◆

## A Puzzle Proof

Since the Pythagorean Theorem concerns the relationships among the sides of a right triangle, first we must become well acquainted with a right triangle—see it, feel it, handle it. In your kit you will find four copies of the same right triangle and one small square. Play with these shapes, move them around, put them together and search for interesting patterns: things that fit, relationships, familiarity. Just see what you find, and remember that no discovery is too small or too inconsequential.

As you play with these shapes, notice that the two nonright (which we jokingly refer to as "wrong") angles of one of the triangles fit together to form a right angle. Just move the triangles around to confirm this fact. You may remember from high school geometry that the sum of the angles of any triangle is 180° and that a right angle has 90°. Thus, the sum of the two remaining angles should be 90° as well—one of the rare occasions when two wrongs actually do make a right.

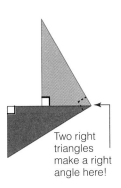

Two right triangles make a right angle here!

## A Hypotenuse Square

Now we begin the proof of the Pythagorean Theorem in earnest. Let's call the length of the hypotenuse of the right triangle $c$, and let's call the lengths of the other two sides $a$ and $b$. We may now state the Pythagorean Theorem as

$$a^2 + b^2 = c^2.$$

We first wish to use the five pieces from the kit to construct a large square with each side of length $c$, which is the length of the hypotenuse. You have four identical copies of a right triangle, and so you have four hypotenuses, one for each side of the square. Assemble your four right triangles and one little square into a square having sides equal to $c$, the length of the hypotenuse.

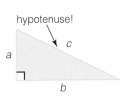

Remember that the hypotenuse of a right triangle is the side opposite the right angle. You may have to flip your triangles over to get them to fit together into a square. Make certain that you keep the hypotenuses toward the outside of the square you are building. Also recall that the angles at each corner of the square must be 90°. Now build this square having sides equal to the hypotenuse, using the four right triangles with the little square in the middle. Here we see in action the credo "doing and trying are powerful means of building understanding."

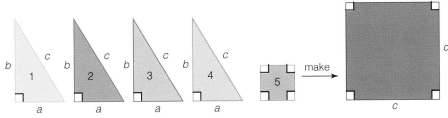

Using the pieces on the left, construct the square on the right.

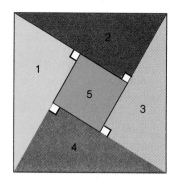

You have now created a large square out of four right triangles and a small square. This big square's area is equal to the length of its base multiplied by the length of its height. But a square has its base equal to its height, and this square has each side equal to the length of the hypotenuse. Therefore, the area of this big square is the length of the hypotenuse of the original right triangle multiplied by itself or $c^2$. So we just discovered that the total area of all five pieces is equal to $c^2$.

## The Two Small Square Challenge

Make two squares.

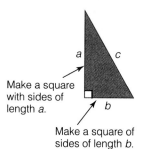

Make a square with sides of length $a$.

Make a square of sides of length $b$.

The goal now is to rearrange the five pieces we just used, the four triangles and the small square, into a new configuration to create two smaller squares: one having sides of length $a$ and the other having sides of length $b$ (remember that $a$ and $b$ are the lengths of the non-hypotenuse sides of the right triangle).

Once we have built these two squares, the area of one of the squares will be $a^2$, and the area of the other will be $b^2$. So, the area of this total new configuration will be the sum of the two areas: $a^2 + b^2$. But remember that we are using the exact same pieces we used previously to make the large square that had area $c^2$. Since the areas of these two configurations must be the same (we just move the pieces around, we don't change their size), we will therefore have shown that $a^2 + b^2 = c^2$. Try it now by moving the triangles around and making shapes.

Keep an open mind to all possible possibilities.

◆  ◆  ◆  ◆  ◆  ◆

## A Hint

The two squares having areas $a^2$ and $b^2$ will not be freestanding squares. They will be parts of an object that looks like a thickened **L**. *Don't read on until you have really tried to solve this puzzle. You can do it!*

## The Answer

We can make the two squares by just moving two triangles from the original large square we constructed.

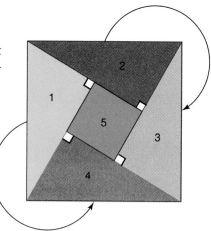

By drawing the line that divides that figure into two squares, one having area $a^2$ and the other having area $b^2$, we see that the total area is $a^2 + b^2$.

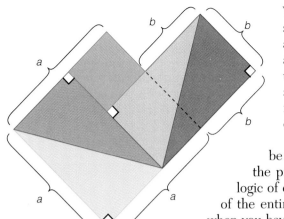

We see that the same four triangles and small square that previously made a large square of area $c^2$ now fit together to make a figure having area $a^2 + b^2$. Thus, for this right triangle anyway, the Pythagorean Theorem holds. Of course, this same proof could have been done starting with any right triangle. So, this geometric demonstration completes a proof of the Pythagorean Theorem.

Although our proof is now complete, it will not be real until we have made it our own. Work through the proof yourself. Attempt not only to understand the logic of each and every step, but try to capture the essence of the entire argument. Say, "Voilà!" or "Eureka!" or "Cool!" when you have drawn the line separating the two squares $a^2$ and $b^2$. Once you have digested the ideas, explain the proof to a roommate, a relative, or a friend. Use puzzle pieces and be patient if they do not immediately see it. It is only after we have explained the proof to several people, and witnessed their moment of understanding, that it is ours for life.

> . . . a theorem as "the square of the hypotenuse of a right angled triangle is equal to the sum of the squares of the sides" is as dazzlingly beautiful now as it was in the day when Pythagoras discovered it. —Lewis Carroll

## A Fascination with Geometrical Ideas

Geometrical issues have fascinated people since the dawn of thought and continue to inspire and tantalize us. The Greeks of antiquity viewed the line and the circle to be the most elegant and fundamental of all geometric forms. Consequently, the Greeks were captivated with the study of geometrical constructions—identifying those shapes and lengths that can or cannot be drawn using just a compass (which makes circles) and a straight, unmarked, edge (which makes lines). The Greeks devised many constructions, but a handful

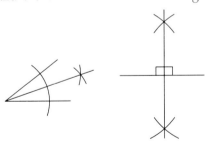

of construction questions remained unanswered for more than 2,000 years, when new mathematical ideas finally settled them. The fact that ancient mysteries were definitively resolved after two millennia of effort can give us all hope that other great problems of humanity will also be solved.

While geometry provides us with an intellectual tie to ancient Greek thinkers that transcends time itself, today computers allow us to see geometrical objects in previously unimaginable ways. These powerful tools are spawning a new generation of creativity and imagination. In the summer of 1995, two eighth-grade students, David Goldenheim and Daniel Litchfield,

*Hard problems can be solved—we should never give up.*

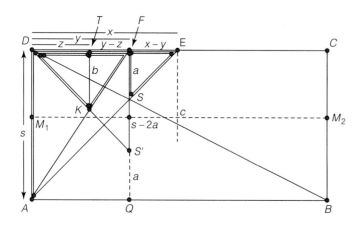

discovered a completely new geometric construction for dividing a segment into equal pieces. Their original construction, which they discovered by playing around with some clever ideas on a computer program known as the Geometer's Sketchpad, is believed to be the first such construction since the time of the ancient Greeks! Indeed, this interweaving of ancient ideas with modern thought forms the detailed fabric we now view as geometry.

**A LOOK BACK**

The Pythagorean Theorem can be proved by physically building with puzzle pieces the square on the hypotenuse of a right triangle and showing how those pieces can be reassembled into a thickened L shape with area equal to the sum of the squares of the other two sides.

The Pythagorean Theorem is an idea that may at first seem abstract, but when we hold and manipulate physical triangles, that abstract idea can become concrete and real. Using several ways of learning about an idea helps us to understand it.

*Experience ideas in as many ways as possible.*

◆ ◆ ◆ ◆ ◆ ◆ ◆ ◆ ◆ ◆ ◆ ◆ ◆ ◆ ◆ ◆ ◆

## MINDSCAPES  Invitations to Further Thought*

### I. SOLIDIFYING IDEAS

1. ***Operating on the triangle.*** Using a straightedge, draw a random triangle. Now carefully cut it out. Next amputate the angles by snipping through adjacent sides.

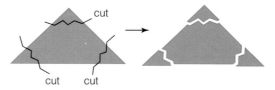

Now move the angles together so the vertices all touch and the edges meet. What do you conclude about the sum of the angles of a triangle? Try this procedure with triangles having different dimensions.

---

*In the Mindscapes section, exercises marked (H) have hints for solutions at the back of the book. Exercises marked (S) have solutions.

2. ***Excite your friends about right triangles.*** Describe the proof of the Pythagorean Theorem to someone who has never seen it before. Try to get him or her inspired and intrigued by math! Record the event and the various reactions.

3. ***Easy as 1, 2, 3?*** Can there be a right triangle with sides of length 1, 2, and 3? Why? Can you find a right triangle whose side lengths are consecutive natural numbers?

4. ***Sky high*** (S). On a sunny warm day, a student decides to fly a kite on the college green just to relax. His kite takes off and soars. He lets all 150 feet of the string out and attracts a crowd of onlookers. There is a slight breeze, and a spectator 90 feet away from the student notices that the kite is directly above her. Unlike a real kite, this math-question kite has the string going in a straight line from the student to the kite. How high is the kite from the ground?

5. ***Sand masting*** (H). The sailboat named Sand Bug has a tall mast. The backstay (the heavy steel cable that attaches the top of the mast to the back, or stern, of the sailboat) is made of 130 feet of cable. The base of the mast is located 50 feet from the stern of the boat. How tall is the mast?

6. ***Getting a pole on a bus.*** For his 13th birthday, Adam was allowed to travel down to Sarah's Sporting Goods store to purchase a brand new fishing pole. With great excitement and anticipation, Adam boarded the bus on his own and arrived at Sarah's store. Although the collection of fishing poles was tremendous, there was only one pole for Adam and he bought it: a five foot, one piece fiberglass "Trout Troller 570" fishing pole. When Adam's bus arrived, the driver reported that Adam could not board the bus with the fishing pole. Objects longer than four feet were not allowed on the bus. In tears, Adam remained at the bus stop holding his beautiful five foot Trout Troller. Sarah, seeing the whole ordeal, rushed out and said, "Don't cry Adam! We'll get your fishing pole on the bus!" Sure enough, when the same bus and the same driver returned, Adam boarded the bus with his fishing pole and the driver welcomed him aboard with a smile. How was Sarah able to have Adam board the bus with his five foot fishing pole without breaking the bus line rules and without cutting or bending the pole?

7. ***The scarecrow.*** In the 1939 movie *The Wizard of Oz*, when the brainless scarecrow is given the confidence to think by the Wizard (by merely handing him a diploma by the way), the first words the scarecrow utters are, "The sum of the squares of the sides of an isosceles right triangle equals the square of the hypotenuse." An isosceles right triangle is just a right triangle having both legs the same length. Suppose that an isosceles right triangle has legs each of length 3. What is the length of the hypotenuse? If the length of each leg is L, then can you figure out the length of the hypotenuse in terms of L?

8. ***Rooting through a spiral.*** Start with a right triangle with both legs having length 1. What is the length of the hypotenuse? Suppose we draw a line of length 1 perpendicular to the hypotenuse and then make a new triangle as illustrated. What is the length of this new hypotenuse? Suppose we continue in this manner. Describe a formula for the lengths of all the hypotenuses.

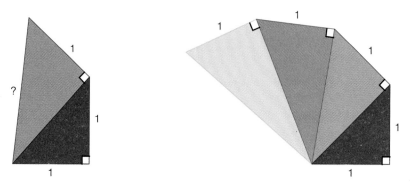

9. ***Is it right?*** (**H**) Suppose someone tells you that she has a triangle with sides having lengths 2.6, 8.1, and 8.6. Is this a right triangle? Why? Is there an angle in the triangle larger than 90°? Justify your answer.

10. ***Train trouble.*** Train tracks are made of metal. Consequently, they expand when it's warm and shrink when it's cold. When riding in a train, you hear the clickety-clack of the wheels going over small gaps left in the tracks to allow for this expansion. Suppose you were a beginner at laying railroad tracks and forgot to put in the gaps. Instead, you made a track 1 mile long that was firmly fixed at each end. On a hot day, suppose the track expanded by 2 feet and therefore buckled up in the middle creating a triangle.

Roughly how high would the midpoint be? Now you may appreciate the click-clack of the railroad track.

## II. CREATING NEW IDEAS

1. ***Does everyone have what it takes to be a triangle?*** Suppose a friend comes up to you and says, "Hey, I just made a triangle with sides of length 2,431; 5,642; and 3,210." How would you respond to

him? What basic fact about triangles do you conclude from this dialogue with your friend?

2. **Getting squared away.** In our proof of the Pythagorean Theorem, we stated that the second figure is actually two perfect squares touching along an edge. Prove that they are indeed both perfect squares. It will be useful for you to use the puzzle pieces from your kit to build the first big square again and carefully notice how the pieces all fit together.

3. **The practical side of Pythagoras.** Suppose you are building a patio and you want to make certain that the sides of your patio meet at a right angle. Give a practical and easy method, using your knowledge of the Pythagorean Theorem, to check that the angle between two adjacent sides is 90°.

4. **Pythagorean pizzas (H).** You have a choice at the local pizza place: For the same price you can get either one large pizza or both a small and a medium. How can you determine which way you get more by using just the Pythagorean Theorem and knowing the diameters of the different sizes of pizza? Describe an easy procedure to figure out which deal to choose.

5. **Natural right.** Suppose $r$ and $s$ are any two natural numbers where $r$ is bigger than $s$. Show that the triangle having side lengths equal to $2rs$, $r^2 - s^2$, and $r^2 + s^2$ is actually a right triangle. Are there infinitely many different right triangles having all sides of integer lengths?

## III. FURTHER CHALLENGES

1. **Well-rounded shapes.** Suppose we have two circles having the same center. The small one has radius $r$, and the large one has radius $R$. Let's now consider two shapes. The first is the doughnut-like region between the two circles. The second is a disk whose diameter is the length of the line segment whose endpoints are on the large circle and whose center point touches the small circle at its north pole. What are the areas of these two shaded regions? How do their sizes compare with each other?

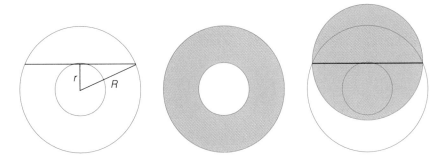

2. **A Pythagorean Theorem for triangles other than right triangles.** Suppose we have a triangle that is not a right triangle. For example, consider:

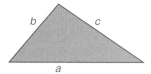

Now, if we drop a perpendicular line from the top vertex down to the side $a$ and we cut $a$ into two pieces, $a'$ and $a''$, we would have:

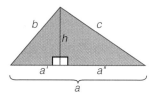

It turns out that $a^2 + b^2 = c^2 + 2aa'$. Use the Pythagorean Theorem with the two right triangles in the preceding picture to produce two equations. Subtract one equation from the other and notice that the $h^2$ terms drop out. Now, deduce the preceding formula (remember that $a = a' + a''$, so you can solve for $a''$). Once you have proved this formula, show what happens if the angle between sides $a$ and $b$ is 90°. What result do you get? Are you surprised?

## IV. IN YOUR OWN WORDS

1. **With a group of folks.** In a small group, discuss and actively work through the proof of the Pythagorean Theorem. After your discussion, write a brief narrative describing the proof in your own words.

# 4.2 A View of an Art Gallery:
## Using Computational Geometry to Place Security Cameras in Museums

*The interior is infinite, all the way to the mystery of the inmost, the charged point, a kind of sum total of the infinite.*

—Paul Klee

We now move from the Pythagorean Theorem, a geometric gem that has been around for over 2,000 years, to a geometric question that has been around for just 20 years. This new geometric question asks how many eyes are necessary to view an entire art gallery—all at once.

### A Look at the Gallery Itself

When we walk around an art gallery, we study and enjoy the works hanging on the walls. Normally we aren't conscious of the walls themselves—that is, we don't pay much attention to the floor plan. The floor plan, however, is exactly what we wish to study now. If we consider an aerial view of the museum, and assume the gallery is one big open area with no interior walls, then we see only the pattern of the museum walls. We will assume that each wall is flat (so the Guggenheim is out).

The walls might fit together to make an interesting shape, such as:

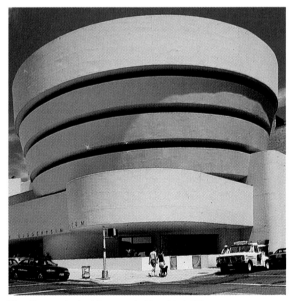

The Guggenheim Museum in New York City.

Top view of an interestingly shaped gallery

sealed!
no crossing

not sealed!
NOT ALLOWED
(not **closed**)

Since we are assuming that the museum consists of one large area with no interior walls or partitions, the surrounding walls form a jagged "loop" that does not cross through itself. In other words, the building is completely sealed, and there is only one gallery on the inside. Let's call any figure that is made up of straight line pieces that are connected end to end to form a loop a *polygonal closed curve*. The corners where walls meet are called *vertices*. So, the floor plan of a gallery made up of straight walls and having just one big open space on the inside is an example of a polygonal closed curve—notice that there may be many nooks and crannies.

Gallery has an inside gallery—NOT ALLOWED. Walls cannot cross through themselves.

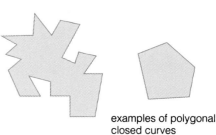
examples of polygonal closed curves

## Finding Points of View

A sad truth is that today museums need security systems to guard their valuable works of art. How can one keep a watchful eye on things? Suppose our goal is to strategically place video cameras such that every point of the gallery will be viewed by at least one camera. In an attempt to be subtle, suppose we wish to mount cameras only in the corners of the gallery. Each camera is equipped with a wide-angle lens that allows the camera to view everything inside the V formed by the two walls that the camera is housed between. It is possible that the corner formed by the two adjacent walls actually sticks into the room. In this case, the camera equipped with a special fish-eye type lens has nearly panoramic vision and has a sweeping view of everything visible between the outside of the V formed by the two walls.

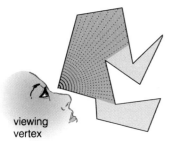
viewing vertex

Shading shows the area that can be viewed from the camera located at the vertex with the eye.

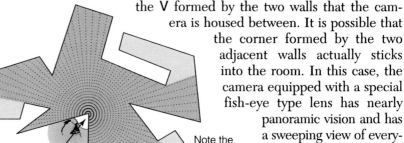
viewing vertex

Note the panoramic vision!

## Our Main Question

How many cameras do we need to make sure every point in the gallery is seen by some camera?

Of course, the answer depends on how complicated the gallery is. However, no matter how many walls and bends it has, if we place one

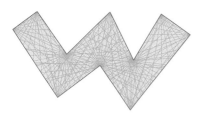

**many** lines of sight with cameras at **every** vertex

camera at every corner, then we are able to keep a watchful eye on every point in the gallery. But cameras are expensive, and perhaps a camera in every corner is not necessary. For example, in the previous illustration, what is the largest number of cameras we can remove so that every point inside is still visible from some camera? This issue leads to the following question.

## Our Modified Main Question

For an arbitrary gallery having $v$ corners (vertices), how many cameras do we need to make sure every point in the gallery is seen by some camera?

Pared down to its geometrical essence, the question becomes the Klee Art Gallery Question.

## The Klee Art Gallery Question

For an arbitrary polygonal closed curve in the plane with $v$ vertices, how many vertices are required such that every point on the interior of the curve is visible (directly in the line of sight) from at least one of these chosen vertices?

This question represents an issue in an area of study known as *computational geometry*—an area of mathematics where one is concerned with efficient methods of computing things involving geometric objects. These methods often lead to improved algorithms, and thus there is much cross-pollination between computational geometry and computer science. Before guessing an answer to the Klee question, let's first explain the famous art name associated with this question. Paul Klee was an imaginative, abstract artist whose work is world-renowned. However, this question is named after Victor Klee from the University of Washington, who in 1973 first posed it casually in conversation. Victor Klee's work in combinatorics

*Open Book (Offenes Buch)* (1930) by Paul Klee.

*Glass Facade* (1940) by Paul Klee.

*Construction under a Waterfall, a Crossing* (1924) by Paul Klee.

and computational geometry is also world-renowned. In fact, Victor Klee's gallery question has led to a tremendous body of interesting work in combinatorial geometry.

Let's now consider various art gallery shapes to build some insights and make some guesses. For each example, mark the fewest vertices required so that every point inside is directly in the line of sight of at least one of your marked vertices.

## An Untangling Example—The Comb Gallery

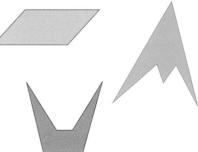

Where would you place your cameras in these museum floor plans?

Let's look at a floor plan that resembles a comb. Each tine is a thin triangle that opens onto one long hall. Notice that each long narrow triangular hallway needs its own camera somewhere. But how many cameras do we need? Well, we could place two cameras in a number of ways to view the entire interior. One camera alone cannot do the job since it is impossible for one camera to see down both triangular hallways. So, the smallest number of cameras needed for this 2-tine comb

2-tine comb gallery

is 2. Notice that the comb has 6 vertices. What is the fewest number of cameras needed for the 3-, 4-, and 5-tine combs? How many vertices does each comb have? What is the relationship between vertices and cameras? What can we conclude about the minimum number of cameras needed for comb-shaped galleries?

If you answered the previous questions, you've discovered that we need exactly 3 cameras for the 3-tine comb, which has 9 vertices; 4 cam-

eras for the 4-tine comb which has 12 vertices; and 5 cameras for the 5-tine comb, which has 15 vertices. Of course, we could make comb-shaped galleries with any number of tines. Such a gallery with $t$ number of tines has $3t$ vertices and requires $t$ cameras. So, from these examples, we conclude that a gallery with $v$ vertices may need as many as $v/3$ cameras. We will now discover that that number of cameras will *always* be enough, no matter what the shape of the gallery—thus answering the Klee Art Gallery Question with the following theorem.

> **THE ART GALLERY THEOREM.** *Suppose we have a polygonal closed curve in the plane with $v$ vertices. Then there are $v/3$ vertices from which it is possible to view every point on the interior of the curve. If $v/3$ is not an integer, then the number of vertices we need is the biggest integer less than $v/3$.*

This result was first formulated and proved by Vasek Chvátal, from Rutgers University, a couple of days after Victor Klee raised his off-the-cuff but interesting question. A few years later Steve Fisk, from Bowdoin College, gave a different proof of the Art Gallery Theorem. It is his argument that we will now think about together.

## CREATIVITY HAPPENS

Creative ideas can arise anytime and anywhere as long as our minds are open. The story of how Steve Fisk came up with the idea for his proof is a wonderful and funny real-life illustration of this maxim. Here Professor Fisk recounts the episode.

> Here's the quick story. My wife and I were traveling across Asia during 1975–1976, and during our one-month stay in Iran I visited the math library at the University of Tehran. I read Chvátal's article—or rather I looked at it, for the proof seemed so complex that I didn't want to read it. I did like the result though. A few weeks later we were traveling on a bus between Herat and Kandahar in Afghanistan. It was hot, dusty, and noisy with squawking chickens on the roof. I was thinking about it—well I was probably dozing off actually—and I thought of the proof.

## Proving the Theorem

Let's see why $v/3$ vertices are always enough to see every point inside a polygonal curve. Where do we begin? The difficulty is that we have no idea what the polygonal curve looks like. That is, we don't know the floor plan of the gallery. It may have hundreds or thousands of complex hallways and corners. What should we do? When faced with a challenging issue in life, we should always try to begin with an easy case and warm up to the more difficult task.

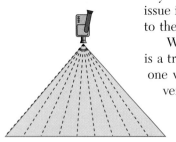

For the simplest gallery—the triangle—any vertex can be used to view the entire museum.

What is the simplest polygonal curve (gallery floor plan)? The answer is a triangle. Notice that the triangle has three vertices and that from any one vertex we are able to view every point in the triangle. So only one vertex is needed to view the entire interior, and, happily, $v/3$ in this case is 3/3 which equals 1 vertex. Thus the theorem is true for any triangular polygonal curve. We also notice that, in this simple case, we can use any of the three vertices as our vertex to install our camera. That was easy. Now that we've tackled the simple, let's move to the more complex.

## Divide and Conquer

Of course we cannot expect our arbitrary polygonal curve gallery to be a triangle. What do we do with a more complex-shaped gallery? The answer is to divide it into a whole bunch of triangles—that is, convert one complicated situation into a collection of easy situations. Let's now see how to divide the inside of the polygonal closed curve into triangles.

We build triangles in the interior of the polygonal curve by adding edges from one vertex to another with the only restriction being that two edges cannot cross. The vertices of these triangles must all be vertices of the original polygonal curve. Notice that there are many different ways to create such triangles.

*Often in life, hard questions are made up of many easy pieces.*

◆ ◆ ◆ ◆ ◆ ◆ ◆

How do we know that we can always divide the polygonal gallery into triangles in this way? In practice, whenever we draw a polygonal curve, it's easy to break its interior into triangles, but we need to describe a method that will always work. Our strategy is first to do something a little easier than dividing our gallery into triangles. Instead, we just divide the inside of the polygonal curve into two pieces by drawing in an edge that spans the inside of the polygonal curve between two vertices. We then have two easier problems to solve—in this case, two smaller polygonal closed curves that we want to divide into triangles. In each of those two

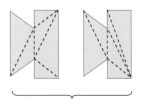

Two examples of two different triangulations of the same polygonal gallery.

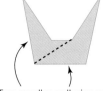

Two smaller galleries made by adding a spanning arc.

smaller curves, we will draw spanning lines again. Pretty soon our pieces will be so small that they must be triangles. Let's see how to find a spanning arc. Remember, if we can break a hard problem into two easier problems, we can solve it.

## *Illuminating the Dividing Line*

Start at any vertex and notice that two edges emanate from it and that they form an angle. Actually they form two angles, one that faces the interior of the polygonal curve and one that faces the outside. From the vertex we point a flashlight along one of the two emanating edges. The flashlight shines against one of the sides and illuminates the vertex at the other end of that side—let's call that vertex STARTING-VERTEX. We now start sweeping the light around the interior angle toward the other emanating edge. At all times we look to see whether the light ever strikes another vertex. The first time it does, we

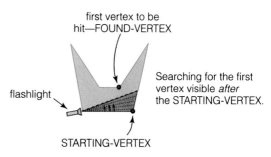

connect that vertex (let's call it the FOUND-VERTEX) to the vertex where the flashlight is located. In this case, we have located a line that spans the interior of the curve. If, as the flashlight swings, the first vertex the light hits happens to be the far vertex of the other emanating edge (let's call it the ENDING-VERTEX), then we connect the STARTING-VERTEX to the ENDING-VERTEX to create a spanning edge.

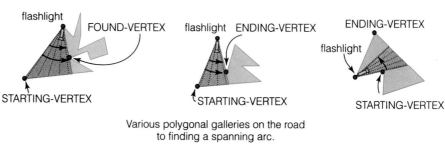

Various polygonal galleries on the road to finding a spanning arc.

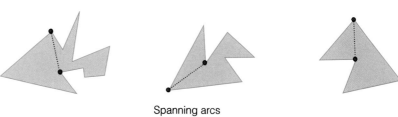

Spanning arcs

## Repeat Until Triangles Abound

We have now broken the interior of the curve into two pieces, creating two polygonal closed curves. If either or both of these curves are not triangles, then we can deal with each of them separately just as before, finding spanning edges until we have broken the whole interior into triangles. Recall two important features of this triangulating process:

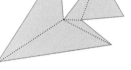
a triangulation

- Each vertex of each triangle is a vertex of the original polygonal curve.
- For any triangle, the entire triangle is visible from any one of its three vertices.

## A Dash of Color

Next, we are going to color each vertex of the polygonal curve gallery one of three colors—red, yellow, or blue. We may color the vertices any way we wish, as long as we abide by one rule: The three vertices of any one individual triangle must have different colors. That is, every single triangle in the triangulation must have one red vertex, one yellow, and one blue. How can we color the vertices following this rule? Try some examples and figure out how to do it.

The answer is just to pick one of the triangles and color each of its vertices a different color. At least one of the edges of that triangle must be an edge of another triangle. This new triangle has two of its vertices colored (the two vertices from the common edge that the triangles share). This leaves one uncolored vertex. We color it the third color. We repeat this procedure, one triangle at a time. Since all the vertices are on the polygonal curve, this process colors all the vertices as required.

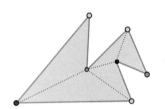

## Red Eye

Suppose we put cameras at every red vertex. Would we be able to see every point inside the polygonal gallery? Think about this question, and try to provide both an answer and a reason.

The answer is, "Yes, we'd be able to see everything." Why? Select a random point inside the polygonal curve. It must be in one of our triangles. We know that that point is visible from each of the three vertices of the triangle. By our coloring rule, there must be a vertex of this triangle colored red (remember that every triangle must have its vertices colored with different colors). From that red vertex we can see the point! Therefore, every point inside the polygonal curve is visible from some red vertex. Would we be able to see every point inside the polygonal curve from just the yellow vertices? Why? Would we be able to see every point inside the polygonal curve from just the blue vertices? Why?

We have now discovered that every point inside the polygonal curve is visible from any set of vertices all having the same color. We are now going to count! How many red vertices are there? How many yellow vertices are there? How many blue vertices are there? The answers can be given succinctly: "We have no idea." Okay, well that isn't too informative. What can we say about the number of vertices of various colors? Let's call the number of red vertices $R$, the number of yellow vertices $Y$, and the number of blue vertices $B$. Even though we do not know what $R$, $Y$, or $B$ equals, we do know something about all of them together. What does $R + Y + B$ equal?

Since every vertex of the polygonal curve was colored one of the three colors and we are told that there are a total of $v$ vertices, we know that $R + Y + B = v$. This is a bit of progress. We have three numbers ($R$, $Y$, and $B$) that add up to $v$, so they can't all three be bigger than $v/3$. That is, at least one of them must be less than or equal to $v/3$. So, we have just discovered that there must be a color that is used less than or equal to $v/3$ times. We have already seen that, if we look just from the vertices colored with that color, we can see the entire interior of the polygonal curve. So, we are always able to see every point inside from less than or equal to $v/3$ vertices—which is exactly what we wanted to prove.

## The Strategy in Review

Let's review the strategy of our proof. We first reduced the difficult issue into an enormous number of less difficult issues (we divided our gallery into triangles). Then, using three colors, we colored the vertices such that every triangle was colored with three different colors. We discovered that, using all vertices of the same color, every interior point was visible. Finally, we counted the number of vertices of each color and realized that we cannot have each of the three colors occurring more than 1/3 of the time. Therefore, one color must appear less than or perhaps equal to $v/3$, and those vertices are where we should place our cameras. The key to this geometry question was to convert it into a counting question. This technique is the main theme of combinatorial geometry: geometrical questions are answered by counting.

## Mirror, Mirror on the Wall . . .

We have considered both ancient and modern questions and their answers. Often it's interesting to look into the future and contemplate those questions that remain unanswered. Here is one such geometry question that relates to our art gallery discussion. No one knows the answer.

Suppose we have a gallery in the shape of a polygonal closed curve, but this time it is a gallery of mirrors. That is, each wall is completely covered by a mirror. Reflect on this image if you will. Since the walls are all mirrored, we can see around some corners by looking through reflections off various walls. Guarding a mirrored gallery might need far fewer video

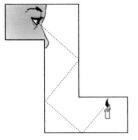

Seeing around corners in a mirrored room.

cameras. In fact, from a single point inside the gallery (not necessarily at a vertex), we might be able to see every other point inside the gallery by using the mirrors. Some people believe there is always at least one point in the gallery from which every other point inside is visible, but no one knows for certain.

**VISIBILITY WITHIN A HOUSE OF MIRRORS QUESTION.**
*For an arbitrary polygonal closed curve having mirrored sides, must there exist a point from which every point inside the polygonal curve is visible?*

We can find such vantage points in some examples. But how can we prove that a point of visibility *always* exists for *all* polygonal curves—no matter how complicated the curve? Will the solution involve some

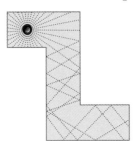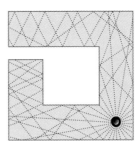

of the analysis used by the Greeks; or will it follow from an insightful observation by some eighth graders using computer graphics software; or will the solution involve new techniques; or will someone find a mirrored polygonal closed curve that contains no point of complete visibility? As with all unanswered questions, we will not know for sure until someone's creative ideas lead us to a complete answer.

## A LOOK BACK

If we have a polygonal closed curve in the plane with $v$ vertices, then there are $v/3$ vertices from which it is possible to view every point on the interior of the curve. This Art Gallery Theorem can be proved by dividing the interior into triangles, coloring their vertices, and then placing cameras at vertices of the same color.

We began our exploration of this topic by drawing and examining examples to develop insights. Then, faced with potentially complicated closed curves, we broke this complicated situation into many manageable pieces—triangles. We looked at examples and then attacked a difficult issue by dividing it into many simpler pieces. These life strategies are among the most potent in our arsenal of weapons against the unknown.

*Ground your understanding in examples.*

*Look for patterns.*

*Divide and conquer.*

# MINDSCAPES  Invitations to Further Thought*

## I. SOLIDIFYING IDEAS

1. ***Klee and friends.*** Explain to a friend the statement of the Art Gallery Theorem and its proof. What are his or her reactions and questions?

2. ***Putting guards in their place.*** For each museum, place guards at appropriate vertices so that every point in the museum is within view.

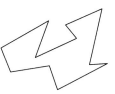  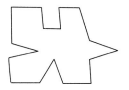

3. **Guarding the Guggenheim.** The Art Gallery Theorem does not tell us how many guards are needed to guard museums that have curved walls. For each museum, place the minimum number of guards at vertices so that every point in the museum is within view.

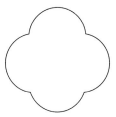  

4. ***Triangulating the Louvre*** (**H**). Triangulate the museums by adding straight segments that do not cross each other yet span the insides and extend from one vertex to another.

---

*In the Mindscapes section, exercises marked (H) have hints for solutions at the back of the book. Exercises marked (S) have solutions.

5. ***Triangulating the Clark.*** Triangulate the museums by adding straight segments that do not cross each other yet span the insides and extend from one vertex to another.

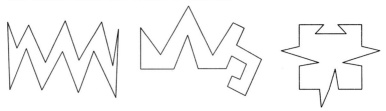

6. ***Tricolor me*** (**H**). For each triangulation, color the vertices red, blue, or green so that every triangle has all three colors.

7. ***Tricolor hue.*** For each triangulation, color the vertices red, blue, or green so that every triangle has all three colors.

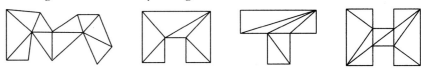

8. ***One third.*** Write the number 6 as a sum of three natural numbers in several different ways, and, in each sum, circle a number that is less than or equal to 2.

9. ***Easy watch.*** Draw the top view of a museum with six sides that needs only one guard to view the entire gallery.

10. ***Two watches*** (**S**). Draw the top view of a museum with 10 sides that needs exactly two guards to view the entire gallery.

## II. CREATING NEW IDEAS

1. ***Mirror, mirror on the wall.*** Consider the aerial views of the rooms drawn below. Suppose that all the walls are made of mirrors. For each room, find a point inside from which all other points are visible. Use illustrations showing the reflections to justify your answers.

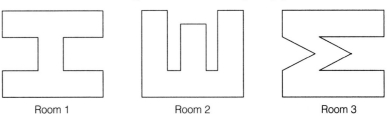

Room 1            Room 2            Room 3

2. **Nine needs three.** Draw a floor plan for a museum having nine sides that needs exactly three guards to watch the entire gallery. Show the placement of the guards in your drawing.

3. **One third again.** If a natural number is written as the sum of three natural numbers, show that one of the numbers in the sum must be less than or equal to one third of the original natural number.

4. **Square museum (S).** If a museum has only right-angled corners, how many guards are necessary? It turns out that you can guard such museums with only one fourth as many guards as corners. In the examples below, where would you place the guards to guard the museum?

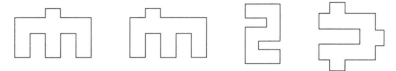

5. **Worst squares (H).** Draw examples of museums with only right-angle corners having 12 sides, 16 sides, and 20 sides that require 3, 4, and 5 guards, respectively.

### III. FURTHER CHALLENGES

1. **Pie are squared.** The circumference of a circle of radius $r$ is $2\pi r$. Suppose you have a round pie, and 1,000 of your best friends come over for dessert. You divide the circular pie into 1,000 equal-size pieces by cutting from the center. Then, before serving them, you rearrange the pieces by putting the first piece with the curved part up, the next piece right next to it with curved edge down, the next one with curved side up, alternating until all 1,000 pieces are arranged.

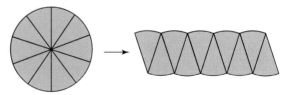

The shape you have constructed is almost an exact rectangle except that its top and bottom edges are each made of 500 just slightly curved tiny segments that came from the edge of the pie. How long is the rectangle? How wide is it? What is its area? Why is this story a convincing demonstration that the formula for the area of a circle is correct?

2. **I can see the light.** Suppose you are in a polygonal mirrored room. Will there always be a point in the room from which, if a light were placed, that light would illuminate the entire room? What is your guess? How does this question compare with the House of Mirrors question mentioned at the close of this section?

## IV. IN YOUR OWN WORDS

1. ***Personal perspectives.*** Write a short essay describing the most interesting or surprising discovery you made in exploring this section's material. If any material seemed puzzling or even unbelievable, address that as well. Explain why you chose the topics you did. Finally, comment on the aesthetics of the mathematics and ideas in this section.

2. ***With a group of folks.*** In a small group, discuss and actively work through the statement and proof of the Klee Art Gallery Theorem. After your discussion, write a brief narrative describing the theorem and proof in your own words.

3. ***Creative writing.*** Write an imaginative story (it can be comical, dramatic, whatever you like) that involves or evokes the ideas of this section.

4. ***Power beyond the mathematics.*** Describe several real-life situations—ideally, from your own experience—for which some of the strategies of thought presented in this section would provide effective methods for approaching and resolving them.

# 4.3 The Sexiest Rectangle:
## Finding Aesthetics in Life, Art, and Math Through the Golden Rectangle

*Geometry has two great treasures: one is the theorem of Pythagoras; the other, the division of a line into extreme and mean ratio. The first we may compare to a measure of gold; the second we may name a precious jewel.*

—Johannes Kepler

On our journeys through various mathematical landscapes we have become conscious of the issue of aesthetics—in particular, the intrinsic beauty of mathematical truths. We're discovering that mathematics is not just a collection of formulas tied together by algebra but is instead a wealth of creative ideas that allows us to investigate, explore, and discover new realms. Now, however, we wonder if mathematics can be used to discover structure behind the aesthetics of art and nature.

### Rectangular Appeal

In our discussion of Fibonacci numbers we asked the following geometrical question that begs to be asked again: What are the dimensions of the most attractive rectangle—the rectangle that we imagine when we think of what a rectangle should look like when we close our eyes on a dark starry night? When someone says "rectangle," we think of a shape. What is it? From the rectangles here, choose the most appealing one:

Given these choices, most people think that the second rectangle from the left is the most aesthetically pleasing—the one that captures the true spirit of rectangleness. That rectangle is referred to as the *Golden Rectangle*. It is the relative length of the base compared to the length of the height of a rectangle that makes it a Golden Rectangle.

What precisely is the ratio of base to height that produces the Golden Rectangle? Recall that, in our conversations about numbers, we found a ratio that was an especially attractive number. That ratio arose in our discussions of the Fibonacci numbers, and we denoted it by the Greek letter phi, $\varphi$. It was called the *Golden Ratio* because it satisfied the symmetrical equation of ratios:

$$\frac{\varphi}{1} = \frac{1}{\varphi - 1}$$

Specifically, we found that the Golden Ratio, $\varphi$, is the number $(1+\sqrt{5})/2 = 1.618\ldots$. You may want to glance back at the Fibonacci discussion in section 2.2 and revisit the relationship $\varphi/1 = 1/(\varphi - 1)$. (The Greek letter $\varphi$ used to denote the Golden Ratio was introduced in the past century to honor the famous ancient Greek sculptor Phidias, much of whose work appears to involve the Golden Ratio.)

The Golden Ratio gives us the satisfying relationship of height to width for those rectangles that many deem extremely pleasing to the eye. The precise mathematical definition of a Golden Rectangle is any rectangle having base $b$ and height $h$ such that

$$\frac{b}{h} = \varphi = \frac{1+\sqrt{5}}{2}$$

We have already discovered how the Fibonacci numbers and the Golden Ratio appear in nature's spirals. Do the proportions of the Golden Ratio make the Golden Rectangle especially attractive and, if so, why? These questions have given rise to heated debate and much controversy. In 1876, Gustav Fechner, a German psychologist and physicist, conducted a study of people's taste in rectangles—a taste test—and found that 35% of the people surveyed selected the Golden Rectangle. So, although the Golden Rectangle seems likely to win an election, we would not expect the outcome to be a landslide.

## The Golden Rectangle in Greece

The Greeks appear to have been captivated by the proportions of the Golden Rectangle as evidenced by its frequent occurrence in Greek architecture and art. As a classic illustration, consider the magnificent Parthenon in Athens, built in the 5th century B.C. The Parthenon today is pretty run down—in fact it's in ruins.

Perhaps you're a step ahead of us, guessing that the big rectangle contained in the Parthenon is a Golden Rectangle.

Actually, if we measure the sides and do the division, we will see that the rectangle is not a Golden Rectangle! So what's the point? Well, when the Parthenon was built, it was much fancier—in particular, it had a roof. Imagine now that the roof is in place. If we form the rectangle from the tip of the rooftop to the steps, we will see a nearly perfect Golden Rectangle.

Another example of the Golden Rectangle in Greek sculpture is the Grecian eye cup. The one pictured is inscribed inside a perfect Golden Rectangle.

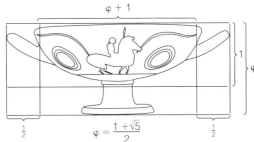

It remains an unsolved question whether Greek artists and designers intentionally used the Golden Rectangle in their work or whether they chose those dimensions solely based on aesthetic tastes. In fact, we are not even certain that such artists were consciously aware of the Golden Rectangle. Although we will likely never know the truth, it is romantic to hypothesize that the Greeks were not conscious of the Golden Rectangle, because it then shows how aesthetically appealing its dimensions are and how we are naturally attracted to such shapes. Some people, however, believe that the occurrence of Golden Rectangle proportions is simply coincidental and random. While some believe that ancient Greek works definitely contain Golden Rectangles, others believe that it is nearly impossible to measure such works or ruins accurately; thus, there is plenty of room for error. In the preceding pictures, all the superimposed rectangles are perfect Golden Rectangles. Was their presence random or deliberate? Are Golden Rectangles really there? What do you think?

## *The Golden Rectangle in the Renaissance*

It appears that mathematicians in the Middle Ages and the Renaissance were fascinated by the Golden Rectangle, but there is much question as to whether this enthusiasm was shared by artists of the time. Leonardo da Vinci was a math enthusiast, but did he know about the Golden Rectangle? Did he deliberately use it in his work? While the historians debate such issues, let's take a look at da Vinci's unfinished portrait of St. Jerome from 1483. In the following reproduction, we have superimposed a perfect Golden Rectangle around the great scholar's body.

Intentional or otherwise, da Vinci selected proportions that were aesthetically appealing, and such dimensions resemble those of the Golden Rectangle. Although we are not certain whether da Vinci intentionally used the Golden Rectangle, we do know that 26 years later he was aware of

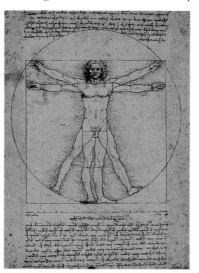

its existence. In 1509, da Vinci was the illustrator for Luca Pacioli's text on the Golden Ratio entitled *De Divina Proportione*. It was famous mainly for the reproductions of 60 geometrical drawings illustrating the Golden Ratio.

The *Divine Proportion* is a synonym for the Golden Ratio. In fact, many people, including Johannes Kepler, referred to the Golden Ratio as the Divine Proportion, or as the *Mean and Extreme Ratio*. Sometimes imaginations ran a bit too wild. Pacioli claimed that one's belly button divides one's body into the Divine Proportion. If you're not ticklish, you can easily check that this is not necessarily true.

## The Golden Rectangle and Impressionism

Let's now leap ahead about 300 years to the creative age of French Impressionism. Painter Georges Seurat was captivated by the aesthetic appeal of the Golden Ratio and the Golden Rectangle. In his painting *La Parade* from 1888, he carefully planted numerous occurrences of the Golden Ratio through the positions of the people and the delineation of the colors.

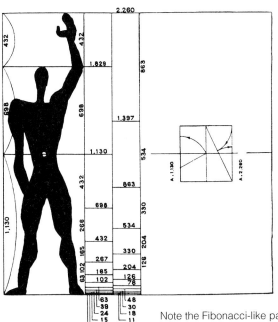

Note the Fibonacci-like pattern in Le Corbusier's 1946 *Modulor* Proportional System: 6+9=15, 9+15=24, etc. . . .

*ABCD, FGHJ, EBIK* are all golden rectangles.

$$\frac{GE}{EA} = \frac{EA}{FE} = \varphi$$

The use of the Golden Ratio in works of art is now known as the *technique of dynamic symmetry*.

## The Golden Rectangle in the 20th Century

In the 20th century, artists were still fascinated with the beautiful proportions of the Golden Rectangle. Abstractionist painter Piet Mondrian was interested in the

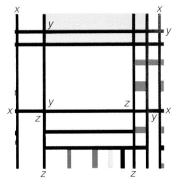

The rectangles marked by *x*'s, *y*'s, and *z*'s are three Golden Rectangles.

"destruction of volume," as illustrated in his *Place de la Concorde* (1938–1943). Hidden in Mondrian's work are at least three Golden Rectangles.

French architect Le Corbusier believed that people are comforted by mathematics. In this spirit, he deliberately designed this villa to conform with the Golden Rectangle.

Le Corbusier was one of the architects involved in the design of the United Nations Headquarters in New York City. Here we again see the influence of the Golden Rectangle in this monolithic structure.

Finally, we note that the Golden Rectangle appears often in other art forms, including musical works. As an illustration, consider the work of French composer Claude Debussy. In his 1894 work "Prelude to the Afternoon of a Faun," he deliberately placed numerous ratios of musical pulses (called *quaver units*) that approximate the Golden Ratio.

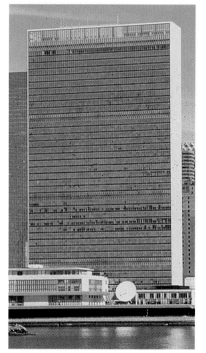

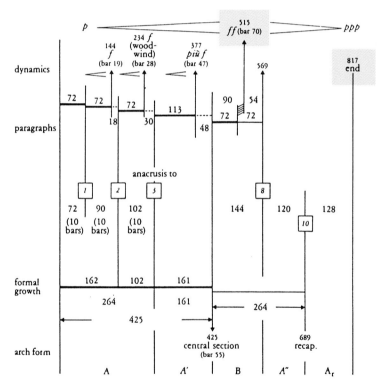

Quaver units for "Prelude to the Afternoon of a Faun."
Note: $\frac{817}{515} = 1.5864\ldots \approx \varphi$

## Why the Appeal?

Why do we see proportions conforming to the Golden Ratio in so many works of art? Let's return to Le Corbusier's villa and notice that the living area creates a large square, whereas the open patio on the left has a rectangular shape. Look what happens when we compare the proportions of the whole villa to the small rectangular patio.

Patio turned on its side and enlarged.

Both are Golden Rectangles!

This rectangular similarity is actually a fundamental and beautiful mathematical property of the Golden Rectangle. This property might explain why the Golden Rectangle is so aesthetically pleasing.

To examine this property in general, let's picture a Golden Rectangle with base equal to $(1+\sqrt{5})/2$ and height equal to 1, so that $b/h = \varphi = (1+\sqrt{5})/2$.

We now divide this Golden Rectangle *aefd* into a square (*abcd*) and a smaller rectangle *befc*. The smaller rectangle is formed by removing that largest square from a Golden Rectangle. We will soon prove that it was the Golden Ratio proportions of the Golden Rectangle that automatically made the smaller rectangle, *befc*, golden!

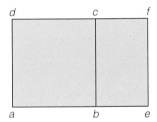

## *An Unexpected Rectangle*

The fact that a Golden Rectangle comprises a square and a smaller Golden Rectangle may well explain its aesthetic appeal. This "self-proliferation" feature represents an attractive regenerating property: If we look at the smaller Golden Rectangle and now remove the largest possible square inside it, we are left with an even smaller perfect Golden Rectangle. Can you visualize continuing this process of removing the square and getting another even smaller Golden Rectangle forever? There is, in some sense, a self-similarity property at work here: At any stage in this process, no matter how small the Golden Rectangle is, when we chop off the biggest square possible, we have created an even smaller Golden Rectangle. We will observe a similar situation when we consider fractals.

Why is this surprising mathematical fact true? It comes from the pleasing algebraic relationship that the Golden Ratio satisfies:

$$\frac{\varphi}{1} = \frac{1}{\varphi - 1}$$

**THE GOLDEN RECTANGLE WITHIN A GOLDEN RECTANGLE.** *If a Golden Rectangle is divided into a square and a smaller rectangle, then the small rectangle is another Golden Rectangle.*

## Proof

Let's begin with our picture of a Golden Rectangle.

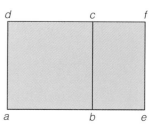

As before, we might as well declare that the length *ad* is 1 unit, and *ae* has length φ. To show that rectangle *befc* is a Golden Rectangle, we must show that the ratio of its longer side to its shorter side, that is, *ef/be*, is φ. So we will need the lengths of the sides of the smaller rectangle. Well, *ef* is easy to figure out: It equals *ad*. So *ef* = 1. What is *be*?

We note that *be* is just *ae* minus *ab*. So,

*be* = *ae* − *ab*.

But *ae* = φ, and *ab* = 1. So,

*be* = φ − 1.

So, the ratio

$$\frac{ef}{be} = \frac{1}{\varphi - 1}.$$

But recall our pleasing identity:

$$\frac{\varphi}{1} = \frac{1}{\varphi - 1}.$$

Therefore, *ef/be* equals φ, and the small rectangle *befc* is indeed a Golden Rectangle. This observation completes our proof.

## Constructing Your Own Golden Rectangle

Perhaps you are now convinced that the Golden Rectangle is aesthetically intriguing and downright cool. You want one for yourself. Sure you can call 1-800-COOL-REC and order one (operators are standing by), but why waste your money? We can make a perfect Golden Rectangle ourselves for free. It may appear that such a perfectly proportioned rectangle would be complicated to create. Not so. In fact, it's easy to construct a perfect Golden Rectangle. Here's how: First we build a square.

Next, we connect the midpoint of the base of the square to the northeast corner of the square with a straight line segment. We then extend the base of the square with a straight line segment off to the east, like a landing strip. We now have a picture that looks like this:

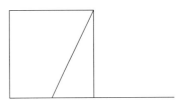

We now let the tilted line fall onto the landing strip by drawing a part of the circle centered at the midpoint of the base and passing through the northeastern corner. We note where the circle portion hits the landing strip. The line segment drawn inside the square from the midpoint to the northeastern corner is actually a radius of the circle arc drawn. We now have the following picture.

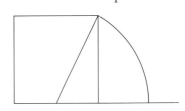

Next, we construct a line perpendicular to the landing strip and passing through the point where the circle hit the landing strip. We then extend the top edge of the square to the right with a straight line until it hits the perpendicular line just drawn. Finally, we erase the excess landing strip to the right of the arc, giving us:

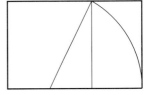

That was pretty easy. Now take a look at that big rectangle we just constructed (we made ours a bit darker). Do you find yourself drawn to that tall, dark, and handsome rectangle? If so, it's all right because that rectangle is a perfectly precise Golden Rectangle.

## Why the Procedure Described Above Produces a Golden Rectangle

We begin by recalling the final picture of our construction and labeling some of the vertices.

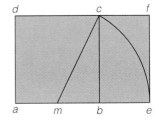

To prove that the dark rectangle, rectangle *aefd*, is really a Golden Rectangle, we must show that the length of *ae* divided by the length of *ad* is equal to the Golden Ratio $(1+\sqrt{5})/2$. So, we want to prove that

$$\frac{ae}{ad} = \frac{1 + \sqrt{5}}{2}$$

Often in life when faced with a difficulty, it is valuable to look for something else that is comparable, but easier to resolve.

◆ ◆ ◆ ◆ ◆ ◆

The size of the rectangle is not important. It is only the ratio of the two sides that matters. We can call the length of *ad* 1 unit and note that this now completely determines the length of everything else in the rectangle. Given this agreement, our goal is to figure out what the length of *ae* is. Notice that *ae* is just *am* plus *me*. If we can find *am* and then *me*, then we will have *ae*, since *ae* = *am* + *me*. Remember that we started with a square, and *m* bisected the bottom side. So *am* = *mb* = 1/2. Great—all we need to do is find *me*.

The truth is that the length of *me* is mysterious. Let's see if we can find another line segment having the exact same length as *me*. Examine the preceding picture and find another line that has the same length as *me*. Try this before reading on.

Did you guess *mc*? If so, great. Note that both *mc* and *me* are radii for the same circle, so the segments must have the same length. Instead of finding the length of *me*, let's find the length of *mc*. Why is this quest easier? The answer is that *mc* is part of a right triangle. In fact, it is the hypotenuse of the triangle *mbc*. Notice that we already saw that *bc* is equal to 1 and *mb* is equal to 1/2. Thus, using the Pythagorean Theorem, we can figure out the length of *mc*. Why not try to figure it out on your own before reading on?

Here we go.

$$(1)^2 + \left(\frac{1}{2}\right)^2 = (mc)^2.$$

That is,

$$1 + \frac{1}{4} = (mc)^2 \quad \text{or} \quad \frac{5}{4} = (mc)^2.$$

Notice the 5 making its debut in this discussion. This development is great news since we want a $\sqrt{5}$ at some point. In fact, note that to solve for *mc* we need to take +/− the square root of both sides, but, because the length of *mc* is positive, we have

$$mc = \frac{\sqrt{5}}{2} \quad (\text{since } \sqrt{4} = 2).$$

This equation is looking good. So, the length of

$$mc = \frac{\sqrt{5}}{2}.$$

Remember that *mc* has the same length as *me*, so,

$$me = \frac{\sqrt{5}}{2}.$$

Therefore,

$$ae = \frac{1}{2} + \frac{\sqrt{5}}{2} = \frac{1 + \sqrt{5}}{2}.$$

Now for the big finish:

$$\frac{ae}{ad} = \frac{\left(\frac{1+\sqrt{5}}{2}\right)}{1} = \frac{1+\sqrt{5}}{2} = \varphi.$$

So, we have a Golden Ratio, which proves we've constructed a perfect Golden Rectangle.

> *Geometry has two great treasures: one is the theorem of Pythagoras; the other, the division of a line into extreme and mean ratio. The first we may compare to a measure of gold; the second we may name a precious jewel.* —Johannes Kepler

## Golden Spirals

We close with one last aesthetically pleasing construction. Let's take a Golden Rectangle and start chopping off the successive squares. As we cut a square, we will draw a quarter of a circle having a radius equal to the side of the square we are cutting. If we do this, we get a spiral.

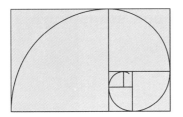

This spiral is called the *logarithmic spiral*, and it occurs in nature in various forms. For example, it occurs in the nautilus sea shell.

The natural and aesthetic beauty of this spiral may be explained mathematically. We first consider the center of the spiral. By the center we mean that point at which the spiral spins around infinitely often—the point that the spiral is heading toward. How can we locate the very center of the spiral?

The location of the center is surprisingly simple to find. We need only draw the diagonal in the largest Golden Rectangle starting from the northwest corner down to the southeast corner and then draw the diagonal in the next largest Golden Rectangle starting in the northeast corner and ending at the southwest corner:

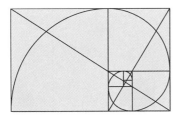

The point where these two diagonals intersect is the precise center of the spiral.

You may also have observed another unexpected fact: All analogous diagonals on all subsequent pairs of Golden

Rectangles lie on the first two diagonals. This follows from the fact that each rectangle has exactly the same proportions. Thus, we see structure and beauty in the construction of the Golden Rectangle and the associated spiral.

What makes the curve of the spiral so appealing? Here is a mathematical fact that may account for its appeal. 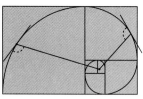 Select any point on the spiral and connect that point with the center of the spiral. Now draw the line that passes through that chosen point on the spiral but just grazes the curve of the spiral (such a line is called a *tangent line*). Notice the angle made by these two lines (the tangent at the point and the line connecting

Angles are always equal!

the point to the center). That angle is always the same, no matter which point on the spiral you selected.

Finally, we note that this beautiful spiral inspired Matisse's 1953 work *L'Escargot*.

We'll now close our discussion of the Golden Rectangle, but not forever. Several other examples of Golden Rectangles occur in most surprising places; but for them we will have to wait until we talk about the Platonic solids.

**A LOOK BACK**

A rectangle is a Golden Rectangle if its ratio of base to height equals the Golden Ratio. If we remove the largest square from a Golden Rectangle, the small remaining rectangle is itself another Golden Rectangle. Thus, we can create a sequence of smaller and smaller Golden Rectangles. This sequence of Golden Rectangles leads to spirals that occur in nature.

We can build a Golden Rectangle by starting with a square and elongating it by using a simple geometric procedure. We can verify that the ratio of base to height is the Golden Ratio by applying the Pythagorean Theorem.

Art, aesthetics, geometry, and numbers all meet in the Golden Rectangle. Its appealing proportions appear in art throughout history and in nature. Do the mathematical properties of the Golden Ratio somehow create the beauty of the Golden Rectangle? Some ideas span the artificial boundaries of subjects—in this case from the algebra of numbers (the Golden Ratio) to the geometry of rectangles (the Golden Rectangle). Seeking connections across disciplines often leads to new insights and creative ways of understanding.

Take ideas from one domain and explore them in another.

# MINDSCAPES  Invitations to Further Thought*

## I. SOLIDIFYING IDEAS

1. ***In search of gold.*** Find at least three examples of Golden Rectangles in your surroundings. If possible, include photographs or sketches and estimates of the ratio of base to height for each example.

2. ***Golden art.*** In the masterpiece *Paris Street; Rainy Day* by Caillebotte (1877), find all the Golden Rectangles you can.

3. ***A cold tall one?*** Can a Golden Rectangle have a shorter base than height? Explain your answer.

4. ***Fold the gold.*** Suppose you have a Golden Rectangle cut out of a piece of paper. Now suppose you fold it in half along its base and then in half along its width. You have just created a new, smaller rectangle. Is that rectangle a Golden Rectangle? Justify your answer.

5. ***Sheets of gold.*** Suppose you have two sheets of paper, an unmarked straightedge, and a pair of scissors. Explain how you can use one of the sheets of paper and the straightedge to construct a perfect Golden Rectangle on the other sheet. (*Hint:* You may cut the first piece of paper.)

6. ***Circular logic?*** (**H**). Take a Golden Rectangle and draw the largest circle inside it that touches three sides. The circle will touch two opposite sides of the rectangle. If we connect those two points with a line and then cut the rectangle into two pieces along that line, will either of the two smaller rectangles be a Golden Rectangle? Explain your reasoning.

7. ***Growing gold*** (**H**). Take a Golden Rectangle and attach a square to the longer side so that you create a new larger rectangle. Is this

---

*In the Mindscapes section, exercises marked (H) have hints for solutions at the back of the book. Exercises marked (S) have solutions.

new rectangle a Golden Rectangle? What if we repeat this process with the new, large rectangle?

8. **Counterfeit gold?** Draw a rectangle with its longer edge as the base (it could be a square, it could be a long and skinny rectangle, whatever you like,

Attach a big square.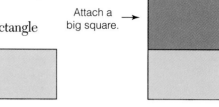

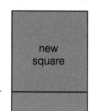

but we suggest that you do not draw a Golden Rectangle). Now using the top edge of the rectangle, draw the square just above the rectangle so that the square's base is the top edge of the rectangle. You have now produced a large new rectangle (the original rectangle together with this square sitting above it). Now attach a square to the right of this rectangle so that the square's left side is the right edge of the

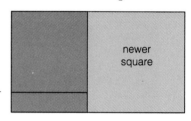

large rectangle. You've constructed an even larger rectangle. Now repeat this procedure—that is, append to the top of this huge rectangle the largest square you can and follow that move by attaching the largest square you can to the right of the resulting rectangle. Start with a small rectangle near the bottom left corner of a page and continue this process until you have filled the page. Now measure the dimensions of the largest rectangle you've built and divide the longer side by the shorter one. How does that ratio compare to the Golden Ratio? Experiment with various starting rectangles. What do you notice about the ratios?

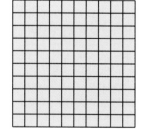

9. **In the grid** (S). Consider the 10 × 10 grid at left. Find the four dots that, when joined to make a horizontal rectangle, make a rectangle that is the closest approximation to a Golden Rectangle. (*Challenge*: What if the rectangle doesn't need to be horizontal?)

10. **A nest of gold.** Consider the figure of infinitely nested Golden Rectangles. Suppose we remove the largest square, and, with the rectangle that remains, we enlarge the entire picture so that its size is identical to the original rectangle. How will that enlarged picture compare to the original figure? Explain your answer.

## II. CREATING NEW IDEAS

1. **Comparing areas.** Let G be a Golden Rectangle having base $b$ and height $h$, and let G' be the smaller Golden Rectangle made by removing the largest square possible from G. Compute the ratio of

the area of G to the area of G'. That is, compute Area(G)/Area(G'). Does your answer really depend on $b$ and $h$ (the original size of G)? Are you surprised by your answer?

2. **Do we get gold?** Let's make a rectangle somewhat like the Golden Rectangle. As before, start with a square; however, instead of cutting the base in half, cut it into thirds and draw the line from the upper right vertex of the square to the point on the base that is one third of the way from the right bottom vertex. Now use this new line segment as the radius of the circle, and continue as we did in the construction of the Golden Rectangle. This produces a new, longer rectangle:

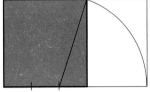

What is the ratio of the base to the height of this rectangle (that is, what is base/height for this new rectangle)? Now remove the largest square possible from this new rectangle and notice that we are left with another rectangle. Are the proportions of the base/height of this smaller rectangle the same as the proportions of the big rectangle?

3. **Do we get gold this time?** (S) We now describe another construction of a different type of rectangle. It is exactly the same as the Golden Rectangle except that, instead of starting with a square, we begin with a rectangle whose base is twice as long as its height. Now connect the midpoint of the base to the upper right vertex with a line, and use this line as the radius of the circle and continue as we did in the construction of the Golden Rectangle. This produces a new, longer rectangle:

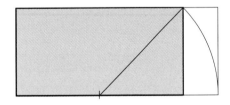

What is the ratio of the base to the height of this new big rectangle (that is, what is base/height for this new rectangle)? Now remove the original rectangle we started. This gives us a new, smaller rectangle. Are the proportions of the base/height of this smaller rectangle the same as the proportions of the big rectangle? Experiment with other starting rectangles.

4. **A silver lining?** (H) Consider the diagonal in the following Golden Rectangle and draw in the largest square possible. Notice that one edge of the square cuts the diagonal into two pieces. What is the ratio of the length of the entire diagonal to the length of the part of the diagonal that is inside the square? That is, compute the length of the entire diagonal divided by the length of the part of the diagonal that is inside the square. Surprised?

5. *Cutting up triangles.* Draw any right triangle. Find a way of cutting up that triangle into four identical triangles such that each one is exactly like the original large triangle except that it is scaled down to one fourth the area.

### III. FURTHER CHALLENGES

1. *Going platinum.* Determine the dimensions of a rectangle such that, if you remove the largest square, then what remains has a ratio of base to height that is twice the ratio of base to height of the original rectangle.

2. *Golden triangles.* Draw a right triangle with one leg twice as long as the other leg. This triangle is referred to as a *Golden Triangle*. Suppose that one leg has length 1 and the other has length 2. What is the length of the hypotenuse? Next draw a line from the right angle of the triangle to the hypotenuse such that the line is perpendicular to the hypotenuse.

   Now cut up the larger of the two new right triangles into four triangles (see "Cutting up triangles," above). Demonstrate that all five triangles are the same size and are Golden Triangles. We will use this neat cutting up of the Golden Triangle in the following section.

### IV. IN YOUR OWN WORDS

1. *Personal perspectives.* Write a short essay describing the most interesting or surprising discovery you made in this section's material. If any material seemed puzzling or even unbelievable, address that as well. Explain why you chose the topics you did. Finally, comment on the aesthetics of the mathematics and ideas in this section.

2. *With a group of folks.* In a small group, discuss and actively work through the geometric construction of the Golden Rectangle. After your discussion, write a brief narrative describing the construction in your own words.

3. *Creative writing.* Write an imaginative story (it can be comical, dramatic, whatever you like) that involves or evokes the ideas of this section.

4. *Power beyond the mathematics.* Describe several real-life situations—ideally, from your own experience—for which some of the strategies of thought presented in this section would provide effective methods for approaching and resolving them.

# 4.4 Soothing Symmetry and Spinning Pinwheels:
## Can the Floor Be Tiled Without Any Repeating Pattern?

*The mathematical sciences particularly exhibit order, symmetry, and limitation; and these are the greatest forms of the beautiful.*

—Aristotle

Understanding the world often comes down to discovering pattern and order. When we perceive that the orbits of the planets are ellipses or that crystals are made of orderly arrangements of molecules, we feel that we have detected something significant about the structure of nature. Our sense of beauty is centered on balance and harmony, but sometimes things affect us through their departure from expected order. People have lived with regular patterns for centuries, but only recently have we created new jumbled patterns that straddle order and chaos. These patterns without symmetry are products of the imagination with a haunting, unsettling beauty of their own. And again we find that an exploration driven by abstract intrigue ends in ideas that have an uncanny ability to describe our real physical world.

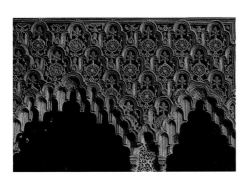

Before considering chaos, let's consider order. What makes the artistic works on the previous page so alluring? The answer, in short, is symmetry.

We begin by describing what symmetry is in general, and then we define its meaning more precisely. By pinning down the meaning of symmetry, we allow ourselves to develop variations of it, including a whole new world of patterns that paradoxically combine order with chaos.

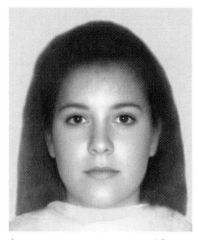

An average computer generated face made from 32 actual faces.

## Symmetry

We are drawn toward the symmetric. Recently psychologists performed experiments to show that facial symmetry is an important part of what makes beautiful people beautiful and what influences us in choosing a mate. How does one study human symmetry and beauty? Judith Langlois of The University of Texas at Austin wanted to know what faces were considered beautiful, so she showed people different photographs of faces and recorded their reactions. We might think that the most beautiful faces would be extreme in some way—extremely thin, or extremely chiseled, or extremely something else. But the faces most often chosen as beautiful were extremely . . . well, average and symmetrical. Average—not extreme. Dr. Langlois created the most symmetrical and most average faces she could by creating composites of many photographs of people. The averaged face was completely symmetrical and was the one most preferred. Thus, the love of symmetry may literally be in our genes.

When we turn from faces to patterns in floor tiles, wall coverings, and paintings, again we are drawn toward the symmetric. In most classical patterns, the symmetry is soothing. However, modern patterns, such as the Penrose Pattern and the Pinwheel Pattern, shown on pages 253 and 254, combine symmetry with chaos to create a disturbing yet hypnotic dissonance.

*Symmetry Drawing E85* by M.C. Escher.

## What Is Symmetry?

We sense symmetry when small portions of a pattern are repeated elsewhere in the overall pattern. Let's imagine that the Escher drawing of devils, frogs, and fish has been extended indefinitely to cover the entire plane. Each fish, frog, and devil has a symmetry of its own, but the symmetry of the whole pattern lies in the regularity with which the devils, fish, and frogs appear over the whole plane. Imagine that

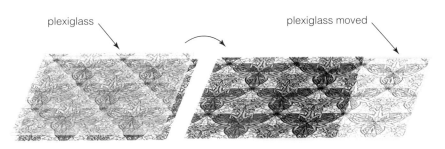

we cover the entire plane with a sheet of rigid plexiglass. Suppose we trace every fish, frog, and devil onto the plexiglass. Now let's imagine sliding the plexiglass over one fish to the right. What would we see? We'd notice that each fish on the plexiglass matches up perfectly with a fish on the plane and each plexiglass frog matches a plane frog, and similarly with those little devils. The plexiglass experiment confirms our intuition that the pattern has symmetry.

To refine this idea, we'll say that *a rigid symmetry* of a pattern in the plane is a motion of the plane that preserves the pattern and does not shrink, stretch, or otherwise distort the plane. In other words, it is a motion of a plexiglass copy of the pattern that ends up in a position that again exactly matches the pattern. A rigid symmetry could be a shift, a rotation, a flip, or any combination of these.

By tracing the different patterns onto transparencies, you can try to discover the various rigid symmetries (or lack thereof) in the illustrations at the opening of this section.

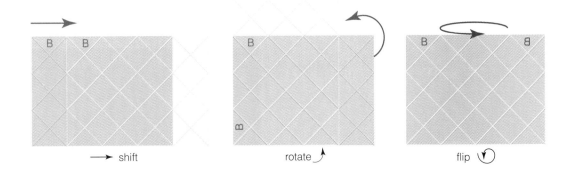

## A Symmetry of Scale

One day in Wonderland, Alice found a cake helpfully labeled, "Eat Me," which she dutifully did. She quickly grew to five times her height. Although it is not explicitly mentioned in *Alice in Wonderland*, surely she was standing on a floor tiled with square tiles. When she had finished her growth spurt, she looked down at the distant floor expecting to feel giddy because of her new altitude. However, much to her surprise, she had no such sensation. Instead, by squinting, she saw that the original square tiles of the floor could be amalgamated in groups of five by five, so that these

bigger super-tiles formed a pattern in the plane that looked, from her new height, exactly like the original pattern.

Taking another bite of the cake, she grew to five times her new height. But again, the super-tiles could be grouped to form super-super-tiles whose pattern of covering the plane looked identical to the original pattern. No matter how much cake Alice ate—and she packed it away—the pattern, when properly grouped, produced an identical pattern but on a larger scale. Being mathematically inclined, Alice noted that this feature of the pattern revealed a symmetry of scale. Being in Wonderland, she wondered if symmetry of scale was related to the cool pictures of self-similar objects that she saw while thumbing through Chapter 6 of this book. She had, however, already learned that, after discovering an idea, one should pin it down with a definition.

We'll say that a pattern in the plane has *a symmetry of scale* or *is scalable* if the tiles that make up the pattern can be grouped into super-tiles that still cover the plane and, if scaled down, can be rigidly moved to coincide with the original pattern.

Several simple patterns possess this new kind of symmetry. The square pattern and the design composed of equilateral triangles (see page 261) clearly illustrate the concept of duplicating the same pattern at a larger scale by grouping the original tiles together to create super-tiles. This duplication of pattern through grouping seems to indicate a high level of organization, regularity, and symmetry. Or at least so it appears.

## A Strange Question

Suppose we have a pattern that is so regular that it has a symmetry of scale. Must that pattern also have rigid symmetries?

Anyone's first attempts at making a pattern with a symmetry of scale involve putting the tiles down in a highly regular pattern. But one of the most remarkable recent developments in geometry shows us that the world is more interesting and varied than we might first guess. Scalable patterns can be so chaotic that they have no rigid symmetries at all! These scalable patterns with no rigid symmetries are bizarre and intriguing modern creations.

## *Chaotic Patterns*

In the mid-1960s, Robert Berger, in his Ph.D. thesis in applied mathematics at Harvard University, constructed some examples of patterns that had no rigid symmetries and yet had a scalable property. These examples used thousands of different shapes of tiles, and they opened the door to an exploration of what types of exotic patterns in the plane might be possible. A decade later, Roger Penrose constructed famous patterns, called *Penrose Patterns*, with no rigid symmetries that used only two tile shapes, referred to as *kites* and *darts*.

 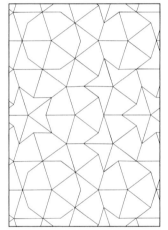

Here are pictures of two patterns created by Penrose tiles. No covering of the plane by Penrose kites and darts has any rigid symmetries; nevertheless, these Penrose Patterns do not look completely chaotic. Part of the reason for their somewhat stable appearance is that, in any Penrose Pattern, every tile occurs in one of 10 possible orientations in the plane. But even that amount of regularity can be absent in some truly exotic patterns.

In 1994, John Conway of Princeton University and Charles Radin of The University of Texas at Austin described another exotic tiling of the plane called the *Pinwheel Pattern* using one single triangular tile. This Pinwheel Pattern not only has a symmetry of scale and has no rigid symmetry, but, in addition, the tiles occur in infinitely many orientations, thereby generating the sense of a pinwheel spiraling out of control from any place you look. The Pinwheel Pattern gives the impression of a bewildering and disturbing jumble. Only by looking carefully can we find the way to group the triangles by fives to produce the super-tiles that demonstrate its symmetry of scale.

## *Construction of the Pinwheel Pattern*

How can we construct such a pattern? Where would we start to build a pattern that has a symmetry of scale but has no rigid symmetry? The answers are that we think ahead and build in both scalability and lack of rigid symmetry during our construction.

For a pattern to have a symmetry of scale, we must be able to group the tiles to create super-tiles of the same shape as the original tiles but larger. In the case of the Pinwheel Pattern, the basic tile is pretty simple. It is a right triangle with one leg twice as long as the other, which we'll call a *Pinwheel Triangle*. By the Pythagorean Theorem we see that a Pinwheel Triangle with legs of lengths 1 and 2 has a hypotenuse equal to $\sqrt{5}$. Those specific proportions are chosen because five identical Pinwheel Triangles can be assembled to form a larger Pinwheel Triangle—that is, another right triangle with one leg twice as long as the other. We'll call that large Pinwheel Triangle made up of five smaller Pinwheel Triangles a *5-unit Pinwheel Triangle*. The phrase "5-unit" reminds us how many of the original-sized tiles were used to make a super-tile. Each of the five triangles in the 5-unit Pinwheel Triangle is in a particular position, and we label those positions 1, 2, 3, 4, and Interior, as illustrated. We will always arrange the five triangles making up a 5-unit Pinwheel Triangle in exactly the arrangement shown, which we call the *T-arrangement*, recognizing the T whose stem is the edge between triangles 4 and Interior, and whose top is an edge of triangle 3.

## Building Super-Tiles

We are now ready to describe how to tile the entire plane with the Pinwheel Pattern. We begin with a single Pinwheel Triangle. Then we surround it with four other identical Pinwheel Triangles to create a 5-unit Pinwheel Triangle in the T-arrangement such that our original triangle is the Interior triangle of the 5-unit Pinwheel Triangle. This strategy specifies for us how to lay the first five triangles. After those five tiles have

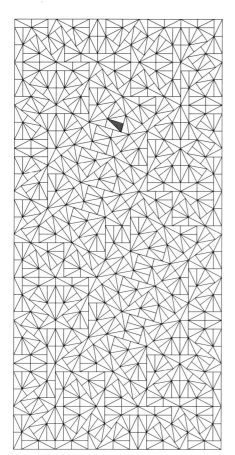

been laid, how would we decide where to lay the next 20 tiles?

The answer requires us first to note that, after we lay the first five tiles, we have created one 5-unit Pinwheel Triangle. We now surround this 5-unit Pinwheel Triangle with four other 5-unit Pinwheel Triangles in the T-arrangement. We have just created a 25-unit Pinwheel Triangle such that our original 5-unit triangle is in the Interior position of the 25-unit Pinwheel Triangle. Each of the four 5-unit Pinwheel Triangles that we just put down are made up of five Pinwheel Triangles in the T-arrangement. Thus, we have just placed down 20 tiles.

We can determine the location of the next 100 tiles by surrounding the 25-unit Pinwheel Triangle that we just created by four other 25-unit Pinwheel Triangles in the T-arrangement, thus creating one 125-unit Pinwheel Triangle. Notice in the illustrations of this process that the successively larger triangles cover increasingly larger regions around the original tile so that the process can be continued forever to cover the entire plane. Also, observe how the increasingly larger Pinwheel super-Triangles spin around in different orientations as they grow, giving the pattern its name. After continuing this process forever, the resulting pattern in the plane is the Pinwheel Pattern.

## *The Pinwheel Pattern Has a Symmetry of Scale*

Let's group the triangles of the Pinwheel Pattern into groups of five using the T-arrangement so that the 5-unit Pinwheel Triangles form an identical Pinwheel Pattern in the plane but at a larger scale. To begin, we hunt down the original triangle. Then we find the four surrounding triangles that

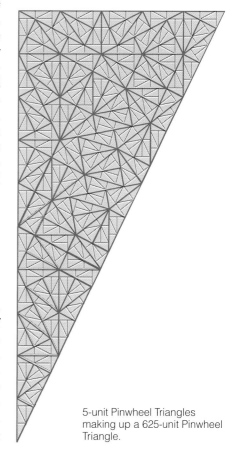

5-unit Pinwheel Triangles making up a 625-unit Pinwheel Triangle.

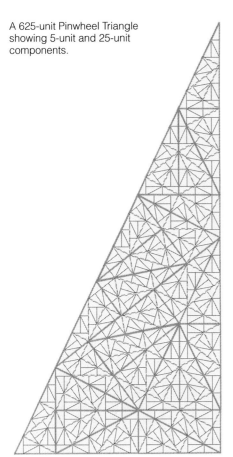

A 625-unit Pinwheel Triangle showing 5-unit and 25-unit components.

make up the first 5-unit Pinwheel Triangle. That 5-unit Pinwheel Triangle is the first of the 5-unit super-tiles that will make up a Pinwheel Pattern in the plane but at a larger scale. The next stage of the construction of the Pinwheel Pattern consists of placing four more 5-unit Pinwheel Triangles in the T-arrangement to construct a 25-unit Pinwheel Triangle. Those four 5-unit Pinwheel Triangles are the next four of the 5-unit super-tiles.

The construction proceeded by surrounding the 25-unit Pinwheel Triangle by four more 25-unit Pinwheel Triangles, each of which was composed of five 5-unit Pinwheel Triangles.

Those five 5-unit Pinwheel Triangles in each of the four new 25-unit Pinwheel Triangles are the next 20 5-unit super-tiles. This process can be continued to amalgamate all the triangles in the Pinwheel Pattern into groups of 5-unit super-tiles. These create a Pinwheel Pattern of the plane that is identical, up to scaling, to the original Pinwheel Pattern.

The Pinwheel Pattern looks jumbled and disorderly. In fact, it's not easy to pick out the super-tiles from the jumble. But they are there, and we ask you now to take a blue marker and trace the super-tiles, each comprising five of the original Pinwheel Triangles. After you have traced these super-tiles, you will have drawn a new, blue pattern of the plane with tiles five times as large as the original ones. Now take a green marker and group the blue tiles in fives to create super-super-tiles, each containing 25 of the original tiles. These green super-super-tiles form yet another Pinwheel Pattern of the plane. Of course, this process of creating larger and larger super-tiles can continue indefinitely. The only way to get a real sense of this amalgamation process and its subtlety is by actually drawing the super-tiles in. Take the time to explore and enjoy the gentle structure of this tundra of triangles.

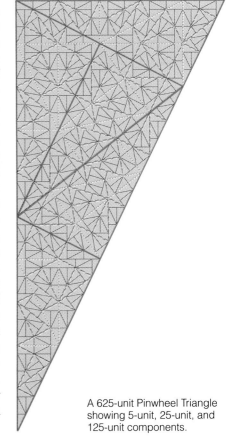

A 625-unit Pinwheel Triangle showing 5-unit, 25-unit, and 125-unit components.

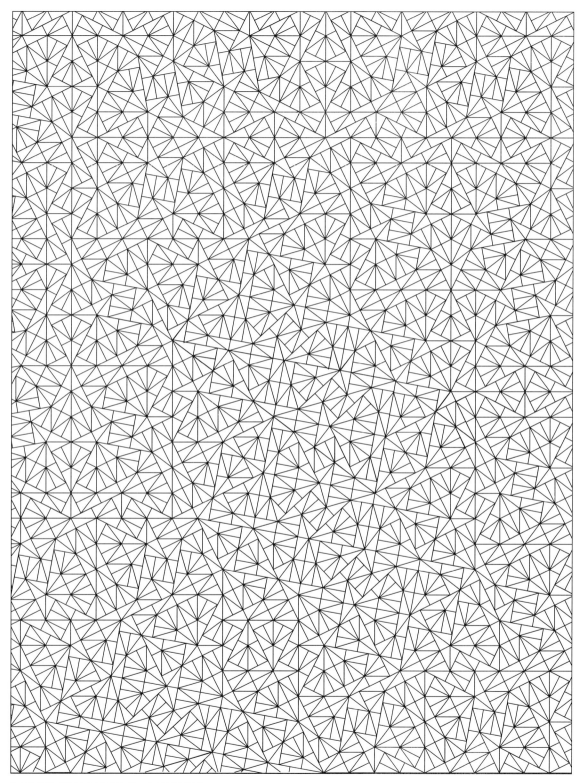

A random piece from the Pinwheel Pattern.

## *Different Groupings?*

In many patterns that possess a symmetry of scale, it is possible to group the tiles in several different ways to demonstrate the symmetry of scale. For example, consider the square pattern. We could group the tiles into a two-by-two square of four red tiles to create 4-unit super-tiles that again tile the plane in the square pattern. However, we could also shift the 4-unit groupings diagonally down one unit, for example, to create a different way of grouping the tiles into 4-unit super-tiles. An upper-right tile in the first grouping would become a lower-left tile in the alternative grouping. Thus, in the case of this pattern of squares, many different ways exist to group the original tiles into 4-unit super-tiles to create a new square pattern that demonstrates its symmetry of scale.

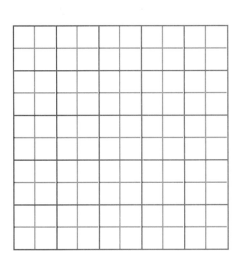
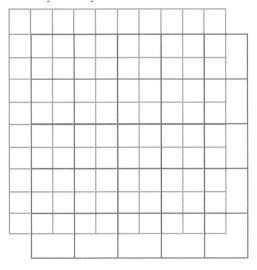

However, as you identified the 5-unit super-tiles in the Pinwheel Pattern, you may have noticed that you had no choice as to how to group the five triangles together to create the super-tile. In fact, there is only one way to group them together.

> **UNIQUENESS OF SCALING.** *There is only one way to group the Pinwheel Triangles into super-tiles to create a Pinwheel super-pattern in the plane.*

The symmetry of scale of the Pinwheel Pattern shows us that there is a way to group the Pinwheel Triangles into fives to create a Pinwheel super-pattern of the plane. We want to show now that it is impossible to group the triangles into 5-unit, T-arrangement super-tiles in any other grouping.

Here we have labeled the five triangles that make up a 5-unit super-tile as 1, 2, 3, 4, and Interior. Would it be possible for an Interior triangle to be part of a 5-unit group in some position other than the Interior position? *Hint*: the answer is no. To see that no 5-unit group could contain an Interior triangle in positions 1, 2, 3, or 4, we simply consider all the possibilities.

Let's look at the Pinwheel Pattern, find any 5-unit super-tile, and locate the Interior triangle of that super-tile. Now let's explore the possibility that, by grouping the tiles differently, the Interior triangle might be a part of a different 5-unit super-tile. That different super-tile would, of course, overlap the first super-tile, and the Interior triangle would coincide with a triangle in position 1, 2, 3, or 4 of the other super-tile. Here we see what would happen if triangle 1 of a supposed purple super-tile B were to coincide with the Interior triangle of a gold super-tile A. As we see, the other tiles do not fit. So we know that the Interior triangle of gold super-tile A cannot be part of another grouping that puts it in position 1 of purple super-tile B.

Could the Interior triangle of gold super-tile A be in position 2 of a supposed purple super-tile B? Again, by overlaying the position-2 tile of purple super-tile B over the Interior tile of super-tile A, we see that the other tiles do not line up, so such an alternative grouping is not possible.

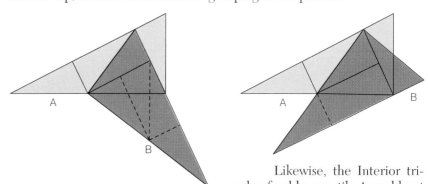

Likewise, the Interior triangle of gold super-tile A could not be part of purple super-tile B in either position 3 or position 4, as seen here. Consequently, each Interior triangle must always be an Interior triangle, and the super-tile of which it is a part must be the super-tile in the original super-pattern of the Pinwheel Pattern. So, no alternative grouping is possible.

We now know that it is impossible to group the Pinwheel triangles into 5-unit, T-arrangement super-tiles in any other grouping. We can use that fact to see that we also cannot group the Pinwheel tiles into 25-unit super-super tiles in any other arrangement. Likewise, there is no alternative grouping of the tiles of the Pinwheel Pattern into 125-unit super-super-super-tiles, and so on. Thus the groupings of the tiles in the Pinwheel Pattern into super-tiles are unique. This observation, in turn, demonstrates that its symmetry of scale is unique and completes our proof.

## *Rigid Symmetries?*

We now turn our attention to the rigid symmetries of the Pinwheel Pattern of the plane. We will soon discover that the uniqueness of grouping result we just proved will be a key insight into resolving the rigid symmetry issue. We have seen that the Pinwheel Pattern enjoys a symmetry of scale. Let's now explore other symmetries that the Pinwheel Pattern might exhibit. Photocopy the Pinwheel Pattern onto a transparency, and move the transparency in search of translational, rotational, or reflectional symmetries of the Pinwheel Pattern. Mark a tile as your base tile. Move the transparency so that the base tile lies above some other tile in the pattern. Now look at the other tiles on your transparency. Do they all lie directly over an identical tile in the picture? If not, try again. You can, in fact, get large groups of tiles around your moved base tile to line up. But do all the triangles on your transparency coincide with a triangle on the page?

Perhaps that's enough trying, because we could try until the proverbial cows come home and we would never get the transparency pattern to cover the entire original pattern. It cannot be done! The Pinwheel Pattern has no rigid symmetries at all!

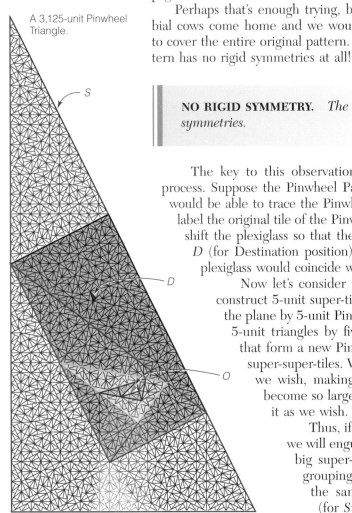

A 3,125-unit Pinwheel Triangle.

> **NO RIGID SYMMETRY.** *The Pinwheel Pattern has no rigid symmetries.*

The key to this observation is the uniqueness of the grouping process. Suppose the Pinwheel Pattern had a rigid symmetry. Then we would be able to trace the Pinwheel Pattern onto a piece of plexiglass, label the original tile of the Pinwheel Pattern *O* (for Original position), shift the plexiglass so that the triangle *O* covers a different triangle *D* (for Destination position), and every Pinwheel Triangle on the plexiglass would coincide with a Pinwheel Triangle on the plane.

Now let's consider the grouping process again. Let's first construct 5-unit super-tiles to form a new Pinwheel Pattern in the plane by 5-unit Pinwheel Triangles. Next let's group those 5-unit triangles by fives to create 25-unit super-super-tiles that form a new Pinwheel Pattern of the plane by 25-unit super-super-tiles. We can continue this process as long as we wish, making the super-tile with triangle *O* in it become so large that it contains as big an area around it as we wish.

Thus, if we continue this process long enough, we will engulf both triangles *O* and *D* in one really big super-duper-tile. Let's do sufficiently many groupings so that both triangles *O* and *D* are in the same super-duper Pinwheel Triangle *S* (for Super-duper).

Notice that our supposed rigid symmetry rotated and slid all the original triangles of the Pinwheel Pattern so they exactly coincided with other original Pinwheel Triangles. So, if we outlined with a marker the super-duper-triangles created by grouping, then the supposed rigid symmetry would take each such super-duper-triangle onto another super-duper-triangle. Therefore, super-duper-triangle S would be shifted to some other super-duper-triangle T. But remember that there is only one way to group the triangles in a Pinwheel Pattern to construct Pinwheel Patterns of a larger scale. If we could move S to T, then we would have constructed two different groupings of the original tiles into super-duper-tiles—the one grouping that contains S and the other grouping that contains T. Since S and T overlap, they could not be two different super-duper-triangles in the same grouping—they would represent two different ways to group the original tiles into super-duper-tiles. This statement contradicts the *Uniqueness of Scaling* observation, and thus our supposed rigid symmetry cannot possibly exist. Therefore, we conclude that the Pinwheel Pattern can have no rigid symmetry.

The previous argument requires a great deal of mental and visual effort to absorb. But once we understand the basic idea of the overlapping large triangles and remember the uniqueness of scaling, the concepts fall into place.

*It requires time and effort to make ideas and strategies our own.*

◆ ◆ ◆ ◆ ◆ ◆

## A Deduction from the Proof

The proof shows us a feature of patterns that possess symmetry of scale and *do* have rigid symmetries: Namely, there must be more than one way to group the tiles to create the super-tiles. We just showed that, if there is only one grouping of tiles into super-tiles, then that pattern would not have any rigid symmetries. Take the square pattern or the equilateral triangle pattern, for example, and see how to group the tiles into super-tiles in several different ways. This variety is shown in the illustrations.

## Exotic Patterns in Nature

Patterns with no rigid symmetries are fascinating objects in their own right, and, as often happens with intriguing mathematical insights, they may also have implications in nature. In this case, these exotic patterns may reflect the geometric structure of a certain class of materials called *quasicrystals*. The connection between patterns without rigid symmetries and the structure of quasicrystals was first noticed in 1984 when scientists observed a diffraction pattern in the quasicrystal alloy $Al_{5.1}Li_3Cu$ that showed an unusual five-fold symmetry.

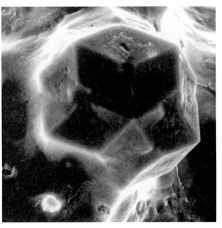

The similar look of this quasicrystal and this aperiodic Penrose Pattern suggested the connection. As with many other developments in mathematics, perhaps these exotic patterns will be among those that at first seem rare and bizarre and later are found everywhere. These exotic patterns will become better understood as more examples are developed and more subtle properties of them are explored.

## *Patterns in Three Dimensions*

As an epilogue to the issues of patterns in the plane, we take a brief, pictorial look at patterns in three-dimensional space. Cubes, of course, will do the job in an orderly manner full of symmetries of all kinds. Here is a surprising pattern constructed by Colin Adams of Williams College that fills space with identically shaped, fattened solid knots. The drawing of the knotted tile on the left is the work of Pier Gustafson.

Finally, here is a pattern in space that has no rigid symmetries. This pattern is a three dimensional analogue of the Pinwheel Pattern in the plane. The picture has a wonderfully jumbled look that may better reflect reality than the orderly, perhaps artificial, neatness of the cubical filling of space.

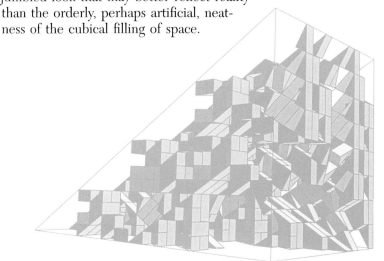

262 ◆ GEOMETRIC GEMS

## A LOOK BACK

Patterns in the plane can display at least two types of symmetries—rigid symmetries and symmetries of scale. Surprisingly, some patterns have symmetry of scale but do not have any rigid symmetries at all. The Pinwheel Pattern is constructed by building increasingly larger super-tiles of the same triangular shape. The tiles can be grouped into super-tiles in only one way. That uniqueness of grouping is the key to showing that the Pinwheel Pattern has no rigid symmetries.

We developed these exotic patterns by starting with a vague idea of symmetry and then pinning it down. Once we defined our concept more precisely, we could identify different types of symmetries and we could then develop new patterns that we had not known before.

*Specifying the meaning of a familiar term or notion can open our eyes to new possibilities and ideas.*

❖ ❖ ❖ ❖ ❖ ❖ ❖ ❖ ❖ ❖ ❖ ❖ ❖ ❖ ❖ ❖ ❖ ❖ ❖

# MINDSCAPES  Invitations to Further Thought*

### I. SOLIDIFYING IDEAS

1. ***Build a super.*** Draw a 1, 2, $\sqrt{5}$ right triangle in the center of a page. Draw four identical right triangles around it to create a similar right triangle that is a 5-unit super-triangle in the T-arrangement.

2. ***Another angle.*** Look at the 5-unit super-tile you created in the last question. Measure the angle between the line determined by the base of the original triangle and the base of the 5-unit super-tile.

3. ***Super-super.*** Surround your 5-unit super-tile with 20 more right triangles, each identical to the original triangle, to construct a 25-unit super-super-triangle.

4. ***Expand forever*** **(H)**. If you continue the process of the last two questions, why can you cover the whole plane?

5. ***Triangular expansion.*** Take an equilateral triangle. Surround it by three other identical equilateral triangles to create another equilateral triangle. Which way is it facing?

6. ***Expand again.*** Take your 4-unit equilateral triangle and surround it with 12 equilateral triangles to create a 16-unit super-triangle. Which way is it oriented?

---

*In the Mindscapes section, exercises marked (H) have hints for solutions at the back of the book. Exercises marked (S) have solutions.

**7. One-answer supers.** Here is a Pinwheel Pattern. For each filled-in tile, outline the surrounding tiles that create the 5-unit super-tile and the 25-unit super-tile of which it is a part. There is only one correct answer.

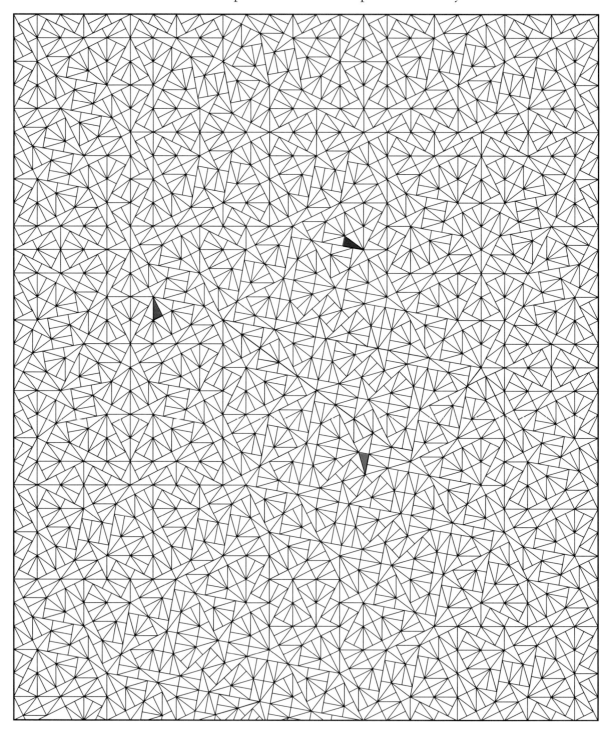

**8. Close to periodicity (S).** If you cover the Pinwheel Pattern with a transparency and shift it to make tile ★ cover tile 1, how many tiles near the moved ★ line up before they don't?

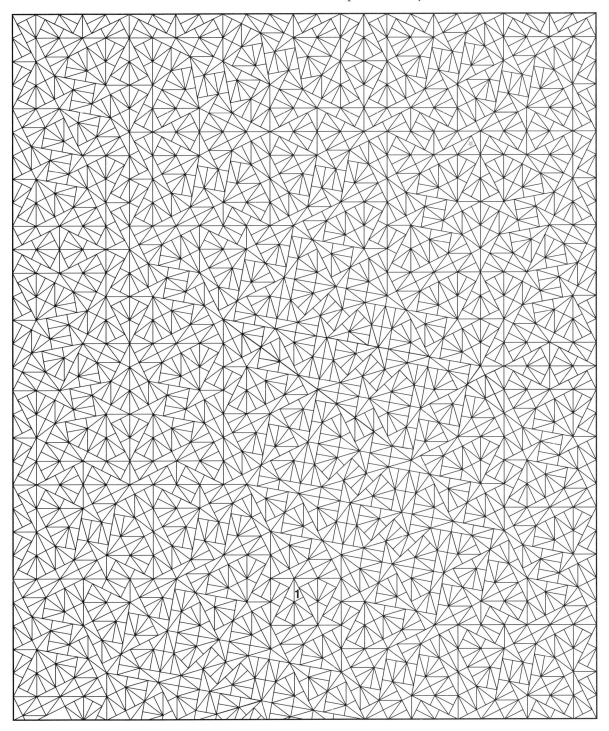

9. ***Golden periodicity*** (**H**). Can you construct a periodic pattern that covers the plane and is made from the 1, 2, $\sqrt{5}$ Golden Triangles?

10. ***Many answer supers.*** Here are pictures of the square and equilateral triangle patterns. For each tile indicated, outline surrounding tiles that create 4-unit super-tiles and 16-unit super-tiles. In each case, show that there is more than one correct answer.

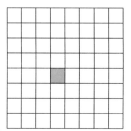
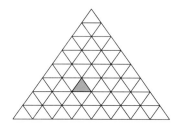

## II. CREATING NEW IDEAS

1. ***Personal Escher*** (**S**). Make your own Escherlike tiles that can cover the plane. Begin with the square pattern and deform the top edge a bit. Why does that require that the bottom edge also be modified? How does that change all the tiles? Note how a change to an edge propagates to every tile in the plane. Distort the square tiles to become Escheresque, fanciful shapes.

2. ***Fill 'er up?*** (**H**) For each set of tiles below, could those tiles be assembled to create a pattern filling the plane? If so, indicate how.

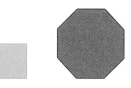
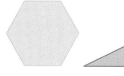
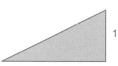

3. ***Friezing.*** In the patterns shown below, can you find a symmetry that shifts every point to the right? Can you find one that flips the right half and the left half? Can you find one that flips the frieze over? or flips and shifts?

4. ***Wallpapering.*** In the patterns shown above, can you find a symmetry that shifts every point to the right? Can you find one that flips the right half and the left half? Can you find one that flips the pattern over? or flips and shifts? Can you rotate the pattern?

5. ***Commuters?*** (S) Take the square pattern covering the plane. Consider two symmetries. One flips the pattern about a diagonal line through the central square. The other rotates the pattern 90° clockwise. If you perform first the flip, then the rotation, do you get the same symmetry as you would get by first doing the rotation and then doing the flip?

### III. FURTHER CHALLENGES

1. ***Penrose tiles.*** Roger Penrose constructed two tiles that can be used to cover the plane only in aperiodic ways. One tile is called a *kite*, one is called a *dart*. Show how to get a tiling.

2. ***Expand forever.*** Why does any shape that can be assembled to form a larger version of itself that surrounds it on all sides give a means for covering the whole plane with tiles of that shape?

### IV. IN YOUR OWN WORDS

1. ***With a group of folks.*** In a small group, discuss and actively work through the construction of the Pinwheel Pattern and prove that it has no rigid symmetries. After your discussion, write a brief narrative describing the construction and proof in your own words.

2. ***Creative writing.*** Write an imaginative story (it can be comical, dramatic, whatever you like) that involves or evokes the ideas of this section.

# 4.5 The Platonic Solids Turn Amorous:
## Discovering the Symmetry and Interconnections Among the Platonic Solids

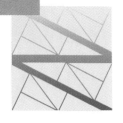

*God eternally geometrizes.*

—Plato

The Calligrapher and Mathematician Johann I. Neudorfer and his son (1561) by Nicolas Neufchatel.

Symmetry and regularity lie at the heart of classical beauty. We have an instinctive affinity for symmetrical objects—things that can be turned or reflected and return to their original shapes. The sphere is the ultimate in symmetry. From any vantage point, the sphere looks the same. If we forgo the graceful constant curvature of the sphere and consider objects with flat sides, then how symmetric and graceful can they be? Here we examine such flat-sided objects and explore their robust symmetries that have intrigued people for thousands of years.

Here we explore symmetric solids—referred to as regular, or Platonic, solids—by thinking about them concretely. We will imagine ourselves holding them, building them, being inside them, cutting them. These "solids" are solid. They are real things for us to hold and enjoy. We study them by first moving from the qualitative to the quantitative—how many edges, faces, vertices does each solid have? Then we record our observations and find coincidences. In life, coincidences are flashing lights signaling us to look for reasons and relationships that show a deeper structure. We find surprising connections among the regular solids that turn these separate objects into a coherent collection.

### Start in the Plane

As always, we consider a simple case with which to ground ideas. Before tackling objects in space, let's quickly consider symmetric shapes in the plane. In the plane, the most symmetrical object is the circle. But if we restrict ourselves to objects with straight sides, the most symmetrical objects are the *regular polygons*—polygons having all sides of equal length and all angles of equal measure. For any natural number *n*, there is a regular polygon having *n* sides. These regular polygons have lots of symmetry.

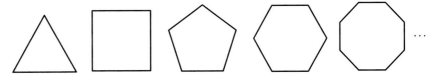

We see that there are infinitely many different regular polygons, and the more sides a regular polygon has, the more it looks like a circle—the most symmetric of all shapes in the plane. Perhaps we could view the circle as an infinitely many-sided polygon.

## *What 3D Objects Are Analogous?*

We now turn from figures in the plane to solids in space. We will discover a surprising difference between the regular polygons in two dimensions and the analogous objects in three dimensions. For now, we will just think about solids in space.

The sphere is the most symmetrical of solids in space because it remains unchanged when revolved about any line through its center or reflected through any plane through its center. Building a sphere isn't easy. If we wanted to construct a solid with lots of symmetry, we might decide to settle for something that is less symmetrical than the sphere but that can be more easily built. Specifically, suppose our solid has flat sides and straight edges. What properties would such solids have in order to be as symmetric as possible?

## *Symmetric Solids with Flat Faces*

Certainly, all the faces should be the same and should be symmetric themselves; that is, the faces should be identical regular polygons. But how will the faces fit together? We would like every corner (vertex) of the solid to look exactly like any other vertex. Thus, we require that the number of edges emanating from any vertex of the solid always be the same. So, for a solid made up of flat sides to be as symmetric as possible, its faces should be identical regular polygons, and the number of edges coming out of any vertex of the solid should be the same for all vertices. Such symmetric solids are called *regular*, or *Platonic, solids*.

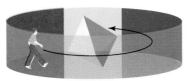   

Regular solid looks the same from different points of view.

Different points of view.

The cube is the most familiar of the regular solids; its classic structure is well known. Every side of the cube is a square and from any vertex of the cube, three edges come together. Can you think of other regular solids? How many other regular solids do you think there are? The Pythagoreans pondered these same questions thousands of years ago. Remember that there are infinitely many regular polygons in the plane. It seems reasonable to guess that there will be infinitely many regular solids in space. The following are 3D pictures of five regular solids.

Can you think of another one? No, you can't. Why? Because even though there are infinitely many regular polygons in the plane, there are only *five* regular solids in space. What a dramatic difference between the geometry of the plane and the geometry of space! Why are there only five? If you cannot wait to learn why, you can take a peek at the Feeling Edgy section in Chapter 5: Contortions of Space. The argument ironically relies on features of solids that are not solid! For now, let's just accept the fact that there are only five regular solids. The regular solids are regular since all the faces of each are regular polygons identical in size to each other face of the regular solid, and the number of edges emanating from each vertex is the same. Each Platonic solid looks identical from the vantage point of each face.

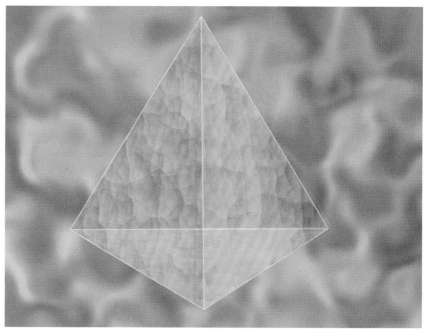

tetrahedron in 3D (Use your 3D glasses from the kit.)

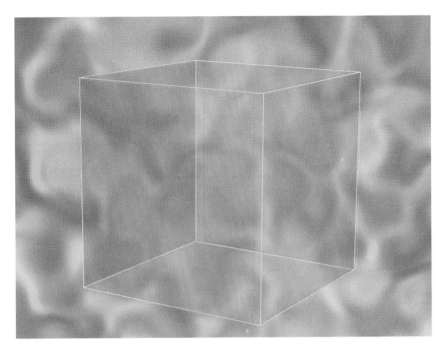

cube in 3D

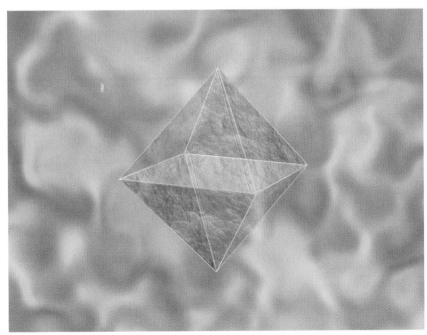

octahedron in 3D

dodecahedron in 3D

icosahedron in 3D

272 ◆ GEOMETRIC GEMS

## A Mystical Allure

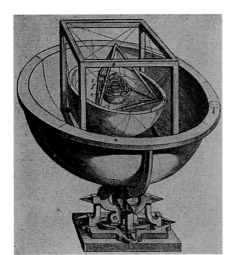

For centuries, the Platonic solids were associated with mystical powers. Even though the great Greek mathematician Euclid wrote about these regular solids, they are named after Plato because he apparently attempted to relate them to the fundamental components that he believed made up the world. The Pythagoreans, who knew there were exactly five regular solids and no others, held them in awe.

Johannes Kepler shared this sense of awe for the Platonic solids. Although Kepler is remembered best for his laws of planetary motion, he was proudest of his book *Mysterium Cosmographicum*. In it he proposed a theory to explain the structure of the solar system. At the time Kepler lived, there were six known planets. Kepler showed that it is possible to take the five regular solids, put one inside the other, and have the sizes of inscribed and circumscribed spheres about these solids reveal the sizes of the orbits of the planets.

Kepler's work was based on a mystical power of the five regular solids that he felt dictated the planets' orbits. Sadly his theory was refuted in 1781 when another planet, Uranus, was discovered. Unfortunately, there still are only five regular solids. We know that the regular solids have no relation to the orbits of the planets, but we also know that the only way to understand and make discoveries is to try new things and not be afraid to be wrong. These qualities cannot be separated from Kepler's genius.

Don't be afraid to make mistakes while creating new ideas.

◆ ◆ ◆ ◆ ◆ ◆

## Hold Them in Your Hands

Before we examine the Platonic solids closely, please explore these shapes a bit on your own. To do that, you must have a set of them. So, before proceeding, you will need to make a complete set of the five regular solids. You will find a set of models and instructions in your kit.

The first and simplest regular solid is the *tetrahedron*. The tetrahedron is made up of four identical equilateral triangles. We assemble them to make a triangular-based pyramid, the tetrahedron. A tetrahedron that we hold and see is imperfect. A real tetrahedron exists only in the mind and is the idea that this physical model suggests. The real tetrahedron is perfect—without blemishes or flaws. It has four identical faces, four vertices, and six edges. We can hold a model with our hands, but we can also hold the real tetrahedron in our minds, turn it about, and see it from the inside out or in other ways.

Think about the tetrahedron and understand its geometry so completely that you could answer any question about it. For example, suppose we were inside a tetrahedron sitting on one face with a vertex directly above our head. What would we see as we looked up? Try to visualize the tetrahedron in several ways. Hold it in your mind. Are the edges sharp or dull? Are the vertices dangerously pointy? Feel it. Cut it up. Explore it. What questions do you ask in getting to know it? Learn to rely on the model in your mind.

Now rotate the tetrahedron and understand all its possible symmetries. Do not be satisfied with general impressions. Move from having a feeling about symmetry to being able to define exactly what symmetry is. Think about how you would describe all symmetries of the tetrahedron.

## *The Other Four*

Next we move to the *cube*—the most familiar of the regular solids. One incarnation of cubes is dice. Their regularity allows us to expect with equal likelihood that any one of the six faces will land up after a die is thrown. Cubes are roughly the shape of some rooms. This fact is helpful in trying to visualize the cube from the inside. If you are in a room right now, look at the corners. You see three sides coming together at the corners. You see the faces meeting at right angles.

The next regular solid is the *octahedron*, which is constructed from eight identical, equilateral triangles. It has the appearance of two bottomless pyramids stuck together.

The *dodecahedron* is the only solid with pentagonal faces. It is made up of 12 identical pentagons.

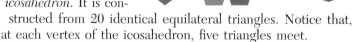

The Platonic solid with the most faces is the *icosahedron*. It is constructed from 20 identical equilateral triangles. Notice that, at each vertex of the icosahedron, five triangles meet.

## *From Qualitative to Quantitative*

After you have made your own set of Platonic solids, you can begin to explore their nuances. Let's leap beyond qualitative impressions and move toward quantitative properties.

In the case of the regular solids, what natural features can we count? How many faces, vertices, and edges are there? How many sides does each face have? How many edges come together at each vertex? How many faces come together at each vertex? Let's be systematic about this exploration by making a table with the regular solids noted down one side and the features to count along the top. Now fill in this chart of features.

> In life, quantitative exploration can often bring into focus hidden texture and richness.

|  | Number of Vertices | Number of Edges | Number of Faces | Number of Faces at Each Vertex | Number of Sides of Each Face |
|---|---|---|---|---|---|
| tetrahedron |  |  |  |  |  |
| cube |  |  |  |  |  |
| octahedron |  |  |  |  |  |
| dodecahedron |  |  |  |  |  |
| icosahedron |  |  |  |  |  |

How can we make certain that we are not miscounting? It is always worthwhile to consider how to avoid or catch errors. One way to minimize the chance for mistakes is to try to do the same thing in more than one way and see if we get the same answer.

## *Double Counting*

We could fill out the table in several different ways. First, we could just count each item. The problem is, how can we be certain that we have not failed to count, say, an edge, or that we have not counted a vertex more than once? Although we can be pretty certain that we are error-free with the tetrahedron, that icosahedron has a heck of a lot of edges, vertices, and faces.

One way to check our answers is to discover relationships among the numbers. Let's begin with the tetrahedron, the simplest figure. Examining simple cases often gives us insights into the more complicated situations and allows us to discover unexpected phenomena.

The tetrahedron has four faces, and each face is a triangle. Once we know that, we can figure out how many edges it must have. How?

Remember that each of the four faces has three edges. So, our first guess is that there must be three edges for each of the four faces for a total of 4 × 3, or 12, edges. Of course, this method of counting is wrong, because each edge of the tetrahedron is counted twice. Each edge touches two different faces, so we've counted each edge twice.

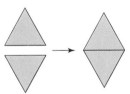

Because each edge has been counted exactly twice, we know that 4 × 3 gives us a number that is exactly twice too big. Hence, we know that the number of edges is (4 × 3)/2 = 6. Of course, with the tetrahedron, it is relatively easy to count the number of edges. Still, we have learned a lesson that can be applied to the more complicated figures.

Let's try this line of thought with the other figures. The cube has six faces. Each face is a square. So there must be (6 × 4)/2, or 12, edges on a cube. We arrive at this answer because there are 4 edges on each of the 6 squares, so there are 6 × 4 edges before we glue the squares together

*Understand simple things deeply.*

❖ ❖ ❖ ❖ ❖ ❖

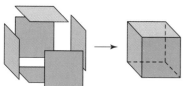

to make the cube. Once we make the cube, we glue one edge of one square to exactly one edge of another square. This gluing then makes the number of edges on the cube exactly half of the number of the 6 × 4 original edges of the squares.

Thus, there are 12 edges on a cube.

The octahedron has 8 faces, each one a triangle. Therefore, an octahedron has (8 × 3)/2 = 12 edges. The dodecahedron has 12 faces, each one a pentagon. Therefore, a dodecahedron has (12 × 5)/2 = 30 edges. The icosahedron has 20 faces, each one a triangle. Therefore, an icosahedron has (20 × 3)/2 = 30 edges.

Now let's look at the table and check the number of vertices in a similar way. Can you devise a way of calculating the number of vertices of a regular solid once you know how many faces it has, how many sides each face has, and how many faces come together at each vertex? After you have devised the method, use it to check your entries for the number of vertices for each figure.

## *Coincidence? We Don't Think So*

> Coincidences are flashing lights alerting us to potential insights.
>
> ◆ ◆ ◆ ◆ ◆ ◆

Look at the table again. Do you notice any coincidences? Coincidences occasionally are just coincidences, but if we see several numbers appearing more than once, we might wonder whether those entries are somehow related. What coincidences appear in our table of vertices, edges, and faces?

## *Observations*

The number of faces of the cube equals the number of vertices of the octahedron; the number of vertices of the cube equals the number of faces of the octahedron; and the number of edges of the cube and octahedron are the same. Why?

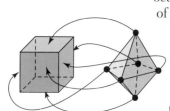

Let's think about these observations while examining our models. The number of faces of the cube equals the number of vertices of the octahedron. Could we put those two apparently coincidental numbers together geometrically? Can we explain geometrically why those coincidental numbers are not at all coincidental?

## *Visualizing the Big Picture*

Let's think big. Stand in the middle of a room—preferably a small square room, since such a room is most nearly a cube. As you stand in this room look up, down, ahead, behind, left, and right. Those six directions are the six faces of the cube in which you stand. We want to understand the coincidence of the octahedron having the same number of vertices as the cube has faces. So, look at each face of the cube—that

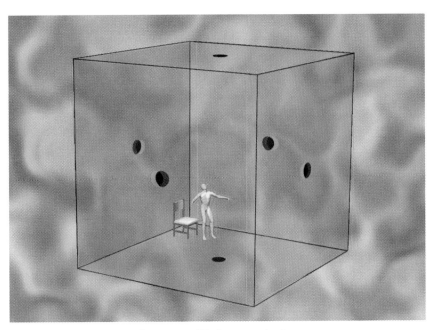

Put on your 3D glasses and enjoy.

is, each wall plus the floor and ceiling. If you placed a dot (vertex) on each face of the cube, where would be the most natural place on each face for it? Look up. Where would you put a vertex? There is really only one natural place—the middle, right in the middle of the ceiling. Imagine a dot in the center of the ceiling. Now look at the center of each of the four walls, and don't forget the floor. Place a dot on each. How many vertices have you placed? Six. One for each face of the cube in which you stand.

Now that you have placed some vertices, try drawing edges between them in an attempt to make the skeleton of a solid. Which vertices would you connect to create an edge of another solid? Look at the center of the ceiling. Remember that, at the center of the ceiling, you have placed a vertex in your imagination. The ceiling is surrounded by four walls, and each wall also has a vertex in its center. It seems natural to connect the ceiling vertex with each wall vertex. Don't you feel compelled to draw those four connecting lines—those edges? The edges connect the ceiling with each wall because each wall abuts the ceiling. It seems reasonable that faces that touch should have their centers joined by an edge. Do you see in your mind's eye the four edges from wall centers to the ceiling center? You are standing so that you see those four edges rising like a tent or a pyramid over your head.

Now join consecutive wall-centers together to create a square encircling your midsection and forming the base of the pyramid. Do you see those four sides going from wall-center to wall-center encircling you in a square embrace? Now complete the picture with edges connecting the wall-centers with the center of the floor where your feet are planted. Do

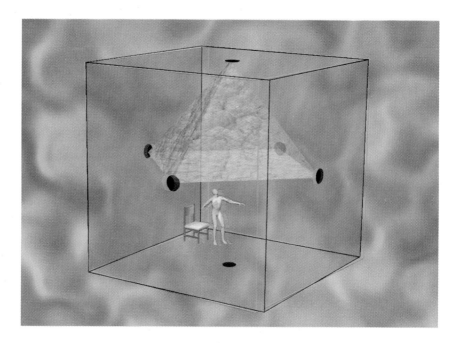

you see those four edges creating an upside-down pyramid on the same central diamond base?

You now have the skeleton—the edges—of the octahedron. Do you see why the edges of the cube are in one-to-one correspondence with the edges of the octahedron? Each edge of the cube is an edge where two faces abut. But each of those two faces has a vertex in its center, and those

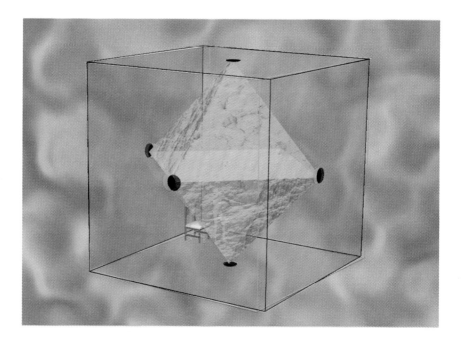

two face-centers are connected with an edge that goes across the direction of the cube edge. Those two edges naturally correspond. Each cube edge has a natural octahedron edge and vice versa. So there are the same number of cube edges as octahedron edges.

Now we have the edges of an octahedron. It remains only to fill in the triangular faces to complete the octahedron. It is easy to fill those triangles in with our mind's eye, but let us think as we do so. Do you see that each corner of the cube is associated with one triangular face of the octahedron? Do you see that the fact that there are three faces coming together at each vertex of the cube leads to the fact that the figure we are creating has triangles for faces?

We have just seen that the very essence of the cube somehow includes in its being the essence of the octahedron. These figures are closely linked. What happens if we start with an octahedron and imagine the same process? Can we float in the center of the octahedron? Of course, we can in our mind's eye. We can move two pyramids of Cheops, one upside down, to create in our mind a wonderful octahedron. We can float inside that octahedron and begin to place a vertex in the center of each of the eight triangular faces. The vertices in the centers of the triangular faces can be joined to create edges to enclose you in what? Try to visualize it! Do you see yourself surrounded by that cube suspended amid the great double pyramid? Do you see the connection between the cube and octahedron?

## *Dualing Solids*

These Platonic solids are paired at the most basic level of their essential defining characters. They are two views of some common essence. The one entails the other. They are our first examples of what we will soon call *duality*, but they will not be our last.

Let's test the limits of the idea and continue to gain insight. In this case, we started with one geometric solid and created another. We put a vertex in the middle of each face. When two faces of the original solid shared a common edge, we connected their centers with an edge. Then each vertex of the original solid became a face of the newly constructed figure. This process will always allow us to create a new solid from an existing one.

Now let's look at the relationship between the old solid and the new one we create in this fashion. We first notice that each face of the old gives a vertex of the new. That is easy to see, because we put a vertex in the middle of each face of the old solid.

Next we notice that there are the same number of edges in the old and

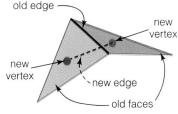

New edge corresponds to the old one right above it.

When we have made an observation, we should think about what other, related ideas it can lead to.

◆ ◆ ◆ ◆ ◆ ◆

the new solids. This equality is true since every edge of the original is shared by two faces. Each of those two faces had a vertex put in it, and the two vertices were connected by an edge. Each edge of the old is crossed by one edge of the new. So they are in one-to-one correspondence.

Finally, we look at each vertex of the original solid. Each face that meets there has a vertex of the new figure placed in its center. These vertices are connected to create a polygonal face of the new solid. For example, look at the corner of a room and imagine it as a vertex of a cube. If we put a vertex in the center of each of the walls that come together at that corner and connect them, we get a triangular face of the new solid we are creating. So, in fact, we know how many sides the polygon has by seeing how many faces come together at the vertex. That is, each face of the solid we are creating from the cube is a triangle. Of course, after we noticed that we had created an octahedron, we knew that each face was a triangle. But now we see a relationship between the triangles of an octahedron and the number of faces that come together at each vertex of a cube.

If we now take our new solid and perform the same process of constructing a new solid from it, it is no coincidence that we get a solid of the same type that we started with.

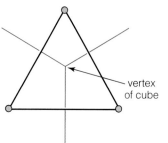

Vertex of cube abuts corresponding octahedron edge.

## *Duality*

This process of creating one solid from another is so appealing that it deserves a name. Since we see great connections between the two solids—the faces of one correspond to the vertices of the other, and the edges

3D art: natural pairings between the platonic solids

280 ◆ GEOMETRIC GEMS

4.5 / THE PLATONIC SOLIDS TURN AMOROUS

correspond—the name chosen to refer to this phenomenon is *duality*. This name suggests that the two solids form a pair. It also suggests that the opposite features of the two solids correspond. That is, the faces of one correspond with the vertices of the other. A polygonal face is the highest dimensional external feature of the solid. It corresponds to the lowest dimensional feature of the dual object, the vertex. The edge is in the middle of the vertex-edge-face hierarchy, and it corresponds to the dual solid's edge because it is also in the middle.

What does duality generate from each solid? The cube gives an octahedron. An octahedron gives the cube back. A dodecahedron gives the icosahedron; and the icosahedron gives the dodecahedron back. What does the tetrahedron give back as its dual? Itself! The tetrahedron is *self-dual*.

### Amorous Solids

For over two millennia, the regular solids have been Platonic. Surely with all this duality going on, the time has come to rename certain pairs the *Amorous Solids*. The cube and octahedron have a physical intimacy that is not purely Platonic. The dodecahedron and the icosahedron are closely entwined. And the self-dual tetrahedron has a relationship with itself that surely goes beyond the bounds of Platonism. We hereby begin a campaign to rename the Platonic Solids the more lively and appropriate Amorous Solids to reflect their physical beauty and their habit of congregating in dual pairs.

### Our Golden Promise

At the end of our journey through the aesthetic world of the Golden Rectangle we said we would see Golden Rectangles occurring in surprising places. It turns out that they occur in Platonic solids. We illustrate this fact by examining the icosahedron. Consider the following drawing and note the rectangle inside the icosahedron. Then examine your physical model of the icosahedron and put your fingers on the opposite sides of the rectangle. Feel the rectangle between your fingers.

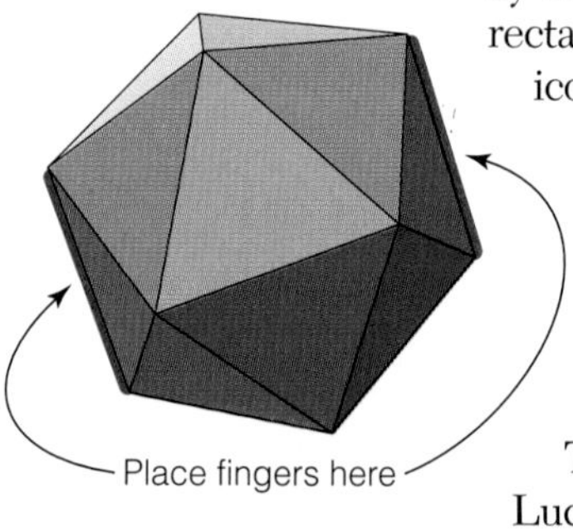

In fact, if that were a perfect icosahedron, that rectangle would be a perfect Golden Rectangle! Can you find other Golden Rectangles inside the icosahedron? Consider the next picture. All three of those rectangles are Golden Rectangles. Do you see how they all appear to be perpendicular to each other?

They actually are! In fact, this is the way that Luca Pacioli constructed the icosahedron in his book *De Divina Proportione*. He wrote that, if we began with the three Golden Rectangles in the position below, then attaching vertices together to make triangles will result in 20 equilateral triangles, thus creating an icosahedron. We ask you to think about this construction for yourself in the Mindscapes.

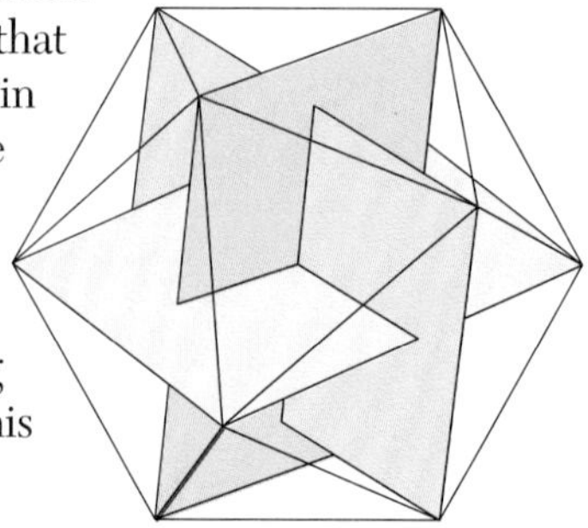

Place fingers here

Notice in the preceding illustration that the rectangles are all linked in an exotic manner. Can you figure out what is unusual about that group of three rectangles? We'll describe this remarkable property in the next chapter.

One last golden fact. We know that the icosahedron and the dodecahedron are dual, that the cube and the octahedron are dual, and that the tetrahedron is dual with itself. Thus, it appears that there is no connection between the octahedron and the icosahedron. We know for sure they are not dual. Well, these two solids turn out to have an interesting connection. We describe this now, but caution you that this special phenomenon is not duality!

Observe that there are 12 edges on the octahedron and 12 vertices on the icosahedron. If we are careful, it is possible to place a small icosahedron inside a larger octahedron in such a way that each vertex of the icosahedron touches exactly one edge of the octahedron. It is a delicate placement. Look at your octahedron and see where you can place the vertices of the icosahedron on the edges of the octahedron. Visualize the icosahedron within the octahedron before reading on.

Notice that the vertices of the icosahedron do not touch the edges of the octahedron exactly in the middle of the edges. The vertices touch the edge somewhere off center. If we take any edge of the octahedron in the preceding figure and cut it where the vertex of the icosahedron touches that edge, we would have two line segments, one longer than the other.

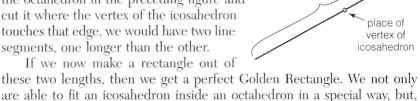

The octahedron surrounding and protecting the icosahedron

If we now make a rectangle out of these two lengths, then we get a perfect Golden Rectangle. We not only are able to fit an icosahedron inside an octahedron in a special way, but, in doing so, we cut up all the edges of the octahedron into pieces that exhibit the Golden Ratio! Mathematics is full of amazing connections!

## A LOOK BACK

The five regular, or Platonic, solids are the tetrahedron, the cube, the octahedron, the dodecahedron, and the icosahedron. No other solids can be created with identical, regular polygonal faces meeting together so that the number of edges emanating from any vertex of the solid is the same. The regular solids come in dual pairs where the vertices of one correspond to the faces of the other. The tetrahedron is self-dual. We saw surprising relationships among the Platonic solids, the Golden Rectangle, and the Golden Ratio. Different mathematical ideas are interconnected. The deep relationships between seemingly distinct mathematical objects are at the heart of mathematics and are the core of its beautiful and delicate structure.

Our strategy for discovering interesting properties of the Platonic solids was to take a quantitative look at their features and then look for patterns. We counted the vertices, edges, and faces of each regular solid and saw some surprising coincidences. Exploring those seeming coincidences led us to discover, among other things, the idea of duality.

To understand ideas or objects, a good first step is to experience them as directly as possible. Going from a qualitative to a quantitative view allows us to see far more detail and opens a whole new world of understanding. Quantitative measurements allow us to see patterns that might not otherwise be apparent. Patterns and coincidences are signposts signaling relationships and interconnections. These techniques can help us analyze the unknown and think of new ideas in every arena of life.

Make it quantitative.

Look for patterns or coincidences.

Seek reasons for the patterns.

## MINDSCAPES  Invitations to Further Thought*

### I. SOLIDIFYING IDEAS

1. ***Build them.*** Using toothpicks and pieces of plastic hose, strings and straws, posterboard, or mahogany, or your kit, make a complete set of the Platonic solids for yourself.

2. ***Unfold them.*** For each of the Platonic solids, draw a picture of how you can unfold it in various ways so it lies flat on the plane, and illustrate how it could be folded to create the solid.

3. ***Edgy drawing*** (H). Draw pictures in the plane that show the edges of each of the regular solids in the sense that each edge in your drawing corresponds to an edge on the solid and each one joins another one whenever the corresponding edges on the solid touch. For example, you could inflate each regular solid until it becomes a sphere with curved edges on it and then draw the stereographic projection.

4. ***Drawing solids.*** Draw each solid by completing the beginnings of the drawings here.

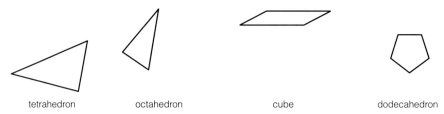

tetrahedron   octahedron   cube   dodecahedron

---

*In the Mindscapes section, exercises marked (H) have hints for solutions at the back of the book. Exercises marked (S) have solutions.

5. **Life drawing.** Draw the regular solids using your physical models. Include shading and perspective. You may title your work *Regular Still Life*.

6. **Count.** For each of the regular solids, take the number of vertices, subtract the number of edges, add the number of faces. For each regular solid, what do you get?

7. **Soccer counts (H).** Look at a soccer ball. Take the number of vertices, subtract the number of edges, add the number of faces. What do you get? This counting can be tricky, so think of a systematic method.

8. **Golden Rectangles.** Take your toothpick or straw model of an icosahedron and place it on the table. Now prop it up so that only one edge is resting on the table. Locate the Golden Rectangle spanning the edge on the table and the top edge. Locate the pair of vertical edges that form two sides of a Golden Rectangle. Locate a pair of horizontal edges halfway up from the table to the top edge that form two sides of a Golden Rectangle. If you are dexterous, carefully weave pieces of string to construct those three Golden Rectangles. Measure the base and height of one of those rectangles and then divide the height into the base and see how it compares to the Golden Ratio.

9. **A solid slice (S).** For each regular solid, imagine slicing off a vertex. What shape is the boundary of the cut? For example, slicing a vertex off a tetrahedron gives a triangular cut.

10. **Siding on the cube.** Suppose we start with the edges of a cube (that is, its skeleton). Now, on each square face, we glue four triangles together as shown:

Sketch a drawing of this figure. Count the number of its vertices, edges, and faces.

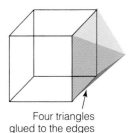

Four triangles glued to the edges of the cube.

## II. CREATING NEW IDEAS

1. **Cube slices (H).** Consider slicing the cube with a plane. What are all the different-shaped slices we can get? One slice, for example, could be rectangular. What other shaped slices can we get? Sketch both the shape of the slice and show how it is a slice of the cube.

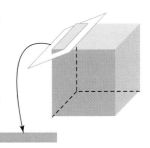

slice is

2. **Dual quads (S).** Suppose you have a cube with edges of length 1. Suppose you construct a dual octahedron inside it whose vertices are in the centers of the faces of the cube. How long are the edges of that octahedron?

3. **Super dual.** Suppose you take a cube with edges of length 1 and construct a dual octahedron around it by making the center of each triangle of the octahedron hit at a vertex of the cube. How long are the edges of the octahedron?

4. **Self-duals.** Suppose you have a tetrahedron having each edge of length 1. Suppose you construct a dual tetrahedron inside whose vertices are in the centers of the faces of the original tetrahedron. How long are the edges of that inscribed tetrahedron?

5. **Not quite regular.** Suppose you allow different numbers of triangles to come together at different vertices of a solid. Then show how to produce solids with arbitrarily large numbers of triangular faces.

### III. FURTHER CHALLENGES

1. **Truncated solids.** Take each regular solid and slice off the vertices to produce new solids that have two different types of sides. Fill in the chart by counting or computing the number of vertices, edges, and faces of each. Also, describe how many faces of each type the truncated solid has.

| Solid (pretruncating) | Truncated Vertices | Edges | Faces |
|---|---|---|---|
| tetrahedron | | | |
| cube | | | |
| octahedron | | | |
| dodecahedron | | | |
| icosahedron | | | |

2. **Stellated solids.** Take each regular solid and replace each face by a collection of triangles (one attached to each edge) to produce new solids that look like stars. For example, here is

the stellated tetrahedron:

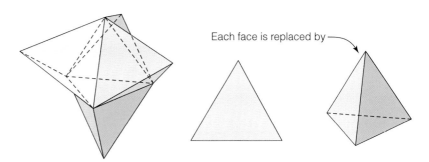

Fill in the chart by counting or computing the number of vertices, edges, and faces of each.

| Solid (prestellating) | Stellated Vertices | Edges | Faces |
|---|---|---|---|
| tetrahedron | | | |
| cube | | | |
| octahedron | | | |
| dodecahedron | | | |
| icosahedron | | | |

## IV. IN YOUR OWN WORDS

1. *Personal perspectives.* Write a short essay describing the most interesting or surprising discovery you made in exploring this section's material. If any material seemed puzzling or even unbelievable, address that as well. Explain why you chose the topics you did. Finally, comment on the aesthetics of the mathematics and ideas in this section.

2. *With a group of folks.* In a small group, discuss and actively work through the idea of duality of the Platonic solids. After your discussion, write a brief narrative describing duality in your own words.

3. *Creative writing.* Write an imaginative story (it can be comical, dramatic, whatever you like) that involves or evokes the ideas of this section.

4. *Power beyond the mathematics.* Describe several real-life situations—ideally, from your own experience—for which some of the strategies of thought presented in this section would provide effective methods for approaching and resolving them.

# 4.6 The Shape of Reality?
## How Straight Lines Can Bend in Non-Euclidean Geometries

*If there is anything that can bind the heavenly mind of man to this dreary exile of our earthly home and can reconcile us with our fate so that one can enjoy living—then it is verily the enjoyment of the mathematical sciences and astronomy.*

—Johannes Kepler

Mathematics can help us understand the cosmic, the unapproachable, and the mysterious. Nothing is more cosmic and mysterious than the entire universe. For thousands of years, people have pondered the fundamental question: What is the shape of our universe?

In any attempt to understand the world around us, it is only natural to wonder about the geometry of our physical existence. Of course, our universe is incredibly vast, and our experience is limited by time and space. Thus our question is by no means easy. Does space bend or curve? What does it even mean for space to bend or curve? Since we do not see space bending or curving around us, our initial sense is that space is flat. However, we exist on such a microscopic scale compared to that of the entire universe, perhaps we don't sense the reality of the "big picture." So let's apply some of our techniques of analysis and try to discover the shape of space.

## A First Sketch of the "Big Picture"

How do we start to understand the geometry of something so large that it is beyond our understanding? Start by looking at something smaller.

*In dealing with life's complex issues, start with the simple and build from there.*

❖ ❖ ❖ ❖ ❖ ❖

First, we look at the ground under and just around our feet. What do we see? Flat.

Thus, it seems reasonable to guess that our world around us is flat like a plane. This guess is not completely crazy, especially to those who live in Kansas. The world around us does tend to look pretty flat. This observation led people throughout history to study the flat plane and its rich geometry. However, it turns out that our Earth is

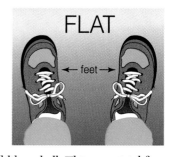

shaped like a ball. This nontrivial fact illustrates two important points. First, there is no pressing need for the Flat Earth Society; and second, what we observe locally may not accurately depict what is occurring on a larger scale.

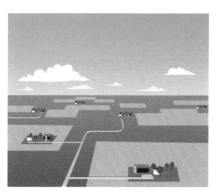

Before taking on the whole universe, perhaps we should consider the geometry of the next simplest realm: the sphere.

## A Next Sketch: The Geometry of a Sphere

What is the shortest distance between two points? In a flat, unobstructed world, that shortest distance is always a straight line. But in New York City the shortest distance from 5th Ave. and 42nd St. to 8th Ave. and 38th St. is not a straight line. What path does the crow take? If the crow drove a taxi, he would find the shortest path, but that path would follow the grid of streets. So the shortest paths between two points—the "straight lines"—depend on the shape of the space where we live. We live on Earth. So what are the "straight lines" on Earth? Let's travel around and see.

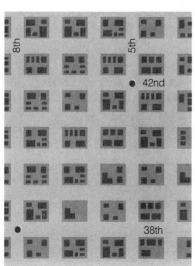

Manhattan

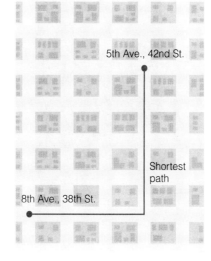

Since the Earth is round, we do not live on a plane; yet many travelers live in a plane a good deal of the time. Pilots have a great attachment to fuel and hate to run out of it at 35,000 feet in the air. Thus, airplanes go from place to place along the shortest routes possible. Pilots, like crows, know the Earth is round and choose their routes accordingly. Let's see what those routes are. The best method for bringing this point home would be for you to now take a nonstop plane trip from Chicago, Illinois to Rome, Italy. We'll wait patiently, but you had better send us a postcard.

## *Traversing Your Travels*

Chicago and Rome are both at the latitude of nearly 40° north. You might think that the shortest route from Chicago to Rome would be to stick to the 40° latitude line the whole way. Let's measure how far that route would be. We will do this by measuring distances on a globe and using the scale to tell the mileage. If we take out a tape measure and place it along the 40° latitude line, we see that the distance is 5,300 miles. Is there a shorter route? If we're flying the plane, we had better find out.

path shown = 5,300 miles

A good, though messy, way to find the shortest route uses a greased globe and a rubber band. We take the globe and grease it until it is so slippery that nothing, including the rubber band, will stick to it. We next put two pins in the globe, one at Chicago and one in Rome, and then stretch a rubber band over the two pins. We first hold the rubber band down so it sits on the latitude line. Then we let go. Did it stay on the 40° latitude line? We don't think so. In fact, it slid up to a shorter route. Instead of flying along the latitude, the rubber band found a genuinely shorter route. Notice that the route heads north and goes over Labrador and Dublin, Ireland before heading back south on its way to Rome.

rubber band's path = 4,800 miles

Is this route really shorter? Let's measure. We placed our measuring tape along our new route and measured about 4,800 miles. This new route saves about 500 miles!

The arc path is part of a "great circle".

Let's take a closer look at this rubber-band route. If we extend the route, we get a *great circle* that is as big as possible going around the whole globe—that is, a circle whose center is at the center of the globe. It is as long as the equator.

Indeed, the shortest path between any two points on the globe is always a great circle that contains them. The segments of great circles are the shortest distance between two points on the globe. Why?

## *Why Great Circles Are the Way to Go*

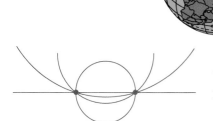

Latitude is longer.

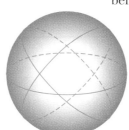

Let's think about why the great circle segments are the shortest paths. If the Earth were hollow, the shortest path from Chicago to Rome really would be a straight line inside the Earth. So for our purposes, let's imagine a straight-line tunnel connecting Chicago and Rome burrowing right through the Earth. The shortest route on the surface of the Earth would deviate as little as possible from that straight, underground Chicago-Rome tunnel.

Let's notice something about circles and lines. If we take two points and make a circle that contains them, then bigger circles are flatter and therefore stay closer to the straight line between the two points. So, among the paths that stay on circles, taking the biggest circle on the globe containing Chicago and Rome—that is, the *great circle*—will stay closest to the straight line. The latitude circle, being smaller, bends out more from the straight line and is therefore longer. "Straight lines," that is, the shortest paths on Earth, are great circle segments.

The larger the circle—the "flatter" the segment.

## *Distances in a Different World*

We live on Earth, which is essentially a ball, but how about a bug on the wall? If our bug doesn't fly, its world is in the shape of the walls. So, when it sees its dining destination on some other wall, it has some serious calculations to perform. What is the shortest distance from here to dinner? Take a guess.

shortest path?

A good guess would be the path shown in the figure on the left.

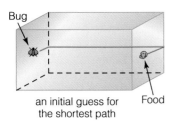

an initial guess for the shortest path

Think about some simple cases.

◆ ◆ ◆ ◆ ◆ ◆

Is this guess the shortest path? Suppose dinner is on the same wall. Then the situation is easy. The bug simply walks in a straight line.

The bug is off to a great start. How about if dinner is on an adjacent wall?

It is pretty clear that the bug needs to go straight to the boundary edge and then straight on the next wall to its dinner. The question is, "Where on that edge should it cross?" Describe a method for locating the best crossing place.

shortest path = straight line

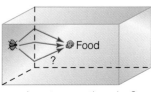

where to cross the edge?

Notice that, if the walls were at some angle other than 90°, the distances from the bug to the crossing point and the crossing point to dinner do not change. So let's consider a different question. Suppose the bug is on an open door, and its dinner is on the wall.

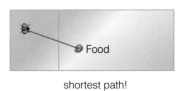

shortest path!

Imagine that, as the bug is considering its shortest path, someone comes along and closes the door. Suddenly the question becomes much easier. Now the bug and its dinner are on the same wall, and the bug simply proceeds in a straight line. Now the bug is on a roll.

Let's now return to the scenario where the bug's dinner is on the opposite wall. How will it figure out the shortest route? Having experienced the closing of the door, surely our bug cannot resist the idea of unfolding the walls. The problem is that there are many ways to unfold the walls.

Can you think of other ways of unfolding the walls, keeping the food and the bug in the same relative locations?

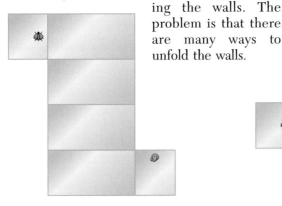

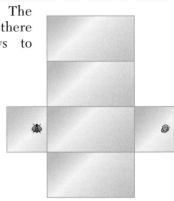

292  ◆  GEOMETRIC GEMS

Which one should the bug choose? The straight lines on some unfolded walls from bug to dinner are different lengths from others.

What to do?

Some straight paths are shorter than others.

One method would be to unfold the room in all possible ways and measure the straight-line distances. We've seen several different room-unfolding possibilities. Notice that different unfolding scenarios result in different placements of the food and the bug. Visualize the reassembly of the flattened rooms and verify that the relative positions of the bug and the food are always the same. Once we find the shortest flattened path, we can draw the straight line on that unfolded version and then reassemble the room. In this case, the shortest path takes the bug over five walls—a dramatic departure from our first guess.

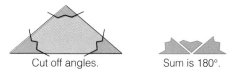

shortest path

> Sometimes, when we are faced with a problem and we don't know what to do, we should just consider everything.
>
> ❖ ❖ ❖ ❖ ❖ ❖

Now we have a better sense of shortest paths and straight lines in various worlds, including our own earthly sphere. Putting these straight lines together allows us to explore some basic geometry that captures the essence of the graceful curvature of the sphere. Let's put three straight lines together to make a triangle.

## *Triangles on the Sphere*

Draw any triangle in the plane. Add up the three angles. The result is 180°.

Cut off angles.   Sum is 180°.

But we've seen that, in different realms, we have different ideas of straightness. Let's now explore the angles of a triangle made out of straight lines on a sphere. We begin by drawing a large triangle on a sphere. For exam-

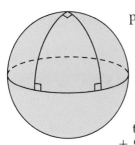

ple, let's put one vertex on the north pole, one vertex on the equator at 0° longitude, and the third vertex on the equator at 90° longitude. The edges of this triangle consist of two longitudinal segments from the north pole to the equator and a segment that goes ¼ of the way around the equator. What are the angles at each vertex? Each one is 90°. So, what is the sum of the three angles of that triangle on the sphere? 90° + 90° + 90° = 270°. Yikes.

This result is slightly disconcerting. The sum of the angles of this triangle on the sphere is not 180°, but 270° (90° too much). Is it possible that all triangles on the sphere have angles that sum to 270°? Let's see.

Take the big triangle above and break it into two by drawing in the longitudinal segment from the north pole down to the equator at 45°.

Now each of the half-sized triangles has angles of 90°, 90°, and 45°. That sum is 225°; 45° more than the 180° we would have in a triangle on the flat plane. This result is stranger still, since not only do the angles fail to add up to the comfortable 180° we know and love, but now we see that on a sphere, different triangles have different sums of angles. As always, we must look for patterns.

Can we find any regularity among our measurements? The big triangle had 270°, 90° degrees too much. When we divided it in half, each half had 225°, 45° degrees more than 180°. Did you notice that the total surplus of angle for the two smaller triangles stayed at 90°? In other words, when we took the big triangle and measured the surplus angle bigger than 180°, we got 90°. Then, when we divided the big triangle into two smaller triangles, each of the halves had a surplus of 45°, or 90° altogether. Suppose we start with any triangle on a sphere and cut it in half by bisecting one of the angles.

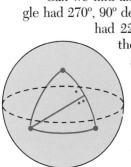

•° + ••° = 180°

What is the relationship between the angles of the original triangle and the angles of each of the two subtriangles? Well, the new angles created add up to 180° since they are on a straight line. So, the total excess of the two triangles must be equal to the excess for the original big triangle.

Notice what happens if we take a small triangle on the sphere. What is the surplus of its angles? Not very much. A small part of a sphere is basically flat, so the angles of a triangle there will have almost exactly the same angles as the angles of a triangle on a flat plane. So, it seems that larger triangles have greater excess in the sum of their angles than small triangles do. Furthermore, if a large triangle is divided into smaller triangles by adding edges, since all the

Sum is just a smidge greater than 180°.

added angles created are along straight lines or divide existing angles, the total excess of all the subtriangles making up a bigger triangle must be the same as the excess of the big triangle. What corresponds to the excess? It turns out that the excess increases as the area of the triangle increases. Thus, we see that the sum of the angles of a triangle on a sphere will always exceed 180° but that small triangles will just barely exceed 180° and large triangles will exceed 180° by a more substantial amount.

### *Extra Degrees Through Curvature*

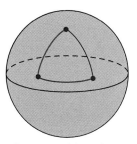

Curvature of the sphere causes "straight lines" to bow out a bit.

The fact that the sum of the angles of any triangle on a sphere exceeds 180° is due to the curvature of the sphere. And, since every triangle has a sum of angles exceeding 180°, we will say the sphere has *positive curvature*. Notice that the curvature on a sphere can be determined by measurements taken on the sphere itself. It is not necessary to see the sphere from outside. For example, suppose we were bugs whose whole universe was a sphere. Perhaps light stayed right along the sphere so that we could see things. We would not see a horizon, because the light would bend around the sphere providing us with ever more distant vistas. Nevertheless, we could determine that our world has positive curvature by drawing a triangle and measuring the sum of the angles. Even though the individual lines would appear completely straight, the sum of the angles would be more than 180°, clinching the positive-curvature claim.

We now have one space, the plane, where all triangles have angles that add up to 180°. That constant sum is our benchmark, so we will say the plane has *zero curvature*—it is flat. We saw another space, the sphere, with positive curvature where all triangles have angles that add up to more than 180°. Surely we cannot resist asking the question, "Is there a space with *negative curvature*—that is, where the sum of the angles of a triangle is less than 180°?"

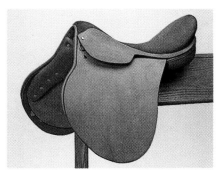

### *Geometry on a Saddle*

Horseback riding provides us not only with a sore bottom but also with an interesting geometrical opportunity. The surface of a saddle has an appealing shape and provides a surface ripe for experiments using rubber

3D picture of a saddle (Use your 3D glasses from your kit.)

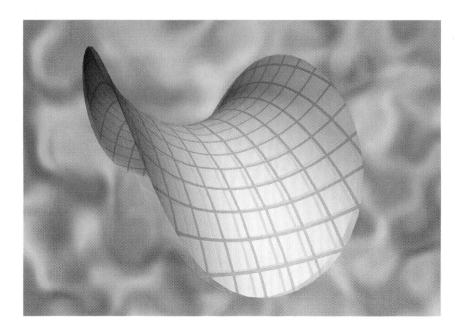

bands and butter. Suppose we place three pins as shown, one near the front of the saddle and two near the stirrups. (This is a poor time to actually sit on the saddle.) We now grease the saddle with butter and put a rubber band around every pair of pins. The rubber bands will slide to the shortest distances between pins. So, we will have a rubber band triangle on the surface of the saddle. We now wish to measure the angles. If you don't happen to have a saddle handy, look at the following picture, estimate the measure of the angles, and compare the sum of the angles to

triangle on the surface of a saddle (Angles sum to less than 180°.)

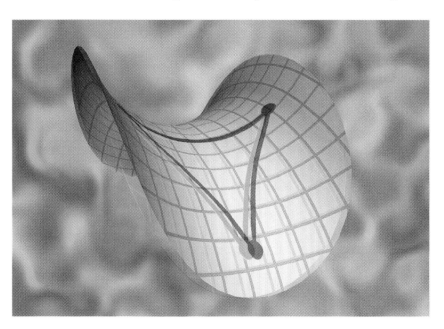

180°. The sum of the angles of this triangle is less than 180°. Of course, the rubber-band method will not always work on a saddle since the line between some pairs of points, like the center front to the center back of the saddle, would leave the surface and float in the air. However, rubber bands are good tools for finding the shortest distances between some pairs of points on the saddle. Whether you use rubber bands or another method for finding the shortest distances between points, triangles on the saddle will have sums of angles less than 180°, because the sides of the triangles curve inward and thus cause those angles to shrink. So this space has negative curvature and is an example of the exotic world known as *hyperbolic geometry*.

Sum of the angles is less than 180°.

We have seen three different types of geometries: plane geometry, which we say has zero curvature (all triangles have angle sums of exactly 180°); spherical geometry, which we say has positive curvature (angle sums vary depending on the size of the triangle, but always exceed 180°); and hyperbolic geometry, which we say has negative curvature (angle sums vary depending on the size of the triangle, but always are less than 180°). It certainly appears as though hyperbolic geometry is exotic and foreign to our real-world existence, which brings us back to our original question: What is the shape of our universe?

## *The Shape of our Universe*

We have just caught glimpses of three types of geometries: planar, spherical, and hyperbolic. Which is our universe? How could we tell which geometry accurately models our universe? Think of an experiment that we could perform to answer this. (*Hint:* What property distinguishes the three?)

Let's measure the angles of triangles. Suppose we make a big triangle and measure its angles. If the sum of those angles equals 180°, then we would conjecture that our universe has zero curvature. If the sum of those angles exceeds 180°, then we'd guess that our universe has positive curvature. If the sum of those angles is less than 180°, then we'd guess that our universe is curved negatively. Would anyone actually attempt this experiment? Yes!

The great mathematician Carl Friedrich Gauss tried this experiment in the early 1800s. He formed a triangle using three mountain peaks near Göttingen: Brocken, Hohenhagen, *und* Inselsberg. He had fires lit and

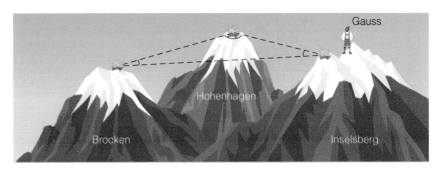

used mirrors to reflect the beams of light to form a triangle having side lengths roughly 43, 53, and 123 miles. He carefully measured the angles of the triangle and added them up. His sum was within 1/180 of a degree of 180°. That small difference could easily have been caused by errors in measurement. This evidence certainly leads us to think that our universe is neither positively nor negatively curved and that the universe is flat. What is the problem with this conclusion? Think about this question in view of our spherical geometry observations.

Recall that, in spherical geometry, if we have a small triangle, then the triangle is nearly flat, and thus the sum of its angles is nearly 180°. Thus, although Gauss's triangle was big, compared to the entire universe it wasn't even a speck. Thus, on such a microscopic scale, it is not surprising to see that the evidence points to a geometry having zero curvature. We would need an enormous triangle to detect the existence of any actual curvature. Is this experiment even practical? And if it were, would anyone actually attempt it?

The answer to the first question is possibly and to the second is yes. Today scientists believe that the universe exhibits two important properties. The first is that it is *homogeneous*, which basically means that any two large sections of space will look the same—of course here "large" needs to be LARGE. The second is that the universe is *isotropic*, which means that, as we look around, things look about the same in every direction. It turns out that we can find geometrical objects that are homogeneous and isotropic that are either planar, spherical, or hyperbolic. This fact leads to a question of great interest to scientists today: Does the universe have zero, positive, or negative curvature?

A large group of scientists now believe that the universe is negatively curved—that is, that the geometry of the universe is actually the exotic hyperbolic geometry suggested by the saddle. In fact, a conference was held in October 1997 at Case Western University that brought together 20 cosmologists and 5 mathematicians to discuss the possible shape of the universe and how to measure its curvature. NASA is scheduled in 2000 to send *MAP*—the Microwave Anisotropy Probe—into space. This probe will measure microwave background radiation, which is a residue of the "big bang." By studying slight variations in the measurements—which are actually temperature measurements—scientists are hoping to discover the exact geometry and curvature of the universe. It will take about two years to map out the heavens with the probe and another four years for scientists to analyze and collect the data. European scientists plan to send up the Planck Probe in about six years. This probe should be able to make even more careful measurements of the variations in microwave radiation. These modern experiments capture the spirit of Gauss's attempts to measure the curvature of the universe.

So, what is the shape of our universe? Although many experts believe it may be hyperbolic and negatively curved, no one knows for certain. However, 21st-century science and technology together with mathematics may enable us one day to measure the curvature of our vast space and understand its subtle and beautiful geometry.

**A LOOK BACK**

Space can have various shapes. We can distinguish how space bends by examining the shortest paths—straight lines, although they may not necessarily be straight. Three different kinds of geometry are planar, spherical, and hyperbolic. The flat plane, round sphere, and saddle are good models for planar, spherical, and hyperbolic geometries, respectively. On a very small scale all look nearly the same, and thus we have not yet been able to determine the shape of our universe by taking measurements of our local environment. However, some scientists now believe that the universe may be hyperbolic, and experiments are being devised to give evidence about the geometry of space.

The distinguishing feature of the three different geometries is their curvature. If a space has zero curvature (the sum of the angles of any triangle is exactly 180°), then the space is flat. If a space has positive curvature (the sum of the angles of any triangle exceeds 180°), then the space is spherical. Finally, if the space has negative curvature (the sum of the angles of any triangle is less than 180°), then the space is hyperbolic.

When we wish to consider big issues it is often valuable to start with simple and familiar models or examples and build from there. By identifying both similarities and differences in our various examples, we can often discover the underlying structure that determines the general case.

*Start with the simple and build from there.*

*When you don't know what to do, consider everything.*

*Look for patterns.*

◆ ◆ ◆ ◆ ◆ ◆ ◆ ◆ ◆ ◆ ◆ ◆ ◆ ◆ ◆ ◆ ◆ ◆

## MINDSCAPES  Invitations to Further Thought*

### I. SOLIDIFYING IDEAS

**1–3. Travel agent.** Get a globe and trace the shortest paths between the following pairs of cities. Match each pair in the left list with the location in the right list that is on the shortest path between them.

| | |
|---|---|
| Austin, Texas—Tehran, Iran | Reykjavik, Iceland |
| Williamstown, Massachusetts—Beijing, China | Denali, Alaska |
| Austin, Texas—Beijing, China | near the north pole |

---

*In the Mindscapes section, exercises marked (H) have hints for solutions at the back of the book. Exercises marked (S) have solutions.

**4–6. *Latitude losers* (H).** Each pair of cities below is on the same latitude. Fill in the following table by measuring the distance from city to city, first staying along the latitude and then measuring the distance taking the great-circle route.

| Pair of Cities | Latitude Distance | Great-Circle Distance |
|---|---|---|
| Beijing, China—Chicago, Illinois | | |
| Mexico City, Mexico—Bombay, India | | |
| Sydney, Australia—Santiago, Chile | | |

**7–11. *Triangles on spheres.*** On a globe, draw triangles whose vertices are the following sets of cities. For each such triangle, measure the sum of the three angles of the triangle.

7. Minneapolis, Minnesota; Austin, Texas; Williamstown, Massachusetts.
8. (S). Panama City, Panama; Nome, Alaska; Dublin, Ireland.
9. Quito, Ecuador; Monrovia, Liberia; Thule, Greenland.
10. Quito, Ecuador; Bangkok, Thailand; the south pole.
11. Wellington, New Zealand; Moscow, Russia; Rio de Janeiro, Brazil.

**12–16. *Spider and bug.*** For each pair of points on the boxes drawn, describe the shortest path from one point to the other.

12.

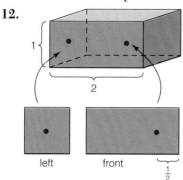

13. on side

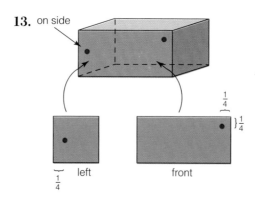

300 ◆ GEOMETRIC GEMS

**14.**

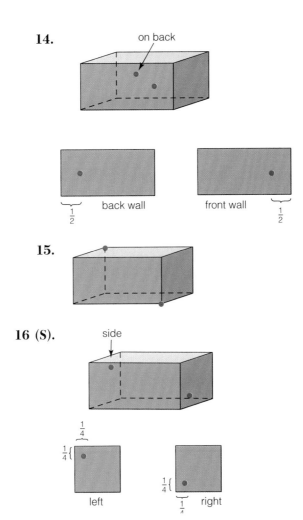

**15.**

**16 (S).**

17. **Becoming hyper.** Professor William Thurston found a neat way to build a model of hyperbolic geometry. Photocopy many equilateral triangles. (Enlarge the following sheet on your copier.)

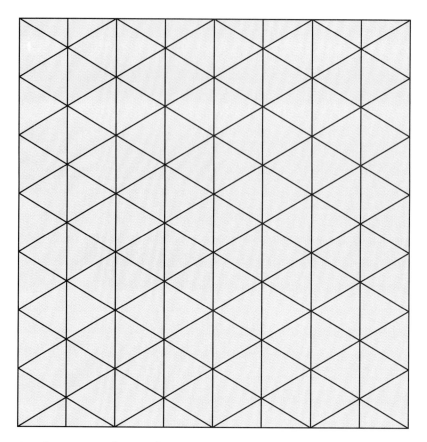

Cut them out and tape them together so that seven triangles meet at each vertex. You will have to bend the triangles to fit together. Continue attaching the triangles so that seven come together at each vertex. You will notice that your model will become floppy. The larger you make it, the floppier and more accurate your model will be.

18. **Deficit angles (H).** Draw a big triangle on your floppy sheet spanning several of the pieces by flattening a section on the ground and drawing a straight line, then flattening another section and drawing another straight line, and then completing the triangle in the same way. There is a lot of squashing involved. Now measure the three angles and add them up. What do you get?

19. **Same old.** Go to a vertex on your floppy plane. Look at the pattern of all the triangles that you can reach from there passing through at most two triangles. Now go to another vertex and do the same. Are the patterns the same or different?

20. **Gauss II.** Try Gauss's experiment. Select three tall objects that are reasonably far away from one another (for example, three buildings or trees) and measure the angles between them. What are your measurements? Sum the angles.

## II. CREATING NEW IDEAS

1. **Big angles (H).** What is the largest value we can get for the sum of the angles of a triangle drawn on a sphere? Experiment with larger and larger triangles and compute the largest sum value.

2. **Many angles (S).** Draw three different great circles on a sphere. How many triangles have you made on the sphere? Compute the sum of the angles of all the triangles. Draw another group of three different great circles and answer the questions again. What do you notice? Make some conjectures.

3. **Quads in plane.** Measure the sum of the angles of the quadrilaterals below. Why is the sum of the angles of any quadrilateral in the plane the same?

4. **Quads on the sphere.** Below are quadrilaterals on the sphere. Measure the sum of the angles of each quadrilateral. Make a conjecture about the relationship between the sum of the angles of a quadrilateral and its area.

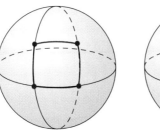  

5. **Parallel lines.** On a plane, if you draw a line and then choose a point off the line, there is one and only one line that goes through that point and misses the line. Suppose we try the same experiment on the sphere. Take a line (which, remember, is a great circle) and take another point. How many lines—that is, great circles—can go through the point and miss the first great circle altogether? Explain your findings.

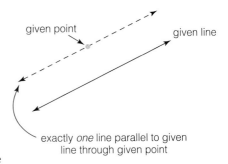

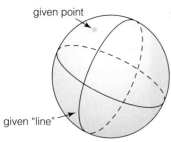

given point

given "line"

6. **Floppy parallels.** On the floppy plane you constructed in question I.17, draw a line and then choose a point some distance off the line. How many lines can go through the point and miss the first line altogether even if they were extended indefinitely?

7. **Cubical spheres.** Take a cube. Put a point in the middle of each face. Now draw the straight lines to the middles of each of the sides of that face. After you have done that for all the faces, how many triangles have you created? The line that goes from the center of one face to the center of an adjacent face is straight since, if you unfolded the cube along the edge, you would get a straight line. Now what is the sum of the angles for each of those triangles? What is the sum of all the angles of all the triangles? Compare your answer to the answer to question II.2. *Many angles*.

8. **Tetrahedral spheres.** Let's do a similar calculation for the tetrahedron. Put a vertex at the center of each face of a tetrahedron and connect adjacent faces over the center of each edge. Answer all the questions in II.7.

9. **Dodecahedral spheres.** This question is the same as the previous ones, except start with a dodecahedron.

10. **Total excess.** Using the observations from the previous questions and question II.2. *Many angles*, make a conjecture about the total excess of sums of angles of triangles that cover up polyhedra as described.

### III. FURTHER CHALLENGES

1–4. **Geometry on a cone** (**H**). Take a piece of paper and cut out a pie-shaped piece of angle $z$. Now put the two ends of the cut-out piece together to construct a cone. Let's see what happens if we look at triangles that surround that cone point. To draw a triangle, we have to decide what is a straight line on the cone. That is fairly easy, since the cone is made from a piece of paper that was originally flat. Take two points on the cone, then flatten the cone in such a way that both points are on the flat part. Then connect them with a straight line. A triangle is just made of three straight lines. First measure the angles in a triangle that does not go around the cone point.

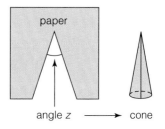

paper

angle $z$ → cone

flattened

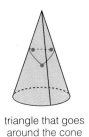

triangle that goes around the cone

1. **What is the sum of those three angles? Why?** Now let's take the more interesting case of a triangle that goes around the cone point. Draw the three sides separately by flattening in different ways.

2. **What is the sum of the angles of your triangle?** Is the sum the same for all triangles that go around the cone point? Let's try some more experiments, this time by removing thinner and fatter pie slices before making our cone.

3. Complete the following table by making the cones, drawing triangles around the cone points, measuring the angles of the triangles, and adding them up.

| Angle of Pie Removed | Sum of Angles of Triangles | Difference with 180° |
|---|---|---|
| 30° | | |
| 60° | | |
| 90° | | |
| 180° | | |

4. Make a conjecture about the relationship between the angle of the slice removed to make the cone and the excess above 180° of the sum of the angles of a triangle that goes around the cone point.

5. **Tetrahedral angles.** What is the sum of the angles around each vertex of the tetrahedron? For each vertex compute 360° minus the sum of the angles at that vertex. Multiply that number by 4 since there are four vertices of the tetrahedron. What do you get? Are you surprised?

## IV. IN YOUR OWN WORDS

1. **Personal perspectives.** Write a short essay describing the most interesting or surprising discovery you made in exploring this section's material. If any material seemed puzzling or even unbelievable, address that as well. Explain why you chose the topics you did. Finally, comment on the aesthetics of the mathematics and ideas in this section.

2. **With a group of folks.** In a small group, discuss and actively work through the relationship between the sum of the angles of a triangle on a sphere compared to that sum on the plane. After your discussion, write a brief narrative explaining that relationship in your own words.

3. **Creative writing.** Write an imaginative story (it can be comical, dramatic, whatever you like) that involves or evokes the ideas of this section.

4. **Power beyond the mathematics.** Describe several real-life situations—ideally, from your own experience—for which some of the strategies of thought presented in this section would provide effective methods for approaching and resolving them.

# 4.7 The Fourth Dimension: Can You See It?

*Listen there's a hell of a universe next door: let's go!*

—e.e. cummings

*Interior in the Fourth Dimension* (1913) by Max Weber.

The very phrase "fourth dimension" conjures up notions of science fiction or even the supernatural. The fourth dimension sounds eerie, romantic, mysterious, and exciting; and it is. Physicists, artists, musicians, and mystics all visualize the fourth dimension differently and for different purposes. The mystique of the fourth dimension appeals to all who contemplate it.

The fourth dimension lies beyond our daily experience. So, visualizing, exploring, and understanding it seem at first impossible. Such understanding would require us to develop an intuition about a world that we will never see. Nevertheless, that understanding is within our reach. We will succeed in building insights without experience. We will become at home in an environment that we cannot touch, see, or otherwise sense. Successful explorations of an unfamiliar realm often begin by delving into the depths of the familiar. The fourth dimension provides a dramatic testament to the power of developing a concept through analogy.

By developing a deeper understanding of what we believe is basic and understood, we are led to new, unforeseen frontiers.

## Building Up from the Familiar

There is no physical direction for us to look in to see the fourth dimension. Our everyday world is too confining. Instead, we must start with what we know and understand. We will get to the fourth dimension by starting with 0 and moving up.

## Making a Point

What is 0-dimensional space? What would a 0-dimensional world be like? In a 0-dimensional space there are zero degrees of freedom. That means there would be no room for anything. No information is required for us to identify a particular location in this 0-dimensional space. What must this space look like? The answer is, a point. In 0-dimensional space, the entire universe is just a point. A point has no length, height, or width—so, there are zero degrees of freedom.

•
⎿ 0-dimensional space.
If you lived there, you'd be home by now.

We now take the extremely confined 0-dimensional world and build a more spacious one. Suppose we had a remarkable ink pad: If we ink up an object and drag it, the ink leaves a trail behind the object. What kind of geometric object would we create if we were to take 0-dimensional space (the point), ink it up on this ink pad, and then drag it in a new direction? We would sweep out a line, 1-dimensional space.

Dragging the inked-up point in a new direction to produce the 1-dimensional line.

## Getting in Line

One-dimensional space is a line. For convenience, let's suppose that the line is marked with numbers. If someone lives at a certain point on the line, what is the minimum amount of information we require to locate that person? The answer is just one number, namely, the number on the line where that person lives. If we think about this number as an address to find where this person is living, we need only the house number, since there is only one street (namely the line). So, we need only one piece of information to identify precisely any point on the line. Therefore, we say that the line is 1-dimensional space.

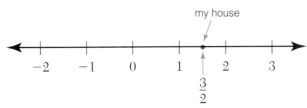

my house

$-2 \quad -1 \quad 0 \quad 1 \quad 2 \quad 3$

$\frac{3}{2}$

My address is $\frac{3}{2}$: one piece of information and you know where I live → 1-dimensional space!

We moved from 0-dimensional space to 1-dimensional space. We now wish to move another dimension higher. Suppose we take the line and ink it up all over and drag it in a new direction. Notice that, if we drag the line east or west, we just place more ink on the line we've already created. We need to drag it in a different direction to create something new. Suppose we now drag the line in the north-south direction (perpendicular to the original line). We sweep out a plane. The plane is 2-dimensional space.

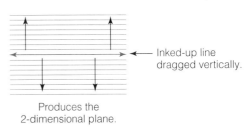

Inked-up line dragged vertically.

Produces the 2-dimensional plane.

## Home on the Plane

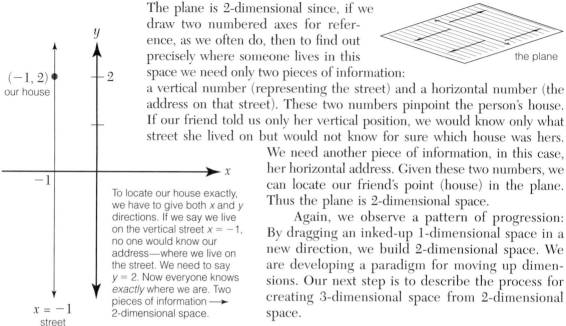

the plane

$(-1, 2)$ our house

$-1$

$x = -1$ street

To locate our house exactly, we have to give both *x* and *y* directions. If we say we live on the vertical street $x = -1$, no one would know our address—where we live on the street. We need to say $y = 2$. Now everyone knows *exactly* where we are. Two pieces of information → 2-dimensional space.

The plane is 2-dimensional since, if we draw two numbered axes for reference, as we often do, then to find out precisely where someone lives in this space we need only two pieces of information: a vertical number (representing the street) and a horizontal number (the address on that street). These two numbers pinpoint the person's house. If our friend told us only her vertical position, we would know only what street she lived on but would not know for sure which house was hers. We need another piece of information, in this case, her horizontal address. Given these two numbers, we can locate our friend's point (house) in the plane. Thus the plane is 2-dimensional space.

Again, we observe a pattern of progression: By dragging an inked-up 1-dimensional space in a new direction, we build 2-dimensional space. We are developing a paradigm for moving up dimensions. Our next step is to describe the process for creating 3-dimensional space from 2-dimensional space.

## Space—The Final Frontier?

Following the analogy, we ink up the entire 2-dimensional plane and drag it in a new direction different from the east-west and the north-south directions. We choose a direction perpendicular to the plane. Suppose we drag the inked-up plane in the up-down direction. Then we sweep out space as we know it. In this space, we need three pieces of information to locate a point, and we call this 3-dimensional space. This space is the world in which we live. We have 3 degrees of freedom: north-south; east-west; and up-down. Once again, we note how a completely inked-up 2-dimensional plane dragged in a new direction results in 3-dimensional space.

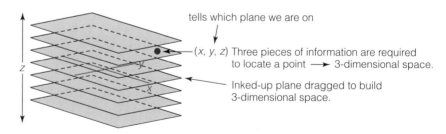

tells which plane we are on

$(x, y, z)$ Three pieces of information are required to locate a point → 3-dimensional space.

Inked-up plane dragged to build 3-dimensional space.

Dragging all of 3-dimensions into a "new" direction. This figure cannot be drawn here!

## The Fourth Dimension

We have now come to the moment we've been eagerly awaiting. Do we now have enough experience to make a first attempt at describing 4-dimensional space? Let's try to follow our established pattern. First we ink up all of 3-dimensional space. Now this inking is both challenging and messy to do in practice, so we'll just perform it in our minds. We ink up every single point in 3-dimensional space; that is, we ink up every molecule of air. It may be useful to think of 3-dimensional space as a sponge that can hold ink everywhere. Now we drag that inked-up space in a new direction—let's say in a direction perpendicular to all three of the original directions of 3-dimensional space. At first glance, this last step may present something of an obstacle.

What does it mean to select a direction perpendicular to all the directions of the space in which we live!? A fourth direction that is different from the three we already know? Where is it? We are unable to point to it, but if we ignore that fact for the moment and drag the inked-up copy of our 3-dimensional space in that new direction, what we sweep out is 4-dimensional space. Each point in that 4-dimensional space can be located by four numbers. To make this foreign idea meaningful, we must develop some insight and clarity into the notion of a "new direction," one that is different from the three we all know and love.

## Understanding Through Analogy

Instead of trying to answer a question that appears too difficult, it is always wise to step back and try to address an easier, but related, question.

◆ ◆ ◆ ◆ ◆ ◆

When the going gets tough, the smart stop going! Thus, instead of searching for this mysterious fourth direction, we will attempt to understand it by looking at the dimensions with which we are comfortable. This fourth direction is a direction that exists entirely in the mind. It is a virtual direction. Why not become familiar with the 4-dimensional world through a process of analogy? We seek to understand how objects fit into this 4-dimensional space just as we understand them in 2-dimensional and 3-dimensional space. Even the basic notion of "inside" will lead to unexpected realizations.

We begin our process of analogy by stepping back and considering the simpler world of two dimensions—the plane of this piece of paper. Imagine that in this world live two authors—Ed and Mike. Thus, they themselves are 2-dimensional beings. They have no thickness at all. They cannot lift their heads to see over lines. So a line, though having no thickness at all, is able to block their way and their view. We, however, can look down onto the page—their world—and see everything. It's, in some sense, an overall panoramic view of their entire universe. Think about this vantage point and then try to draw what the two 2-dimensional authors would look like to you from your view. Make a quick sketch before reading on.

Here is how Ed and Mike may look to us.

how authors appear to us

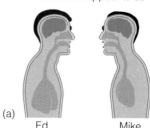

(a)

Ed    Mike

A line blocks the authors' view of each other.

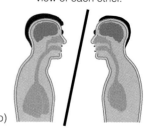

(b)

We immediately notice that, looking at this 2-dimensional world from our vantage point above it, we can see inside each 2-dimensional person and examine his or her inner organs. To a 3-dimensional being, the insides of a 2-dimensional person are exposed and appear open. Notice, however, that, when Ed and Mike look at themselves, they are unable to see each other's insides; their bodies are completely sealed by their 1-dimensional skin. Their 1-dimensional skin is able to separate their insides from their outside world.

Why are 3-dimensional people able to see the 2-dimensional people's inner organs? The answer is that we are able to look down at their entire 2-dimensional universe from a third direction and thus are able to see everything: insides and outsides simultaneously. They cannot hide anything from us. In fact, if they wanted us to perform internal surgery, and give them their much needed lobotomies, we could perform it without even cutting their skin open. We could simply reach in from the third dimension and touch their insides.

Imagine the great frustration we would experience in failed attempts to describe our extra degree of freedom to these 2-dimensional authors. Three-dimensional people certainly have an overall aerial view of the entire 2-dimensional world that 2-dimensional people could not easily comprehend.

## *An Analogy in Our World*

Suppose now that a 4-dimensional creature is looking "down" at us. Since it would be looking at our 3-dimensional world from a new and different direction, it would have a view of our world analogous to what we had of that 2-dimensional world. It could see everything—inside and outside—simultaneously. Our inner organs would be open and in full view of the 4-dimensional creature.

Let's call the 4-dimensional creature Fredddd (notice Fredddd is 4-D! ... don't groan, it's cute). Now is it really possible for Fredddd to see our insides since they are completely covered all around by our skin? Remember the analogy of the 2-dimensional beings. They appeared to themselves as completely sealed all around *in their world*. However, their 1-dimensional skin does not block the view of 3-dimensional eyes. Similarly, although our 2-dimensional skin separates our insides from our 3-dimensional outside world, it would not block Fredddd's view from his 4-dimensional vantage point. Think again about the 2-dimensional creatures with their 1-dimensional skin that did not block our view.

We can think about Fredddd's incredible "panoramic" view and attempt to develop a sense of 4-dimensional space from our 3-dimensional world via analogy to our attempts to describe our world to the 2-dimensional plane-folk.

## Are Safes Safe Across Dimensions?

Suppose now that the 2-dimensional authors, Ed and Mike, wrote a most successful and incredibly well-received 2-dimensional math book, *The Spleen of Mathematics*. Instead of cash royalties, in their contract negotiations they agreed to receive 2-dimensional bars of gold (actually Golden Rectangles, but that is another story). Now they wish to put their gold bar in a sealed vault so it will safely sit until Ed's and Mike's retirement. What would a sealed vault look like in 2-dimensional space?

Here is their 2-dimensional sealed vault. Notice that, when the 2-dimensional plane-authors look all around it, they see it is completely sealed up. All the sides are solid, and it is impossible for anyone to steal their gold. To make sure, they even stand guard and watch the vault and surrounding area intensely.

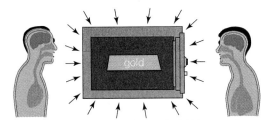

Our 2-dimension vault—sealed ALL around!

A 3-dimensional person could trivially remove the gold without breaking the sealed vault and without the authors seeing it happen. We can simulate this theft with a Post-It™ note playing the role of their gold. The 3-dimensional person just reaches down to the page right over the Post-It™ note and peels it off the 2-dimensional page. Suppose the authors decide to forget about their retirement and take out the gold now. They break the sealed vault, and what do they see? Their nest egg has vanished into thin 2-dimensional air! Think about how they must feel. They saw with their own eyes that the vault was completely sealed *all around the entire time*; there was no way to get in. The disappearance appears simply impossible.

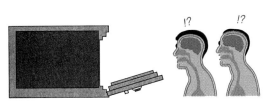

Our gold is gone! But the vault was *never* broken! Impossible!

Let's return to our 3-dimensional world. Imagine that we placed this book (a 3-dimensional object that is worth its weight in gold) into a sealed box. We examine the box along all its sides and convince ourselves that it is completely sealed all around. Now, working by analogy, we invite you to describe how Fredddd can remove our hidden treasure without opening the sealed box and without our seeing him take it. By doing so, we are building some insights and intuition into the abstract realm of the fourth dimension.

Our book inside a COMPLETELY SEALED box. Can't get to it.

Don't close your mind to ideas that first appear counterintuitive.

◆ ◆ ◆ ◆ ◆ ◆

## *Lassoing in Low Dimensions*

Returning to the 2-dimensional world of the authors, suppose that one of the authors (the loud one) annoys the other author to his breaking point. Almost crazed, the aggravated author decides to rid himself of his chattering co-author once and for all. As he does not believe in murder, he decides to take some rope in two dimensions and encircle his co-author with a lasso. Although the aggravated author can still hear the rantings of the loud author, it is impossible for the lassoed author to escape (he has no sharp objects to cut the rope). Think now of ways to help the talkative author to escape without touching him. Remember that the lasso is made of pliable rope.

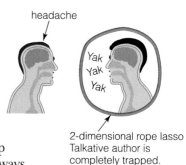

2-dimensional rope lasso Talkative author is completely trapped.

To rescue the author, we reach down and lift a little bit of the rope up into the third dimension (out of the plane). What would the quiet author, who is watching all this, see from his perspective? What would it look like to him? It would appear as though a piece of the rope was cut off and removed. Is the rope really cut? No, it's just that we moved a piece of it out of sight into the third dimension, and thus it appears that a section of the rope disappeared. If we now allow the rope to fall back into the plane, it would seem as if the missing piece of rope had magically reappeared.

Now, back to our 3-dimensional world. Suppose we

Lift a piece of the rope out of the plane and into the third dimension.

"Part of the rope has vanished or was cut!?"

I'm out!

Use your 3D glasses to really see the lifting rope.

312 ◆ GEOMETRIC GEMS

have a piece of real rope in our hands. The rope's ends are glued together so it forms a loop. However, on closer inspection, we notice that there is a knot in the loop. So, we are holding a piece of rope that has been knotted and then glued. Can you explain how Fredddd can help you unknot the rope in 4-dimensional space without ever actually cutting it? Visualize what you

knotted rope loop in three dimensions

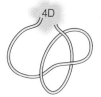

Fredddd lifts a piece of the rope up into the fourth dimension. Is the rope cut? No! But from our 3-dimensional vantage point, it's open. So, ...

 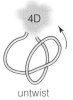 

move loop     untwist

would see and what is actually happening. Remember to argue by analogy. When you are done, you should be holding the rope, still a sealed loop, but without any knot in it.

Now Fredddd lowers back the piece of rope he was holding in the fourth dimension ⟶ We see the rope "fuse" together magically.

Unknotted rope! No cutting.

These mind experiments allow us to develop a sense of the structure of 4-dimensional space. This type of analogy appears in the classic short book *Flatland* by Edwin Abbott, written in 1884.

## *Visualizing Cubes*

To illustrate further the notion of dimensionality, we now consider the geometry of space by constructing and visualizing a 4-dimensional cube. We proceed, as always, to warm up by building cubes in all the lower dimensions first. This method will allow us successively to build a cube in the next higher dimension using the cube constructed in the previous dimension. Notice that we are again examining dimensionality sequentially.

The 0-dimensional cube is pretty easy—remember *everything* is just that one point. Thus, the 0-dimensional cube is a point. We now ink up the 0-dimensional cube and drag it one unit in a new direction. This dragging produces a line segment that is actually a 1-dimensional cube. If we ink up the line segment and drag it one unit in a direction perpendicular to the previous one, we get a 2-dimensional cube (also known as a square). If we ink up the entire 2-dimensional cube and drag it one unit in a direction perpendicular to the previous ones, we construct a 3-dimensional cube (also known as a cube).

0-dimensional cube    drag    1-dimensional cube    drag    2-dimensional cube    drag    3-dimensional cube

The last cube is actually just a drawing—an artist's rendition—of a 3-dimensional cube. It is impossible to draw a perfect and complete 3-dimensional cube on 2-dimensional paper. Instead, we draw a perspective picture, which our 3-dimensional eyes understand and interpret correctly. For example, consider the following drawing, but view it solely as a figure on a piece of paper and not as a perspective picture. What do we literally see?

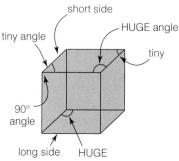

A figure in the plane—certainly not symmetric!

Notice that some lines of this drawing are longer than others, and some of the angles are huge while others are very small. These properties are certainly not properties of a 3-dimensional cube. On a 3-dimensional cube, all the edges have the same length, and all the angles are perfect right angles. So, why does the drawing look so funny? Again, it's just an artist's rendition of the perspective. Because we are creatures who see and understand 3-dimensional objects instinctively, we make the adjustments and interpret the picture in the appropriate manner. We see all the angles as right angles and all the edges as equal even though they are not drawn that way. Our mind adjusts automatically to what our eyes actually see.

*Look at things in new ways.*

◆ ◆ ◆ ◆ ◆ ◆

Can you now describe a process to construct a 4-dimensional cube and draw a picture of it? Again, as always, we proceed by analogy. Let's take the 3-dimensional cube (note that it is a solid object) and ink it up. We don't just ink up its faces, but we ink up every point of the cube, even the points inside. Perhaps it is useful to think of the solid 3-dimensional cube as a cubical sponge. This way we may be able to visualize the ink being inside all the points of the solid cube. We now drag it one unit in a direction perpendicular to the previous ones, and we have a 4-dimensional cube. Since we are moving the 3-dimensional cube in a direction different from the previous ones, the 3-dimensional cube never hits itself as it is dragged. Of course, this dragging is performed entirely in our imagination. Here is one view of the 4-dimensional cube, also known as a *hypercube*.

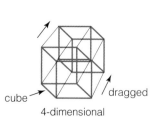

4-dimensional

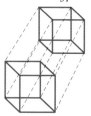

4-dimensional cube with new edges dotted to illustrate the "dragging" in the "new" direction.

The picture looks complicated, and indeed it is. It is a 2-dimensional picture of a 4-dimensional object. We have to squash that object down two

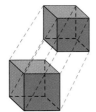

dimensions to have it fit on the page. The drawing contains many different angles and lengths, and it doesn't appear to be a cube made of right angles. But remember, it is drawn in perspective. In four dimensions, that cube is perfect: All the edges are the same length, and all the angles are exact right angles. Our difficulty is that we do not have eyes that perceive four dimensions, so we cannot automatically see the perspective as we did earlier with the 3-dimensional cube. A better picture would be a 3-dimensional picture of a 4-dimensional cube.

"Top" view of a 3-dimensional cube

Here is another 2-dimensional picture of the 3-dimensional cube and the analogous perspective of the 4-dimensional cube in three dimensions made out of soap film. We must try to create a 3-dimensional picture in our minds to get a better sense of its geometrical shape.

## Beyond the Fourth

We can now even build a 5-dimensional cube. How? Just remember: When life gets you down, just ink it up and drag it! Ink up every point of the 4-dimensional cube and drag it one unit in a direction perpendicular to the previous ones, and, voilà, we have a 5-dimensional cube. We can

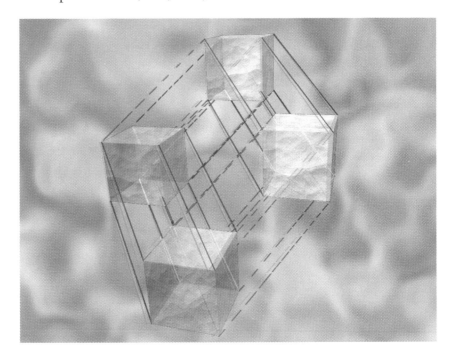

5D cube: Just a 4D cube dragged in a new direction. Dotted edges are the new ones added.
Put on your 3D glasses to get a hint of 5D!

4.7 / THE FOURTH DIMENSION ◆ 315

use this successive reasoning to consider the geometrical structure of dimensions greater than four. Through careful analysis we are able, within our mind, to see geometrical insights in worlds that we are unable to see with our eyes.

## *Cubes—Unplugged and Unfolded*

Before leaving these cubes, we return to the issue of the perspective drawing of the 3-dimensional cube on the paper and the inability to draw a perfect 3-dimensional cube. The truth is we can depict the 3-dimensional cube on paper in such a way that all edges are exactly the same size and that all angles are perfect right angles. This seems impossible since we need all three dimensions to build the 3-dimensional cube. The secret is to draw the cube *unfolded*. Thus, we could draw it as:

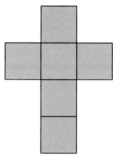

Unfolded 3D cube.
Now all the sides are
perfect squares.

We must understand how to assemble it: We fold it up along the drawn edges and glue the outer edges together in the correct manner. Here is the unfolded 3-dimensional cube; we have marked the pairs of edges that need to be glued together.

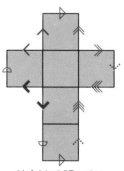

Unfolded 3D cube
with instructions as to
how to glue edges
together.

This construction seems to indicate that it would be possible to unfold the 4-dimensional cube in 3-dimensional space in such a way that all edges

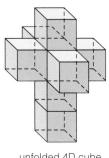

unfolded 4D cube in three dimensions

are of the same length and all angles are right angles. All we would have to do is join and glue pairs of faces together. Indeed, here is an unfolded 4-dimensional cube in 3-dimensional space. To assemble it, the external faces would have to be paired up and glued. This unfolded 4-dimensional cube was the inspiration for Salvador Dali's 1954 painting *The Crucifixion, Corpus Hypercubicus*, in which Dali attributes religious implications to the fourth dimension.

## *Yes, Virginia, There Is a Fourth Dimension*

Finally, we close with the question: Is there a fourth dimension? What would it mean for the fourth dimension to exist? From a practical point of view, higher dimensional space definitely does exist and is used frequently, for example, in big business and high finance. We illustrate this fact by answering the question, How do I use the fourth dimension to make money? Suppose we were in the business of manufacturing dental floss, and we wish to make lots of profit through wholesome dental hygiene. What decisions would we have to make? Here are some simple ones.

- *Where do we have our factory?*
  If it's in the middle of nowhere, the rent is inexpensive, but then we incur high shipping costs. If we're in the big city, we have high rents but almost no shipping costs.

- *How big should our factory be?*
  The bigger the factory, the faster we can produce dental floss but the greater the rent.

- *How many workers do we hire?*
  If we hire a few, we achieve less productivity; if we hire a million people, they would just squeeze into the factory, couldn't move a bit, and thus would not produce anything.

- *How much capital should we borrow?*
  There are tax advantages in realizing some debt and also in having capital on hand, but if we incur dramatic debt, we may go bankrupt in our attempts to pay off our loans.

Even though this scenario is oversimplified, there are already four degrees of freedom, or four different variables. We have to juggle all four

Open your mind to the practical utility of abstract notions.

❖ ❖ ❖ ❖ ❖ ❖

variables until we find levels for each that maximize overall profit. This problem lives within the fourth dimension. Higher dimensions do exist—even in the world of big business.

## Is It Time?

Many believe that time is perhaps the fourth dimension. Although time does play a dimensionlike role, we do not enjoy the freedom of motion through time that we have in the other dimensions; for example, it is difficult for us to move backward in time. In fact, any attribute may be used to represent dimensions in a model of space. For example, we may use sound, color, or temperature as dimensions. That is, to locate a unique point in 6-dimensional space we can give you a point in 3-space (three coordinates) and then tell you a pitch (say C sharp), a color (purple), and a temperature (35°F). These attributes precisely describe a point in a 6-dimensional world. Our discussion of the fourth dimension has been spatial, and all four dimensions have equal geometrical status.

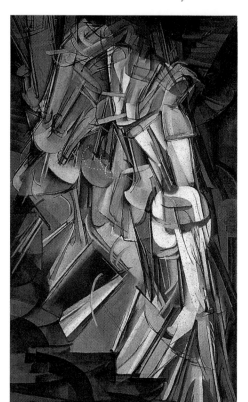

Marcel Duchamp captures the fourth dimension as time in *Nude Descending a Staircase #2* (1912).

## Is It Physical?

Okay, but does the fourth dimension exist physically? Certainly the fourth dimension may exist physically, but, just like the 2-dimensional beings who are unable to see the third dimension, perhaps we are unable to see that extra direction of freedom. Perhaps we are living on just a slice of a 4-dimensional universe. In fact, modern scientific theories describe the universe as having many physical dimensions beyond three. At this point, however, we are unable to perceive them directly. But we have already seen that the fourth dimension does exist in our minds. We can describe it, draw it, navigate around in it, and enjoy it. The fourth dimension is a real world that we can explore and view with aesthetic appreciation all beyond the sensory world but within the world of the mind.

Before moving on, we pose one last question that leads to the next stop on our mathematical tour: What happens if we are allowed to *bend* space?

## A LOOK BACK

Four-dimensional space is a world we construct in our minds by following the pattern created by the 1-, 2-, and 3-dimensional spaces that we already know. Four-dimensional creatures could pluck this text from a 3-dimensional locked safe as easily as we can lift a paper clip off a table top. To enclose valuables in four dimensions, we need a hypercube—the 4-dimensional analog of a cube. We can visualize 4-dimensional space by looking at 2- or 3-dimensional "pictures" or by unfolding objects like hypercubes.

We developed our ideas of the fourth dimension by looking at the familiar and finding patterns and analogies. The fourth dimension as developed here is not a vague wonderland, ill-defined and idiosyncratic for each individual. It has a specific shape and definite form. We followed the pattern of sweeping a lower dimensional space to create the next higher dimension. This pattern among the known spaces was sufficiently clear that we could proceed to follow the pattern even when physical models were no longer available to confirm our understanding.

One method for opening our minds to ideas imperceptible to our immediate senses is the use of analogy. Starting with familiar objects or ideas, we can look for patterns and then extend those patterns to create new ideas. We can ask ourselves, If this pattern were extended, what properties would the next object on the list have? Once we have described that next object, we have created a new idea, whether the next object can actually be constructed physically or not. Reasoning by analogy is a powerful method for creating new ideas.

*When a question seems too difficult to answer,*
*try to answer an easier related question.*

*Find patterns among familiar ideas.*

*Create new ideas by analogy.*

◆ ◆ ◆ ◆ ◆ ◆ ◆ ◆ ◆ ◆ ◆ ◆ ◆ ◆ ◆ ◆ ◆ ◆ ◆ ◆

# MINDSCAPES   Invitations to Further Thought*

## I. SOLIDIFYING IDEAS

1. ***On the level in two dimensions.*** Here are level slices of objects in two dimensions. That is, we took a 2-dimensional object and made several parallel slices with a line at different levels. What are the objects?

|            | Level 1 | Level 2 | Level 3 | Level 4 | Level 5 |
|------------|---------|---------|---------|---------|---------|
| Object 1:  | •       | • •     | •   •   | • •     | •       |
| Object 2:  | ———     | •   •   | •  •    | • •     | •       |
| Object 3:  | ———     | •   •   | •   •   | •   •   | ———     |
| Object 4:  | •       | •       | •       | •       | •       |

2. ***On the level in three dimensions*** (S). Here are level slices of objects in three dimensions. That is, we made several parallel slices with a plane at different levels. What are the objects?

|            | Level 1 | Level 2 | Level 3 | Level 4 | Level 5 |
|------------|---------|---------|---------|---------|---------|
| Object 1:  | •       | ○       | ○       | ○       | •       |
| Object 2:  | •       | △       | △       | △       | ▲       |
| Object 3:  | ●       | ○       | ◯       | ○       | ●       |
| Object 4:  | •       | ⬯       | ○  ○    | ⬯       | •       |

3. ***On the level in four dimensions.*** Here are level slices of objects in four dimensions. That is, we made several parallel slices with 3-dimensional space at different levels. What are the objects?

|            | Level 1 | Level 2 | Level 3 | Level 4 | Level 5 |
|------------|---------|---------|---------|---------|---------|
| Object 1:  | •       | ⊖       | ⊖       | ⊖       | •       |
| Object 2:  | ▪       | ▢       | ▢       | ▢       | ▪       |
| Object 3:  | •       | •       | •       | •       | •       |
| Object 4:  | •       | △       | △       | △       | ▲       |

---

*In the Mindscapes section, exercises marked (H) have hints for solutions at the back of the book. Exercises marked (S) have solutions.

4. **Tearible 2's.** In the pictures, just by bending and moving objects in the 2-dimensional plane, describe how you would remove the gold bar from the various barriers. Indicate which containers require tearing to remove the bar.

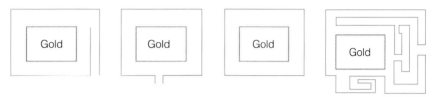

5. **Dare not to tear?** For the figures in the Tearible 2's that required tearing to remove the gold, describe how you could remove the gold without tearing its container by using the third dimension.

6. **Unlinking (H).** Using the fourth dimension, describe how you would unlink the pictured pairs of objects.

7. **Unknotting.** Using the fourth dimension, describe how you would unknot the pictured knots.

8. **Latitude.** Some 4-dimensional person has an object. She shows you 3-dimensional cross-sectional slices of it. The slices look like a circle in one level of 4-space with increasingly smaller circles at each level above and below until they end at a point at the top and a point at the bottom. What is the object?

9. **Edgy hypercubes (H).** Produce drawings of the regular cube, the 4-dimensional cube, the 5-dimensional cube, and the 6-dimensional cube.

10. **Hypercube computers.** Parallel processing computers use 4, 5, and higher dimensions by locating a processor at each vertex of the cube. One processor sends information to processors that are attached to it by an edge. The distance between two vertices is

defined to be the minimum number of edges required to create a path from one of the vertices to the other. What is the longest distance between vertices on a 3-dimensional cube? 4-dimensional cube? 5-dimensional cube? In general, an *n*-dimensional cube?

## II. CREATING NEW IDEAS

1. *N-dimensional triangles* (**H**). We saw how to build cubes in all dimensions; how about triangles? A 0-dimensional triangle is just a point. A 1-dimensional triangle is a line segment; you know what a 2-dimensional triangle looks like; a 3-dimensional triangle is a tetrahedron. What is the pattern? We take the triangle we just created and then add a new point in the next dimension "above" the triangle. If we draw new edges from the vertices of the triangle to our new point, then we have a next higher dimensional triangle. Sketch a 4-dimensional triangle and then a 5-dimensional triangle. Fill in the following table:

| Dimension of the Triangle | Number of Vertices | Number of Edges | Number of 2-Dimensional Faces | Number of 3-Dimensional Faces |
|---|---|---|---|---|
| 1 | | | | |
| 2 | | | | |
| 3 | | | | |
| 4 | | | | |
| 5 | | | | |
| *n* (in general) | | | | |

2. *Doughnuts in dimensions.* Suppose we have a mysterious object in four dimensions. If we slice the object in half with a 3-dimensional slice, we'd see the surface of a hollow doughnut.

If we take a 3-dimensional slice just above or below that first slice, we'd see the surface of another doughnut—this one thinner.

As we slice higher (and lower) we see doughnuts whose waistlines are getting thinner and thinner, until finally we just see a circle.

What is the 4-dimensional object if we know the level 3-dimensional slices? Why is the answer "a circle of spheres"?

3. **Assembly required (S).** In the preceding section we discussed unfolding cubes and promised to give you a chance to glue together a 4-cube. Here it is. Draw a picture of the unfolded 4-dimensional cube as eight 3-dimensional cubes. Indicate in your drawing which faces get glued together to reassemble the 4-cube in 4-dimensions.

4. **Slicing the cube.** Take a 3-dimensional cube balancing on a vertex and imagine slicing it with many parallel planes starting with this one. Sketch the various types of level curves we'd see. For example, the first few levels would look like triangles that are increasing in size. Continue sketching the slices and make sure to include all the shapes we'd see.

Move plane down and slice cube.

5. **4-D swinger.** The plane is just a half line swung around a point. Three-dimensional space is a half plane swung around a line.

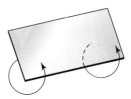

Swing a half plane around
a line and make a 3D space.

Suppose we make a circle with a dot inside it on the half plane. When we swing the half plane around to make 3-dimensional space, the circle and the point produce objects in 3-dimensional space. Describe the objects. What pair of objects do we produce if we swing only the point around but keep the circle fixed?

Swing around and
sweep out a plane.

Now what is 4-dimensional space? It is half 3-space swung around a plane. Hard to see? Yes—but try. What do you get if you take a pair of linking circles in the half 3-space and swing it around to make 4-dimensional space? What if we take a sphere in the half 3-space with a point inside. Swing the point around, but leave the sphere still. Do you get a sphere linked with a circle? Explain your answer as best as you can.

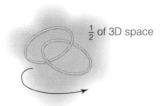

Spin around to make 4D space.
What do the circles make?

### III. FURTHER CHALLENGES

1. ***Spheres without tears.*** A sphere is the set of points at a fixed distance from a given point. A sphere in 1-dimensional space is just two points. To make a sphere in 2-dimensional space, we take two copies of the sphere in 1-dimensional space, fill each one in (color in all the points between the points), and then glue the outer edge of one of the filled-in spheres in 1-dimensional space to the outer edge of the other (this requires us to bend each sphere out a bit). This process produces a circle (a sphere in 2-dimensional space). Generalize this procedure to produce a sphere in 3-dimensional space (include pictures), and then use this method to describe how to construct a sphere in 4-dimensional space.

   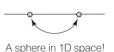
   A sphere in 1D space!

   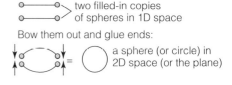

2. ***Linking.*** Start with two linked circles in 3-dimensional space. Make one of the circles the equator of a sphere in 4-dimensional space by constructing increasingly smaller circles as we rise and descend in levels of 4-dimensional space. Can you pull the sphere and circle apart in 4-dimensional space, or are they linked? Explain your answer.

### IV. IN YOUR OWN WORDS

1. ***Personal perspectives.*** Write a short essay describing the most interesting or surprising discovery you made in exploring this section's material. If any material seemed puzzling or unclear or even unbelievable, address that as well. Explain why you chose the topics you did. Finally, comment on the aesthetics of the mathematics and ideas in this section.

2. ***With a group of folks.*** In a small group, discuss and actively work through how to untie a knot in 4-dimensional space. After your discussion, write a brief narrative describing the process in your own words.

3. ***Creative writing.*** Write an imaginative story (it can be comical, dramatic, whatever you like) that involves or evokes the ideas of this section.

4. ***Power beyond the mathematics.*** Describe several real-life situations—ideally, from your own experience—for which some of the strategies of thought presented in this section would provide effective methods for approaching and resolving them.

# CHAPTER FIVE

# Contortions of Space

*The whole of mathematics is nothing more than a refinement of everyday thinking.*
—Albert Einstein

### 5.1

*Rubber Sheet Geometry*

### 5.2

*The Band That Wouldn't Stop Playing*

### 5.3

*Feeling Edgy?*

### 5.4
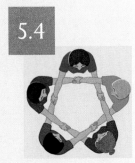
*Knots and Links*

Most objects in our everyday world are more or less rigid. Cars, buildings, and tables are of a fixed size and shape. Consequently, the common building blocks that we use to describe the shapes in our world include the firm templates of the straight line, triangle, circle, and sphere. However, some objects in our world do bend, stretch, or distort—rubber bands, waterbeds, and our waistlines. In our mind's eye, we can imagine the rigid objects of our world as stretchable and amorphic. That image gives us a whole new potential reality to explore. In such an unreasonably contortable universe, we discover things that surprise and amuse us. These discoveries lead us to new insights into our real world, including unraveling some of the mysterious behavior of DNA, the stuff of life itself.

We begin our exploration of contortable space through some classic examples of flexible possibilities, including removing a stretchable vest without first taking off one's coat. Such examples lead us to a notion of equivalence by distortion. This idea of equivalence captures some geometrical aspects of objects but is far more flexible than the rigid equivalence of "congruence" that is discussed in high school geometry. Some features of objects do not change with bending or stretching, and it is these we describe and explore.

Counting basic components of surfaces leads us to intriguing, twisted objects that initially defy our intuition. For example, a relationship among vertices, edges, and faces of random doodles on rubber surfaces helps us prove the rigid fact that there are only five regular solids. Another application of this flexible geometry is meteorological: some pair of points opposite each other on Earth must right now have the exact same temperature and pressure. Intrinsically intriguing ideas always seem to find applications.

As we have mentioned previously, a great way to stumble on new insights is to first find different ways of describing a situation. Also, seeking the essential and ignoring the irrelevant, we can represent the very core of an issue free from distracting and unnecessary baggage. Even in an abstract world created in our imagination, we can look for patterns and describe similarities and differences. In addition, whether grappling with abstract ideas or concrete ones, we can often deal effectively with complicated situations by describing a means for measuring the complexity of the situation. Then we can seek incremental progress rather than being submerged in an ocean of complication. The tools of thought that are illustrated in this contortable world are most powerfully applicable in the real world of our everyday lives.

❖ ❖ ❖ ❖ ❖ ❖ ❖ ❖ ❖ ❖ ❖

## 5.5

**Fixed Points, Hot Loops, and Rainy Days**

# Rubber Sheet Geometry:
## Discovering the Topological Idea of Equivalence by Distortion

*The moving power of mathematical invention is not reasoning but imagination.*

—Augustus de Morgan

Have you ever noticed that a rubber band can be stretched to resemble a circle, a square, a triangle, and, in fact, any polygon or any distorted version of a polygon? From a "rubber geometry" point of view, all these shapes are equivalent since we can deform any one into any other. Compare this new view of geometric equivalence with that of clas-

 same as  same as  same as

sical, Euclidean geometry in which two objects are basically the same if one can be rigidly moved to the location of the other, with the two matching perfectly.

As much as one may try, a rubber glove cannot be stretched to look like a Ping-Pong ball or a hollow inner tube. So, let's imagine that all objects are made of material more malleable than the rubber glove. Then, rubber geometry will provide us with an idea of geometric equivalence that is more flexible than Euclidean geometry yet captures some of the geometric character of objects. This "rubber geometry" has evolved into an entire new branch of

*The Persistence of Memory* (1931) by Salvador Dali.

are congruent, ... They can be lined up perfectly.

mathematics known as *topology*. The word "topology" was coined in 1847 and derives from the Greek word *topos*, which means *surface*. Topology is a mind-stretching subject that frees us from conventional geometry and allows us to appreciate geometric characteristics of objects that remain unchanged even when the objects are stretched, shrunk, distorted, or contorted.

Seek the essential.

◆ ◆ ◆ ◆ ◆ ◆

## Fun with Rubber

Fooling around with rubber can be fun; but besides its entertainment value, it can stretch the mind. To stir the imagination and introduce the essence of elastic space, we present three short tales. (Well, since they are rubber stories, they really can be stretched into three tall tales.) Solve each of the challenges before reading the answer.

## First Challenge—The Investment

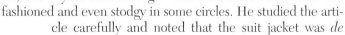

One of the authors frequently wore a natty suit complete with vest. However, one day he read in *Vogue* that vests were considered old-fashioned and even stodgy in some circles. He studied the article carefully and noted that the suit jacket was *de rigueur*, whereas vests were definitely out. Not wishing to be out of fashion for even a moment, he wanted immediately to remove his vest without removing his jacket. Given that his vest was rather large and flexible, was this feat possible? After in*vest*ing some thought in this puzzle, he wondered if he could accomplish it without unbuttoning either the vest or the jacket—if his vest were made of an extremely flexible and stretchable rubberlike material. Experiment in your mind's eye or in your clothes closet.

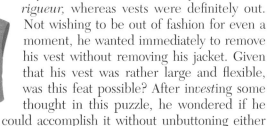

## Second Challenge—Grandma's Flat Tire

While the first author was taking off his vest, the other author happened on an elderly lady whose car had been incapacitated by a flat tire. He first carefully advised her about how to use the tire iron. Soon the 86-year-old

grandmother had pulled off the tire and removed the tube—being guided throughout by the author's numerous and helpful suggestions. She showed him the rather large puncture in the tube. In fact, the hole was about 5 inches across.

The silver-haired matron was patting the perspiration from her brow and catching her breath from the exertions of jacking up the car while our math hero said, "Do you think it is possible to turn the inner tube completely inside out? Just reach in through that hole and pull on the insides and see what happens."

The grandmother dutifully did as instructed. At last she stood, covered with grime from head to toe. "You have just discovered a topological fact about the torus!" cried our impassioned author with glee. "Is my tire almost fixed?" asked the grandmother. "Whoops, got to get to class," said our fearless author as he nimbly dodged the tire iron that the grandmother apparently did know how to use after all. Challenge: Was she able to turn the inner tube inside out?

### Third Challenge—The Ring

We are given a thin sheet of extremely flexible and stretchable rubber with two holes. An expensive gold ring passes through the two holes. Is it possible to remove the ring from one of the holes? Your challenge is two-fold: Try to answer this question, and then try to write a funny story that uses the question as the punch line.

### The Investment Solution

To illustrate the notion of topology, we first turn to The Investment story and examine how, if the vest is made of very distortable rubber, we can remove the vest without even unbuttoning the jacket or the vest.

Since we are not allowed to unbutton the vest, the vest has the same characteristics as a pullover sweater vest made of rubber. As the vest is extraordinarily flexible, we can shrink the bottom part of the vest upward without changing its topological properties. (Remember, stretching or shrinking will not change an object in this new type of geometry.) In fact, we can shrink the bottom of the vest so that its bottom seam comes within an inch of the head hole and the arm holes. The vest now appears to be more like a stretchable tank top. Notice that all this shrinking can take place without the vest ever leaving the cozy confines of the sports jacket.

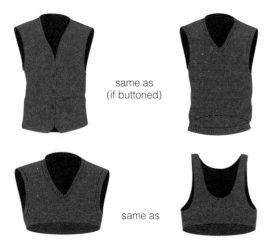

We can now stretch one shoulder strap down through the jacket sleeve to our hand, pass the strap around our hand, and shrink the strap back up the sleeve to its original size. If we then move the other shoulder strap down and around the other arm, we see that the straps are no longer over

our shoulders. Thus by wiggling, we can slide the entire rubber garment down to our feet right through the bottom of the jacket. We can then pick up the tank top and stretch it all out again to make the unfashionable, original vest.

Despite the fact that we bent, stretched, and shrunk the vest, at every stage of the "devestment" the vest remained *equivalent by distortion* to our original one.

## Grandma's Flat Tire Solution

To refine this notion of equivalence by distortion, we return to Grandma's flat tire and show how we can turn an extremely distortable inner tube, with a hole, inside out. In the sequence of figures below, notice how we begin with an inner tube that is red on the inside and blue on the outside. We then carefully stretch and deform that hole and, once the dust settles, we are left with the red surface on the outside and the blue surface on the inside.

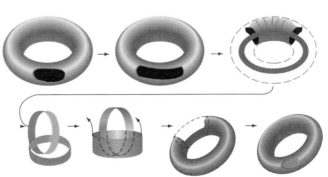

By the way, this feat can physically be accomplished with a regular inner tube having a 5-inch diameter puncture. If you can get your hands on an inner tube, give it a try.

## The Ring Solution

The counterintuitive fact is that the ring *can* be removed from one of the holes. Here is a sequence of stretches that accomplishes this surprising feat.

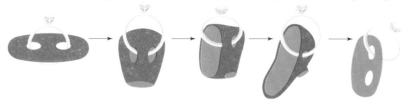

## Equivalence by Distortion

Let's further develop our intuition regarding rubber sheet geometry by comparing various objects and determining what kinds of deformations are allowable and what are not. Since we wish to preserve, protect, and defend the fundamental structure of our object during a distortion, we will not allow breaking or tearing. We want to maintain the integrity of the object in the sense that points that were close to one another in one position remain close throughout the distortion. As an analogy, let us think of the object as being made of molecules that grip adjacent molecules. If the object is stretched, each molecule still grips the same molecules. However, if we cut a rubber band and straighten it out, the continuity of the cycle is broken; a bond between connecting molecules has been destroyed. So, cutting a rubber band to get a line segment is not allowed, since cutting would change the fundamental structure of the rubber band. (The rubber band can hold a bunch of index cards together, whereas a rubber line segment cannot.) The rubber band and the rubber line are topologically inequivalent objects.

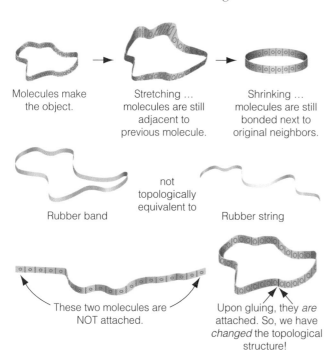

In a similar spirit, we do not wish to allow gluing together points that were not connected to each other before. For example, suppose we take a rubber line segment and glue the ends together. We have just created a rubber band that, as we have already noted, is structurally different from the segment. When we attach the ends, we are bringing together points that were not touching in the original segment. In the analogy of molecules, we are not allowed to create connections between molecules that did not exist before.

So, summing up, we'll say that two objects are *equivalent by distortion* if we can stretch, shrink, bend, or twist one, without cutting or gluing, and deform it into the other.

## *Undisturbed by Distortion*

What are some features of an object that remain intact no matter how much it is distorted? A circle has the property that, if we remove any point from it, what remains of the circle is still one piece. Any distortion of the circle retains this property. This feature of a circle allows us to deduce that we could not stretch it to make it look like a circle with a feeler attached, because there is a point on the circle with a feeler that, if removed, actually breaks the object into two pieces.

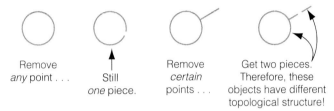

Remove *any* point . . .   Still *one* piece.   Remove *certain* points . . .   Get two pieces. Therefore, these objects have different topological structure!

However, if we remove any two points from a circle, the circle falls into two pieces. This feature allows us to deduce that a circle cannot be distorted to look like a theta curve—that is, an object in the shape of the capital Greek letter *theta*—because there are pairs of points that, when removed, leave the theta curve in one piece.

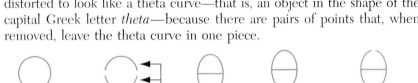

Remove *any* pair of points . . .   We get TWO pieces.   theta curve   Remove *certain* pairs of points . . .   Still have one piece. Thus, the circle and theta curve are topologically inequivalent!

Let's imagine ourselves as a small bug crawling along a circle; our entire sensory apparatus allows us to detect only the immediate neighborhood of the circle on which we crawl—not the space off it. So, wherever we stop, we can feel ahead and behind, and that's all. Our neighborhood feels the same as a slightly distorted line segment.

Suppose now that we are crawling along a curve, but we are not told whether we are crawling around on a circle or on a theta curve. Could we tell on which curve we are marching, purely by sensing our immediate neighborhood?

At most places, we could not tell the difference between the neighborhoods of a point on the circle versus a point on the theta curve; however, if we were standing at one of the two points where the horizontal bar attaches to the circle on the theta curve, we would feel a different surrounding. We would notice that there are three directions in which we could travel from that point. The number of segments emanating from a point would be preserved under any distortion. Therefore, the circle and the theta curve are not equivalent by

Where are we?

What are we on?

distortion, because the theta curve contains a point (in fact, two points) from which three different segments emerge, whereas every point in the circle is like every other point on the circle in having only two directions. So, this local difference in features distinguishes these two curves.

Three different directions to move. There's no such intersection on a circle. So, this can't be a circle!

## *Feeling Our Way on Surfaces*

To gain a deeper understanding of the idea of equivalence by distortion, we now examine various surfaces. Let's consider a sphere—that is, the boundary of a ball—and attempt to find features of it that would be preserved under distortions. If we remove one or two points from a sphere

Remove a few points → still one piece! → Remove a loop → we get *two* pieces.

(or even more, of course), the object left over remains in one piece. But, if we cut out any loop, the sphere will fall into pieces.

Observe that, if we were a bug on a sphere and could sense only the neighborhood immediately around us, every neighborhood would seem like every other one. In fact, every point on a sphere has a neighborhood surrounding it that is equivalent by distortion to a neighborhood of a point in the flat plane. Any object with the property that around every point there is a small neighborhood equivalent to a small neighborhood on the plane is called a *surface*.

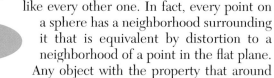

Neighborhood looks like

## *What Is a Torus?*

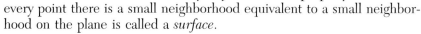

torus

A *torus* is the boundary of a doughnut. This inner-tube-shaped object has some similarities to the sphere, as well as some differences from it. Suppose we were a bug and we lived on a torus. As we crawl over the surface sensing only the surface itself, we would notice that every neighborhood felt the same as any other. In fact, we could not tell the difference between crawling around on the torus and crawling around on the sphere.

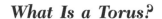

inside is hollow

334 ◆ CONTORTIONS OF SPACE

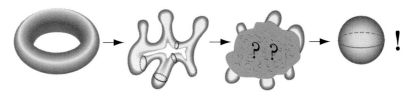

Perhaps this is actually how the Earth is.

There is no difference on the local level—they are both surfaces. In fact, this question raises an amusing issue. How do we know that the surface of the Earth is equivalent by distortion to a sphere? Everyone learns this fact at an early age, but how can we know this and prove this by personal experience? Sure, we see photographs of the earth taken from outer space, but perhaps they were altered by computer. The point is that, as we travel along the surface, we cannot be certain that we are walking around on a sphere rather than on a torus! Is there a difference between a sphere and a torus? If so, how can we tell the difference?

Probably we believe that a torus is different from a sphere because it has a hole. But how can we be certain that the hole cannot be distorted by some unthought-of maneuver such that the torus could be stretched to look like the sphere? The answer is to find some feature of the torus that is different from the sphere and yet would be preserved under any distortion.

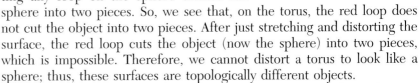

## Is a Torus a Sphere?

Look at the red circle on the torus pictured here. Suppose we cut the torus around that circle. Notice that the cut torus is still one piece; it's become a tube. We have not separated the torus into two pieces. Suppose now that we could somehow distort the torus to make it look like the sphere. Then, after the distortion that red circle would become a red loop on the sphere. But cutting any loop on the sphere divides the

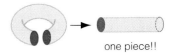

one piece!!

sphere into two pieces. So, we see that, on the torus, the red loop does not cut the object into two pieces. After just stretching and distorting the surface, the red loop cuts the object (now the sphere) into two pieces, which is impossible. Therefore, we cannot distort a torus to look like a sphere; thus, these surfaces are topologically different objects.

We close this section with two intriguing questions. We hope that you first spend some time drawing pictures, visualizing objects in your mind's eye, and talking about these questions with others before reading the answers.

## Puzzle 1—Holy Doughnuts

hollow inside

The torus we considered above had one hole. Let's now consider tori with more than one hole. A two-holed torus is just the surface of a two-holed doughnut. Notice that the inside is hollow. It is easy to see that the

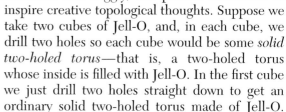

following figures are all equivalent by distortion to the original two-holed torus. Our question here is, What happens if the two holes are linked? Specifically, is the linked two-holed torus equivalent by distortion to an unlinked two-holed torus? If so, draw a sequence of pictures showing how to stretch and distort one to get the other. If not, provide a reason why the linked holes cannot be unlinked.

## *Puzzle 2—Knotted Jell-O Cubes*

Jell-O is beautifully translucent and wiggly, the perfect substance to inspire creative topological thoughts. Suppose we take two cubes of Jell-O, and, in each cube, we drill two holes so each cube would be some *solid two-holed torus*—that is, a two-holed torus whose inside is filled with Jell-O. In the first cube we just drill two holes straight down to get an ordinary solid two-holed torus made of Jell-O.

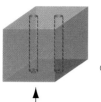

solid Jell-O with two holes drilled

NOT equivalent by deformation to

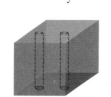

Romantically tied together...

?
=

However, in the second cube, one of the holes is knotted. Although we will not prove it, these two Jell-O cubes are not equivalent by distortion; there is no way to unknot that knotted hole. Here we wish to consider a third Jell-O cube with two holes, one of which is knotted and circles around the unknotted hole. Question: Is this knotted Jell-O solid two-holed torus equivalent by distortion to the ordinary (unknotted) Jell-O solid two-holed torus? If so, draw sketches that show how to deform one into the other. If not, provide a reason why they are not equivalent. (You may use the previous fact about the first knotted Jell-O mold.)

## *Solutions—A Picture Is Worth a Thousand Words*

The surprising and counterintuitive answer to both questions is, yes, they are equivalent! We now illustrate a sequence of moves that depict the distorting to get from one object to the other.

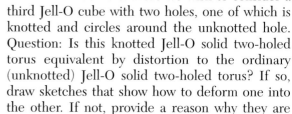

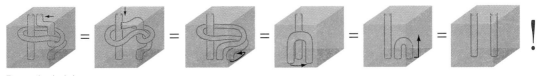

Down the hole!

## A Twisted Final Thought

Each of the surfaces we've examined thus far has two sides, an inside and an outside—as we expect. If you are standing on the outside surface of a tube, for example, and you walk along the surface and return to your starting point without crossing an edge, then you will still be on the same side of the surface. Is it possible to make a surface so strange that we could start at one point, walk along a path never crossing an edge, and return to the same point, but find ourselves on the other side of the surface? We might think that is impossible. But, as usual, questioning our intuition and our expectations reveals a whole world of fascinating possibilities. We will grapple with these disorienting ideas in the next section.

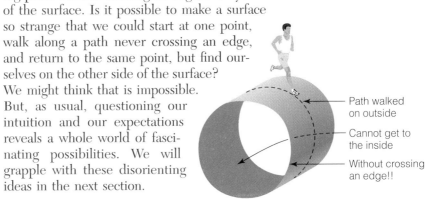

Path walked on outside

Cannot get to the inside

Without crossing an edge!!

**A LOOK BACK**

Rubber sheet geometry, the nickname for topology, centers on a flexible notion of equivalence. Two objects are equivalent by distortion if we can stretch, shrink, bend, or twist one, without cutting or gluing, and deform it into the other. These distortions alter some geometric properties of the objects but preserve others. Looking for these invariant, or unchanged, features is an effective way to determine when two objects cannot be made equivalent by distortion. If we cut along any loop on a sphere, we divide the sphere into two pieces; however, there are some loops on a torus that do not separate the torus by cutting. Thus, the torus is not equivalent by distortion to the sphere. However, surprising pairs of objects are equivalent by distortion.

Our primary strategy for investigating this idea of rubber sheet geometry is to try things physically when possible or to draw lots of pictures. We must be open-minded to avoid narrow thinking and must seek specific reasons when we believe some distortion is impossible. Rubber sheet geometry forces us to identify essential features of objects and to ignore others.

In real life, we group people or objects together on the basis of characteristics they share. However, a great way to find new ideas and new connections is to choose alternative ways to identify things. In this way, we can see old objects and ideas in a new light. We can explore this new

Seek the essential.

Ignore the irrelevant.

Explore uncharted means of comparing and contrasting objects and issues.

◆ ◆ ◆ ◆ ◆ ◆ ◆ ◆ ◆ ◆ ◆ ◆ ◆ ◆ ◆ ◆ ◆ ◆ ◆

# MINDSCAPES   Invitations to Further Thought*

## I. SOLIDIFYING IDEAS

1. *That theta* (S). Does there exist a pair of points on the theta curve whose removal breaks the curve into three pieces? If so, the existence of those two points would provide another proof that the circle is not equivalent by distortion to the theta curve. Why?

2. *Your ABCs* (H). Consider the following letters made of 1-dimensional line segments:

   A B C D E F G H I J K L M N O P Q R S T U V W X Y Z

   Which letters are equivalent to one another by distortion? Group equivalent letters together.

3. *Puzzled?* Find the topology puzzle in your kit. Read the kit instructions for the puzzle and then solve the puzzle. Describe your solution in words and pictures.

4. *Half dollar and a straw.* Suppose we drill a hole in the center of a silver dollar. Would that coin with a hole be equivalent by distortion to a straw? Explain why or why not.

5. *Drop them.* Is it possible to take off your underwear without pulling down your pants? You may assume you are wearing rubber undies. Explain why or why not. Include pictures.

6. *Coffee and doughnuts* (H). Is a standard coffee mug equivalent by distortion to a solid doughnut? Explain why or why not.

---

*In the Mindscapes section, exercises marked (H) have hints for solutions at the back of the book. Exercises marked (S) have solutions.

7. ***Lasting ties.*** Handcuff a person with a thin rope tied around his wrists, and then handcuff yourself with another rope, with the rope handcuffs linked as pictured. Can you unlink yourselves without cutting the ropes? Why or why not? (This challenge is one you should definitely make physical!)

8. ***Will you spill?*** **(S).** Suppose you rest a glass of water in the palm of your hand. Is it possible to rotate the palm of your hand 360° without spilling the water or dislocating your arm?

9. ***Grabbing the brass ring.*** Suppose a string passes through a metal, nondistortable frame and then is tied to a brass ring and fastened to a ceiling, as illustrated in the picture. Is it possible to remove the brass ring?

10. ***Hair care.*** Is a regular comb equivalent by distortion to a regular hair pin? Explain why or why not.

11. ***Three two-folds.*** Take three pieces of paper and fold each in half and then half again, as shown in the figure. Suppose from the first paper you cut out a semicircle from the edge opposite the two folds, and from the second you cut out a small semicircle from the one-fold edge, and from the third you cut out a semicircle from the two-fold edge. Are those cut sheets of paper equivalent? Explain.

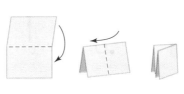
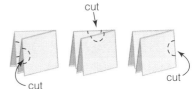

12. ***Equivalent objects.*** Consider this photograph, and group the objects into collections that are equivalent by distortion.

13. ***Clips.*** Is a paper clip equivalent to a circle? If not, to what other small stationery products is a paper clip equivalent by distortion?

14. ***Pennies plus.*** Consider the two pictured objects. One is made of two pennies, cut and glued together as shown. The other is a thickened plus sign. Are these two objects equivalent by distortion?

**15. *Starry-eyed.*** Consider the two stars. Are they equivalent by distortion?

**16. *Learning the ropes.*** Here are two ropes, coiled differently. Are the ropes equivalent by distortion?

**17. *Holy spheres.*** Consider the two spheres. Each has four holes on its surface. Through the holes, two ropes are looped in different ways. Is the first sphere with its two rope loops equivalent by distortion to the second sphere with its rope loops?

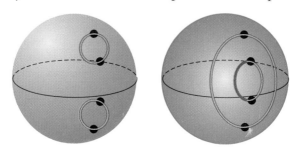

**18. *From sphere to torus.*** The following sequence of drawings takes a sphere and deforms it into a torus. Does this sequence describe an equivalence by distortion? Why or why not?

**19. *Half full, half empty.*** One glass is half filled with cranberry juice as illustrated. Another glass is also half filled with juice, although it is the top half (ignore such pesky issues as gravity). Are the two half-filled glasses with their contents equivalent by distortion? Explain.

**20. *Male versus female.*** Consider the male and female symbols. Assuming they are made out of 1-dimensional lines and curves, are the symbols equivalent by distortion? Explain.

## II. CREATING NEW IDEAS

1. ***Holey tori.*** Are these two objects equivalent by distortion? If so, demonstrate the distortion with a sequence of pictures; if not, explain why not.

   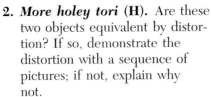

2. ***More holey tori*** (**H**). Are these two objects equivalent by distortion? If so, demonstrate the distortion with a sequence of pictures; if not, explain why not.

   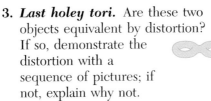

3. ***Last holey tori.*** Are these two objects equivalent by distortion? If so, demonstrate the distortion with a sequence of pictures; if not, explain why not.

   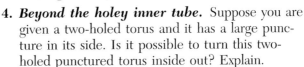

4. ***Beyond the holey inner tube.*** Suppose you are given a two-holed torus and it has a large puncture in its side. Is it possible to turn this two-holed punctured torus inside out? Explain.

   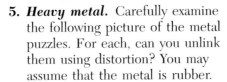
   Puncture hole

5. ***Heavy metal.*** Carefully examine the following picture of the metal puzzles. For each, can you unlink them using distortion? You may assume that the metal is rubber.

6. ***Rings around the ring.*** Return to The Ring challenge given at the beginning of this section. Draw a red circle around one of the holes on the rubber sheet and a blue circle around the other.

   Now redraw the sequence of moves that unlinks the ring from one of the holes, but on each figure now include how the two colored circles get deformed. What is the result? Are both colored circles still looped around the ring?

7. ***The disk and the inner tube.*** Suppose you have a rubber disk with two holes and you glue a tube from one hole to another. Is that object equivalent by distortion to an inner tube with a puncture? Explain.

   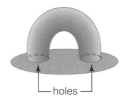
   holes

   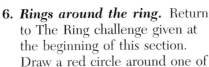

8. ***Building a torus*** (**S**). Suppose you are given a rectangular sheet of rubber. How could you glue the various edges together to build a torus? Indicate which edges get glued to which and how the edges match up.

9. **Lasso that hole.** Consider the two tori on the left. Both have two punctures on their sides. On the first torus, a rope is looped through the two holes but does not go around the hole of the torus. On the second, the rope is looped around the hole of the torus. Is it possible to distort the first object to look like the second? How about if the rope looped around the hole twice as shown on the right?

10. **Knots in doughnuts.** We are given two solid doughnuts: one with a worm hole drilled as shown and the other with a knotted worm hole drilled as shown. Are these two-holed doughnuts equivalent?

### III. FURTHER CHALLENGES

1. **From knots to glasses.** Take the thickened knot and then add a solid tube, as illustrated. Is it possible to distort this new object into a pair of eyeglass frames?

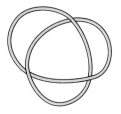 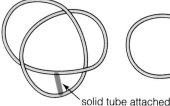 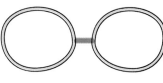

solid tube attached

2. **More Jell-O.** Suppose we take a cube of Jell-O, drill two holes in it, and then glue a Jell-O tube between them.

Is this Jell-O object equivalent to the Jell-O cube with two holes?

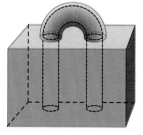 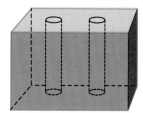

3. **Fixed spheres (H).** We are given two spheres that are made of glass and thus are not distortable. They are rigidly fixed in space: one inside the other as indicated, and they cannot be moved. The middle sphere floats miraculously in midair without any means of support. A rubber rope

hangs from the inside ceiling of the big sphere to the roof of the smaller sphere, as illustrated. The rope has a knot in it. Is it possible to unknot the rope without moving or distorting the two spheres?

4. *Holes.* Is a torus equivalent to a two-holed torus? If not, carefully justify your answer. (Consider cutting the objects more than once.)

5. *More holes.* Is a two-holed torus equivalent to a three-holed torus? If not, carefully justify your answer. Can you generalize your observations and arguments?

### IV. IN YOUR OWN WORDS

1. *Personal perspectives.* Write a short essay describing the most interesting or surprising discovery you made in exploring this section's material. If any material seemed puzzling or even unbelievable, address that as well. Explain why you chose the topics you did. Finally, comment on the aesthetics of the mathematics and ideas in this section.

2. *With a group of folks.* In a small group, discuss and actively work through the reasoning for how a two-holed torus with its holes linked can be distorted into an unlinked two-holed torus. After your discussion, write a brief narrative describing the method in your own words.

3. *Creative writing.* Write an imaginative story (it can be comical, dramatic, whatever you like) that involves or evokes the ideas of this section.

4. *Power beyond the mathematics.* Provide several real-life situations—ideally, from your own experience—for which some of the strategies of thought presented in this section would provide effective methods for approaching and resolving them.

## 5.2 The Band That Wouldn't Stop Playing:
### Experimenting with the Möbius Band and Klein Bottle

*If a man's wit be wandering, let him study mathematics.*

—Francis Bacon

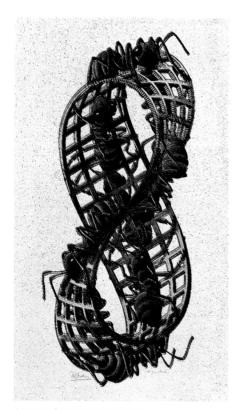

Mobius Strip II by M.C. Escher.

A sphere, the surface of a doughnut, and the surface of a two-holed doughnut are all different surfaces; it is impossible to distort any one to look like the others. However, each of these surfaces does have two sides—an inside and an outside—in the sense that, if we were bugs crawling on the outside surface and we walk along the surface and return to our starting point without crossing an edge, then we will still be on the outside of the surface. Does every surface have two sides in that sense? Maybe any surface automatically has two sides. But maybe not.

To help us think about the two sides of this sidedness issue, we will build a physical model and then explore it through various different experiments. In particular, we'll see what happens when we cut it up in various ways. We may encounter some surprising results. That feeling of surprise, however, means that there is more to understand. So, we'll keep looking at the situation in different ways until the results change from mysterious to simply neat. Even after we understand it completely, the Möbius band remains intriguing.

- Path walked on outside
- Cannot get to the inside
- Without crossing an edge!!

## The Möbius Band

The Möbius band has been a crowd-pleaser for decades. After we have held a Möbius band and explored its endless edge, we experience an eerie sense of oneness: one edge, one side, one, one, one.

Looking at a picture is not good enough. You need a Möbius band, and you need it now. To make a Möbius band, get a strip of paper about 11 inches long and about 1 inch wide. First, bring the two ends of the strip together to make a loop. Now give one end of the strip a half twist and again bring the ends together. Tape the ends together. You now have a band with a half twist in it. You have a *Möbius band*.

What makes the band you hold so Möbius? It's that twist, of course. The half twist creates unexpected wonders. Let's explore. Hold the Möbius band; touch its edge; notice its simple elegance. As you slide it through your hands you may feel the urge to explore it a little further. What has the twist done besides create such haunting beauty? Some basic physical experiments will lead us to some startling discoveries.

### Experiment 1—Life on the Edge

Let's start with an easy experiment. Take a red marker and begin coloring around the very edge of the Möbius band. We are envisioning your marker as soaking a little of the band as you slide along. Keep going until you get back to where you started. By the way, how many edges are there? Let's count them. There is the one you just colored, and . . . Surprise! Why do we see such a strange phenomenon?

### Experiment 2—A Walk on the Wild Side

Let's now use a blue pen to count the number of sides. Put the pen in the middle of a side of the Möbius band and start drawing a line right down the middle all the way around the band. Keep drawing without lifting the pen. Keep going until you return to where you started. You have now drawn a circle on the Möbius band, and, of course, that circle, being one piece, all lies on the same side of the band. But wait. Peek on the opposite side of the band. What do you see? How could that circle be on the back side, too? The back side is the same side as the front side. How is that possible?

### Experiment 3—Making the Cut

So, the Möbius band seems to be deficient in edges and sides—only one of each, but abundant in mystery and intrigue. What is going on here? Before uncovering any explanations, let's try another experiment. Take a pair of scissors and cut the band lengthwise down the center core. (In fact, your blue pen line provides a guide for cutting.) Before physically executing this experiment, try it first in your mind's eye and make a guess as

Make guesses.

◆ ◆ ◆ ◆ ◆ ◆ ◆

to what you will see. *Any guess is fine, but please guess something!* You may want to record your guess because guesses and even mistakes often lead to insights and an appreciation for hidden subtlety. Now cut the Möbius band along the center blue core line. What do you see?

## *Experiment 4—Hugging the Edge*

Cut, staying $\frac{1}{3}$ inch away from the right edge of the strip with the scissors.

Build another 11″ × 1″ Möbius band. We're going to want you to cut it again lengthwise. But this time, instead of cutting right down the center core line, cut so that your scissors hug (stay close to) the right edge as you slide the Möbius band along while you cut. That is, cut so that the scissors are always ⅓ inch away from the right edge. Before proceeding, think about the result of this experiment in your mind's eye and make a guess as to what you think you will have. Remember: First think, then guess, then cut. Now perform the experiment. Cut the band lengthwise, but keep the scissors ⅓ inch away from the right edge. *No matter what happens, don't give in to the temptation to deviate from that right edge where your scissors currently are cutting!* The outcome is surprising!

## *Some Experimental Results*

How many pieces did you end up with in experiment 3? Did you guess *one*? One, one, one. One edge, one side, and one piece after you cut it down the middle. Notice how the new narrow band is longer than the original one. How many edges does your long twisted strip now have? Fortunately, one edge is already marked in red. The other edge is basically marked in blue if you stayed right in the middle of the blue marked line while you cut. So, how many edges does this new band have? Two edges. In fact, check for yourself that the twisted strip you now hold has two full twists, two edges, and, you guessed it, two sides. Just one cut and we doubled everything! Why? Given all these observations, were you surprised by the outcome of experiment 4?

How can we understand why these peculiarities occur? We have physically seen that the Möbius band behaves as it does, but we can understand why by looking at the band in different ways.

Look at alternative representations of things.

◆ ◆ ◆ ◆ ◆ ◆ ◆

When trying to understand an object or an idea, it is often helpful to consider different descriptions of it. Frequently it is helpful to create a model—not necessarily a physical model but a representation that helps us to see the object from a certain point of view. In this case, let's think about how the Möbius band is constructed and use that knowledge to help us construct a flat representation of the twisted band.

Recall that we construct the Möbius band using a strip of paper. We attach the ends together in a special way, namely, with a half twist. To represent this process, let's draw a strip and indicate how the ends of the strip are to be glued to create the Möbius band. We'll mark the short sides with arrows to indicate that, to make the Möbius band, we must bring those short sides together and glue them to each other in such a way that the

edge identification

Glue so arrows line up.

A Möbius band!
(assembly required...)

arrows are pointing in the same direction. Notice that the twisting is now captured in the directions of the arrows. In fact, as long as we understand that those arrows have to line up, we can think about the Möbius band without actually gluing the ends. We just understand that our picture means that those ends are, in fact, to be glued in the appropriate manner to create the Möbius band. The picture can be viewed as a kit with instructions to build the Möbius band (some assembly required).

Using this representation of the Möbius band, let's take another look at some of the results we observed.

## *Making the Cut—Experiment 3 Revisited*

Cutting down the center line would look like this:

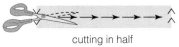

cutting in half

As the scissors reach the right edge, the cut returns where it started. How can we now see why we get only one piece after cutting? Let's place a small bug on the surface and have it walk around. As the bug crawls toward the right and crosses the upper part of the right side, it doesn't fall off— there is no cliff. In fact, it doesn't even see an edge; it just sees a seam. Remember the gluing. It just sees

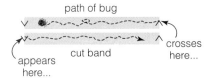

path of bug

appears here...

cut band

crosses here...

the band continuing. As it crosses that upper-right edge, where does it reappear? Visualize this activity and remember the arrows.

The bug steps out on the lower-left edge (on the back) and continues to walk across until it reaches the lower-right edge and then continues on the upper-right edge (now in front) until it returns to where it started.

How many pieces are there? One, since there is no piece of the band that was not touched by the bug! This demonstrates why cutting a Möbius band down the middle results in only one piece. In fact, our analysis allows one to

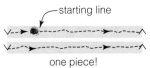

starting line

one piece!

determine the length of the new band and the number of twists it has. (If you don't believe us, just take a look at the Mindscapes.) Who said a math proof shouldn't have bugs in it? Okay, so much for our cutting sense of humor, let's see if this method can be used to analyze the cutting experiment in which we hugged the right shoreline.

## *Hugging the Edge—Experiment 4 Revisited*

In this cutting experiment, we cut keeping the scissors 1/3 inch off the right (bottom) edge. As we cut, we come to the right edge of the strip. Where do the scissors move beyond

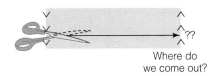

??

Where do we come out?

this point? The scissors appear at the upper-left side and continue to cut to the right. However, now, as the band is twisted, the rule of staying 1/3 inch off the bottom side requires us, as in our diagram, to stay 1/3 inch off the top side. The key lies in the twisting. We continue to cut and when we reach the right edge of our strip we return back to the lower left where we started our cut. So how many pieces do we have? Imagine a bug on one of the pieces and let it wander around. What do you see?

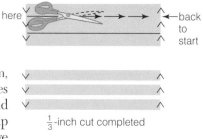

$\frac{1}{3}$-inch cut completed

If we place a ladybug on the upper strip and have it walk to the right, it will walk until it gets to the right edge, and then it emerges . . . where? Once it moves beyond the lower-left edge, it continues to travel, passing the lower-right edge, which brings it back to the upper-left edge and then back to its starting place. So, the top and the bottom strips are connected to make one large strip. Did the ladybug traverse everything? No! That middle piece was untouched by the ladybug and thus is a second piece. This model demonstrates that, when we cut the Möbius band in this manner, we get two pieces. This analysis also shows much more, as you will again discover in the Mindscapes.

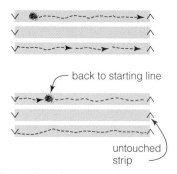

back to starting line

untouched strip

## *Möbius Bands Abound*

Although Möbius bands have a great many wondrous properties that may surprise us, it turns out that, in our everyday world, we are surrounded by Möbius bands. Where? Just look at any recycling symbol. Why do you think the standard logo is a Möbius band?

Möbius bands appear in some surprising places. Factories sometimes use Möbius bands as conveyor belts. Some conveyor belts have a half twist underneath them so that the belt wears evenly on both "sides" at the same time. This twist extends the life of the belt. In fact, such a Möbius belt was patented by the B. F. Goodrich Company.

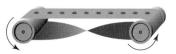

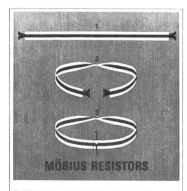

348   ◆   CONTORTIONS OF SPACE

She can never get to the inside from the outside; he can never get to the outside from the inside. The sphere has no edges but two sides!

## One-Sided Surfaces

The mysterious features of the Möbius band arise from its single-sidedness, but, of course, the Möbius band has an edge to it. Is it possible to construct a surface that has only one side, with no edge at all? A sphere has no edge, but, if we walk on one side of the surface, we can never get to the other side. Now we are asking if there is a surface without any edges such that we can start at one point, walk along the surface, and arrive at the opposite side of our starting point—just as we did on the Möbius band. It is worthwhile to think about this disorienting question before moving on.

The answer is yes. One strategy for constructing such a surface would be to start with a Möbius band and somehow get rid of its edge. Since it is not clear how to accomplish this task, we examine similar, related situations in the hopes of discovering a useful idea.

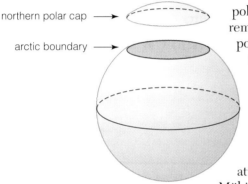

Suppose someone gave us a sphere with its northern polar cap removed. The sphere with its northern polar cap removed has a boundary, namely, the arctic circle. Also, the polar cap itself has a circle boundary. We could eliminate the arctic circle boundary by pulling the northern polar cap back in. Could we do something similar to eliminate the boundary of the Möbius band, thus creating a surface with no boundary?

Well, let's try to add a cap as we did on the arctic circle. Recall that the Möbius band has only one edge. Let's just take a distortable disk (a truly floppy disk) and attach the disk to the Möbius band along the edge of the Möbius band. To make it physical, let's suppose we have a cloth Möbius band and a cloth disk whose boundaries have the same lengths. Now we simply start sewing the edge of the disk along the edge of the Möbius band. We do fine for a while, but eventually something gets in our way. In fact, we will always get stuck if we remain in our pedestrian 3-dimensional world. If, however, we are more liberal in our choice of habitat, we can take our cloth disk and sew its boundary on to the boundary of the Möbius band while keeping the inside of the disk in the fourth dimension out of the 3-dimensional space in which the Möbius band lives. This construction does eliminate the boundary of the Möbius band. However, this surface, known as the *projective plane*, is a bit unsatisfying in that it is too abstract and hard to visualize in our mind's eye. Is there a one-sided surface that is easier to visualize?

Start with what you know.

◆ ◆ ◆ ◆ ◆ ◆ ◆

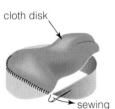

Eventually we get stuck.

## The Klein Bottle

Klein bottle

Although we do not justify it here, the fact is that no one-sided surface without an edge can be constructed entirely in 3-dimensional space. Nevertheless, we can effectively describe an elegant one-sided surface known as the *Klein bottle*. The rules for constructing a Klein bottle are simple, and the resulting surface can be attractively modeled. To construct the Klein bottle, take a rectangle and glue one pair of opposite sides together without a twist, creating a tube. Next glue the other pair of opposite sides to each other, putting in a half twist as in the Möbius band.

   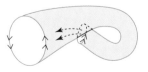

gluing ...

Arrows don't match up!

Move inside so that arrows line up.

When we follow the directions for constructing a Klein bottle, we soon discover that the Klein bottle passes through itself. We can construct a non-self-intersecting Klein bottle in the fourth dimension, or we can construct the Klein bottle in three dimensions having one circle-shaped self-intersection. The neat property of a Klein bottle is that its inside is the same as its outside! In the rectangular-edge matching model, we can see the one-sidedness by just following a path over the twisted edge and seeing that we switch sides. Alternatively, we can look at the 3-dimensional model and trace our finger lengthwise along the tube and see that, when we continue around, we find ourselves on the inside. What would happen if you tried to fill a Klein bottle with your favorite beverage? Is it inside? Is it outside? Cheers!

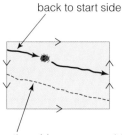

back to start side

on other side ... so, one side!

Besides the amazing hidden secrets we uncovered in the mysterious Möbius band and the convoluted Klein bottle, we also saw the value of the diagrams with the arrows. They enabled us to create, understand, and explain phenomena that first appeared to be mystifying. In fact, similar diagrams can be used to describe and analyze any surface.

The Klein bottle has many interesting mathematical properties, but it is also an especially intriguing work of art. Stone, glass, and metal renditions of the Klein bottle have added grace and beauty to many museums. Perhaps this section will inspire some artists to make a Klein bottle of their own.

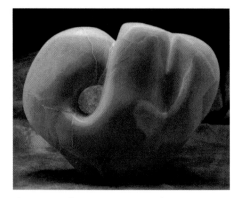

*Eine Kleine Rock Musik III* (1986) by Helaman Ferguson.

**A LOOK BACK**

The Möbius band has one edge, one side, and remains in one piece after being cut lengthwise down the middle. It demonstrates that a surface can be one-sided. A Klein bottle is a surface without an edge that also has only one side; however, the Klein bottle cannot be constructed entirely in 3-dimensional space without passing through itself.

To better understand the Möbius band and the Klein bottle it is useful to represent those surfaces as a construction model with construction rules. Each can be described as a rectangle or a square with edges glued together according to the assembly instructions. Examining these deconstructed versions of the Möbius band and the Klein bottle lets us visualize relationships that may be harder to see in the assembled versions.

Often what first appears surprising and mysterious can be explained clearly by thinking about the situation in a completely different manner. By doing so, we see another side to the issue, and we begin to build greater understanding and deeper insights.

*Building physical models and performing experiments is a powerful method for making new discoveries.*

*Look for alternative means of describing a situation to help explain surprising outcomes and build new insights.*

❖ ❖ ❖ ❖ ❖ ❖ ❖ ❖ ❖ ❖ ❖ ❖ ❖ ❖ ❖ ❖ ❖ ❖

# MINDSCAPES  Invitations to Further Thought*

### I. SOLIDIFYING IDEAS

1. ***Record reactions.*** Explain to a friend how to make a Möbius band and how to cut it up in the various ways described previously. Make your friend guess the outcomes before he or she does the cutting. Explain why they work as they do. Record their reactions and thoughts.

2. ***The unending proof.*** Take a strip of paper and write on one side: "Möbius bands have only one side, in fact while." Next turn it over on its long edge and write "reading *The Heart of Mathematics*, I learned that." Now tape the strip to make a Möbius band. Read the band. This provides another proof that the Möbius band has only one side.

3. ***Two twists.*** Take a strip of paper, put two half twists in it, and glue the ends together. Cut it lengthwise along the center core line. What do you get? Can you explain why?

---

*In the Mindscapes section, exercises marked (H) have hints for solutions at the back of the book. Exercises marked (S) have solutions.

4. ***Two twists again.*** Take a strip of paper, put two half twists in it, and glue the ends together. Now cut it lengthwise while hugging the right edge. What do you get?

5. ***Three twists*** (H). Take a strip of paper, put three half twists in it, and glue the ends together. Cut it lengthwise along the center core line. What do you get? Find an interesting object hidden in all that tangle.

6. ***Möbius length*** (S). Use the edge identification diagram of a Möbius band (page 347) to find out how long a band we get when we cut the Möbius band lengthwise along the center line. Give the length in terms of the length of the original Möbius band.

7. ***Möbius lengths.*** Use the edge identification diagram of a Möbius band to find the lengths of the two bands we get when we cut the Möbius band by hugging the right edge. Give the lengths in terms of the length of the original Möbius band.

8. ***Squash and cut.*** Take a Möbius band and squash it flat on the table. Cut it like a buzz saw. What do you get?

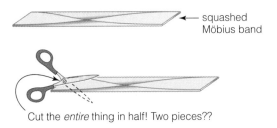

Cut the *entire* thing in half! Two pieces??

9. ***Two at once.*** Take two strips of paper and put them on top of each other. Twist them together as though you were making a Möbius band, tape the tops together, then tape the bottoms. What do you have? What do you get if you cut it lengthwise down the center core if you keep the two bands together?

Tape each pair!

10. ***Parallel Möbius.*** Is it possible to have two Möbius bands of the same length situated parallel in space so that one hovers over the other exactly 1/8 inch away? Explain why or why not.

Part of another Möbius band $\frac{1}{8}$ inch away from the front ... Possible??

11. *Puzzling.* Suppose you have a collection of jigsaw pieces as shown. They can be put together to form a strip. Can they be assembled into a Möbius band? Can you explain why or why not?

12. *Möbius triangle.* Make a 1-inch wide Möbius band, lay it like a circle on a table, and carefully flatten it (thus making three folds). What shape do you have? What is the shortest flattened Möbius band that you can make? Please build one.

13. *Thickened Möbius.* Take a Möbius band and thicken it so the edge becomes as thick as the side. We'll call this a *thickened Möbius band*. How many edges does it have?

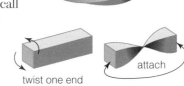

14. *Thickened faces.* How many faces (sides) does a thickened Möbius band have (see question 13)?

15. *Thick then thin.* Suppose we take a Möbius band, thicken it to make a thickened Möbius band (see question 13), and then shrink the original face to make it an edge. Is this new object a Möbius band?

16. *Drawing the band.* Suppose you take a Möbius band made of cloth with a drawstring around the edge. Can you draw the drawstring completely together? Why or why not?

17. *Tubing* (H). Suppose we take two Möbius bands and make a small hole in each. We then glue each hole to a tube connecting the two bands. How many edges does this new object have? How many sides?

18. *Bug out.* Suppose you are a ladybug on the outer surface of the Klein bottle. Describe a path on the surface of the bottle that you can travel that would get to the inside where a special gentleman bug is waiting.

19. *Open cider.* Consider the Klein bottle half filled with apple cider, as illustrated. Describe how you could pour out a glass of cider without opening the bottle.

20. *Rubber Klein* (S). Suppose you have a rectangular sheet of rubber. Carefully illustrate how you would associate and then glue the edges of the sheet together to build a Klein bottle. Draw a sequence of pictures illustrating the construction.

## II. CREATING NEW IDEAS

1. **One edge.** Using the edge identification diagram (p. 347) of the Möbius band, prove that the Möbius band has only one edge.

2. **Twist of fate (S).** Using the edge identification diagram of the Möbius band, prove that, when you cut a Möbius band lengthwise down the center, you have two half twists.

3. **Linked together.** Using the edge identification diagram of the Möbius band, prove that, when you cut the Möbius band by hugging the right edge, the two pieces you get are interlocked.

4. **Count twists.** Using the edge identification diagram of the Möbius band, determine the number of half twists each band has after you cut the Möbius band by hugging the right edge.

5. **Don't cross.** Can you draw a curve that does not intersect itself and that goes around a Möbius band three times without crossing over the edge? Experiment and explain.

6. **Twisted up (H).** Suppose you are given a band of paper with a lot of twists in it. How can you tell without counting whether you have an even number of half twists or an odd number? Can you deduce a general fact about what happens if there are an odd number of twists and what happens if there are an even number of twists? Try. (*Hint*: Experiment by drawing.)

7. **Klein cut.** Consider a rubber rectangular sheet with the edges associated as to create a Klein bottle. Suppose we make two cuts as indicated. How many pieces would we have? What would those pieces look like?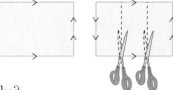

8. **Find a band.** Find a Möbius band on the surface of a Klein bottle. [The answer to the Rubber Klein story (Mindscape I.20) may help.]

9. **Holy Klein.** Show that the figure on the left is the same as a Klein bottle with a hole.

10. **Möbius Möbius.** Show that the Klein bottle is two Möbius bands glued together on their edges.

## III. FURTHER CHALLENGES

1. **Attaching tubes.** Consider a Möbius band with two small holes removed. Suppose we connect these holes with a tube. We could glue the tube to the holes in two different ways, as illustrated. Are these pictures equivalent? (*Hint*: Slide one end of the tube around the band.)

2. *Möbius map* (**H**). Draw two maps on a transparent Mobius band, or, if not transparent, then make each of your countries soak right through the Mobius band. One map should have five countries and the other six countries where each country shares a length of border with each of the other countries in the map.

3. *Thick slices.* Thicken a Möbius band and then carefully cut (slice) the band along two adjacent sides straddling a common edge. How many pieces are you left with? If there is more than one piece, what is their relationship to one another?

Slice off edge with a knife.

4. *Bagel slices.* If we take a bagel and slice it in the usual way, we notice that the newly cut face is equivalent to an untwisted looped strip. Suppose we use a peculiar cutting method whereby, instead of cutting along a closed untwisted loop, we cut along a Möbius band. Into how many pieces will our bagel fall? Would it fit in a toaster?

5. *Gluing and cutting.* Consider the following rectangular sheet of rubber with its edges identified. Suppose we make the two cuts as indicated. How many objects would we be left with? Describe the objects.

## IV. IN YOUR OWN WORDS

1. *Personal perspectives.* Write a short essay describing the most interesting or surprising discovery you made in exploring this section's material. If any material seemed puzzling or even unbelievable, address that as well. Explain why you chose the topics you did. Finally, comment on the aesthetics of the mathematics and ideas in this section.

2. *With a group of folks.* In a small group, discuss and actively work through the reasoning for how we get two interlocked strips when we cut the Möbius band a third of an inch off the right edge. After your discussion, write a brief narrative describing the method in your own words.

3. *Creative writing.* Write an imaginative story (it can be comical, dramatic, whatever you like) that involves or evokes the ideas of this section.

4. *Power beyond the mathematics.* Provide several real-life situations—ideally, from your own experience—for which some of the strategies of thought presented in this section would provide effective methods for approaching and resolving them.

# 5.3 Feeling Edgy?
## Exploring Relationships Among Vertices, Edges, and Faces

*Inspiration is needed in geometry, just as much as in poetry.*

—Aleksandr Pushkin

In our discussion of the regular (Platonic) solids, we boldly asserted that there were only five of them. This statement seems just plain wrong given the fact that there are infinitely many regular polygons in the plane. We said we would later prove that there are exactly five regular solids, and later is now. The regular solids are hard, rigid bodies. Their essence includes their flat faces, each a regular polygon. Surprisingly, we can prove that there are only five regular solids without even thinking about them as solids. The realm of the distortable contains all the ideas that limit the number of regular solids. Thinking about stretching and distorting allows us to uncover hidden structure within the rigid world of the regular solids.

We discover properties of rigid solids by first doodling in the plane and looking at what we draw. Looking carefully and quantitatively at something as simple as a random doodle gives a key insight about the regular solids. Once again, the key to unlocking a hard idea is to observe a common phenomenon clearly and quantitatively.

## Discovering by Doodling

"Do I have to draw you a picture?!" This exclamation is often heard in conversations involving an exasperated and impatient pontificator trying to get across some idea to a poor soul who would rather be somewhere else. Nevertheless, the answer is yes.

Few methods for communicating fundamental ideas are more productive and successful than drawing a picture. Pictures not only reveal structure and pattern but often show more than what the artist intentionally put into them. The discovery of such new information is an important feature of the power of pictures. To illustrate this idea and to move us toward better understanding the Platonic solids, let's try a simple example. Please take out a blank piece of paper, and remember the artistic adage: "A picture is worth a thousand words."

Put a piece of paper in front of you and place a pen on the paper. Close your eyes. Now draw sweepingly, randomly on the page without lifting your pen or running off the edge. Don't draw too much, because later we're going to ask you to count various parts of your work of art. Once you've had enough, stop drawing, lift the pen from the paper, and open your eyes. You have created your own masterpiece—you're another Picasso. Here is a sample of the genre:

We, the authors, will now attempt to demonstrate our ability to predict your actions across the void of time and space. Specifically, we will divine some features of the random drawing you just created. Please give us a moment to concentrate. Something is coming into focus . . . yes . . . yes . . . We see a squiggly object. Right? *Amazing.*

Please now accentuate each place where the curves cross each other in your drawing by placing a big dot at each crossing point. Also, draw dots at the points where you started and stopped. We now confidently assert that your drawing divides the paper into various regions.

## *Count On It—Detecting a Pattern*

Okay, you may not yet be impressed with our clairvoyance, but hold on. We are detecting a pattern. We notice that each of your big dots has exactly four curves coming out from it—except for the starting and ending dots. Are we right? If we aren't, then we must wonder if you really closed your eyes and drew randomly. If not, please start over again.

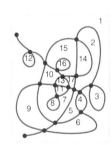

Hold on—we see something else. In our minds, we see that you have not just drawn a random squiggle after all. You have drawn exactly two fewer edges than dots and regions combined. Now why in the world did you do that?

Please check our clairvoyance: Count the number of dots, regions, and edges (an edge is a segment, curved or not, connecting two dots). Don't forget to count the one region on the outside—that is, the one containing the edge of the paper. (How did we know there was exactly one of these exterior regions?) Notice that the number of dots plus the number of regions is exactly two more than the number of edges!

We have duly demonstrated our powers to read your mind and see through your eyes. However, these magical illusions are really mathematical feats. When you drew your squiggly curve, you unknowingly created some relationships among parts of your picture. These relationships are consequences of the topology of the paper and its idealized counterpart, the 2-dimensional plane. These relationships are topological in nature in that they are not associated with the particular size or shapes in your drawing. Let's discover why the number of dots plus the number of regions is always two more than the number of edges. As always, we begin with an easy case.

## Building Up Pictures

On a sheet of paper, let's draw a dot (also known as a *vertex*) and then draw a line segment from there, making a dot at the end of the segment. Now count the number of segments we have drawn. That is not too difficult—it is one. How many dots have we made? Two. How many regions are there on the page? One—everything around the segment. In this simple example, the number of dots plus the number of regions is three, which is indeed two more than the number of edges.

> Hidden insights often come to light by first analyzing easy cases.

So far, so good. Now draw another line segment that starts at one of the dots and goes in any direction. Draw a dot at its end. Count again:

vertices = 3

edges   = 2

regions = still 1

Next, join up the two end dots to make a triangle. Count again.

vertices = still 3

edges   = 3

regions = 2 (inside the triangle and outside the triangle)

Continue adding segments that start at an existing dot and either go to an existing dot or just stick off some way and end with a new dot. New edges may not cross or touch any point of the existing diagram except for the vertices. Notice that, if we add a segment that connects two existing vertices, we create another region as we add an edge. If the new edge just sticks out somewhere,

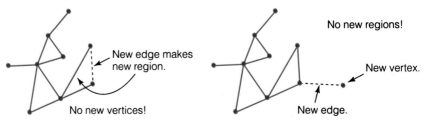

New edge makes new region.

No new vertices!

No new regions!

New vertex.

New edge.

we do not create another region, but we do add another vertex. So, whenever we add an edge, we also add either one more region or one more vertex. We will continue to have two fewer edges than vertices and regions combined no matter how many edges we add to our picture. The objects we are drawing, consisting of edges connected together, are called *connected graphs*.

## The Pattern Revealed

By adding one edge at a time, we can draw connected graphs as complicated as we wish. We could in theory draw one with 1,234,867 edges having 746,346 vertices. The thing we now know for sure is that any connected graph with 1,234,867 edges and 746,346 vertices will have exactly 488,523 regions. We are certain of the number of regions because, as we draw the figure one edge at a time, the number of vertices minus the number of edges plus the number of regions always equals 2, even if we draw millions of edges.

We were not the first to notice this fact about doodling. About 350 years ago, Rene Descartes observed this same phenomenon; however, he did not prove it. It took another 200 years until Leonhard Euler (pronounced *"oiler,"* although most of his *tackling* was within math) came along and provided a complete and rigorous justification why this conjecture is always true. Euler figured out how to take the observation and prove it is fact—it is a theorem. Let the letter $V$ stand for the number of vertices in the graph, $E$ the number of edges, and $F$ the number of regions into which the graph divides the plane. (Why do we use the letter $F$? We'll *face* that question soon.) The outside region does count as a region.

> **THE EULER CHARACTERISTIC.** *For any connected graph in the plane, $V - E + F = 2$, where V is the number of vertices; E is the number of edges; and F is the number of regions.*

All this formula says is that, when we add an edge, we also add either one vertex or one face. So, we have already shown why this theorem is true. Writing our insight in this form allows us to give it a fancy name that will memorialize Euler through the ages. The formula $V - E + F$ is called the *Euler Characteristic*. The result seems like a simple insight, and it is a simple insight. Its significance and importance are that we can build great conclusions on a simple rock.

## Only Five Regular Solids

In our discussion of the Platonic solids, we stated that the universe contains only five regular solids (the tetrahedron, cube, octahedron,

dodecahedron, and icosahedron). We can now give a complete justification of this statement. As we will discover, this result can be deduced as a surprising consequence of the Euler Characteristic.

Remember the table of data you filled in regarding the regular solids in Section 4.5 (p. 275). The table listed the numbers of vertices, edges, and faces in each solid. Now let's add another column to the table. The new column will compute $V - E + F$, vertices minus edges plus faces for each of the regular solids. (This connection with the solids clears up why we used $F$ for regions.) For each of the regular solids, let's compute $V - E + F$, giving us the following table.

|  | Number of Vertices | Number of Edges | Number of Faces | $V - E + F$ |
|---|---|---|---|---|
| tetrahedron | 4 | 6 | 4 | 2 |
| cube | 8 | 12 | 6 | 2 |
| octahedron | 6 | 12 | 8 | 2 |
| dodecahedron | 20 | 30 | 12 | 2 |
| icosahedron | 12 | 30 | 20 | 2 |

Our answer to $V - E + F$ is 2 for every solid. This answer is like a mystical mantra for the universe: 2. Remember that we already showed that the Euler Characteristic for any connected graph in the plane is 2. Is this repeated answer of 2 a coincidence? Unexpected similarities in seemingly different settings suggest hidden relationships.

## A Plane-Sphere Correspondence

Although the sphere, the boundary of a solid ball, and the plane are quite different from one another in some ways, in other ways they are similar. For example, on any surface equivalent by distortion to a sphere, any connected graph drawn on that surface having $V$ vertices, $E$ edges, and $F$ faces has the property that $V - E + F = 2$. As we will soon see, this fact is a neat consequence of the Euler Characteristic Theorem.

Suppose we draw a connected graph on a rubber balloon. Let's further suppose that the air hole, where we blow up the balloon, is located in the middle of a face (so the air hole does not pass through an edge or vertex).

Let's place our rubber balloon surface on a flat plane so that it is resting with the air hole (tied in a knot) sticking up. Now let's unknot the air hole and carefully stretch out the balloon (remember it's stretchable rubber) so that it's flat on the plane. This activity takes

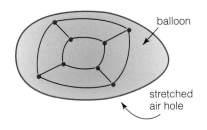

a great deal of visualization in our mind's eye. Another way to think about this procedure is to project the balloon onto the plane using a stereographic projection. (Recall our discussion of stereographic projection from our geometric look at infinity in Section 3.5.) We now have a connected graph on the plane, and the number of vertices, edges, and faces of the connected graph are equal to the number of vertices, edges, and faces of the original surface on the balloon. Notice that the face that had the air hole is now the region that entirely surrounds the graph.

We know from the Euler Characteristic Theorem that the number of vertices minus the number of edges plus the number of faces of the connected graph in the plane is equal to 2. But since these numbers are exactly the same as the corresponding numbers on the balloon surface, we must have that $V - E + F = 2$ for any surface that is equivalent by distortion to a sphere. This stereographic projection process explains why all those numbers from our table combined to give 2: All the regular solids are equivalent to a sphere. Notice how we used the notion of a one-to-one correspondence here to see that the number of vertices, edges, and faces on the balloon equal the number of vertices, edges, and regions of the corresponding graph on the plane.

## Only Five Regular Solids and No More

We now know something about the relationships among vertices, edges, and faces of connected graphs drawn on a sphere or anything distortable to a sphere, such as the regular solids. Our challenge now is to use this Euler Characteristic relationship to show that there are only five regular solids.

**FIVE PLATONIC SOLIDS.** *There are only five regular solids.*

## The Proof

The question is, Could there be a regular solid that we have not thought of? If there were, then that *MYSTERAHEDRON* would satisfy the formula $V - E + F = 2$ just as all the regular solids do. That fact will lead to its demise and show us that the mysterious *MYSTERAHEDRON* is actually a *NONEXISTAHEDRON*. In other words, it cannot exist!

## Part 1—Finding Relationships

Let's suppose that the *MYSTERAHEDRON* has $V$ vertices, $E$ edges, and $F$ faces. Besides the Euler Characteristic formula, $V - E + F = 2$, some additional relationships arise among $V$, $E$, and $F$, because we are supposing that the *MYSTERAHEDRON* is a *regular* solid. Recall that in a regular solid all faces have the same number of sides, and all vertices have the same number of edges emanating out.

Let's use the letter $s$ to stand for the number of sides each face has. Notice that $s$ has to be at least 3, because a face must have at least three sides; otherwise, it would not be a real face. (Remember that the face with the smallest number of sides is a triangle.) Next we can count the number of edges $E$ in a clever way. Multiply the number of faces, $F$, by $s$, the number of sides on each face, and notice that each edge is counted exactly twice because each edge is on exactly two faces. So $sF$ divided by 2 gives the number of edges; that is, $E = sF/2$, or $F = 2E/s$. As an illustration, for the cube we see $s = 4$, $F = 6$, so

$$\frac{4 \times 6}{2} = 12 = \text{number of edges}$$

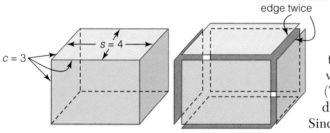

In a regular solid, the number of edges that come together at any vertex is the same as the number that come together at any other vertex. Let's call that number of emanating edges out of a vertex, $c$. Notice that $c$ must be at least 3. (Why must this be? Think about it and draw some pictures to convince yourself.) Since each edge has two ends, we could count the number of edges by multiplying the number of vertices, $V$, by $c$ and dividing the result by 2 since $cV$ is the number of edge ends, rather than the number of edges. So, $E = cV/2$, or $V = 2E/c$. For the cube, we see that $c = 3$ and $V = 8$, so

$$\frac{3 \times 8}{2} = 12 = \text{number of edges}$$

These two formulas, $F = 2E/s$ and $V = 2E/c$ result from the idea of carefully counting things.

## Part 2—Using the Relationships

Now we are in business, because we know that $V - E + F = 2$, and we know what $V$ and $F$ are equal to. Substituting, we get

$$\frac{2E}{c} - E + \frac{2E}{s} = 2$$

or, equivalently,

$$E\left(\frac{2}{c} - 1 + \frac{2}{s}\right) = 2$$

or, equivalently,

$$E\left(\frac{2}{c} + \frac{2}{s} - 1\right) = 2$$

The positive number $E$ multiplied by $(2/c + 2/s - 1)$ equals the positive number 2. So, $(2/c + 2/s - 1)$ must also be a positive number. Therefore, $2/c + 2/s$ must be larger than 1.

Since both $c$ and $s$ are 3 or greater, there are not many possible values for $c$ and $s$ that ensure that $2/c + 2/s$ is larger than 1. Notice that if either $c$ or $s$ were as large as 6, the number $2/c + 2/s$ would be 1 or less. So, neither $c$ nor $s$ can be as large as 6. Therefore, the only possibilities are as follows.

- $c = 3$, $s = 3$ (which gives the tetrahedron)
- $c = 3$, $s = 4$ (which gives the cube)
- $c = 3$, $s = 5$ (which gives the dodecahedron)
- $c = 4$, $s = 3$ (which gives the octahedron)
- $c = 4$, $s = 4$ (which makes $2/c + 2/s = 1$, which is not larger than 1)
- $c = 4$, $s = 5$ (which makes $2/c + 2/s = 9/10$, which is not larger than 1)
- $c = 5$, $s = 3$ (which gives the icosahedron)
- $c = 5$, $s = 4$ (which makes $2/c + 2/s = 9/10$, which is not larger than 1)
- $c = 5$, $s = 5$ (which makes $2/c + 2/s = 4/5$, which is not larger than 1).

This proof demonstrates that there are only five regular solids in the entire universe and no others.

## Concluding Thoughts

There are many proofs of this theorem. This proof has the advantage that it does not use geometry in the usual sense. For example, it does not refer to the interior angles of regular polygonal faces. So, in fact, we have proven that we cannot build even a distorted solid with faces all having the same numbers of sides and the same number of edges emerging from each vertex.

Icosahedral shell of a polio virus.

Perhaps we can empathize with Kepler for believing that this limit on the number of these aesthetically pleasing regular solids has a mystical significance for the structure of the universe. In fact, these solids do have physical implications that influence such things as the structure of certain molecules and some living organisms as well.

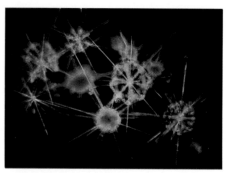

Radiolaria—microorganisms whose skeletons often are regular solids.

**A LOOK BACK**

The Euler Characteristic records a relationship among the number of vertices, edges, and regions created by a connected graph in the plane, namely, $V - E + F = 2$. Connected graphs on a balloon, a sphere, or a solid share the same property since we can create a one-to-one correspondence between figures on a balloon, sphere, or solid and those in the plane. Other relationships among the vertices, edges, and faces of a Platonic solid arise from their regularity. These relationships combined with the Euler Characteristic prove that there are only five regular solids.

We can discover the Euler Characteristic relationship by counting features of random doodles in the plane. Then we confirm our observations by building connected graphs one edge at a time and seeing that the $V - E + F = 2$ relationship continues to hold with each new added edge. Thinking about stretching and distorting allows us to see why drawings in the plane correspond naturally in a one-to-one manner with figures on a balloon or sphere, or the regular solids.

Looking at simple objects in a quantitative way lets us see patterns that might not otherwise be apparent. When we have found a pattern in one domain, we can increase its scope and power by seeking the same or related patterns in different settings. Great ideas are rare and should be applied in as large an arena as possible.

Never underestimate the power of simple counting.

Expand an idea to different settings.

Cultivate ideas and reap great harvests.

◆ ◆ ◆ ◆ ◆ ◆ ◆ ◆ ◆ ◆ ◆ ◆ ◆ ◆ ◆ ◆ ◆

# MINDSCAPES    Invitations to Further Thought*

### I. SOLIDIFYING IDEAS

1. ***Bowling.*** What is the Euler Characteristic of the surface of a bowling ball? Why?

2. ***Making change.*** We begin with the graph pictured at right. For each of the graphs that follow, compute the change in the number of vertices, edges, and regions.

3. ***Making a point.*** Take a connected graph and add a vertex in the middle of an edge, making two edges out of the one. What happens to $V - E + F$? Why?

4. ***On the edge*** (H). Is it possible to add an edge to a graph and reduce the number of regions? Why? Is it possible to add an edge and keep the same number of regions? Why?

5. ***Soap films.*** Consider the following sequence of graphs generated by soap bubble films. As the vertical film shrinks, it passes through the unstable position in the third picture and then jumps to the last. Compute $V - E + F$ at each stage. What are the changes in the V, E, and F counts as we move from the second to the third picture?

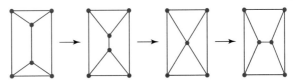

6. ***Dualing.*** What is the relationship between the Euler Characteristic for a regular solid and its dual? (See Chapter 4, section 6.)

7. ***No separation*** (S). Suppose we have a connected graph that does not separate the plane into different pieces. Prove that the number of edges is one less than the number of vertices.

8. ***Lots of separation.*** Suppose we are told that a connected graph cuts the plane into 231 regions. How many more edges than vertices are there?

---

*In the Mindscapes section, exercises marked (H) have hints for solutions at the back of the book. Exercises marked (S) have solutions.

9. *Regions* (H). Suppose we are given a connected graph made up of 151 edges. Is it possible that the graph cuts the plane into 153 regions? If not, what is the largest number of regions possible? (You may use loops.)

10. *Psychic readings.* Someone is thinking of a connected graph in the plane. It is made of 36 edges and cuts the plane into 18 pieces. How many vertices are there?

11. *An odd graph.* Is it possible to draw a connected graph in the plane with an odd number of faces, an even number of vertices, and an even number of edges? If so, draw one; if not, explain why not.

12. *Random coloring.* Draw a random squiggle with your eyes closed, as you did at the beginning of the section. Suppose now that the outside (unbounded) region around the squiggle is an ocean and that the squiggle is a map of a large island cut into different regions. Given the rule that regions that share an edge must have a different color, what is the fewest number of colors required to color the different regions of the island? (Regions that touch only at a vertex are allowed to have the same color.) Does your answer hold for any squiggle? Can you explain a procedure for how to color the map?

13. *Coloring America.* Given the rule that states that share an edge must have a different color, is it possible to color the map of the United States with two colors? If not, how about three? If not three, how about four? When the answer is no, please explain.

14. *Circles on a sphere* (S). Suppose we take a sphere and draw two latitudinal circles on it. Place one dot on each circle. Count the number of faces, edges, and vertices on this sphere. Compute $V - E + F$. Did you get the answer you expected? Why? What hypothesis about the Euler Characteristic Theorem does this question pertain to?

15. *More circles.* Consider the sphere described in Mindscape 14 but add an edge that connects the two dots. Now count the number of faces, edges, and vertices. Compute $V - E + F$ again. Why is this answer different from the previous calculation? What hypothesis about the Euler Characteristic Theorem do questions 14 and 15 illustrate?

16. ***In the rough.*** Count the number of facets, edges, and vertices of this diamond. Compute the Euler Characteristic for this gem.

17. ***Cutting corners.*** The following collection of pictures shows the regular solids with their vertices cut off. Such objects are called *truncated solids*. For each truncated solid, count the number of vertices, edges, and faces and verify that the Euler Characteristic is correct.

18. ***Stellar.*** The following collection of pictures shows the regular solids where each face was replaced by a vertex that sticks out and that attaches to the solid with edges to each of the vertices of the original face. Such objects are called *stellated solids*. For each stellated solid, count the number of vertices, edges, and faces, and verify that the Euler Characteristic is correct.

19. ***A torus graph.*** The Euler Characteristic $V - E + F$ can be applied to other surfaces besides spheres. We can compute the Euler Characteristic for tori, two-holed tori, and yet more complicated figures; however, for those surfaces the 2 in $V - E + F = 2$ is replaced by other numbers. Draw a connected graph on a torus that both circles the hole and goes through the hole of the torus. Compute $V - E + F$ for that graph. Try several other different graphs that go around the torus both ways and again compute $V - E + F$.

Graph must go around both ways, as marked.

20. ***Regular unfolding.*** Each graph below represents some regular solid that has been "unfolded," or projected down onto the plane. For each graph, determine which regular solid it represents.

## II. CREATING NEW IDEAS

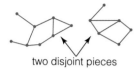
two disjoint pieces

1. *A tale of two graphs.* Suppose we draw a graph that has exactly two pieces; that is, it is not connected but instead consists of two connected parts. What is $V - E + F$ for that two-component graph?

2. *Two graph conjectures* (S). Can you conjecture a new formula for $V - E + F$ for graphs with exactly two pieces? Try various examples first.

3. *Lots of graphs conjecture.* Can you conjecture a new formula for $V - E + F$ for graphs with exactly three pieces? Try various examples first. Can you conjecture a new formula for $V - E + F$ for graphs with exactly $n$ pieces? Search for a pattern.

4. *Torus count.* Carefully count the number of vertices, faces, and edges for this prismatic torus. Compute the Euler Characteristic for this torus.

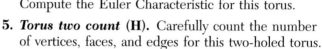

Remove both faces and glue edges together.

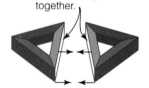

5. *Torus two count* (H). Carefully count the number of vertices, faces, and edges for this two-holed torus. One way to view this two-holed torus is as two copies of the previous picture, with one side removed from each and then the open edges glued together. This operation is called the *connected sum*. Compute the Euler Characteristic for this two-holed torus.

6. *Torus many count.* Using the preceding calculations and the notion of a connected sum, compute the Euler Characteristic for a three-holed torus. (First think about how to put together a three-holed torus using a two-holed and a one-holed torus.) Make a conjecture as to what the Euler Characteristic is for a torus with four holes, five holes, and in general, $h$ number of holes.

7. *Torus tours.* Consider the following rubber rectangular sheet with the edges identified to make a torus. Assuming that the edges are glued together as indicated, how many different edges would there be (*different!*)? How many different vertices would there be? How many faces? Compute $V - E + F$. (Make sure you are not counting the same things more than once!)

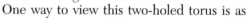

8. *Tell the truth.* Someone said that she made a two-holed torus by gluing 60 triangles together along their edges. Once it was all glued, this person claimed that there were 30 vertices on the object. Is she telling the truth? Explain why or why not.

9. *No sphere.* Suppose we have a sphere built out of 60 triangles. Why can't all the vertices have the same number of triangles coming into them? (*Hint*: Count edges by multiplying by something and then dividing by two, and count vertices by a similar multiplying and dividing technique.)

10. **Soccer ball.** A soccer ball is made of pentagons and hexagons. Each pentagon is surrounded by five hexagons. Use the Euler Characteristic to figure out how many pentagons and how many hexagons are necessary to construct a soccer ball.

### III. FURTHER CHALLENGES

1. **Klein bottle.** Using the diagram on the left for building a Klein bottle (as described in the previous section), count the number of vertices, edges, and faces. Use this data to conjecture the Euler Characteristic for the Klein bottle.

2. **Not many neighbors.** Show that every map has at least one country with five or fewer neighbors—that is, countries that share a border (oceans count as countries). (*Hint*: Put a vertex in the center of each country and join two vertices if the two countries share a border. This procedure produces a graph with one vertex for each country. The number of edges emanating from a vertex equals the number of neighboring countries. Suppose each country had six or more neighbors. Then we would know that there are at least 6V/2 or, equivalently, 3V edges. How many faces would there have to be to make the Euler Characteristic hold?)

3. **Infinite edges.** Suppose we consider a connected graph in the plane, but now we allow edges to go from a vertex out to infinity. What would $V - E + F$ be for such graphs? Why? (Experiment and discover the pattern.)

4. **Connecting the dots** (H). Suppose we are given five points in the plane and we wish to connect each point to the other four points. Prove that it is impossible to draw such a graph without having edges cross each other. (*Hint:* Assume we could draw the graph without any edge crossing. Then how many vertices would we have? How many edges? Given those numbers, figure out how many regions there must be. What's the smallest number of edges that must surround each region? Use that number and the Euler Characteristic to compute the minimum number of edges needed, and show that exceeds the actual number of edges—thus, we must have crossings!)

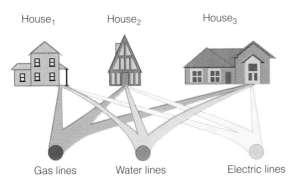

5. **Gas, water, electric.** There are three houses, and each has a gas line (running from the gas company), a water line (running from the water company), and an electric line (running from the electric company). Prove that it is impossible to draw all the gas, water, and electric lines in the plane without having some of the lines cross one another.

## IV. IN YOUR OWN WORDS

1. ***Personal perspectives.*** Write a short essay describing the most interesting or surprising discovery you made in exploring this section's material. If any material seemed puzzling or even unbelievable, address that as well. Explain why you chose the topics you did. Finally, comment on the aesthetics of the mathematics and ideas in this section.

2. ***With a group of folks.*** In a small group, discuss and actively work through the reasoning for why any connected graph in the plane satisfies the relation that the number of vertices minus the number of edges plus the number of regions is equal to 2. After your discussion, write a brief narrative describing the method in your own words.

3. ***Creative writing.*** Write an imaginative story (it can be comical, dramatic, whatever you like) that involves or evokes the ideas of this section.

4. ***Power beyond the mathematics.*** Describe several real-life situations—ideally, from your own experience—for which some of the strategies of thought presented in this section would provide effective methods for approaching and resolving them.

# 5.4 Knots and Links:
## Untangling Ropes and Rings

*God is like a skilful Geometrician.*

—Sir Thomas Browne

When we think of flexible, bendable, and contortable objects, at some point we think of string. Loops of string can contain knots or can be looped together with other loops of string. This mundane, everyday observation actually leads to interesting discoveries with important consequences. Here we investigate the twisted world of knots and links and find that knots, at the most fundamental level, play a role in the creation of life itself.

Physical experimentation is an excellent way to become acquainted with the variety and intrigue of knots. As we hold and try to disentangle a loop of knotted string, we sometimes wonder whether we are making matters better or worse. This question leads us to the important idea of measuring the complexity of a position of the string. By deciding on some way to gauge the complication of a knot, we can determine whether we are making progress toward disentangling it or not. The strategy of specifically measuring complexity is applicable broadly in and out of mathematics. It allows us to tackle complicated issues a step at a time.

## A Twisted DNA Tale

The fiber of life itself is encoded in strands of DNA entwined intricately in its famous double helix. These double strands herald life's message in the watery world of our cells. They jostle among molecular building blocks floating in their surroundings and attract pieces with their geometrical and magnetic charms. At the supreme moment in the life of a DNA molecule, it begins its magical splitting. The

*Alexander the Great Cutting the Gordian Knot* by Perino del Vaga.

Double helix in 3D—Put your 3D glasses on.

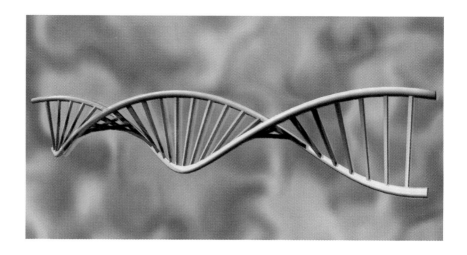

double helix splitting

DNA twisted and knotted

ladder rungs break, and one strand peels away from the other, beginning the process of duplicating itself, thereby manifesting the glorious wonder of self-replication. This image of the two connected sides of the ladder peeling apart to go independently off on their own to gather materials and ultimately create duplicates is amazing.

Unfortunately, it can't be true; the picture we just painted is impossible. Strands entwined in a double-helical fashion would be linked, tangled, and crumpled. For example, take two pieces of string and twist one around the other to form a double helix. Remember that the helix is twisted thousands of times. If we start at one end and attempt to pull the two strands apart, the other end would have to spin around thousands of times, and the complicated tangle would disrupt the process in midstream. The DNA strands would have to be strong and agile enough to untangle and untwist themselves, and chemists assure us that such acrobatics are beyond the physical abilities of DNA molecules. Some DNA molecules have their ends attached to one another, thereby forming a double-helical loop. In this case, the two strands are linked, and they would be impossible to pull apart without breaking some bonds.

What can we conclude? We are forced to conclude that, during the replication phase of DNA, the two long strands do not stay individually connected. They must frequently break apart, cross over the other side of the ladder, and reassemble themselves. The geometrical linking of the two sides of DNA requires this process of cutting, moving, and regluing during the

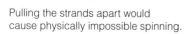

Pulling the strands apart would cause physically impossible spinning.

372 ◆ CONTORTIONS OF SPACE

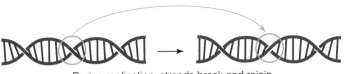

During replication, strands break and rejoin.

process of splitting and separating. Bolstered by the mathematical certainty that a process of breaking and regluing must be happening, molecular biologists are seeking and have isolated enzymes whose special function is to allow unlinking. DNA and its reproductive habits provide a powerful, current application of the theory of knots and links.

The theoretical study of knots and links has been useful to biologists in deducing more subtle conclusions about DNA as well. An enzyme acts on DNA locally, meaning its influence is restricted to the part of the DNA strand that it touches. One application of the abstract study of knots occurred when Dewitt Sumners, a knot theorist, was able to prove in the abstract that only certain types of knots in DNA could result from the localized action of an enzyme. Substantial amounts of money were spent to do molecular biological experiments to show the same thing that Sumners' theoretical work implied without any laboratory experimentation. Such revelations have led scientists and knot theorists to pursue joint projects in the hopes that such investigations may influence one another's fields of expertise.

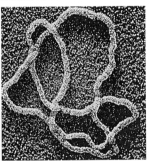

Actual photograph of knotted DNA.

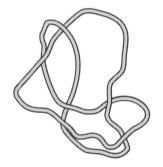

## *The Gordian Knot*

Knots have not been confined to the scientific side of human curiosity. Knots and links have long played their parts in human history in legend as well as science. Perhaps the intricate tangles, crossings, and complexities of knots speak to a gnarled core of the human psyche. Perhaps the single most famous knot in history is the Gordian Knot.

The Gordian knot held a mythical sway over the Phrygian people in the 4th century B.C. "The one who can untie the Gordian knot will be king," the legend maintained. When Alexander the Great arrived in Phrygia and was confronted with the fiendish knot, he took his great sword and cut the knot in two, simultaneously declaring the prophecy fulfilled and himself

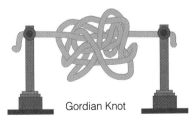

Gordian Knot

king. Had he studied knot theory instead of conquering the known world, he might well have saved his edge and been able to untie the knot without breaking the spirit of the unknotting legend—a breach of etiquette that, no doubt, weighed heavily on his mind. Although we may never be asked to untie a knot to become king of the known world, we remain fascinated by knots, which play a significant role in the structure of life.

## Knots We May Know

We begin this discussion with pictures of some knots we may know. Notice that each example is a closed curve rather than a strand having free ends, because a rope with free ends can always be untied simply by pushing a

unknot

trefoil knot

figure-eight knot

square knot

granny knot

free end through any knottiness in the string. So, although we talk about taking a length of string and putting a knot in it, unless we glue the free ends together, we have not actually created a real knot. Let's return to the pictured knots.

The first of these is the simplest knot of all. It is a round circle, and its only problem is that it is not really knotted. In fact, it is unknotted. Mathematicians no doubt thought long and hard about what to call that circle in the context of knot theory. After deep consideration, it was decided that any unknotted curve would henceforth be referred to as *the unknot*. Although a round circle is an unknot, more complicated curves that can be untangled to become a round circle are also examples of the unknot. For example, we can untangle this curve to see that it is in fact the unknot in disguise.

push through — so not a real knot

A free end means "not knotted."

Looks knotted, but really isn't!

Once we have gotten beyond the unknot, we can start looking at some knotted knots, the simplest of which has three crossings and three lobes when pictured. The 3's involved give it its name, the *trefoil knot*. No matter how hard we try to move the trefoil knot around, we cannot just bend and stretch it, without cutting it, to make it become the unknot. It is genuinely knotted.

The next knots in the sequence of pictures are the *figure-eight knot*, the *square knot*, and the *granny knot*. The figure-eight knot is called the figure-eight knot because it can be drawn to look somewhat like the numeral 8. The square knot is called a square knot because it can be drawn to look somewhat square. The granny knot is called a granny knot because it apparently looks somewhat like someone's Granny. These knots are relatively simple in the sense that each can be drawn with just a few crossings. However, more complicated knots can require any number of crossings.

## An Unsolved Knotty Challenge

An unsolved challenge is to distinguish one knot from another. Suppose someone drew a complicated picture of a knot. Could we give a clear strategy for rearranging it to put it in a position in which it has the fewest possible number of crossings? In particular, if it is the unknot, could we give a strategy for disentangling it to make it look like the circle? No one has yet figured an effective way to answer these questions in general. Perhaps we could think of some way to count crossings or count the pattern of overs and unders that could tell us definitely whether a pictured knot is actually the unknot. No one knows how to do that yet. Many questions about knots have yet to be answered, but other results about knots are known and some are quite surprising. We consider one such result in the following story.

## *Not Math to Knot Math*

The road to mathematical understanding is not without its major impasses and potholes. This story is about a student who hit one pothole too many.

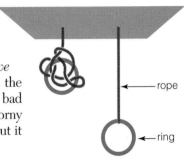

Two gym rings: one tangled, one not.

Chris was a conscientious student in a course that used this book as a text. Sure, Chris wasn't a real math fan (yet), but Chris would read *The Heart of Mathematics: An invitation to effective thinking* and would try to work through the assigned homework. The text was not as bad as Chris had expected. Yes, some of the corny jokes and comments were slightly lame, but it made for some unusual reading. Chris' mind was spinning with notions of prime numbers, different sizes of infinities, the fourth dimension, and the Klein bottle. Chris was experiencing mathematical thinking. Then there was Chris' roommate, Pat.

Pat was in the same math course as Chris, but gave up and used the text to prop up the CD player. Pat had lots of interesting tattoos and was pierced in a variety of unusual places (including Detroit). Pat's mind was spinning, but sadly the spinning was not caused by intellectual, new ideas. Pat could not see the practical value of discussing, arguing, and debating various counterintuitive mathematical notions with friends. Pat, who was unable to participate in these lively gatherings (remember, Pat's text remained under the CD player, unread), decided instead to go to the gym to work out on the rings.

There Pat was faced with a practical problem: Although one ring hung straight and free, the other rope had become hopelessly knotted and entangled through the ring as it hung from the high ceiling. Pat could not reach the ceiling to remove the snared rope, unknot, untangle, and then reattach it. So, the only hope of practicing the iron cross was to untangle the knotted rope without removing it from the ceiling. At this moment, Chris, having finished a lively math discussion, walked in and was inspired to help Pat work out this knotty problem. Chris began encouraging Pat by yelling out some of the life lessons Chris had read in *The Heart of Mathematics*.

Experiment to discover new insights.

Measure the complexity of a problem.

Never give up.

◆ ◆ ◆ ◆ ◆ ◆ ◆ ◆ ◆ ◆ ◆ ◆ ◆ ◆ ◆ ◆ ◆ ◆ ◆ ◆ ◆

Pat thought a bit and saw that one of the most obvious measures of the complexity of this situation was the number of crossings appearing in the rope on its tangled way to the ceiling. If one could always reduce the number of crossings by one, then, Pat realized, it would be possible to untangle the rope; because, if Pat could reduce the number of crossings first by one, then Pat could further reduce the new number by one more and so on until there were no crossings left. After untangling the rope without removing the rope from the roof or the ring from the rope, the rings hung straight. Pat was so happy to discover such a creative idea that Pat felt uplifted, returned to the math course, removed the text from under the CD player, read it, aced the course, got a great job, and got a new tattoo that read "Math Rocks."

## *Untangling a Knotted Rope— Reducing Complexity*

Reduce the complexity of a problem.

◆ ◆ ◆ ◆ ◆ ◆

Upon first inspection, it may appear that it would be impossible to unknot and unlink a rope that is really tangled up and knotted without removing it from the ceiling. However, Pat's general method of attack is an effective one, and it works often in many settings: Namely, measure the complexity of a situation and work to reduce it.

Suppose we have a complicated tangle as described in the story. Further suppose that a way exists to remove one of the crossings of the tangle and that this method does not depend on the idiosyncrasies of the tangle itself. By removing this crossing, we would have another tangle, but it would not be quite as complicated, because it has one less crossing. If we repeat this procedure, then we reduce the number of crossings again by one. Repeating this process until there are no crossings left leaves us with an untangled rope with a loop.

How do we know we can always remove a crossing from our tangled and knotted rope? Start where the rope is attached to the ceiling and move down toward the ring. Find the first place where that hanging rope crosses over or under some other part of the rope or the ring.

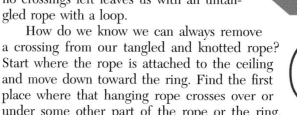

The crossing is made of two parts: the part on the rope from the ceiling and the other part that crosses it. Notice that we can remove that crossing by pulling that crossing piece up toward the ceiling and then all the

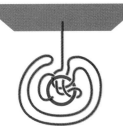

way around the bottom of the tangled mess. Now the rope has one less crossing. We repeat. We find the first crossing we come to as we proceed down the rope from the ceiling and remove that crossing in the same manner as before. We repeat again and again; this leads us to an untangled and unknotted rope. In the Mindscapes section we ask you to physically try this with some string. You will discover that it is actually easier to do by just playing with the string, rather than trying to strictly follow this procedure.

## *Links*

A link is like a chain. It is an object that is constructed from some number of closed loops that may be knotted either individually or about one another. A trivial example of a link is a collection of round circles that do not interact at all. This link is boring; it's called the *unlink*.

The simplest example of a link where some real linking goes on is the two-component link that looks like two pieces of a chain. Notice that each piece is an unknotted curve, but we cannot disassemble the links to become a collection of two circles that do not interact. This example is a genuine link, but links among curves are not always so straightforward and present us with many challenges. Here is one.

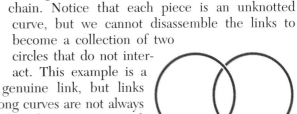

the unlink

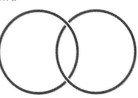

## *The Linking Challenge*

Is it possible to link three rings together in such a manner that they are indeed linked; yet if we remove any one of the links, the other two remaining links become unlinked?

It seems that the answer must be no. Consider, for example, this link. The three circles are linked. If we remove the middle link, the other two become unlinked; however, if we remove the right or left circle, the other two circles remain linked. Our question asks if we could link the three rings so that the removal of *any* (not just a particular) ring leads to unlinking of the remaining two. Therefore, the answer must be no. Do you agree?

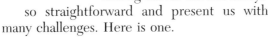

## *The Borromean Rings—They Either Hang Together or They Don't Hang at All*

The answer to the linking challenge is yes. There does exist a way of linking three rings such that the removal of any one will automatically unlink the other two. An example of such a linking is known as the *Borromean*

Borromean rings

*rings*. The Borromean rings slightly resemble the Olympic symbol. Notice that, if we pretend that the red ring is gone, then the blue ring is completely behind the green ring. Therefore, the blue ring and the green ring are unlinked if the red ring is removed. Similarly, notice that we get unlinking if the blue ring is removed and also if the green ring is removed.

another view of the Borromean rings

The delicate linking of the Borromean rings turns out to be a natural wonder. The Borromean rings are not just some abstract idea of the mind; here nature provides us with a tactile example in the regular solids—in particular, the icosahedron, the 20-sided solid. Recall that, if we take two edges of the icosahedron that are parallel and directly across from each other, they form two sides of a Golden Rectangle. The Golden Rectangle is constructed simply by using the parallel sides as the short sides of the rectangle and completing the rectangle with line segments that go directly through the icosahedron.

the Olympic rings

Let's take an icosahedron and balance it on a table on one edge of the icosahedron. At the top, we will find an edge parallel to the bottom. Also, there is a pair of horizontal edges exactly half way from the table to the top edge. Finally, there is a third pair of edges that are vertical. So, we have located three pairs of parallel edges. Now let's complete each pair of edges into a Golden Rectangle. How are these three

Golden Rectangle

icosahedron

related? Please look at them.

The three rectangles form a set of Borromean rings. The icosahedron has in it a natural set of three mutually perpendicular Golden Rectangles that form the Borromean rings. Three ideas—the Golden Rectangle, the Platonic solids, and the Borromean rings—all come together in the icosahedron to form an example of the interconnectedness and beauty of our geometric universe.

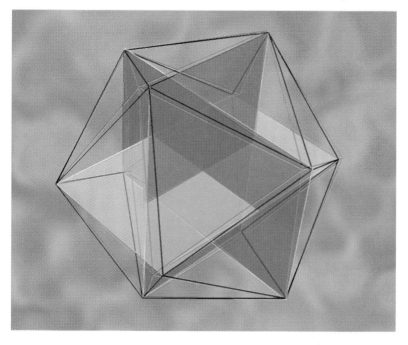

3D version of the icosahedron with golden Borromean Rings

## *Other Knotty and Linking Challenges*

The Borromean rings are three closed curves such that the removal of any one curve allows the remaining two to be separated from each other. Is it possible to construct a set of four, five, or more closed curves such that as a group they are all linked, and yet, if we remove any one curve from the group, the remaining curves all become unlinked? Try it in the Mindscapes section.

Knot theory is a vast subject, full of challenging, unsolved problems. We end this section by repeating the most basic question in knot theory that remains unanswered. We hope you will enjoy turning your mind toward this simple-sounding question that has stumped mathematicians for many years.

## *Unanswered Question*

Given two pictures of knots, is there a practical procedure, for example, a computer program, that would tell us whether one could be deformed into the other?

**A LOOK BACK**

Knots and links made of string or rubber exist naturally in the world of bendable, distortable space. DNA molecules with their twisted, linked structure demonstrate that studying knots and links is important for understanding ourselves. Knots and links surprise us with what can be done and what cannot. A knotted-up noose can be untangled, whereas the Borromean rings cannot. The Borromean rings form a group of codependency: Together they cannot be disentangled; yet, after any one of them has been removed, they become independent—that is, they completely fall apart.

Our basic strategy for dealing with knots and links is first to experience them directly with physical models and then to think about how to measure their complexity. To untangle a knotted noose we measure its complexity by counting crossings and then demonstrate a procedure for removing a crossing. By repeatedly simplifying the knot a little at a time, we can eventually untangle the noose altogether.

The technique of incremental solution allows us to deal with complex situations. Instead of being frozen into inaction by the sheer size and difficulty of an issue, we can look for a way to measure progress step by step. Small, but definite, forward motion, repeatedly applied, will get us anywhere.

Get direct experience.

Measure complexity.

Reduce complexity.

# MINDSCAPES  Invitations to Further Thought*

## I. SOLIDIFYING IDEAS

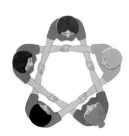

1. ***Human knots.*** Find four friends. Have the five of you make the following knot by joining hands with the appropriate crossings.

   Without unhanding anyone, have the group move around and attempt to unknot. Once the group has tried in earnest, have everyone regroup and take their original positions. Now, however, switch two crossings, as shown.

   Try to unknot. Write about the event and your findings.

2. ***Human trefoil.*** What is the minimum number of people you need to make a human trefoil knot? Get some close friends and try it. Draw a diagram showing the human knot.

3. ***Human figure eight.*** What is the minimum number of people you need to make a human figure-eight knot? Get some close friends and try it. Draw a diagram showing the human knot.

4. ***Stick number.*** What is the smallest number of sticks you need to make a trefoil knot? (Bending sticks here is not allowed.)

5. ***More Möbius.*** Make a Möbius band with three half twists. Cut it lengthwise along the center core. What kind of knot have you made?

6. ***Slinky.*** Take a Slinky, lengthen one of its ends, and push it through and attach it to the other end. Is that a knot?

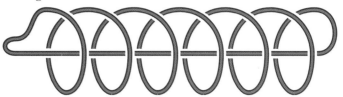

7. ***More slink.*** Take a Slinky, and this time weave an end up and down through the twists and then attach it to the other end. Is this knotted?

8. ***Make it.*** Use a piece of string or an extension cord to determine if this is a knot or the unknot.

---

*In the Mindscapes section, exercises marked (H) have hints for solutions at the back of the book. Exercises marked (S) have solutions.

9. ***Knotted*** (S). Take an unknotted loop. Tie a knot in the middle as illustrated. Is that a knotted loop? Explain why or why not.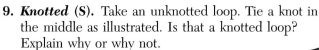

10. ***Slip.*** Take an unknotted loop and put a *slip knot* on it: Why does the knot go away without cutting? Should it be called a *slip unknot*?

11. ***Dollar link.*** Take two paper clips and a dollar and fasten them as illustrated. Now pull the ends of the dollar so as to straighten it out. Report what happens to the paper clips.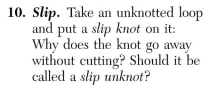

12. ***Knotted loop.*** Take some string and put a loop at the end of it. Now knot the string through and around the loop as much as you want, and tie the unlooped end to a chair or table. Attempt to untangle the string without untying it from the chair. (*Hint:* It is easier to just play around with the knot than to follow the mathematical algorithm presented in the story.)

13. ***Borromean knot.*** Is it possible to switch some number of crossings in the picture of the Borromean rings to make it a trefoil knot? If so, what is the minimum number of crossing switches that you need? A crossing switch just changes an undercrossing to an overcrossing or an overcrossing to an undercrossing.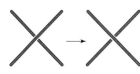

A crossing switch

14. ***Unknotting knots*** (H). In each of the following two knots, unknot the knot by switching exactly one crossing. Show the crossing you switched and how the new object becomes the unknot.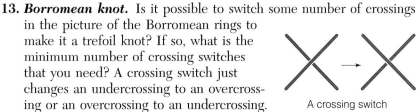

15. ***Alternating.*** A picture of a knot is an *alternating picture* if as you follow along the string, you see an undercrossing, then an overcrossing, then an undercrossing, and so forth until you return to where you started. Of the five simple knots, which are alternating?

unknot    trefoil knot    figure-eight knot    square knot    granny knot

16. ***Making it alternating.*** Consider the knot on the left. Notice that the crossings do not indicate if they are overcrossings or undercrossings. Select the crossings so that the picture of the knot is alternating.

17. ***Alternating unknot.*** Draw an alternating picture of the unknot that has at least two crossings.

18. **One cross (H).** Prove that any loop with exactly one crossing must be the unknot.

19. **Two loops (S).** Is there a picture of two linked loops that has exactly three crossings? Could all three crossings be created by one loop crossing the other?

20. **Hold the phone.** Disconnect the wire from the phone to the wall. Tangle it up and plug it back in. Can you untangle it without unplugging the wire?

## II. CREATING NEW IDEAS

1. **More unknotting knots.** In these two knots, find the fewest number of crossings you need to switch so that the new object is unknotted. Illustrate the crossing changes and how the new loop is unknotted.

2. **Unknotting pictures (S).** Suppose you are given a complicated picture of a knot. Show that you can always change that knot to the unknot if you are allowed to switch as many crossings as you wish. How would you proceed systematically?

3. **Twisted.** Suppose we are given a figure consisting of two twisted strings that are connected by noninteracting curved arcs, as shown. What property of the twist would determine if it is a knot or a link? Experiment!

4. **More alternating.** First reread question I.15. For each knot on the right, re-draw the picture, without changing the knot, so as to produce an alternating drawing of the knot.

5. **Crossing numbers.** Suppose you are given pictures of two knots. If they have a different number of crossings, then must the knots be different knots? If so, explain; if not, provide different pictures of the same knot with different numbers of crossings.

6. **Lots of crossings.** Suppose you are given a picture of a knot. Can you move it, without cutting, so that it has more than any specified number of crossings? If so, how; if not, why?

7. **Torus knots (H).** Can you draw a trefoil knot on a torus without having the loop cross itself? If so, draw a picture and then show that the loop you drew is a trefoil knot.

8. **Two crosses.** Prove that any loop with exactly two crossings must be the unknot.

9. *Hoop it up.* Show that every knot can be positioned such that all of it lies in the plane as disjoint curves except for some unknotted, semicircular hoops all the same size.

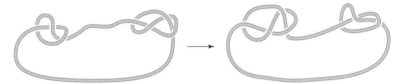

10. *The switcheroo.* Pictured here is a way of combining two knots by putting one after the other. Show that you can deform this combined knot so that the rightmost knot will now be on the left side.

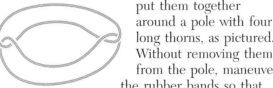

### III. FURTHER CHALLENGES

1. *4D washout.* Why is the study of knots and links vacuous if you allow yourself to move in 4-dimensional space?

2. *Brunnian links* (**H**). Link four loops together in such a way that they are all linked and yet the removal of any one loop will unlink all the others. This generalization of the Borromean rings is called the *Brunnian links*. Can you link five loops together in this fashion? Can you link any number together in this way? (*Hint:* The loops need not be round circles. Make the Borromean rings out of rubber bands, cutting and re-gluing one. Now move them around.)

3. *Fire drill.* A fire starts in your fifth-floor room. All you have available is a collection of slippery 3-foot loops. They are strong, but short. You have a post in the room around which you can link one. How can you arrange the loops so that you can climb down? (The loops are too small and slippery to tie together by doubling them up and using them as if they were small pieces of rope.)

4. *Bing links.* Put two rubber bands together as pictured (this is known as a *Bing link*). Now make another identical pair and put them together around a pole with four long thorns, as pictured. Without removing them from the pole, maneuver the rubber bands so that each rubber band crosses at most one of the four thorns. (In the figure, notice that each band crosses two thorns.)

5. *Fixed spheres again.* We are given two spheres that are made of glass and thus are not distortable. They are rigidly fixed in space,

one inside the other as indicated, and they cannot be moved! The middle sphere floats miraculously in midair without any means of support. There is a rubber rope that hangs from the inside ceiling of the big sphere to the roof of the smaller sphere, as illustrated. The rope has a knot in it. Show that it is possible to unknot the rope without moving or distorting the two spheres. (This question appeared in section 5.1, question III.3.) Suppose now we add another straight rope as shown. Can you unknot both ropes?

### IV. IN YOUR OWN WORDS

1. *Personal perspectives.* Write a short essay describing the most interesting or surprising discovery you made in exploring this section's material. If any material seemed puzzling or even unbelievable, address that as well. Explain why you chose the topics you did. Finally, comment on the aesthetics of the mathematics and ideas in this section.

2. *With a group of folks.* In a small group, discuss and actively work through the reasoning for the solution of the tangled ring story. After your discussion, write a brief narrative describing the ideas in your own words.

3. *Creative writing.* Write an imaginative story (it can be comical, dramatic, whatever you like) that involves or evokes the ideas of this section.

4. *Power beyond the mathematics.* Provide several real-life situations—ideally, from your own experience—for which some of the strategies of thought presented in this section would provide effective methods for approaching and resolving them.

# 5.5 Fixed Points, Hot Loops, and Rainy Days:
## How the Certainty of Fixed Points Implies Certain Weather Phenomena

*What science can there be more noble, more excellent, more useful . . . than mathematics.*

—Benjamin Franklin

Times are changing. Things are moving. The world is in flux. With all this movement, it is comforting to find settings where stability is absolutely required—not by statute or custom but by the higher law of mathematics. Here we'll examine various instances where things must remain fixed and consider some surprising applications and consequences regarding the climate on Earth.

*Starry Night* (1889) by Vincent Van Gogh.

Often the road to insight begins with intentional failure. For example, suppose we seriously attempt a truly impossible task such as constructing an example that contradicts a true theorem. We will fail. But looking carefully at why we fail can lead us to understand why the theorem is true. This strategy of intentional failure is a potent generator of ideas.

Sometimes the ideas are just small pieces of a larger puzzle. We can analyze what is happening there, or there, or there, but we do not have a global picture that incorporates all the information. We now need to step back and seek some organizing structure that puts the local information together to form a global whole.

# The Rubber Break-Up Challenge

Consider two disks: One is red and one is blue, but they are identical in size. The only difference between the two disks, besides their color, is that the red one is made of extremely stretchable, shrinkable and distortable rubber, whereas the blue one is made of rigid material, so it is unbending and cannot be distorted. We are handed these disks, the red on top of the blue. Since they are the exact same size, each point of the red disk touches one and only one point of the blue disk, and, similarly, each point of the blue disk touches one and only one point of the red disk. So, we have paired the points of the red disk with the points of the blue disk—everyone has a partner. In other words, we have found a one-to-one correspondence between the points of the two disks.

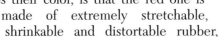

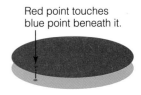

Red point touches blue point beneath it.

*Discovering a global structure can be a considerable challenge but often is the key to true understanding.*

Of course, the world is in flux. Suppose the red points start to dance around. That is, suppose we now lift up the red disk, stretch, shrink, fold, bend, twist and otherwise distort it (but no cutting), and then place it back on top of the blue disk. The only rule is that the red disk cannot hang off the blue one. That is, the red disk must sit completely on the blue, but in any way, shape, or form that you wish.

Plainly, we have mixed up and destroyed many original pairs of partners of points. In fact, there may be some blue points that have no red points above them (they are points without partners), and there may be some blue points that have several red points above them (those blue points have several partners). There may be other blue points that have exactly one red partner, but that partner may not be its original corresponding one.

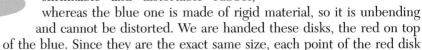

Here we see a blue point without a red partner.

A single pairing of a red and a blue point—but not an *original* pairing.

Here we see 3 red points on top of one blue . . . competition!

Here is a challenge. Can you find a distortion of the red disk so that, once it is placed on top of the blue disk, *no* red point is paired with its original blue partner, that is, each original pair of partners had a break-up? Can you do it? Try to distort the red disk in an attempt to have every point on the red disk move from its original position.

Henri Matisse captures the dancing about of the red points in his 1910 *Dance*.

## Some of Our Attempts

We thought we'd share a couple of our simple attempts at finding a solution to the preceding challenge. Our first try was to twist the red disk on the blue one like a record on a record player (check a history book for a description of a "record player"). The problem is that the center point of the red disk would remain fixed and so that center pair of corresponding points would remain. Foiled!

Red disk is twisted from original position...
but center point is fixed!

Next we tried folding the red disk in half and then half again to make a four-fold quarter piece. We then placed the piece smack-dab in the middle of the blue disk. Notice that the original center of the red disk (now the corner of the quarter piece) is not touching the center of the blue disk. Also, lots of points have

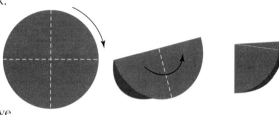

been shuffled around. It may appear that we had met the challenge. But suppose we connect the center of the blue disk to the corner of the quarter red piece with a line segment and mark the midpoint of that segment. Because we folded the red disk twice, there are four red points below that midpoint. If we now rotate the quarter red piece around that midpoint, it matches up perfectly with the upper-right quarter of the blue disk and, therefore, one of the red points under that midpoint was never moved; it remained fixed. Thus, that pairing of points is an original pairing. Foiled again!

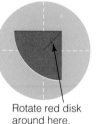

Rotate red disk around here.

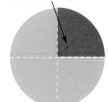

One of these four red points at the midpoint remained fixed!

Finally, in frustration, we just crumpled up the red disk randomly and dropped it on the blue one. Certainly that must do it!?

Crumpled red mess ... fixed points??

## It's Hard to Accomplish the Impossible

Give up? So did we. The annoying part is that the challenge sounded so darned doable. Amazingly enough, however, it is not. That is, it is impossible to stretch, rotate, distort, fold, and squash the red disk and place it back on top of the blue disk in such a manner that each point on the red disk moves. This surprising fact is a mathematical result known as the *Brouwer Fixed Point Theorem*.

**THE BROUWER FIXED POINT THEOREM.** *Suppose two disks of the same size, one red and one blue, are initially placed so that the red disk is exactly on top of the blue disk. If the red disk is stretched, shrunken, rotated, folded, or distorted in any way without cutting and then placed back on top of the blue disk in such a manner that it does not hang off the blue disk, then there must be at least one point on the red disk that is fixed. That is, there must be at least one point on the red disk that is in the exact same position as it was when the red disk was originally on top of the blue one.*

This result has many applications within the abstract mathematical realm of topology and other areas. In fact, recently scientists are realizing that such a result has applications to human physiology. The Brouwer Fixed Point Theorem may help us understand the behavior of heart muscles and heart attacks.

Why is the Brouwer Fixed Point Theorem true? Since we have no idea what distorting we will do, we need to demonstrate that this property holds without knowing what the distortion looks like. This task sounds nearly impossible, but a clever idea saves the day. (*Caution:* The proof ahead is challenging.)

## *The Proof*

In the beginning, we have one disk on top of another, and the points of the blue disk are paired up in a one-to-one manner with the points on the red disk. Each blue point gazes up and memorizes its special red partner point. As the red disk is stretched, bent, folded, rotated, compressed, and otherwise distorted, each blue point carefully watches the movements of its special red point. When the distorting process finally terminates and the mangled red disk is laid to rest on the blue, each blue point sees exactly where its special red partner is now located.

So strong is the attraction of each point on the blue disk toward its original red mate that each blue point has an invisible pull toward its red point's current location. This magneticlike pull can be detected with a special compass. When we place this little compass at any particular point on the blue disk, it points in the direction of where its original red partner is now located. In fact, we may imagine placing a tiny compass on every single blue

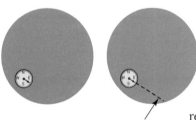

Means that the blue's original red partner is now located somewhere in this direction.

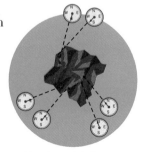

Nearby points have compasses pointing in roughly the same direction.

point; each compass would point in the direction of its corresponding red point. Notice that since we didn't cut the red disk, nearby blue points will have their compasses pointing in similar directions.

## *Moving a Compass Along Circular Paths— The Winding Number*

Let's consider what happens to the compass arrow as we move around the circular boundary of the blue disk. At each point around the boundary circle, the compass arrow points toward the destination of its associated red point. Of course, we don't know where that destination red point is; however, we do know that the destination is somewhere in the blue disk. Consequently, if we start at the top of the boundary circle and move once around the circle, the compass arrow will wind around exactly once. Following the same procedure, do this experiment for yourself. A good way to see it is first to just keep the compass arrow constantly pointing toward the middle of the circle. This experiment shows that the compass arrow circles once around the compass as we move around the boundary of the circle.

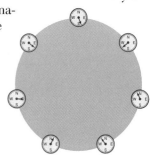

Arrows along the boundary circle always point into the blue disk.

Now what happens when we do the same process but on a circle that is not quite the boundary circle of the blue disk but is just barely inside it? Well, notice that at each position the compass arrow is pointing in almost exactly the same direction as it was pointing from the nearby point actually on the boundary. So, as we go around, we will still see the compass arrow circling exactly once. We call the number of times the compass arrow winds around as we travel along a circle the *winding number*. So, we see that, along the circular boundary of the blue disk and circles just within the boundary, the winding numbers are all 1; the arrow circles once around the compass.

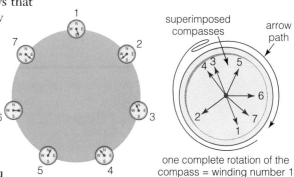

one complete rotation of the compass = winding number 1

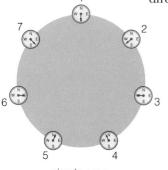

simple case

Let's make another observation. If we take a tiny circle around the center of our blue disk, then, as we move around that tiny circle, all the destination

5.5 / FIXED POINTS, HOT LOOPS, AND RAINY DAYS ◆ 389

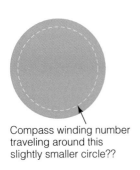

Compass winding number traveling around this slightly smaller circle??

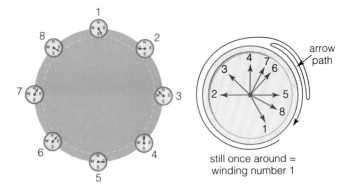

still once around = winding number 1

points stay in almost the same location. Consequently, the compass arrow will be pointing in nearly the same direction for all the points in that tiny circle. So, in going around the tiny circle while pointing to the destination points, the compass arrow does not wind around the compass at all. It has winding number 0.

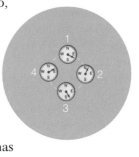

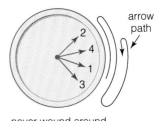

never wound around the compass equals winding number 0

## *Various Winding Numbers*

The question now is, what happens as we consider the winding numbers for the circles of different radii, all centered at the center of the blue disk, as we move from the boundary circle toward the center of the blue disk? Near the boundary, the winding number is 1. Near the center, the winding number is 0. So, somewhere along the way, the winding number had to change. But how is that possible? Whenever a particular circle has a certain winding number, all nearby circles must have the same number since, at each location nearby, the compass arrow would be pointing in nearly the same direction.

The only way the winding number can change is if around some circle, the winding number *does not exist at all*. In

Winding number along here is 0.

Winding number along here is 1.

Also winding number 1.

Similarly, winding number 0.

Where does the winding number change from 1 to 0? Can it?

390 ◆ CONTORTIONS OF SPACE

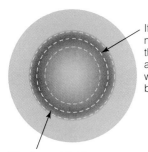

If you know the winding number here, you know the winding numbers for all circles near it—the winding number would be the same!

Winding numbers along circles in the shaded region are all equal!!

other words, at some point on that circle, the compass does not know which direction to point. That situation would occur only if the red point is exactly above its original blue point—if that original red/blue pair of points remained together. This romantic and happy ending demonstrates that there must be a fixed point, and our theorem is proven to be true.

## *Meteorological Fixed Points*

We now illustrate the power of this circle of ideas with an application to the weather. The north pole and the south pole are always cold; however, they are not always equally cold. Air pressure varies from time to time and from place to place. These sentences just about exhaust our personal knowledge of meteorology. But we do know something about the weather that practicing meteorologists might not know.

> **THE METEOROLOGY THEOREM.** *At every instant, there are two diametrically opposite places on Earth with identical temperatures and identical barometric pressures.*

Let's clarify this statement. At any instant in time, every location on Earth has both a temperature and a pressure. Every point on Earth has exactly one point that is diametrically opposite, or antipodal. The *Meteorological Theorem* says that, if we have a globe before us with the temperatures and the pressures written at each point, we can always find a pair of antipodal points on the globe where the temperature and pressure at one point are identical to the temperature and the pressure at the other point.

Although we will not give a proof of the Meteorology Theorem, the fundamental idea is to adopt the winding number argument used in the Brouwer Fixed Point Theorem. In trying to understand this theorem, however, the first question we may ask is, Forget the pressure—I have enough pressure as it is. Why is there a pair of antipodal points where the temperatures are the same? The answer to this more basic question is contained in the proof of the *Hot Loop Theorem*.

> **THE HOT LOOP THEOREM.** *If we have a circle of variably heated wire, then there is a pair of opposite points at which the temperatures are exactly the same.*

## Warm-Up Exercise

Before we read the proof of the Hot Loop Theorem, let's try to defeat the theorem by building a counterexample. Draw a circle and label the points around the circle with temperatures. Try to label it in such a way that no opposite pair of points has the same temperature. Do this before reading on, because your attempt provides the essence of the idea behind the result.

## Result

When you undertake this challenge, you will learn several things.

1. You cannot label each point since there are infinitely many (in fact, uncountably many) points on a circle. So, you label just some of the points and assume that the temperatures between labeled points vary smoothly.
2. Temperatures vary continuously. That is, if it is 50° at one point, it cannot be 100° right next to it.
3. You will learn that you cannot defeat the Hot Loop Theorem.

## The Proof of the Hot Loop Theorem

78° here, so we write temp (12) = 78°

Let's label the points on the circle as a clock is labeled—that is, 12 at the top, then 1, 2, 3, and so on. Assume that points in between these numbers are labeled with decimals. For each point $x$ on the circle, we'll call the opposite point $x'$. So, for example, if $x = 1.5$, then $x' = 7.5$.

For each point $x$ on the circle, let us call the temperature at the point $x$ temp($x$). In other words, temp($x$) is the temperature at point $x$. For example, if the temperature at point 12 is 78°, then we would write temp(12) = 78°. So, we seek to locate a point $x$ on the circle where temp($x$) = temp($x'$) (the temperature at point $x$ is equal to the temperature at the point opposite $x$).

We will start at 12 and consider the difference between the temperature there and the temperature at the opposite point—that is, temp(12) − temp(6). As we move around the clock, we consider temp(1) − temp(7), temp(2) − temp(8), and so on.

Notice what happens to this difference as we move around the circle. Suppose that temp(12) − temp(6) is a positive number. That means that temp(12) is bigger than temp(6). Points near 12 will still be hotter than points near 6. Perhaps, when we get to 1, temp(1) − temp(7) will still be positive; that is, the temperature at 1 is larger than the temperature at 7. Keep going. At 2, perhaps temp(2) − temp(8) is still positive; that is, temp(2) is greater than temp(8). Likewise, temp(3) might be larger than temp(9). Keep going. It might even be that temp(4) is larger than temp(10). Maybe temp(5) is still larger than temp(11).

But look out! Could temp(6) be larger than temp(12)? No! Remember that we started the whole discussion assuming that temp(12) is larger than temp(6). But, as we move half way around the circle, we see that we are switching the order of subtraction. So, if temp(12) − temp(6) is positive, then temp(6) − temp(12) is negative.

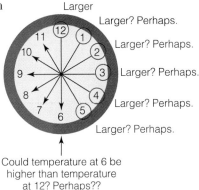

Larger
Larger? Perhaps.
Larger? Perhaps.
Larger? Perhaps.
Larger? Perhaps.
Larger? Perhaps.

Could temperature at 6 be higher than temperature at 12? Perhaps??

Therefore, somewhere between 12 and 6 the difference must change from positive to negative, and, when it does, it is 0. At that point $x$, temp($x$) = temp($x'$). That means that the temperature at $x$ equals the temperature at the opposite point $x'$, and we have established the validity of the Hot Loop Theorem!

## *From Loops to Spheres—The Meteorological Theorem*

The Hot Loop Theorem tells us that there are lots of opposite (or antipodal) points on Earth where the temperatures are the same. Take any great circle at all—for example, the equator. There must be a pair of antipodal points on the equator that have the same temperature. Likewise, for every longitudinal circle there is a pair of opposite points that have the same temperature.

These red dots have the same temperature
These red dots have the same temperature
These red dots have the same temperature
These red dots have the same temperature

By the same reasoning, on the equator there is at least one pair of opposite points where the pressures are the same. So have we proved the Meteorological Theorem? No. A point on the equator whose temperature equals the temperature of the opposite point may not be a point whose pressure is the same as the opposite point's pressure. The temperatures and the pressures form independent sets of numbers around the equator. We require a slightly more elaborate line of reasoning to deduce the Meteorological Theorem, but we won't present that proof here.

We close this section with the secure feeling that, even in an ever-changing world, we are certain, using abstract mathematical ideas, that some things will remain the same. In contrast to this peaceful and comforting thought, the disturbing realm of chaos and fractals will be our next realm.

## A LOOK BACK

The Brouwer Fixed Point Theorem reveals the surprising fact that no matter how we bend, stretch, or deform a disk and put the result within the bounds of where it started, there will always be at least one point that has not moved. The Hot Loop Theorem implies that are always two opposite points on the earth that have the identical temperatures or pressures.

The key to understanding these issues involves examining the situation point by point but then reasoning about the whole. In the case of the Fixed Point Theorem, we consider the winding numbers of a compass arrow as it travels from point to point around various circles. In the Hot Loop Theorem, we consider the differences in temperatures of opposite points. We can try to turn and fold a disk to contradict the Brouwer Fixed Point Theorem. We can try to put numbers on a circle in an attempt to contradict the Hot Loop Theorem. Our failures give us insights into why the theorems are true. These theorems are difficult to prove because they require us to put together many bits of information into a structured whole. In one case, we must look at how circles of points behave, and, in the other, we must look at opposite points on the loop.

Failures can be excellent teachers. If we believe something is true, a good method to find reasons for that truth is to try hard to show that it is false and see where the arguments break down. Looking at the cause of a failure often points the way to success. Sometimes understanding a complicated situation requires us to gather and organize many small bits of information to see a global pattern. When we have mounds of microscopic information, we can step back, squint, and see whether the details coalesce to paint a coherent picture.

*Attempt impossible tasks and carefully observe the failure.*

*Learn from failed attempts.*

*Collect information at the micro level and analyze it at the global level—Think locally, act globally.*

# MINDSCAPES  Invitations to Further Thought*

## I. SOLIDIFYING IDEAS

1. **Fixed on a square.** Does the Brouwer Fixed Point Theorem hold if the disks are replaced by two square sheets? Explain why or why not.

2. **Fixed on a circle.** Does the Brouwer Fixed Point Theorem hold if the disks are replaced by two circles (that is, just the boundaries of the disks without including the inside)? Explain why or why not.

3. **Winding curves (H).** In each drawing is a disk with a marked green circle. The disk was then stretched, twisted, shrunk, and generally distorted and then restored to its original form. Of course, the green circle was stretched, twisted, shrunk, and generally distorted as well. We depict the green circle after the distortion by the analogously marked yellow curve. For each picture, compute the winding number of the yellow image as you travel once around the original green circle.

4. **Winding arrows.** In each drawing we have a circle (equator) drawn with arrows as indicated. For each picture, compute the winding number of the arrows as you travel once around the circle.

5. **Under pressure (S).** Must there be two antipodal points on the equator with the same barometric pressure?

6. **Not the equator.** Must there be two opposite points on any small circle drawn on Earth with the same temperature?

7. **Home heating (H).** Prove that there are two points somewhere in your room that are exactly 5 feet apart and have *precisely* the same temperature.

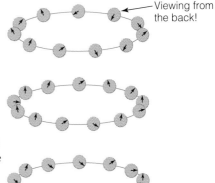

---

*In the Mindscapes section, exercises marked (H) have hints for solutions at the back of the book. Exercises marked (S) have solutions.

8. **...ar populations.** Suppose at each point on Earth you count the number of people within 1,000 miles of that point. Are there two ...tipodal points that have the same number of people within 1,000 ...iles?

9. **Lighten up.** Must there exist two antipodal points on Earth where ...e amounts of sunlight (lightness) are identical?

10. **Shot disk.** Suppose we have two disks, one red and one blue, and we remove the center point from the red and rest that punctured disk on top of the blue. If we now distort the red disk and place it back on the blue, must there be a point on the punctured red disk that remains fixed?

## II. CREATING NEW IDEAS

1. **Lining up.** Suppose we have two line segments having identical length: one red and one blue. We begin with the two segments ...ght on top of each other. We then take the red segment, stretch, ...nd, shrink, and distort it without cutting, and place it on the blue line so that it does not hang off at either end. Must there be a point on the red line that is in its original position?

2. **A nice temp.** Must there be two antipodal points on Earth where the temperature is 62°? Explain.

3. **Off center.** Suppose we have two disks, one red and one blue, and we remove one point from the red (not necessarily the center point) ...d rest the punctured red disk on top of the blue. If we then distort the red disk and place it back on the blue, must there be a point on the punctured red disk that remains fixed?

4. **Diet drill.** Suppose someone weighs 160 lbs and decides to go on a ...et. After 3 months, the person weighs 149 lbs. Did this person ...h 154.5 lbs. after 1.5 months? Did the person weigh 154.5 lbs ...e time within that 3-month period?

...**y (S).** You enter a tollway and are given a toll card at 12:00 ...The speed limit is 65 m.p.h. You travel 140 miles on the ...ay and then exit at 2:00 P.M. You give the card to the attendant, ...looks at it and immediately calls the police. Why did she call the police? How can she prove you broke a law?

## III. FURTHER CHALLENGES

1. **The cut core.** Suppose we have the red and blue disks as before. Now, ...vever, on the ...lisk we cut out ...ll circle from the

cut

Two pieces!

center of the red disk. Thus, the red disk is cut into two pieces: a small disk and a disk with a hole. Is it possible to deform these two red pieces and place them back on the blue disk in such a way that every red point is moved?

2. **Fixed without boundary.** Do you think that the Brouwer Fixed Point Theorem would hold if we replace the two disks by two interiors of disks? That is, we have two disks, but neither contains their boundary circle.

3. **Take a hike.** A hiker decides to climb up Mount Sanitas. There is only one trail to the top. He starts at the base of the mountain at 8:00 A.M. Saturday. He climbs, stops, rests, backtracks a bit, but finally gets to the summit by 5:00 P.M. that evening. The next morning at 8:00 A.M. he begins to hike down. Again he stops, rests, backtracks (in fact returns back to the top because he left his tent up there), but finally gets back to the bottom by 5:00 P.M. that evening. Must there exist a precise point (altitude) on the trail with the property that, at the very moment he crossed that point, his watch showed the exact same time going up as it did coming down? Carefully explain why or why not.

### IV. IN YOUR OWN WORDS

1. **Personal perspectives.** Write a short essay describing the most interesting or surprising discovery you made in exploring this section's material. If any material seemed puzzling or even unbelievable, address that as well. Explain why you chose the topics you did. Finally, comment on the aesthetics of the mathematics and ideas in this section.

2. **With a group of folks.** In a small group, discuss and actively work through the arguments involved in the proofs of the Brouwer Fixed Point Theorem and also the Hot Loop Theorem. After your discussion, write a brief narrative describing the ideas in your own words.

# CHAPTER SIX

# Chaos and Fractals

> *God has put a secret art into the forces of Nature so as to enable it to fashion itself out of chaos into a perfect world system.*
> —Immanuel Kant

**6.1** Images

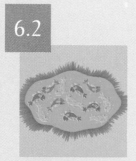

**6.2** The Dynamics of Change

**6.3** The Infinitely Detailed Beauty of Fractals

**6.4** The Mysterious Art of Imaginary Fractals

Fractal images are intricately—in fact, infinitely—detailed pictures. From an artistic point of view, fractals liberate us from the confines of finite detail. Fractals are objects of the mind whose complexity continues to provide revelations and additional wonders at ever-increasing magnification. All the detail is in place, and it is ready for us to explore visually with improved printers and enlarged images. Fractals provide us with whole worlds of beautiful images whose richness and detail are literally infinite.

The images that follow form the starting point for our investigations into fractals and mathematical chaos. We will examine the images and look for patterns. The patterns we will find are ones of self-similarity; that is, small parts of the picture look like the whole picture only on a smaller scale. Once we see the pattern of self-similarity at different scales, we seek processes that could generate such features, and we find that repeated applications of simple transformations result in these beautiful images.

Fractals have significance beyond the images we can produce. They show us that simple processes, repeated many times, lead to objects of great complexity. Many mathematical models of phenomena in our world use simple processes that are repeated many times. These models attempt to predict the future of the weather, the stock market, and the population. The chaotic complexity of the fractal images is reflected in these models and suggests that our ability to predict the future is severely limited.

◆ ◆ ◆ ◆ ◆ ◆ ◆ ◆ ◆ ◆ ◆ ◆

**6.5**

*Predetermined Chaos*

**6.6**

*Between Dimensions*

## 6.1 Images: Viewing a Gallery of Fractals

*Not chaoslike, together
  crushed and bruised,
But, as the world
  harmoniously confused;
Where order in variety we see,
And where, though all things
  differ, all agree.*

— Alexander Pope

A natural first step in the study of fractals is simply to look at some. Look for patterns and similarities within each picture. But mainly enjoy each image for itself.

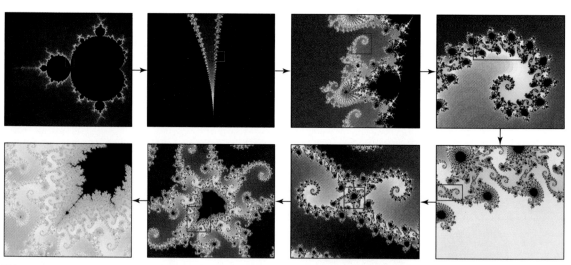

This spectacular Mandelbrot Set provides an infinitely detailed tribute to Benoit Mandelbrot's contribution to the development of a whole new field of study. If graphic lines were infinitely fine and microscopes were infinitely powerful, you could increase the power indefinitely, focus on increasingly small sections, and see at each magnification new intricate details that possess a structure similar to the larger sections. As you will see in section 6.4, the Mandelbrot Set is an intellectual triumph that spans mathematics and art.

Real or fake? These spectacular, cratered vistas never saw the light of day. Instead, these images were created by a simple, repeated mathematical process. The images are so compelling we must ask ourselves whether nature itself relies on similar processes to produce itself.

Real, fake, healthful, or tasty? Answer: Real and healthful, certainly. Tasty, to some. Self-similar, definitely. The whole bunch of broccoli looks much like a sub-bunch, which looks much like a sub-sub-bunch.

bunch looks like

sub-bunch which looks like

sub-sub-bunch

Here they are together—note the relative size differences although they look similar in the above photographs.

In the Sierpinski Triangle, the whole is identical to a part, which is identical to a sub-part and so on —*forever*.

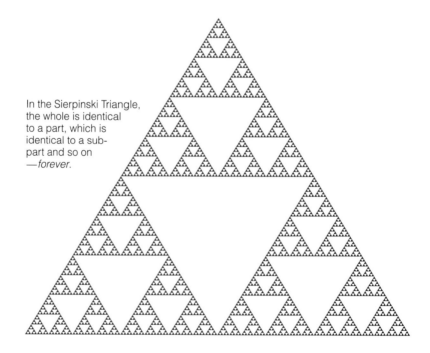

402 ◆ CHAOS AND FRACTALS

How many Quakers do you see? This Quaker Oats box gives us an example of a process of a picture within a picture within a picture, which is the basic construction of many incredible fractal images.

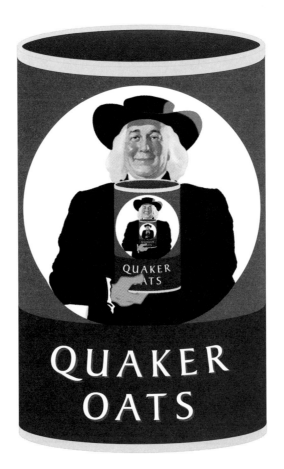

Could this image be on a Quaker Oats box? Yes. The whole fern is not a fern at all but merely a Quaker-esque picture within a picture.

Gaston Julia never saw graphic renderings of the infinitely complex sets he conceived more than 70 years ago.

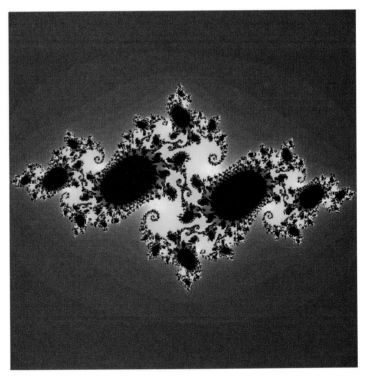

More of Julia's sets.

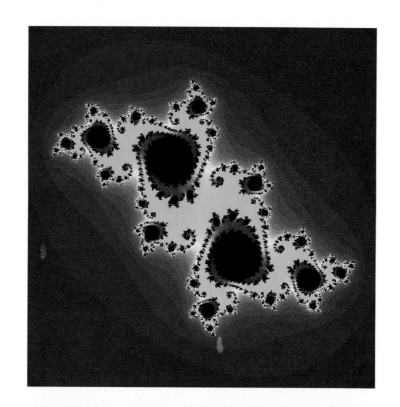

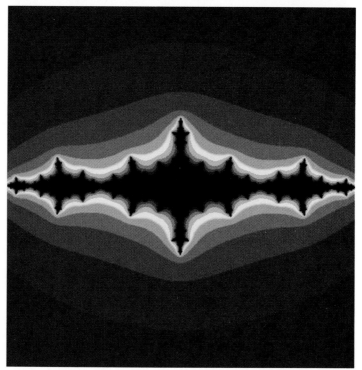

Still more of Julia's sets.

**A LOOK BACK**   These images are all infinitely intricate but arise from simple processes. They are the icons of chaos and fractals. We now proceed to discover and to understand their origins, structures, and nuances.

Look for patterns.

◆ ◆ ◆ ◆ ◆ ◆ ◆ ◆ ◆ ◆ ◆ ◆ ◆ ◆ ◆ ◆ ◆ ◆

## MINDSCAPES  Invitations to Further Thought*

### I. SEEING THINGS

***Self-seeking.*** Each image in the preceding fractal gallery exhibits self-similarity to some extent. Here we ask you to uncover it.

1. ***The incredible shrinking man.*** On the Quaker Oats box, outline the sub-picture that is an identical, but reduced, copy of the whole picture. Roughly what fraction of the height is the reduced picture compared to the whole?

2. ***Multiplicity (S).*** In the Sierpinski Triangle, outline three sub-figures that are identical but reduced copies of the whole figure. For each sub-figure you outlined, compare its width, as a fraction, to that of the whole Sierpinski Triangle.

3. ***Different sizes.*** In the fern, find three reduced copies of the whole picture—one about 85% as large as the whole figure, the other two much smaller—that make up the whole picture except for the stem.

4. ***Blooming broccoli.*** In the bunch of broccoli, find three sub-bunches that look nearly identical to the whole bunch. Which would you rather eat, the whole bunch or the sub-bunch?

5. ***Not quite cloned.*** In the Mandelbrot Set shown, the whole is not identical to any sub-part. Find some sub-parts that nevertheless look similar to some yet smaller sub-parts.

6. ***Julia's descendants.*** Some of the Julia Sets pictured hang together in one piece, and some are made up of many pieces. For the Julia Set that is not connected, find two reduced copies of the whole Julia Set that make up the whole picture.

7. ***Maybe moon.*** What features of the fractal forgeries of the cratered vista make it look realistic?

8. ***Exposing forgeries.*** What features of the fractal forgery of the cratered vista expose it for a fake?

---

*In the Mindscapes section, exercises marked (H) have hints for solutions at the back of the book. Exercises marked (S) have solutions.

9. ***Nature's way.*** Find some examples in nature of self-similarity.
10. ***Do it yourself*** (**H**). Draw a figure that contains several reduced copies of itself in it.

## II. IN YOUR OWN WORDS

1. ***With a group of folks.*** In a small group, discuss the images in this section. After your discussion, write a brief narrative describing themes that go through the images and how they might be produced.
2. ***Creative writing.*** Write an imaginative story (it can be comical, dramatic, whatever you like) that involves or evokes the images and ideas of this section.

# 6.2 The Dynamics of Change:
## Can Change Be Modeled by Repeated Applications of Simple Processes?

*The mathematical phenomenon always develops out of simple arithmetic, so useful in everyday life, out of numbers, those weapons of the gods; the gods are there, behind the wall, at play with numbers.*

—Le Corbusier

The best predictors of where you will be are where you are now and which way you're going.

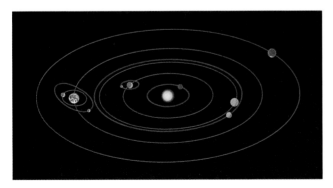

Our world changes. Populations grow and shrink. The weather changes from day to day, season to season, and year to year. The positions of planets, moons, and stars evolve over time. Often we have some understanding of the influences that cause change. Applying this knowledge, we can fairly accurately predict the population, temperature, or positions of planets for the next day. Then, using those predicted conditions as the assumed starting points, we could predict the population, the weather, or the positions of planets for two days from our starting point. Once more, taking the results of previous predictions as the starting point, we could take one more step of prediction. In this fashion, we can predict many days or years into the future. This process is a dynamic model of incremental prediction.

Dynamic models of incremental prediction are natural ways to describe change, and many important mathematical applications use them. As we will soon see, this technique of step-by-step change has intriguing and unexpected consequences. Among them are games, images, and a world of insight about our ability to accurately predict the future.

409

We will clarify our ideas about these repeating processes by looking at several different examples—bank accounts, positions of planets, populations, and games. From these disparate examples, we abstract and develop an idea of systems that change in a step-wise fashion.

## *Money*

Suppose we deposit $1,000 in a savings account that pays 5% interest compounded annually. After one year, our account will have a balance of 1.05($1,000) = $1,050. If we leave the money in another year, we will accrue 5% interest on all the money in the account. So, after two years we will have 1.05($1,050) = $1,102.50. If we leave our money in the account yet another year, we will earn 5% on what we now have; so, after three years we will have 1.05($1,102.50) = $1,157.63. The pattern is clear—not exciting, but clear. Take what we have at the beginning of the year, multiply that number by 1.05, and we have the amount of money we will end up with after one more year. We can just sit back and watch our money grow. Simple enough.

## *Planets*

Suppose we know the locations of all the planets, asteroids, and other matter in the solar system. We know the speed and direction of each object; we know the masses of the sun, planets, and asteroids; and we know about gravitational forces. Can we compute the locations and velocities of all those bodies after one revolution of the earth? Sure, pretty closely at least. How about after two revolutions? We could take our knowledge of the positions and velocities of the bodies after the first revolution and do the same calculations to compute the state of the solar system after two revolutions. And so on. Theoretically, we could continue year after year for thousands, millions, or even billions of years. This method can be used to make predictions, although how accurate our predictions will be is a matter for us to contemplate when we consider chaos. The key is that we are able to take what we have, apply a rule, compute what we have next, and then repeat.

## *Population*

How can we predict the population size after one year? We can take the current population size, estimate the number of births and deaths expected during the next year, and then compute the population size after one year. How can we predict the population after two years? We can just take our estimate for the population after one year, again estimate the number of births and deaths, and compute the new population size after one more year. How can we predict the population size many years from now? Repeat, repeat, repeat.

## *Repeat, Repeat, Repeat*

As we are seeing, models of the world often boil down to:

- Start somewhere
- Apply a process and get a result
- Apply the same process to that result to get a new result
- Apply the process again to the new result to get a newer result
- Repeat patiently and persistently, forever.

The computation of compound interest, predicting the positions of the planets after time has passed, and the estimation of future population size are all examples involving the common underlying theme of repetition. We have encountered this iteration theme before in our discussion of the Fibonacci numbers. Recall that each Fibonacci number was obtained by adding together the previous two Fibonacci numbers. We took the results of what we had done before and used them to get the next number. We are now crystallizing the notion of an iterative procedure for predicting the future of changing, dynamic systems—a procedure that repeatedly uses a feedback process whereby the output from one step is used as the input to produce the next step.

## *Modeling*

To build a mathematical model that reflects reality, we examine the real-world situation and cull from it some features that we believe are important. We then attempt to capture their essence through mathematical relationships or processes that are analogous. As an illustration, let's consider models for population size.

How many people will there be on Earth in the year 2050? Where would we begin? We might try to devise some reasonable predictions about how the population size will change from one year to the next given the number of people living at the beginning of a year, as well as their living conditions. Of course, developing realistic, predictive models is an extremely complicated issue, since so many factors influence population change. Therefore, as always, let's start simply.

Start simple.

One of the recurring themes of this book is the power of starting with a simple or an idealized case, building up insights, definitions, ideas, and methods, and then using that experience to handle the more challenging cases. In analyzing population growth, let's start with a game that reflects some features of what makes populations grow or shrink.

## *The Game of Life*

Life was not invented in the 20th century, but the *Game of Life* was—by a mathematician, John Conway of Princeton University. Although the Game of Life models only one abstracted basic feature of real life, it does reflect some of the dynamic development of populations over time. It may even be more interesting as a model of other phenomena, such as how fires propagate in a forest.

Conway's Game of Life is played on an infinitely extended grid—an infinite checkerboard. At each time interval, a square is either alive or not, depending on how many living squares surrounded it in the previous time interval. Thus, this model of population growth is focused on an organism's need for the right number of companions within its immediate vicinity. Here are the rules for determining which squares will be alive during the next generation.

- A living square will remain alive in the next generation if exactly two or three of the adjoining eight squares are alive in this generation; otherwise, it will die.

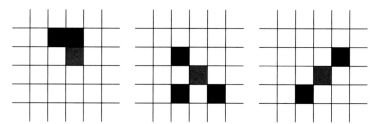

The red square will be alive in the next generation.

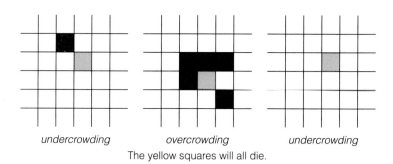

undercrowding        overcrowding        undercrowding

The yellow squares will all die.

- A dead square will come to life if exactly three of its adjoining eight squares are alive; otherwise, it will remain dead.

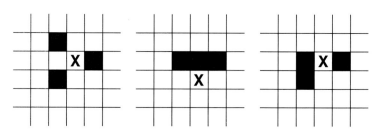

The currently dead **X** squares will come to life in the next generation.

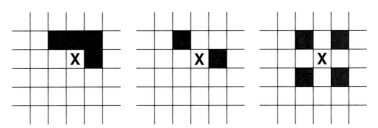

The currently dead **X** squares will remain dead in the next generation.

Let's look at the future generations for some initial sets of organisms.

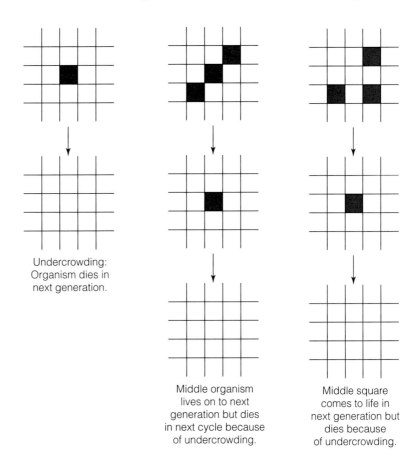

Undercrowding: Organism dies in next generation.

Middle organism lives on to next generation but dies in next cycle because of undercrowding.

Middle square comes to life in next generation but dies because of undercrowding.

Slight differences in the initial sets of living cells sometimes evolve into dramatically different outcomes. Some initial settings give rise to a population explosion that grows without bound. Other populations die out quickly. Some become periodic—growing and shrinking in an infinitely repetitious, predictable way—or become stable. Others move across the grid like a migrant colony marching off forever onward. Here we see all these possibilities at play. For each figure, decide whether it gives rise to (1) a population explosion, (2) an extinction, (3) a stable pattern, (4) a periodic pattern, or (5) a migratory pattern. You can use your CD-ROM from your kit to simulate the Game of Life. Have fun!

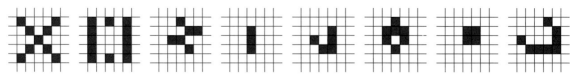

Choices: (1) population explosion; (2) extinction; (3) a stable pattern; (4) a periodic pattern; (5) a migratory pattern

The Game of Life provides a rough model of population development. Even with these simple rules, the potential patterns of development are intricate and fascinating. This game suggests that different initial population distributions give rise to completely different types of future behavior. Mathematical biologists study various models of growth that try to reflect different aspects of actual populations. The mathematical behavior of these models sometimes points to unexpected population patterns that occur in nature. Let's discover this phenomenon in action by considering a different population *pool*.

## Scrod

Fish in a pond, like the birds and the bees, fall in love, marry, and subsequently spawn little fish who, after proper schooling, themselves become amorous and produce little fish of their own. This romantic life cycle results in ponds full of fish and leads to the question: How many fish might we expect in the pond after several years of good breeding?

The simplest model of population growth might assert that the population doubles each year. However, such a model is grossly unrealistic, because the pond has a maximum sustainable population. If the fish population doubled each year, soon the population would far exceed the pond's capacity. Thus, we need to develop a more refined model of population growth that predicts that the population will shrink if the fish population exceeds the maximum number that the pond can sustain.

In the 1840s Pierre Francois Verhulst developed a population model with this refinement in mind. The Verhulst Model represents each year's population not as a number of fish but instead as a density of fish in the pond—that is, as a *fraction* of the maximum sustainable population of the pond. So, for example, if there are half as many fish as the maximum sustainable number, we say the population density is .5. If the number of fish is equal to the maximum sustainable number, then we say the population density is 1.0. If in some year the population temporarily exceeds the maximum sustainable population, the population density might be 1.2, but then the population would drop the next year—there is an overpopulation of fish.

*There are many ways to say the same thing. Choose one that's useful.*

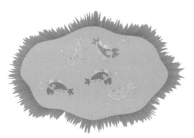
population density = .5

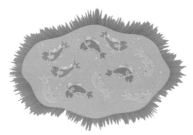
population density = 1.0

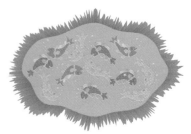
population density = 1.2

## *The Rate of Change of Population*

Verhulst reasoned that, if the population is small relative to the maximum sustainable population, then the fish will have lots of available room and resources and thus will happily reproduce at will. Consequently, the population will grow at a faster rate than it would if it were near the maximum sustainable population (in which case overcrowding and limited resources would inhibit amorous urges). In the Verhulst Model, the rate at which the population increases (or declines) is proportional to how far away the population size is from the maximum sustainable number. Let's take a closer look at how the Verhulst Model works.

We first need to figure out the rate at which the population grows (or shrinks). To do this, let's write $P_n$ for the population density at year $n$ given as the fraction

$$P_n = \frac{(\text{the number of fish at year } n)}{(\text{the maximum sustainable population})}.$$

Then $P_{n+1} - P_n$ is the change in population density from year $n$ to the next year $n + 1$. That change divided by $P_n$ gives the *rate at which the population is increasing or decreasing*. So, the rate at which the population is increasing or decreasing from year $n$ to year $n + 1$ is equal to

$$\frac{P_{n+1} - P_n}{P_n}.$$

For example, suppose that, in a certain pond, the maximum sustainable population of fish is 5,000. If in year 2 the fish population of the pond was 2,500 and in year 3 the population was 3,000, then we can compute the population densities as follows:

$$P_2 = \frac{2{,}500}{5{,}000} = .5 \quad \text{and} \quad P_3 = \frac{3{,}000}{5{,}000} = .6.$$

Thus, the change was

$$P_3 - P_2 = .6 - .5 = .1.$$

If we now take that change (.1) and divide it by the population we started with ($P_2 = .5$), we get .2 as the *rate at which the population increased*.

Stated in English: The population increased by 20% from year 2 to year 3. Recall Verhulst's basic three-part assumption:

- If the population is far from the maximum sustainable population, we expect a large increase in population;
- If the population is not as far from the maximum sustainable population, we expect a lesser increase in population; and finally,
- If the population exceeds the maximum sustainable population, we expect a drop in population.

Specifically, we notice that $1 - P_n$, the difference between the maximum sustainable population density, 1, and the population density at year $n$, $P_n$, measures how close the current population is to the maximum sustainable population. So, Verhulst's Model states that the rate of population change $[(P_{n+1} - P_n)/P_n]$ will be large when $1 - P_n$ is large, and the rate of population change will be small when $1 - P_n$ is small.

We can phrase this observation in mathematical terms as follows: There is some fixed number such that $(P_{n+1} - P_n)/P_n$ will equal the fixed number multiplied by $1 - P_n$. For example, if the fixed number is, say, .7, we can write this idea as an equation:

*Quantify an idea after you formulate it.*

$$\frac{P_{n+1} - P_n}{P_n} = .7(1 - P_n).$$

Multiplying through by the $P_n$ term, we get

$$P_{n+1} - P_n = .7P_n(1 - P_n), \text{ or}$$
$$P_{n+1} = P_n + .7P_n(1 - P_n).$$

So, knowing the population density of one year allows us to find the population density for the next year. This formula, with all the $n$s and $n + 1$s running around, is confusing. Let's solidify the idea by working through a few specific examples.

## Example 1: 1,000 Fish

Let's assume that our pond has a maximum sustainable fish population of 5,000. Suppose that the population this year is 1,000 fish. Thus, we see that

$$P_1 = \frac{1000}{5000} = .2.$$

Since .2 is far from the maximal sustainable population density (which is 1), we would expect a good-sized increase in population for the next year.

In fact, we can compute and see that next year's population density ($P_2$) will be

$$P_2 = .2 + .7 \times .2(1 - .2) = .2 + .7 \times .2 \times .8 = .2 + .112 = .312.$$

So, the population for next year will be

$$.312 \times 5{,}000 = 1{,}560 \text{ fish}.$$

The increase from the population level of 1,000 to 1,560 represents an increase in population of more than 50%—a pretty hefty increase.

## Example 2: 4,500 Fish

Suppose now that the initial population is 4,500 fish. Thus, we have that

$$P_1 = \frac{4500}{5000} = .9.$$

In this case, the population density is very close to the maximum sustainable number, so we would expect the population growth to slow down. Computing, we discover that next year's population density ($P_2$) will be

$$P_2 = .9 + .7 \times .9(1 - .9) = .9 + .7 \times .9 \times .1 = .9 + .063 = .963.$$

So, the population for next year will be

$$.963 \times 5{,}000 = 4{,}815 \text{ fish}.$$

The population increase from 4,500 to 4,815 represents an increase of only 7%—a much smaller increase, just as we had expected.

## Example 3: 7,000 Fish

Finally, suppose the initial population is 7,000 fish. So, we see that

$$P_1 = \frac{7000}{5000} = 1.4.$$

In this case, we have a population that exceeds the maximum sustainable population, so we would expect a decrease in population. In fact, next year's population density ($P_2$) will be

$$\begin{aligned} P_2 &= 1.4 + .7 \times 1.4(1 - 1.4) \\ &= 1.4 + .7 \times 1.4 \times (-.4) \\ &= 1.4 - .392 \\ &= 1.008. \end{aligned}$$

So, here we see that next year's fish population will be down to 5,040. Notice that next year's population is smaller than the population of the year before, just as we had predicted. The population decrease from 7,000 to 5,040 represents a decrease of 28%.

## *Desperately Seeking the Constant*

One key to accurately predicting future population numbers using the Verhulst Model is finding the constant number that best fits the situation. Finding a constant that leads to an accurate predictor is a difficult task. In our examples we used a constant of .7, but, for a real pond, the constant 2 or .01 or some other number might have given more accurate predictions of future populations.

Here are two instructive experiments we can perform with this little population model. The first is to fix a constant, like .7, experiment with different initial populations, and see what happens to the population in future years. The second experiment is to see what happens when we adjust the constant. In 1976, Robert May, a mathematical biologist, published a paper in the journal *Nature* entitled "Simple mathematical models with very complicated dynamics." In it, May explored the surprising results of such experiments and helped to usher in a whole new perspective on predictability and chaos. Before we explore those issues, we will first turn to one of the most intriguing consequences of the repeated application of simple processes: infinitely intricate pictures and beautiful fractals.

## A LOOK BACK

How can we describe change? One method is to take it a step at a time. We can describe the current state and predict the next state. Then we can take that result and predict the next one. In this step-wise way, we can describe change over many steps—that is, over a long period of time. The Game of Life and the Verhulst population model are two illustrations of this basic theme.

The real world is far too complex to understand and describe completely. Often our method of describing the world or predicting the next step is a simplification of reality. We just pick one or two important features and ignore the rest. Although we cannot expect such simplified models to be completely accurate, they may well reveal some aspects of how we can expect the world to change. We start with an oversimplified model with the expectation that we can adjust and augment it later.

To develop ideas, it is important to look at several examples of related phenomena in different settings and find common features. Often there are many ways to describe the same situation, so we can choose a method that is useful to us. By isolating essential features, we can develop descriptions that free us from distractions and allow us to focus on the core issues.

Look at a variety of examples.

Start with simple cases.

Model the same situation to emphasize different features.

◆ ◆ ◆ ◆ ◆ ◆ ◆ ◆ ◆ ◆ ◆ ◆ ◆ ◆ ◆ ◆ ◆ ◆ ◆ ◆

# MINDSCAPES  Invitations to Further Thought*

## I. SOLIDIFYING IDEAS

1. **Call your shots.** For each picture, trace the path of the billiard ball in the sketch until it lands in a pocket. Assume that, when the ball strikes a wall, it bounces off, making the same angle as shown. What pocket do you call?

Destination: Side pocket    Destination: _____    Destination: _____

2. **Getting cornered.** Starting from the spot indicated on a billiard table, sketch two trajectories that result in the ball getting to the back left corner.

3. **Double your money.** Suppose you deposit $1,000 in a savings account that pays 5% interest compounded annually. So, after one year, your account will have a balance of $1.05(\$1{,}000) = \$1{,}050$. By repeating this process, estimate the amount of time required to double your money.

4. **Too many (S).** The surface of the Earth contains approximately $2{,}500{,}000{,}000{,}000{,}000 = 2.5 \times 10^{15}$ square yards including the oceans. The current population of the Earth is about $5{,}700{,}000{,}000 = 5.7 \times 10^9$ people. If the population were to grow at the rate of 1% per year, approximately how soon would there be a person for each

---

*In the Mindscapes section, exercises marked (H) have hints for solutions at the back of the book. Exercises marked (S) have solutions.

square yard on Earth (including the oceans)? What can you deduce about the future rate of population growth?

5. **Rice bowl** (**H**). One day long ago, the Emperor of China wished to reward a clever servant and asked the servant what he wanted. The modest servant said he would be satisfied with one grain of rice on the first square of a checkerboard, two on the next, four on the next, eight on the next, and so on over the 64 squares of the checkerboard. Why did the Emperor behead the servant?

6. **Nature's way.** Find some examples, other than those mentioned in this section, of iterative processes at work in nature.

7–10. **The game of life.** For each initial population in the Game of Life, determine which cells will be alive after 1, 2, 3, and 4 generations. You can use your CD-ROM from your kit to simulate the Game of Life if you wish. (**I.7:S**)

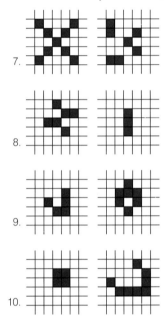

11–14. **Life cycles.** For each initial population in the previous exercise, guess whether that initial population gives rise to (1) a population explosion, (2) an extinction, (3) a periodic pattern, (4) a stable pattern, or (5) a migratory pattern. You might try one or two more generations to help you confirm or reject your guess.

15. **Life on the Web.** Visit our Web site or your CD-ROM to experiment with the Game of Life. Our Web site is: **http://www.heartofmath.com**. Record your experiments.

16. **Explosion.** Devise a new initial population in the Game of Life that leads to a population explosion. (*Hint*: Consider taking an initial population you know will explode and adding a small community far off that will not interfere with the explosion.)

17. **Extinction.** Construct a new initial population in the Game of Life that survives for several generations and then becomes extinct. (*Hint*: Consider taking an initial population you know and adding a small community far away that won't hurt.)

18. **Periodic population (H).** Construct a new initial population in the Game of Life that is *periodic*, that is, that returns to its original position after several generations. (*Hint*: Consider putting together a couple of periodic initial populations you know.)

19. **Programmed population.** On a computer or programmable calculator, enter this simplified version of the Verhulst logistic equation, $f(x) = 4x(1 - x)$, into its function memory: Start with $x = .245$ and compute $4x(1 - x)$ by pressing the function key. Take the result and again press the function key. Repeat 25 times and put the answers in the first column of a table. Now do the whole thing again, but this time start with the slightly different number, $x = .246$. Enter the resulting answers in the second column of the table. The resulting answers start by being similar. Do they remain similar? We will explore your observation more in the Chaos section.

20. **Programmed population: the next generation.** Using a computer or programmable calculator, input this simplified version of the Verhulst logistic equation, $f(x) = 3.5x(1 - x)$: Start with $x = .245$ and compute $3.5x(1 - x)$ by pressing the function key. Take the result and again press the function key. Repeat 25 times and put the answers in the first column of a table. Now do the whole thing again, but this time start with the slightly different number, $x = .246$. Enter the resulting answers in the second column of the table. Are you surprised at the difference between questions 19 and 20?

## II. CREATING NEW IDEAS

1. **How many now?** Suppose that a population is modeled using the Verhulst logistic model, $P_{n+1} = P_n + cP_n(1 - P_n)$. Suppose the initial population measure $P_1$ is 1/4, and the next year's population measure $P_2$ equals 1/2. Find the value of the constant $c$. Using your answer, compute $P_3$.

2. **Fibonacci.** Fibonacci numbers are constructed using an iterative process. Explain how the results of one step in the process become the input for the next iteration of the process. Make a table with two columns. In the first column write the Fibonacci numbers where you start the process as usual with the first two numbers being 1 and 1. In the second column, do the same process of going from generation to generation; however, now start with 1 and 3. How are the columns similar and different?

3. **Fibonacci again.** For each of the two columns you created in question 2, make a new column. Next to the $n$th entry $F_n$ in your existing column, enter $F_{n+1}/F_n$ in the new column. What do you notice about

the two new columns you created? Did the different starting seeds make a difference? What further experiments might you consider?

4. **Alien antenna.** Start with a **V**. Suppose each end sprouts a smaller **V**. Then each end sprouts an even smaller **V**. Draw the first four generations of this alien antenna. How many end points are there after four generations? How many would there be after $n$ generations?

5. **Cobweb plot.** Here are some graphs with the diagonal of the square drawn in. Start with any point on the horizontal axis. Trace a path by going straight up or down until you reach a point on the graph. From there trace a path horizontally left or right until you hit a point on the diagonal line. Now repeat the process again and again. You will generate a path with right angle turns.

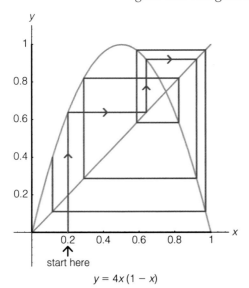

$y = 4x(1 - x)$

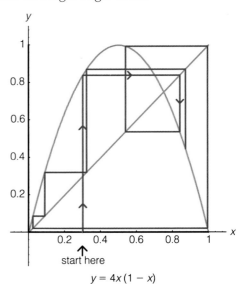

$y = 4x(1 - x)$

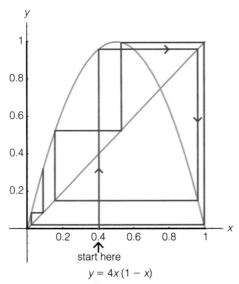

$y = 4x(1 - x)$

The path is called a *cobweb plot*. The cobweb plot records the results of a repeatedly applied transformation—namely, taking a value, applying $f(x)$, then taking that result and again applying $f(x)$, and so on. Now consider the graph of $f(x) = 2.7x(1 - x)$ given below. Start on the horizontal axis at .5 and carefully draw five iterations of the cobweb plot using a straight edge. Does the cobweb plot form a spiral? (*Hint:* To help make the drawing accurate, you should compute $f(.5)$, then take that value and plug it back into the formula, repeat . . .)

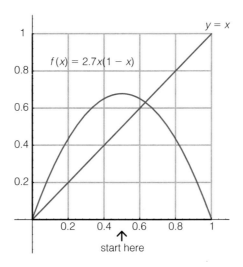

6. **More spiders (S).** Here is the graph of $f(x) = 3.18x(1 - x)$. Start on the horizontal axis at .25 and draw five iterates of the cobweb plot as described in the previous question. Does the path appear to be periodic or not?

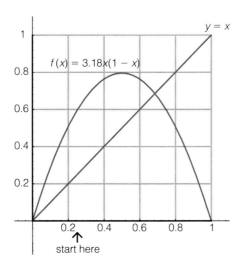

**7. Arachnids.** Here is the graph of $f(x) = 4x(1 - x)$. Start on the horizontal axis at .25 and draw five iterates of the cobweb plot as described in the previous question. Does the path keep going, or does it get stuck?

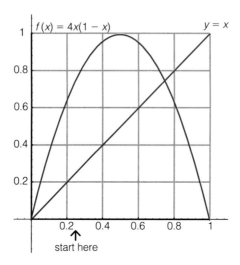

**For 8–10. Making dough.** Many delicious French desserts, such as Napoleons, are made of numerous thin layers of pastry. Once upon a time a French chef began, as usual, to make a masterpiece with a segment of dough one unit long. This time, however, he noticed that three grains of colored sugar—one red, one white, and one blue—were embedded in the pastry. He stretched the dough to twice its length and then folded it in half to produce a double layer of thinner pastry still one unit long. He again stretched the dough to double its length and folded it to produce four layers again one unit long. He repeated the process: Stretch to twice its length, fold to produce eight layers. He stretched and folded again and again.

As our chef stretched and folded, he became fascinated with the movement of those three grains of sugar. In fact, he noticed that the red grain always returned to the identical place where it started. The white grain started

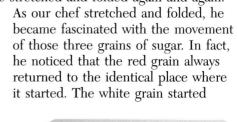

somewhere, went to a different location after a stretch and fold, and then returned to its original starting location after one more stretch and fold. The blue grain rotated among three different places. The chef was so fascinated with these infinitely recurring itineraries of these grains of sugar that he stretched and folded the dough to such an extreme thinness that his dessert creation lived up to the high expectations for light, puffy, French desserts and became the Napoleon of pastries.

8. **Red.** Show where the red grain of sugar may have been located initially.

9. **White** (**H**). Show where the white grain of sugar may have been located initially.

10. **Blue.** Show where the blue grain of sugar may have been located initially.

### III. FURTHER CHALLENGES

1. **More cobwebs** (**H**). Consider the inverted "V" graph below and describe the points whose cobweb path leaves the unit square. (*Hint*: Any point on the diagonal above the interval (1/3, 2/3) leaves immediately. How about points above intervals (1/9, 2/9) and (7/9, 8/9)? Are there others that leave after more steps?)

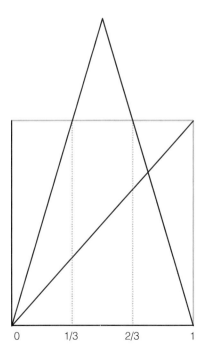

2. **Yet more cobwebs.** Given the inverted "V" graph below, find a starting point by trial and error with pictures such that the cobweb path intersects the diagonal in every 1/6th of the diagonal.

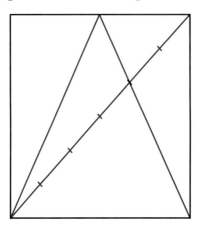

3. **Cantor's cuts.** Start with the unit interval [0, 1]. Remove the middle third of that interval and you will have [0, 1/3] and [2/3, 1]. Next, remove the middle third of each of those intervals. You will have [0, 1/9], [2/9, 1/3], [2/3, 7/9], and [8/9, 1] remaining. If you repeat the process of removing the middle third forever, the points that remain constitute the *Cantor Set*. Describe infinitely many points that remain in the Cantor Set.

4. **How much is gone?** In the construction of the Cantor Set, various intervals were removed. At the first step, the interval (1/3, 2/3) was removed. Then the intervals (1/9, 2/9) and (7/9, 8/9) were removed, and so on. Estimate the total length of the removed intervals. Use your answer to estimate the total "length" of the Cantor Set.

5. **How much remains?** Consider numbers that are not decimal (that is, base 10) but instead are represented in base 3. That is, every digit of the number is 0, 1, or 2. The first digit after the "decimal" point tells how many $1/3^1$'s you have; the next tells how many $1/3^2$'s you have; the $n$th place after the decimal point tells how many $1/3^n$'s you have. So, 0.212 (base 3), for example, represents $2/3 + 1/9 + 2/27$ or 23/27. Show that the points that remain in the Cantor Set are exactly those numbers whose base 3 decimal expansion can be written with only 0's and 2's. (Note that, just as .0999 ... = .1 base 10, .0222 ... = .1 base 3, so put all numbers in the ending 2's

form rather than in the ending 0's form.) Since any sequence of 0's and 2's corresponds to a "decimal" number in the Cantor Set, show that there are more numbers in the Cantor Set than there are natural numbers. In fact, the cardinality of the Cantor Set is the same as the cardinality of the real numbers.

### IV. IN YOUR OWN WORDS

1. *Personal perspectives.* Write a short essay describing the most interesting or surprising discovery you made in exploring this section's material. If any material seemed puzzling or even unbelievable, address that as well. In your brief essay, explain why you chose the topics you did. Finally, comment on the aesthetics of the mathematics and ideas in this section.

2. *With a group of folks.* In a small group, discuss the idea of iterative systems, particularly population models. After your discussion, write a brief narrative about systems that are described by repeated applications of a transforming process.

3. *Creative writing.* Write an imaginative story (it can be comical, dramatic, whatever you like) that involves or evokes the ideas of this section.

4. *Power beyond the mathematics.* Provide several real-life situations—ideally, from your own experience—for which some of the strategies of thought presented in this section would provide effective methods for approaching and resolving them.

# 6.3 The Infinitely Detailed Beauty of Fractals:

## How to Create Works of Infinite Intricacy Through Repeated Processes

*I coined* fractal *from the Latin adjective* fractus. *The corresponding Latin verb* frangere *means* to break: *to create irregular fragments . . . how appropriate for our needs!*

— Benoit Mandelbrot

Some artists are renowned for their attention to detail, producing pictures that are incredibly intricate. But only within the realm of mathematics can one create images that are literally infinitely intricate. What does it even mean to speak of an image that is *infinitely detailed*? Could such an image be drawn with the finest penpoint? No. Instead, we describe such images so precisely that, for any point on our canvas, we can determine whether that point is in the design or not. These images can be drawn to any degree of detail that our finest printers can muster.

But the totality of their intricacy is fully present only in our mind's eye. Every power of magnification reveals yet further detail. In some cases, we will understand the object so well that we can literally draw a blown-up version of the tiniest parts magnified a million, a billion, or a trillion times. Other images reveal unthought-of variations and surprises at ever-increasing levels of magnification. These mysterious images give us new vistas to discover as we explore them to an ever-increasing depth, but they also hold deeper secrets for others to uncover later.

These images of the mind are created by following a pattern that is repeated over and over, forever. Barbers and hair stylists make us all familiar with this

In *The Great Wave* (1823–1829) by Hokusai we see waves within waves . . .

kind of process when they use two mirrors to see the backs of our heads, and we see hairdo after neatly cut hairdo receding into the distance. Feedback loops of video images taking pictures of a TV screen are another example. But other feedback processes also create great pictures of infinite detail and intrigue. Let's begin our exploration of infinitely intricate images with some specific illustrations.

## Example—Parallel Mirrors

Take a hand mirror and hold it facing a regular mirror. What do you see?

What you are seeing is the result of a repeating process. When you look in the big mirror, you see yourself holding a small mirror. But what can be seen within that small mirror? You see a reflection of itself in the big mirror. But what is the reflection of the small mirror reflecting?

Light travels at finite speed; there are limits to reflecting capabilities of mirrors; and eyes have limited acuity. So, real mirrors are not infinitely complex. However, we know what image we would see in an idealized set of mirrors. That image would have infinite detail, because each image of the hand mirror would contain in it an image of a yet smaller hand mirror, and so on forever. Infinitely reflecting mirrors provide a good, concrete way to think about an image with infinite detail.

Parallel mirrors are a real-world example of a repeating process that has a self-similarity property.

## Example 2—Video Feedback

Here's another physical example. Take a video camera and set it up so that a TV displays the image that the video camera sees. Video stores often have the cameras set up in this fashion. Now point the camera at the TV. What do you see? The camera is taking a picture of the TV. But the TV is displaying a picture of what is being seen by the video camera. But the video camera is seeing the TV, so the TV displays a picture of the TV, whose picture is the picture of the TV, whose picture is a picture of the TV, and so on . . .

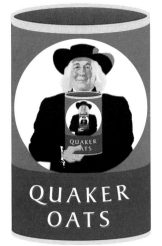

Again, the physical constraints of pixels prevent video feedback from actually having infinite complexity, but in the abstract we can think of an infinitely detailed feedback image displayed on the TV.

## Example 3—Eat Your Oatmeal

Sometimes your day starts with infinite complexity—especially if you eat Quaker Oats. On the cereal box, we see a Quaker holding a box of Quaker Oats, in which we see a smaller picture of the Quaker holding a smaller box of Quaker Oats, in

This 14th-century Vajradhatu Mandala from Tibet, used for meditation, captures some essence of self-similarity long before Quaker Oats ever existed.

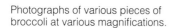

Isolate ideas from examples.

which we again see the Quaker holding the Quaker Oats, and so forth . . . The parallel mirror, video feedback, and the Quaker Oats box all represent infinitely detailed themes. In each of these illustrations, parts of the picture look identical to larger parts of the picture, but at a different scale. These pictures display self-similarity.

We have now seen several examples of a similar theme. Our next step is to isolate the core ideas from these illustrations.

Photographs of various pieces of broccoli at various magnifications.

Same three broccoli pieces, but now with a more tasty treat to show the scaling.

## Self-Similarity

Consider broccoli. Broccoli is a food possessing a legion of virtues for health and well-being. We know, however, that some people do not like broccoli, and we are going to help those people by pointing out a feature of broccoli that will demonstrate that a little broccoli goes a long way. In fact, a little piece of broccoli is similar in structure to the entire head. The upper left picture shows a bunch of broccoli, and underneath it is an enlarged picture of a small piece of it.

430 ◆ CHAOS AND FRACTALS

Underneath the enlarged picture is an even more enlarged picture of an even smaller piece. Notice how similar those three pictures look. So, if you want to convince your mother or doctor (or both) that you are eating your broccoli, just eat the tiny, untastable piece of the last picture, send them the blown-up photo, and they will be impressed with your diligent adherence to a healthful diet. *Bon Appetit!*

Coastlines are wiggly. When we look at the coastline of England on a world globe, we will see a wiggly outline. Now look at a large map of England alone. We discover that each of the wiggles of the globe-level view has been magnified to reveal a finer construct of wiggles at a smaller scale. Now let's look at a map of Wales. Again, we see that the smaller wiggles comprise finer wiggles still. At each level of magnification, the wiggles have some resemblance to the image at previous magnifications—a real-world illustration of self-similarity across scales.

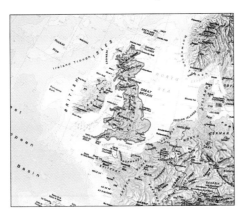

At each magnification we see wiggles on wiggles—at all scales we see self-similarity.

Coastlines and broccoli both capture the flavor of self-similarity—that is, the feature that something looks the same or similar under increasing magnification as it does before the magnification. A set that has this distinctive self-similarity feature is referred to as a *fractal*. Coastlines and broccoli are examples from nature. We will soon see some purely geometrical fractals that so eerily resemble natural objects, we will have to wonder whether the physical processes that produce self-similarities in nature might be similar to iterative processes that produce self-similarities in geometric examples.

## A Process of Repeated Replacement

Pictures with infinite intricacy can be produced by a process of repeated replacement. We start with a picture made of line segments, replace each line by a particular squiggle made up of several shorter segments, then replace each of those shorter segments by the same, but smaller, squiggle, and so on. This procedure, if continued forever, produces objects of infinite detail. In 1904, Helge von Koch first described this type of spikey replacement. Let's see this process in action.

## *Koch's Kinky Curve*

Let's start with a line segment and replace the segment with four equal-length segments each one third as long as each original. Thus we have created a line segment with one kink in the middle. But, in this chapter, we never stop. Let's do it again. Replace each segment with four segments

creation of the Koch Curve

each one third of the length of the previous segments, as before. We now have bumps on the bumps. Let's continue and continue and continue, forever. The object we end up with cannot be drawn completely, because it has infinitely many wiggles. Let's look at this *Koch Curve* under increasing magnification. Take the part of the curve that came from one of the first four parts, and blow it up to be three times as large. The blown-up version is identical to the entire Koch Curve. The Koch Curve exhibits exact self-similarity. Looking at a part of the Koch Curve, you cannot tell if you are seeing it actual size or magnified a billion times. Either way, it looks completely, exactly the same.

The Koch Curve is one example of an infinite replacement process. Let's ground our understanding of this process by looking at other specific examples that use infinite replacement—namely, the Sierpinski Triangle, the Menger Sponge, and, more generally, a special collage-making procedure.

## *Sierpinski Triangle*

Waclaw Sierpinski first described his fractal triangle in 1916. We start with a filled-in triangle and replace it by a triangle with the upside-down middle triangle removed, as illustrated. This process leaves three filled-in triangles, each one of which is a reduced version of the original triangle. We

stage 0

stage 1

stage 2

stage 3

stage 4

have therefore established a process of replacing a filled-in triangle with three smaller triangles contained in it. So, we can take the result of that first replacement process and replace each of the three smaller triangles with three yet smaller triangles. To each of the nine triangles now present, we can apply the same process. We continue forever.

Notice that the resulting object, known as the *Sierpinski Triangle*, looks exactly like any one of the three sub-triangles that made it up, except for scaling. It also looks exactly like any one of the nine triangles at the next stage, and so on. The construction guarantees this self-similarity because the process of replacement is identical for each sub-triangle as it was for the whole triangle.

*Sierpinski Garden* (1997) by Khaldoun Khashanah. Studying and teaching fractals inspired Professor Khashanah to paint his interpretaion of the landscape along the Delaware River.

## *Menger Sponge—A 3-Dimensional Example*

Self-similar objects need not be confined to the 2-dimensional world of the plane. In 1926, Karl Menger described a neat example of a 3-dimensional fractal that has come to be known as the *Menger Sponge*. Here a solid cube is replaced by the 20 solid sub-cubes shown. It's as

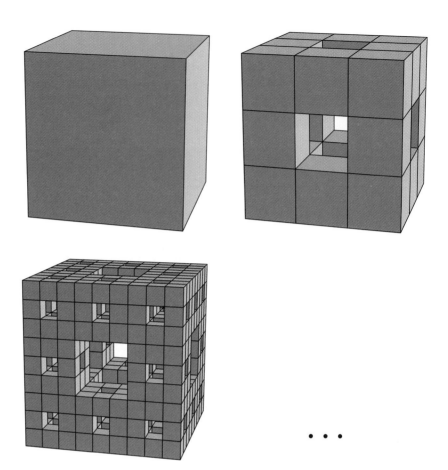

construction of the Menger Sponge

*The Kitchen Knife Cuts Through Germany's First Weimar Beer-belly Culture* (1919) by Hannah Hoech.

though we got frustrated with a Rubik's cube and punched it out. Continuing by replacing each of those 20 cubes by 20 sub-cubes yields 400 yet smaller cubes. This process can be continued forever, and the resulting object is known as the *Menger Sponge*. Notice that the whole Menger Sponge is precisely the same, up to scaling, as each of the 20 original sub-cubes, and it is also precisely the same as each of the 400 sub-sub-cubes at the next stage, and so on, forever.

## Getting a Collage Education

Aren't collages attractive? A collage is made up of separate images that are artistically grouped together. Although the collage gathers together many individual pictures, these pictures give one total impression. Here we are going to do an experiment in collage making that illustrates the ideas of self-similarity and repeating processes.

*Explore ideas carried to the extreme.*

◆ ◆ ◆ ◆ ◆ ◆

In making collages, some people prefer to use just one picture, make several reduced copies of it, and assemble them to create a collage based on one initial theme. In other words, they take their favorite picture (probably of themselves), make several smaller copies of it, and then assemble them artistically on the page.

Some people (like us) go to extremes. The collage they just created is so attractive that they decide that it should be the initial picture to be used to craft yet another collage. So, they take a photograph of the entire collage and basically start the process all over. They make the same number of copies as before, each copy reduced in size as before. They then assemble the copies into a collage by configuring the copies in the way they did previously, and lo and behold, they have a new collage.

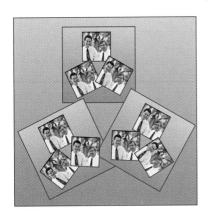

Once started on this road to collage making, who could stop? The next step is to take the new collage and repeat the whole process. Then make reduced copies as before, assemble the copies as before, and create yet another collage. This process could be repeated 100 times, 1,000 times, or a million times. But why stop there? Why not repeat the process forever? The resulting object will capture the essence of that collage. Let's pin this idea down and explore it.

## *Instructions for Creating Repeated-Image Collages*

1. Start with a picture.
2. Make some number of copies, each reduced by a specified amount.
3. Position each reduced picture on a page in a specified location, creating a new picture on the page (which is really a collage of pictures).
4. Start with the resulting picture and return to step 1, repeating this cycle forever.

We'll elaborate and illustrate each of these collage-making instructions so that anyone following them will create the same work of art.

1. Start with a picture.

2. Make a specific number of copies, 5 copies, or 10 copies, or any number. Each copy is reduced by a specific amount. For example, copy 1 may be 1/3 the original size, copy 2 may be 1/2 the original size, copy 3 could be 3/4 the original size, copy 4 could be 1/1000 the original size, and so on. Each of the copies is reduced by a specific amount and will be reduced by that same amount each time.

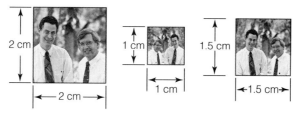

3. Position each reduced copy of the picture on a page in a particular place at a particular angle. It even could be flipped over. For example, the center of copy 1, positioned vertically, might be at the point 3 inches to the right and 4 inches up from the bottom of the page. Copy 2, rotated 20° counterclockwise, might be centered at the point 5 inches to the right and 6 inches up. Copy 3 might be centered . . . and so on. Notice that some of the pictures may even overlap one another.

4. Put all the pictures down to create a new picture comprising a bunch of copies of the original picture. This new picture is then used to start the whole process again. That is, exactly the same number of copies of this picture are made, each reduced to the same size as before. They then are positioned in exactly the same places as the first copies were positioned before. The result is another picture that is a collage of a collage. As you see, it really consists of reduced copies of reduced copies of the original picture. Repeat this whole process again and again, each time using the result of the previous creation as the initial picture for the next collage.

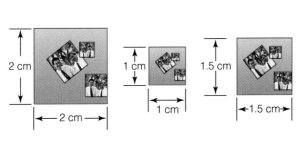
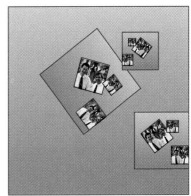

After you repeatedly apply these instructions, you will find that the resulting collages start to settle down in the sense that you see slighter and slighter differences between the resulting collage and the previous one. That is, the difference between the picture we see after the 100th step will look almost identical to the picture we see after the 101st step, However, the full intricacy of the whole infinite process is not actually present except in our minds as we imagine that the process has been repeated infinitely many times.

3rd stage

Intriguing images can be constructed using this collage-generating mechanism. In addition, this method of generating images has some surprises to offer that allow even artistically challenged people to create spectacular images. Before we explore such tantalizing possibilities, let's see some of the images we can construct using this infinitely iterated collage method. As a familiar illustration, we return to the Sierpinski Triangle.

## Sierpinski Triangle Encore

We start with a filled-in equilateral triangle and then make three smaller copies, each half as tall and half as wide as the original. We position them to create a three-triangle collage as shown. These three reduced copies and their positions specify this collage-generating instruction set. We take the resulting picture, make three reduced copies, and place them in the same arrangement again. Continue forever. The resulting image, the Sierpinski Triangle, has infinitely many holes. Notice if we take the Sierpinski Triangle and perform the operation on it (make three reduced copies and arrange them as specified), we will exactly reconstruct the Sierpinski Triangle.

## Barnsley's Fern

Perhaps the images and objects constructed so far appear somewhat formal and sterile. So, here we see how Michael Barnsley developed a fractal that looks uncannily lifelike. This beautiful and lifelike fern is actually created by using our collage-making process, and, in fact, it is rather simple in that only four reduced copies of an original image are needed to produce this intricate and naturalistic result. It differs from our previous examples in that some of the reduced copies shrink the length and the width by different amounts; in fact, one of the four reductions squashes the width down to zero. Let's see this process in action.

We'll begin with a simple caricature of a fern. First, draw the four reduced copies separately so we can see how they are shrunk. The first copy is .85 as large as the original in height and in width. The second copy shrinks the width to .3 its original width and the height to .34 its original height. Copy 3 shrinks the width to .3 its

*Once you have an idea, look for extensions and variations.*

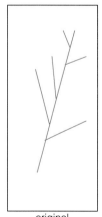
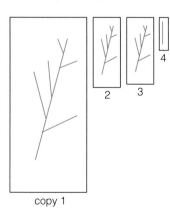

original width and the height to .37 its original height. Copy 4 shrinks the width to 0 and shrinks the height to .16 the original height. These copies are positioned on the page as shown. Notice that copy number 3 is flipped over as well as rotated. Also notice that the copies overlap, which is allowed.

As always, let's take our original picture and repeat our collage-making process (make four reduced copies and position them) to make a collage. Next, as always, let's take that image and apply the collage-making instructions to it, repeating the process once, twice, three times, and infinitely often. The resulting object of performing this operation infinitely is *Barnsley's Fern*. Notice that, as before, if we take this intricate image and perform the operation on it (make four reduced copies and arrange them as specified), we will exactly reconstruct Barnsley's Fern again.

Barnsley's Fern

## *Artistic Ability Unnecessary*

One of the most surprising features of this method of creating images by following the collage-making process is that the final outcome, after infinitely many iterations, does *not* depend on the original, initial image. If we have a specific list of collage-generating instructions, the final collage picture will be the same whether we start with our whole page blackened, a single dot, or the *Mona Lisa*. The form, beauty, and substance of the final delicate picture is completely contained in the instructions of how many copies to make, how much to shrink each one, and how to position each one rather than in the initial image. You might be somewhat skeptical of this assertion, so let's see why it's true.

## *Sierpinski Triangle—The Final Bow*

When we created the Sierpinski Triangle using the collage method, we started with an equilateral triangle, we made three smaller copies, each half as tall and half as wide as the original, and we

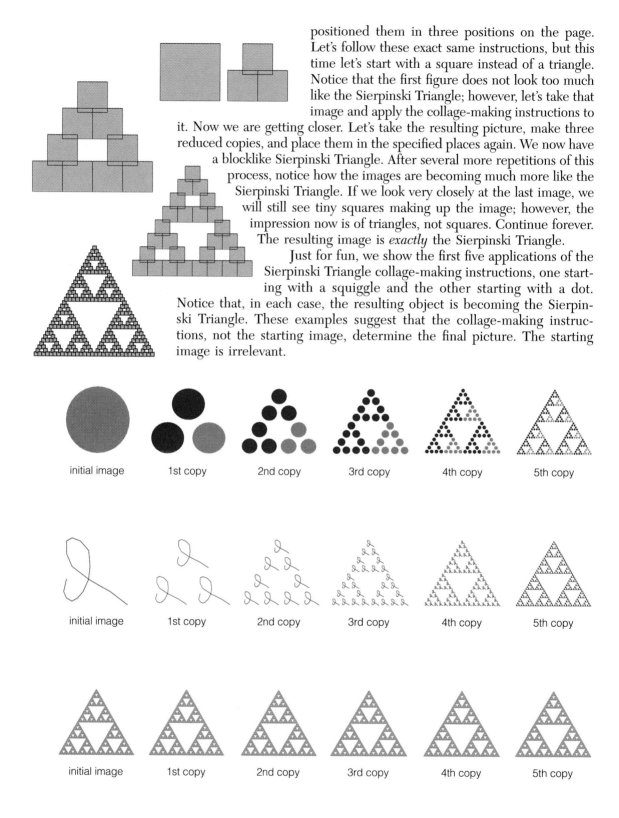

positioned them in three positions on the page. Let's follow these exact same instructions, but this time let's start with a square instead of a triangle. Notice that the first figure does not look too much like the Sierpinski Triangle; however, let's take that image and apply the collage-making instructions to it. Now we are getting closer. Let's take the resulting picture, make three reduced copies, and place them in the specified places again. We now have a blocklike Sierpinski Triangle. After several more repetitions of this process, notice how the images are becoming much more like the Sierpinski Triangle. If we look very closely at the last image, we will still see tiny squares making up the image; however, the impression now is of triangles, not squares. Continue forever. The resulting image is *exactly* the Sierpinski Triangle.

Just for fun, we show the first five applications of the Sierpinski Triangle collage-making instructions, one starting with a squiggle and the other starting with a dot. Notice that, in each case, the resulting object is becoming the Sierpinski Triangle. These examples suggest that the collage-making instructions, not the starting image, determine the final picture. The starting image is irrelevant.

## Barnsley's Fern Regrown

Let's reconstruct Barnsley's Fern from a different initial drawing. Let's try it starting with something that doesn't resemble a fern at all—for example, a car. We make the four reductions exactly following the instructions as before, arrange those reduced images as specified, and repeat the process. Notice how that mechanistic symbol, a car, transforms itself into the ecologically sensitive fern. If we were philosophically inclined, we would say that the study of fractals is helping humanity create a bridge between industrial coldness and natural beauty—lucky for you, we are not so inclined.

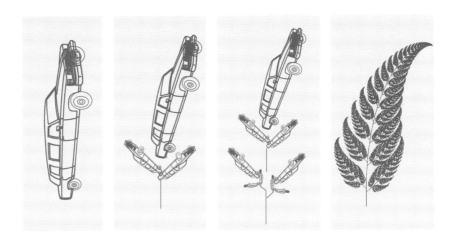

> **IT'S NOT WHERE WE BEGIN, IT'S HOW WE GET THERE.**
> *If one repeatedly applies a specific set of collage-making instructions, starting with any initial image, the infinite collage process will produce the same result.*

## The Proof

It is heartening to know that we cannot go wrong. To understand why any starting image produces the same result, let's just carefully go through the collage-making process and see what happens to the initial image. The first step is to make several copies of the starting image, each one reduced in size, and place those copies on the page. Each one of those reduced copies of the original image, being part of the first collage, is copied and further reduced in the process of making the next collage. Each copy is further reduced in making the next collage, and so on. So, at any stage, the collage we create is made of some number of reduced copies of the original image. Let's be more specific.

## *Idea of the Proof—Example 1*

If the instructions call for making four copies, for example, then the first collage will have four copies of the initial image, although they may overlap one another. After following the instructions again, we will have 16 copies of the original image, since the first collage was made of four copies of the original image and that whole collage was copied four times. After the next stage, there will be 64 copies, and in general there will be $4^n$ copies of the original image making up the $n$th stage collage. But since at each stage the copies are reduced in size, all copies of the original image become tinier and tinier with subsequent repetitions of the collage-mak-

ing process. Soon they become so tiny that what we see after many steps is not determined by the original images, each copy of which is so tiny that we can't see it anyway; the picture is determined by the locations of $4^n$ greatly reduced copies of the original image. The position of each tiny, multiply reduced copy of the original image is determined by the instructions that tell us where to put each copy.

## *Idea of the Proof—Example 2*

As a matter of fact, if we start with a single point, even the shrinking factors are not necessary. In the case where the collage-making instructions call for four copies, the first "collage" would simply consist of four points positioned according to the location instructions. Those four points would be duplicated four times to create 16 points positioned per instructions, then duplicated four times to create 64 points, and so on. Notice that after 100 steps, there will be $4^{100}$ points, each placed where a reduced image would have been placed had we started with an image. But, after reducing any image 100 times, it would be indistinguishable from a point anyway. So, we've just shown that any starting picture produces the same final collage.

## *Modeling the World*

If we want to model a picture with an infinitely detailed fractal approximation, there is a simple way to do it. Just sketch the goal picture and cover it with a set of images that are reductions of an initial picture to define the collage-making instruction set. This process of putting the pieces together generates some amazing artificial images that suggest reality. Examine the two images below, and guess which is an authentic photograph by Scott Woolums and which is a fractal impostor by Ken Musgrave and Benoit Mandelbrot. The fractal fake is strongly convincing, and repetition of simple rules lies at the heart of the fractal image. In fact, if

we throw in a bit of randomness every once in a while, then the resemblance to reality and nature is eerie. Such fractal images are so striking that we can't help but wonder whether the processes by which natural wonders are created might similarly involve repetitious applications of simple rules together with a bit of randomness. This theme of randomness suggests an alternative method of constructing fractals. (By the way, Woolums' photograph *Sunset over Mt. Annapurna* was the image on the right.)

## The Chaos Game

Repeating the collage-making instructions creates intricate images, but another way to generate such images is by random luck. We illustrate this idea next.

Let's number the three vertices of an equilateral triangle 1, 2, 3. We start at any vertex and randomly choose a number: 1, 2, or 3. (For example, we could roll a regular die and let 1 or 4 mean 1; 2 or 5 mean 2; and 3 or 6 mean 3.) Whatever number we roll, we move halfway from where we are toward that numbered vertex and make a dot and remain there. We then roll again and move halfway from where we are toward that numbered vertex and make a dot. We repeat, repeat, repeat, creating a sequence of dots. Computers can easily be programmed to repeat this process quickly. As we make the dots we will notice that the dots do not become a mere random collection, but instead they begin to describe a pattern. If we continue for many steps, we will see the Sierpinski Triangle materialize before our eyes. As we continue forever, the dots we draw will fill in the Sierpinski Triangle in increasingly finer detail. Let's think about why this process will generate the Sierpinski Triangle.

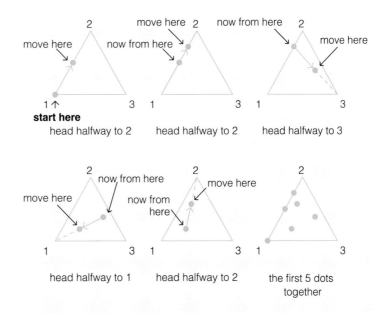

First five stages if the random numbers were 2,2,3,1,2,...

Remember that each of the three sub-triangles in the Sierpinski Triangle is identical to the whole Sierpinski Triangle, except reduced in size. So, taking the whole Sierpinski Triangle and moving each point to the point halfway toward the top vertex, for example, just reduces the entire Sierpinski Triangle to the upper sub-triangle. Similarly, if we start with the

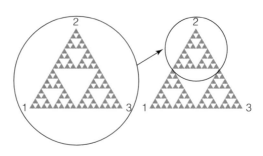

whole Sierpinski Triangle and move halfway toward either of the other two vertices, the whole Sierpinski Triangle gets shrunk to become exactly one of the three sub-triangles. This observation tells us that, if we start with a point on the Sierpinski Triangle and move halfway toward any vertex, we will land on another point of the Sierpinski Triangle. So, our process of generating dots is random, but at least the dots we make are part of the Sierpinski Triangle.

The dots will tend to fill the whole Sierpinski Triangle in the sense that, as we follow the sequence of dots, we will expect to eventually get arbitrarily close to any point in the whole Sierpinski Triangle. By getting close to each point of Sierpinski's Triangle, the pointillistic drawing we are creating will evolve into the whole Sierpinski Triangle in ever-increasing detail. So, we need to ask ourselves why this random sequence of points will tend to draw the whole Sierpinski Triangle.

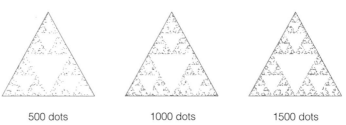

500 dots    1000 dots    1500 dots

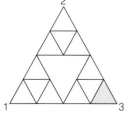

How will we visit points in the shaded lower, lower-right triangle?

Let's begin with an easy question. Suppose we start with the lower-left vertex. Is it possible that we will *never* visit a point in the upper triangle? Well, that is possible if we never randomly pick a 2. But, if we are randomly picking infinitely often and the numbers 1, 2, and 3 are equally likely to arise, as a practical matter, we will eventually pick a 2. As soon as we do, no matter where we were before, we will land in the upper triangle. Similar arguments convince us that we will eventually land somewhere in each of the three sub-triangles. But notice that each sub-triangle is itself made of three sub-sub-triangles. Can we be certain of visiting each of those nine sub-sub-triangles? For example, will we visit the lower, lower-right sub-sub-triangle? What kind of throws would get us into that small sub-sub-triangle? Well, all we have to do is pick two 3's in a row. If we are picking 1, 2, or 3 infinitely often, sometime along the way we will pick two 3's in a row, and, whenever we do that, we will find ourselves in the lower, lower-right sub-sub-triangle.

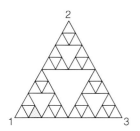

Let's just consider one or two more levels, and we will see the reason why our dots fill the whole Sierpinski Triangle. For example, why are we basically certain to sometimes land in the lower-right sub-sub-sub-triangle in the lower lower-left sub-sub-triangle in the lower left sub-triangle? Or, another way to ask this question is, What sequence of random numbers will guarantee that we will land in that sub-sub-sub-triangle? We can see that any sequence of the form 3, 1, 3 will land us in that sub-sub-sub-triangle. But, if we pick numbers infinitely many times, that specific sequence will eventually come up. We will consider other notions of randomness in the next chapter . . . stay tuned.

By this time, we can see two things:

- For any sub-sub-sub- . . . -sub-triangle in the whole Sierpinski Triangle, some specific sequence of numbers will take us there; and
- If we pick numbers randomly infinitely many times, that specific sequence of rolls will eventually happen.

So, every tiny nook and cranny of the Sierpinski Triangle will eventually be visited, and the Sierpinski Triangle will be drawn to that precision.

## *Even Random Roads Lead to Fractals*

This example shows how a random process can lead to a definite picture. Similar random processes give an alternative way to construct infinitely detailed models of the world and other fractal images arising from the collage-making process. In fact, there are theories that the stock market and even one's heart rate exhibit this type of behavior. Thus we see that chance, together with some simple rules, leads us to the infinitely intricate world of fractals, a world that quite possibly overlaps with our own physical world. Some of the most beautiful examples of fractals, however, are those that arise out of abstract mathematical ideas; more appear in the next section.

## A LOOK BACK

Infinitely detailed images can be created by repeating simple processes infinitely many times. These fractal images can be created by feedback loops, by iterated collage constructions, or by random processes. Following a collage-making instruction set will lead to the same image no matter what picture we start with. Starting at a point and randomly moving halfway toward one of three points sketches the Sierpinski Triangle pointillistically. All these fractal images are infinitely detailed in the sense that they display more intricacy at every increased magnification. These models lead to compelling pictures that seem to capture some of nature's vistas.

Examples of fractals in everyday life suggest the idea of pictures that can have infinite detail. Parallel mirrors, video feedback loops, and Quaker Oats boxes all create images that would be infinitely detailed if they were carried to their logical extremes. All these are generated by repeated applications of a process. We can repeat other processes, for example, repeatedly replacing part of a picture (such as a line segment) with a modified version (such as a bent segment) or repeatedly replacing a collage by a collage of itself. By looking at variations of the idea of repeating a process, we create intriguing images.

Pushing ideas to an extreme often leads to new and unexpected insights and often clarifies essential features. Looking at extremes is a means of using ordinary objects or ideas in uncommon ways. After we have an idea, we can pin it down, and explore it by looking for extensions, variations, and alternative methods.

*Simple repeated processes can lead to complex and interesting outcomes.*

*Take ideas and look for extensions, variations, and alternatives.*

*Explore ideas carried to the extreme.*

❖ ❖ ❖ ❖ ❖ ❖ ❖ ❖ ❖ ❖ ❖ ❖ ❖ ❖ ❖ ❖ ❖

# MINDSCAPES  Invitations to Further Thought*

### I. SOLIDIFYING IDEAS

1. ***Nature's way.*** Find several examples of objects in nature that, like broccoli, display self-similarity at different scales.

2. ***Who's the fairest?*** Can you position three mirrors in such a way that in theory you could see infinitely many copies of all three mirrors? Either describe the positions of the mirrors or describe why it is not possible.

3. ***Billiards and mirrors.*** On an idealized, square billiard table, you want to hit a shot that traces a perfect diamond. To line up your shot, you take four mirrors and position them in the centers of each side wall. Before you remove the mirrors, you make your shot directly along the line from the center of one mirror to the center of the next. As you look into the facing mirror, how many billiard balls do you see?

---

*In the Mindscapes section, exercises marked (H) have hints for solutions at the back of the book. Exercises marked (S) have solutions.

4. **MTV.** You've become a rock star and consequently feel that there can never be too many of you. So you make a video of infinitely many appearances of yourself. Do it! Arrange a TV, a video camera, and yourself as described in  Example 2—*Video Feedback* (p. 429) in this section to produce a picture of infinitely many copies of yourself and your guitar admiring infinitely many copies of yourself and your guitar. If you have a decent voice, you can sing on the video, otherwise call Milli Vanilli.

5. **Photo op.** Suppose you arrange two mirrors facing each other at a slight angle, as shown. Place a camera parallel to one of the mirrors. Snap the picture. The picture will contain many increasingly smaller pictures of the camera. Will they be arranged going off to the right, the left, or up?

6. **How many mes?** Arrange mirrors and camera as in exercise 5, but put yourself in the picture right above the camera. Suppose the mirrors are about 5 feet apart, and the shutter is opened 1/1000th of a second. Light travels at about 186,000 miles per second, and a mile is 5,280 feet. If the photographic paper were sufficiently detailed and sensitive, roughly how many of you will you see in the photo? Is that enough?

7. **Quaker,** *Quaker,* Quaker**.** Suppose you had the job of shooting the photograph for the next Quaker Oats box, and the picture was to be one of you holding a Quaker Oats box. Of course, the picture on the Quaker Oats box you are holding should be of you holding a Quaker Oats box, and so on. Suppose you have a camera but no mirror; however, you do have a photocopier that can reduce images to any size. How can you create the picture for the box?

8. **Sierpinski hexed (S).** Take an equilateral triangle and remove a hexagon from the center as shown, leaving three equilateral triangles with sides 1/3 the length of the sides of the original triangle. Repeat the process of removing a hexagon from each sub-triangle. Repeat, repeat, repeat, forever. Make a sketch of what remains after one step, two steps, and three steps. Describe the locations of infinitely many points that are left after infinitely many steps. Let's call the remaining points *Sierpinski Dust*.

9. **The bends.** Koch's kinky curve is created by starting with a straight segment and replacing it by four segments, each 1/3 as long as the original segment. So, at the second stage the curve has three bends. At the next stage, each segment is replaced by four segments, and so on. How many bends does this curve have at the third stage? the fourth stage? the $n$th stage?

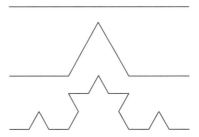

10. **Four times (H).** Draw a picture on a square piece of paper. Now make a collage by reducing your picture to 1/4 its length and width and making four copies. Divide another square sheet of paper into four squares, and center each copy of the picture in each square to make the second stage of the collage. Making the four reduced copies and positioning them in the four squares constitutes the collage-making instructions. What does your collage look like at the third stage? the fourth stage? Sketch your guess of what the collage looks like at the final infinite stage.

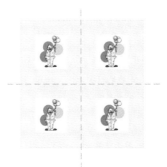

11. **Burger heaven (S).** Sketch a picture of a hamburger, and make three reduced copies, two reduced to 1/4 the original length and width, one reduced to 1/2 its original size. Position them as shown to create the second stage of the collage. Follow these same instructions and sketch the third stage and the fourth stage, and sketch a guess of what the infinite stage looks like.

second stage of burger collage

12. **Ice cream cones.** Draw a picture of an ice cream cone, and follow the same instructions as in the burger heaven question (#11). Sketch the third stage and the fourth stage, and sketch a guess of the infinite stage. How similar are your results compared to the burger heaven results?

13. **Sierpinski boundary.** Take the boundary of a triangle. Make three reduced copies, each reduced to 1/2 the length and width. Arrange them as you did when making the Sierpinski Triangle. Repeat the process and draw the result.

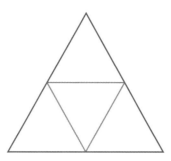

14. **Catching Z's.** Take a Z. Put in nine smaller Z's as shown to create the second stage. If the smaller Z's are 1/6 as long as the large one, roughly how long is the line through the Z's at the third stage if the line through the original, big Z is 24 cm. long?

15. **Replacement pinwheel.** Take a 1, 2, $\sqrt{5}$ right triangle. Divide it into five identical similar right triangles, as shown. Repeat the process in each triangle. Draw the picture that you get after three steps.

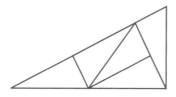

16. **Koch Stool.** Start with a line segment, mark it into three equal pieces and replace the middle piece with a small square missing its bottom. Now, for each of the five smaller line segments, repeat this procedure. Draw the next two iterations of this process. If we repeat this process forever, we can call the final result a *Koch Stool*.

17. **Koch Collage Stool.** Given the Koch Stool described in exercise 16, explain how you can create the figure using the collage method.

18. **Sierpinski shooting.** Suppose that you were playing the Chaos Game to create the Sierpinski Triangle and you rolled 1, 1, 3, 2. Shade in the sub-sub-sub-sub-triangle in which you would land. How about if you rolled 1, 3, 3, 1? How about 3, 2, 2, 2?

19. **Sierpinski target practice (H).** What sequence of numbers will land you in sub-sub-subtriangle labeled $a$ if you are playing the Chaos Game to create the Sierpinski Triangle? What sequence would land you into $b$? How about $c$?

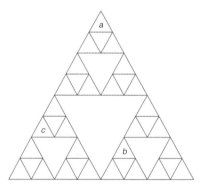

20. **Cantor Set.** Start with the interval [0, 1]. Build a 1-dimensional collage by making two copies of the interval, each shrunk to 1/3 the original length, and putting them at the ends of the interval. Using these collage-making instructions, draw the next two stages of the collage process. The result of doing this process infinitely many times is called the *Cantor Set*.

## II. CREATING NEW IDEAS

1. **Cantor luck (H).** Start with the point 0. Flip a coin. If it comes up heads, move 2/3 of the way toward 1, if it comes up tails, move 2/3 of the way toward 0 from wherever you are at the time. Repeat forever. The points you find are drawing a picture of a set you constructed in the question from the previous section called Cantor's Cuts (Section 6.2, III.3). Verify that any point in the Cantor Set will move to another point in the Cantor Set under the coin flipping and moving process. Notice that the points will approximate the whole Cantor Set as you continue the process.

2. **Cantor square (S).** Take a square. Make four copies, each reduced to 1/3 the length and width of the original. Position these four copies in the corners of the original square. Repeat these collage-making instructions. Draw the next two stages of the construction. We will call the result of doing this process infinitely many times the *Cantor Square*.

3. **Cantor square shrunk.** Take the Cantor Square. Describe why applying the collage-making instructions to it (making four reduced copies and putting them in the corners of the original square) results in the exact same Cantor Square that you started with.

4. **Cantor squared.** Draw the four corners of a square, label them 1, 2, 3, 4, and start at one corner. Randomly choose a number from 1 to 4. Move 2/3 of the way toward the corner whose number comes up from wherever you are at the time. Repeat forever. Verify that, with this process, any point in the Cantor Square will move to another point in the Cantor Square. Notice that, as you continue the process, the points will approximate the whole Cantor Square.

5. **Hexed again.** Suppose you start with the three vertices of an equilateral triangle labeled 1, 2, 3. Start at any vertex and randomly choose numbers 1, 2, or 3. When you choose a number move two thirds of the way toward that numbered vertex from where you are. Show that the points you generate will result in the Sierpinski Dust described in the Sierpinski Hexed exercise (I.8).

6. **Pinwheel spun.** Take a 1, 2, $\sqrt{5}$ right triangle. Divide it into five identical right triangles. Repeat the process in each triangle. Draw the picture that you get after three steps. Describe the relationship between this fractal construction and the construction of the Pinwheel Pattern that was a non-periodic covering of the plane described in Section 4.4.

7. **Antoine's necklace.** *Antoine's necklace* is a delicate necklace indeed. The first stage is a solid doughnut, and the second stage results from replacing it by eight linked sub-doughnuts. Each of these eight sub-doughnuts is in turn replaced by eight sub-sub-doughnuts, and so on, forever. Draw the next stage of Antoine's necklace.

8. **Menger jacks.** For the game of jacks, let's imagine constructing a six-pointed metal jack whose ends are square. Begin with a solid cube. Make seven reduced copies, each reduced to be 1/3 the size in each dimension, and position the seven little cubes in the original cube to create a jack. Draw the result of performing the collage-making instructions again.

9. **A tighter weave.** Look back at the Tight Weave story in Chapter 1 (p. 10). There you constructed a pattern by dividing a purple square into a 3 × 3 grid of nine squares, making the central square gold, and repeating the process in each of the remaining eight squares. The purple remaining after repeating that process infinitely often is sometimes called a *Sierpinski carpet*. Here let's modify the process. Instead of dividing the original square into a 3 × 3 grid and making the inner one gold, suppose we divide the square into a 5 × 5 square and make the central one gold. Then take each of the 24 remaining squares, divide them into a 5 × 5 grid and make the middle one gold, and continue this process forever. Starting with a square as stage 1, draw the first three stages of this replacement process.

10. **A looser weave.** Let's modify the carpet-designing process yet again. Instead of dividing the original square into a 3 × 3 or 5 × 5 grid and making the inner one gold, suppose we divide the square into a 4 × 4 square and make the central four squares gold. Then take each of the 12 remaining squares, divide each of them into a

4 × 4 grid and make the central four gold, and continue this process forever. Starting with the square as stage 1, draw the first three stages of this replacement process.

### III. FURTHER CHALLENGES

1. *From where?* Look at this fractal. What collage instructions would produce it?

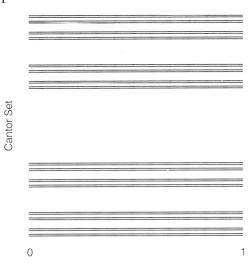

2. *Treed.* Describe collage-making instructions that would result in a fractal collage that resembles a tree.

3. *Flaky* (H). Describe collage-making instructions that would result in a fractal collage that resembles a snowflake.

4. *How big a hole?* In each of the Tighter Weave (II.9) and Looser Weave (II.10) questions, how much of the total pattern is gold after infinitely many steps?

5. *4D fractal.* Describe a fractal in 4-dimensional space that is analogous to the Menger Sponge. How many sub-hypercubes would you divide a hypercube into? How many would you remove to create the next stage?

### IV. IN YOUR OWN WORDS

1. *Personal perspectives.* Write a short essay describing the most interesting or surprising discovery you made in exploring this section's material. If any material seemed puzzling or even unbelievable, address that as well. Explain why you chose the topics you did. Finally, comment on the aesthetics of the mathematics and ideas in this section.

2. **With a group of folks.** In a small group, discuss the infinite collage-making process. After your discussion, write a brief narrative describing the process and why any starting picture yields the same result if you use the same collage-making instructions.

3. **Creative writing.** Write an imaginative story (it can be comical, dramatic, whatever you like) that involves or evokes the ideas of this section.

4. **Power beyond the mathematics.** Provide several real-life situations—ideally, from your own experience—for which some of the strategies of thought presented in this section would provide effective methods for approaching and resolving them.

# 6.4 The Mysterious Art of Imaginary Fractals:
## Creating Julia and Mandelbrot Sets by Stepping Out in the Complex Plane

*... what the imagination seizes*
*as beauty must be truth—*
*whether it existed before or not.*

—John Keats

The infinite iterative processes we will examine here do not involve the collage-generating theme described in the previous section but instead an imaginative motion of points in the plane. The mathematical objects resulting from these moving points were first studied more than 70 years ago—long before computer technology allowed us to generate accurate physical pictures of them. They were first discovered by mathematicians as an abstract idea. Only in the past 20 years, with the development of powerful computers and graphics software, have we become able to see pictures of these abstract conceptions. The pictures not only surprised and excited people around the world, but they led to a new branch of mathematical inquiry.

Visualization is a powerful means of discovering structure. These images are like a Rorschach Test of mathematics. We see patterns within patterns, nuances, similarities, and differences with new intricacies emerging at every level of magnification. Questions about what we see or don't see are still among the unsolved problems of mathematics.

## Julia and Mandelbrot Sets

The first collection of objects we examine are called *Julia Sets*, named after the French mathematician Gaston Julia. In the 1920s Julia discovered these sets as natural objects in his investigation of imaginary numbers. He studied

*God as the Architect of the Universe* (1230). Is the universe a Benoit Mandelbrot-like fractal?

Gaston Julia

some of the many unusual properties these Julia Sets possess even though he could not create or imagine the incredibly rich and intricate pictures of them that tantalize us today. The physical pictures of these sets were created by sophisticated computers more than 50 years after Julia's initial work. These pictures startle and delight all who see them. The beauty and complexity of the Julia Sets are great surprises that add to the interest in Julia's original mathematical work.

Here are some images of various Julia Sets. When we magnify one part of the set, we see a smaller set with a structure similar to that of the original one. This self-similarity produces images of endless intrigue,

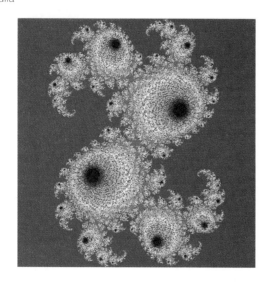

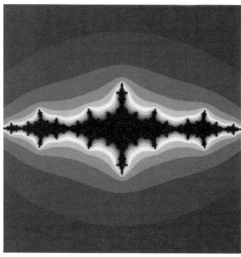

458 ◆ CHAOS AND FRACTALS

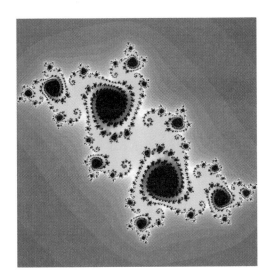 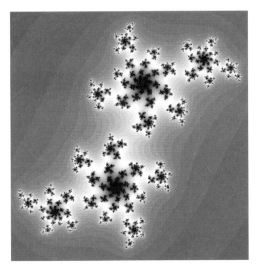

because every magnification exposes new wonders. While you enjoy these spectacular pictures, we invite you to think about how one could ever describe or create such infinitely detailed objects.

In the late 1970s Benoit Mandelbrot discovered a revolutionary means of studying an aspect of all such Julia Sets. He described what is now known as the *Mandelbrot Set*, and in 1980 the image of his set was first created using special computer graphics software. The Mandelbrot Set still conceals mysteries and is the object of continuing investigation. You may well join the many who consider the Mandelbrot Set a supremely intricate and beautiful icon of mathematical art.

Don't be afraid to look and think before you understand.

Benoit Mandelbrot—the father of fractals.

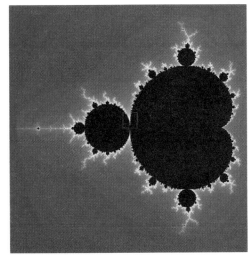

the Mandelbrot Set

## Fractals

Benoit Mandelbrot pioneered a whole new area of study concentrated on fractals, images of infinite detail, such as these Julia Sets and his Mandelbrot Set. He truly is the father of fractals, since, as we have mentioned, he invented the word *fractal*, which he coined from the Latin *fractus*, meaning broken or irregular. He applied the term to all sets of infinite detail, such as the collage-generated sets discussed in the last section and the Julia Sets we will look at here. *Fractal* does not have a rigorous mathematical definition, but it conjures up the ideas of similarity at different scales and of infinitely detailed intricacy.

We now turn our attention to the basic question: How were these incredibly detailed and imaginative sets created? To understand their origin, we follow the fertile imagination of some great thinkers and enter the world of imaginary numbers.

## A Process in the Plane

We first describe a process by which we start with one point in the plane and use it to move to another. This procedure is used to generate the elegant Julia and Mandelbrot Sets. The process is not difficult, but it may seem artificial and strange at first. However, we will see that it is a natural example of performing arithmetic with imaginary numbers.

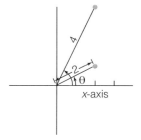

Suppose we take a point in the plane and draw the line segment from that point to the origin. We then measure the length of that segment and the angle going counterclockwise starting from the positive $x$-axis to the segment. Using this point we will now describe a new point. The new point we get will have distance from the origin equal to the square of the length of the segment—that is, the square of the distance of the original point to the origin—and will be in the direction at twice the angle of the original segment. So, we double the angle and square the length. Sounds like an unnatural procedure to us . . . for now. We'll describe this procedure more fully in a few paragraphs.

Points in the plane can be interpreted as imaginary numbers, and the process described in the last paragraph actually corresponds to squaring an imaginary number. We next present a brief look at imaginary numbers and how they are multiplied and added together. For the purpose of constructing Julia and Mandelbrot Sets, however, the geometric process just mentioned is really all we need to know. Interpreting the process as imaginary arithmetic provides the inspiration and motivation for the significance of this particular process of taking one point in the plane and getting another.

*If something is an anomaly in your world view, expand your view to make it fit.*

◆ ◆ ◆ ◆ ◆ ◆

## Imaginary Numbers

Perhaps you have already been introduced to the imaginary number $i$. The unreal number $i$ is defined to be the number satisfying the following amazing property:

$$i^2 = -1.$$

If we take any real number from the number line and square it, we never get a negative number. Thus we immediately see that this new number $i$ is *not* a real number, and so we call it an *imaginary number*. Since $i^2 = -1$, many write $i$ as

$$i = \sqrt{-1}.$$

More generally, we define *complex numbers* to be numbers of the form

$$a + bi,$$

where $a$ and $b$ are real numbers. So, for example, the following are all examples of complex numbers:

$$2 + 3i, \; 5.7i, \; 1 + i, \; 2\pi - 12i, \; -5 + \sqrt{3}i.$$

All we need to do with the complex numbers is add them and multiply them. First we'll just figure out how to perform these operations; then we'll see a way to visualize both addition and multiplication. These two operations are all that are needed to understand how the beautiful Julia and Mandelbrot Sets are made.

## An Addition Complex

A natural guess as to how to add two complex numbers is correct. For example, $2 + 5i$ plus $11 + 4i$ is $13 + 9i$.

Here is the rule for how to add complex numbers:

$$\begin{array}{r} (a + bi) \\ +(c + di) \\ \hline (a + c) + (b + d)i \end{array}$$

So,

$$\boxed{(a + bi) + (c + di) = (a + c) + (b + d)i.}$$

In other words, the approach is not exotic. We add the first two numbers together and then add the two numbers in front of the $i$'s together.

## A Multiplication Complex

Multiplying complex numbers takes a bit more time to figure out. Let's try an example together—the key is to make certain that every term in one complex number gets multiplied by every term in the other. It is similar to the "FOIL" (first, outer, inner, last) method used in some algebra classes. We'll multiply $2 + 5i$ and $11 + 4i$ (remember the crucial fact that $i^2 = -1$ and notice how we use it in the fourth line):

$$
\begin{aligned}
(2 + 5i)(11 + 4i) &= (2 + 5i)11 + (2 + 5i)4i \\
&= (22 + 55i) + (8i + 20i^2) \\
&= (22 + 55i) + (8i + 20i^2) \\
&= (22 + 55i) + (8i - 20) \\
&= (22 + 55i) + (-20 + 8i) \\
&= (22 - 20) + (55 + 8)i \\
&= 2 + 63i.
\end{aligned}
$$

Let's adapt the previous example to find a formula for multiplying complex numbers in general. Notice how all the steps below parallel the steps in the previous example.

$$
\begin{aligned}
(a + bi)(c + di) &= (a + bi)c + (a + bi)di \\
&= (ac + bci) + (adi + bdi^2) \\
&= (ac + bci) + (adi - bd) \\
&= (ac + bci) + (-bd + adi) \\
&= (ac - bd) + (bc + ad)i.
\end{aligned}
$$

So we see:

$$\boxed{(a + bi)(c + di) = (ac - bd) + (bc + ad)i.}$$

The final answer to the multiplication looks strange because $i^2 = -1$; but remember that we arrived at that answer simply by multiplying $(a + bi)$ by $(c + di)$ just as we learned in algebra class, and then we used the fact that $i^2 = -1$ to simplify the answer. Therefore, it is easier to remember the *procedure* for multiplying complex numbers than to try to memorize the *formula*. For the purpose of making Julia and Mandelbrot Sets, all we care about is multiplying a complex number by itself. So, let's state this squaring as a special case:

$$\boxed{(a + bi)^2 = (a + bi)(a + bi) = (a^2 - b^2) + (2ab)i.}$$

Understanding is rock. Memory is sand. Build on rock.

◆ ◆ ◆ ◆ ◆ ◆

## Visualizing Complex Numbers

Recall that we visualized the real numbers by studying the real number line. The complex numbers can be visualized as a plane—the *complex plane*. Here is the complex plane:

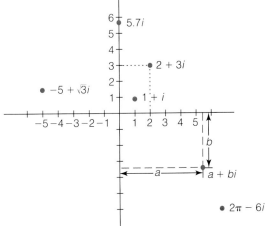

Notice that we have two axes: a horizontal ($x$-axis) and a vertical ($yi$-axis). The number $2 + 3i$ can be found by moving over 2 in the horizontal direction and then moving up 3 in the vertical direction. Here are the numbers $2 + 3i$, $5.7i$, $1 + i$, $2\pi - 6i$, $-5 + \sqrt{3}i$ in the complex plane. So, in general, to find the point $a + bi$, we just need to go over to the number $a$ on the $x$-axis and then move up or down to $b$ on the $yi$-axis. We can now think of a different way to understand addition and multiplication of complex numbers.

Here is an easy way to perform addition of complex numbers geometrically in the complex plane. Draw a line segment from each of the two numbers to the origin ($0 + 0i$). Then make a parallelogram using these two lines as two adjacent sides. The vertex of the parallelogram that is across from the origin is the point that represents the sum of the two numbers. For example, returning to $2 + 5i$ plus $11 + 4i$, we draw the diagram below. Notice that the vertex across from the origin represents the complex number $13 + 9i$, which is the sum. So, the parallelogram gives a geometric way to add complex numbers—cool.

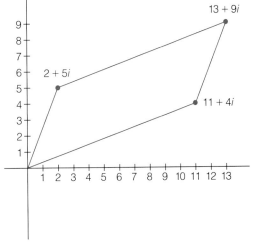

We can also multiply a complex number by itself geometrically in the complex plane, and, as you may have guessed, we will discover that the *Process in the plane* described earlier actually corresponds to multiplying a complex number by itself. Let's look at this geometric view of multiplication. To multiply a number by itself, we first plot the number in the complex plane and connect it by a straight-line segment to the origin. We will take note of the pitch (or slope) of this line segment and its length.

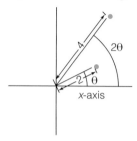

As we described in the *Process in the plane,* we can determine the line's pitch by considering the angle made by the line and the right half of the *x*-axis. Specifically, we consider the angle made by starting at the right half of the *x*-axis and moving counterclockwise until we reach the line segment.

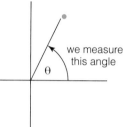

We now consider a new direction: We start at the right half of the *x*-axis and move counterclockwise to an angle that is twice the angle we just found.

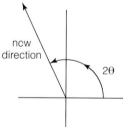

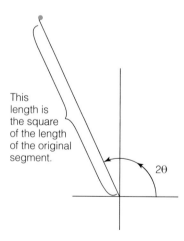

This length is the square of the length of the original segment.

In that direction we move out from the origin to a length that is equal to the square of the length of the original line segment and place a point at that location. That is, we swing around counterclockwise by an angle that is double the angle of our line segment. This angle gives us a new pitch from the origin. We now place a point in the plane such that the line from the origin to this point is in this new direction and the length of this new line is equal to the length of our original line segment multiplied by itself. A little trigonometry would show that this new point is the precise location of the square of the original complex number—that is, the result of multiplying the complex number by itself.

Let's illustrate this elaborate procedure with an example. Using the arithmetic method just outlined, we can compute that $(2 + 3i)(2 + 3i)$ equals $-5 + 12i$. Now notice how we get to the same point from the geometric procedure just outlined. So, to find the product of a complex number with itself geometrically, we look along a direction from the origin that is *double* the angle made by the original number and the right half of the *x*-axis and place a mark that is located at a distance from the origin that equals the *square* of the distance of the original number to the origin.

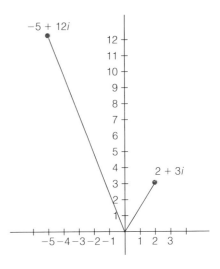

## Squaring Practice and a Julia Set Warm-Up

We now have all the necessary mathematics to describe how the wonderfully intricate Julia and Mandelbrot Sets are created. Those pictures result from an iterative process whereby we perform an operation and then repeat, repeat, repeat . . . forever. This process, however, is different from the collage method. To get us in the mood for this process and also to practice the arithmetic of complex numbers we just outlined, let's consider some experiments and make some simple observations.

Let's take the number $2 + 3i$, square it, take that new number and square it, take that result, square it, and repeat, repeat. . . . What is the list of numbers we get? We already computed $(2 + 3i)(2 + 3i)$ and saw it equaled $-5 + 12i$. If we repeat this process with our new answer we get $(-5 + 12i)(-5 + 12i) = (25 - 144) + [2(-60)]i = -119 - 120i$. If we continue to repeat this process, here are the numbers we get (verify the third one for yourself):

**Starting with $2 + 3i$, each number is the square of the number that preceded it.**

$-5 + 12i$
$-119 - 120i$
$-239 + 28560i$
$-815616479 - 13651680i$
$665043872449535041 + 22269070348069440i$
 . . .
($\pm$ a really huge number) + ($\pm$ another huge number)$i$
 . . .

Here are those same numbers plotted. Notice how small the scale needs to be. Why? It's because those numbers are moving farther and farther from the origin.

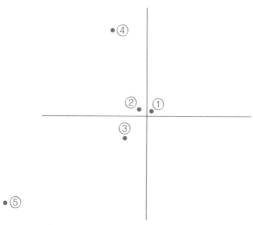

Let's try the same experiment starting with the number $0.5 - 0.8i$. Here is what we get (verify the first entry):

**Starting with 0.5 + 0.8i, each number is the square of the number that preceded it.**

$-0.39 + 0.8i$
$-0.48 ... - 0.62 ... i$
$-0.15 ... + 0.59 ... i$
$-0.33 ... - 0.18 ... i$
$0.07 ... + 0.11 ... i$
. . .
($\pm$ a really tiny number) + ($\pm$ another tiny number)$i$
. . .

Here are those complex numbers plotted. Notice how large the scale is now; that's because we see that these complex numbers are zooming toward the origin, $0 + 0i$.

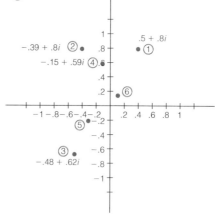

466 ◆ CHAOS AND FRACTALS

Finally, let's try this iterative squaring procedure starting with 3/5 + 4/5i.

**Starting with 3/5 + 4/5i, each number is the square of the number that preceded it.**

$-7/25 + 24/25i = -0.28 + 0.96i$
$-0.84\ldots - 0.53\ldots i$
$0.42\ldots + 0.90\ldots i$
$-0.64\ldots + 0.76\ldots i$
$-0.17\ldots - 0.98\ldots i$
$-0.94\ldots + 0.33\ldots i$

$\ldots$

$(\pm \text{ a modest number}) + (\pm \text{ another modest number})i$

$\ldots$

The graph of these points is quite interesting. Notice that they neither head off to the outer reaches of the complex plane nor spiral into the origin. In fact, they seem to stay along the circle centered at the origin having radius 1. Why? By using the Pythagorean theorem we can compute that the distance between the original point 3/5 + 4/5i and the origin is equal to 1. Therefore, when we square it, the distance away from the origin of the new number will be $1^2$, which is still 1. Similarly, all other squares will remain on the circle of radius 1.

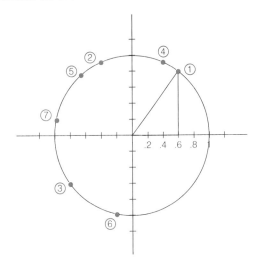

## Insights from Experiments

So, what do we discover from these experiments? We see that if a point outside the circle centered at the origin of radius 1 is repeatedly squared and squared again, the values drift off to infinity. If the point is inside the circle, then successive squaring makes the numbers head toward the origin. Finally, if we start with a point on the circle, then the subsequent squares remain on the circle. So, the circle is the boundary between those points that do not drift off to infinity and those points that do go off to infinity as they are repeatedly squared. This circle is actually a simple and uncharacteristically plain example of a Julia Set. We are now ready to face and understand some incredible, imaginative fractals.

## *From the Complex Plane to the Infinitely Beautiful*

Let's consider the iterative process of starting with a complex number, squaring it, taking that answer and repeating the process (so square that number, then square the new number, and continue to square successively). What happens to these numbers? It all depends on what the initial number was; that is, it all depends on the initial condition. If our starting number is a distance greater than 1 from the origin, then our list of numbers, starting with that first one, will head off to infinity. If our starting number is a distance less than or equal to 1 from the origin, then our list of numbers will not head off to infinity. If we think of all the points in the complex plane as all the possible starting numbers, then we see that the circle of radius 1 centered at the origin is the interface between those starting numbers that result in their lists going off to infinity and those starting numbers that result in their lists not going off to infinity. This interface, or boundary, between these two classes of numbers is called a *Julia Set*. Here is a picture of the "filled-in" Julia Set, where we color in all the starting numbers whose lists do not head off to infinity.

## *Adding Some Spice*

To generate infinitely intricate Julia Sets we need to add one more twist to our process. Suppose we pick a particular complex number: let's call it $a + bi$. So, $a + bi$ is a *fixed number*. Now suppose we begin with a starting complex number, square it, and then add our fixed number $a + bi$. We can now iterate this process: Take the answer we get, square it and then add $a + bi$, take that new answer, square it and add $a + bi$, and so on, forever. For example, suppose $a + bi = \mathbf{-1 + 0i}$. If we start with the initial number $0 + 0i$, here is what we generate:

$$0 + 0i$$
$$(0 + 0i)^2 + (-1 + 0i) = -1 + 0i$$
$$(-1 + 0i)^2 + (-1 + 0i) = 0 + 0i$$
$$(0 + 0i)^2 + (-1 + 0i) = -1 + 0i$$

and so on . . .

If we start with the initial number $2 + 3i$, then we generate:

$2 + 3i$
$(2 + 3i)^2 + (-1 + 0i) = (-5 + 12i) + (-1 + 0i) = -6 + 12i$
$(-6 + 12i)^2 + (-1 + 0i) = (-108 - 144i) + (-1 + 0i) = -109 - 144i$
$(-109 - 144i)^2 + (-1 + 0i) = -8856 + 31392i$

and so on . . .

## Julia Sets Exposed

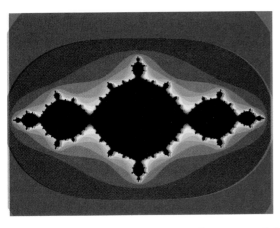

Notice that this procedure is exactly the same as our original squaring procedure, except for the slight difference of adding on the number $a + bi$ at each stage after we square our number. It turns out that this slight difference leads to a world of difference. As before, given a certain starting number, our list of iteratively generated numbers will either drift off to infinity or not. In the preceding example, notice that the starting number $0 + 0i$ does not drift off to infinity, whereas the starting number $2 + 3i$ does. If we think of all points in the complex plane as all possible starting numbers, we see that some of them lead to lists that go off to infinity and some lead to lists that do not. The interface between these sets is called a *Julia Set*. Here is a picture of the filled-in Julia Set when $a + bi$ equals $-1 + 0i$. The boundary around this set is the Julia Set associated with the shifting number $-1 + 0i$.

Notice that, when we did not add any number after squaring, our Julia Set was a boringly symmetric circle. However, when we added $-1 + 0i$ after each squaring operation, all of a sudden the Julia Set became an exotic fractal set. Remember what this set represents.

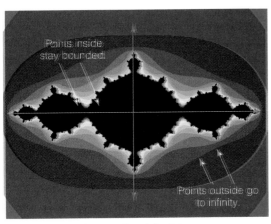

$(-1 + 0i)$ – Julia Set

It turns out that, if we change the $a + bi$ (the number we add after we square at each stage), then the associated $(a + bi)$-Julia Set (the boundary between the starting numbers that result in unboundedly larger and larger values and the starting numbers that result in values that remain bounded) changes dramatically. Here are some different Julia Sets associated with different values of $a + bi$.

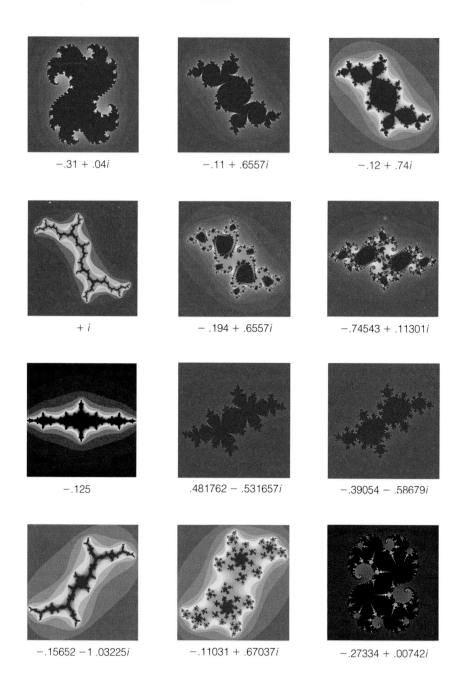

## Color-Coded Retreat

In the pictures, the area outside the filled-in Julia Sets is depicted by several different colors. Remember that each starting point outside the Julia Set generates a sequence of complex numbers that eventually become larger and larger without bound. The colors indicate how quickly the numbers get larger. So, for example, starting with a complex number in the red or orange regions, its iterative list will go off to infinity faster than if the starting number was selected from the blue or yellow areas. So, the colors give a sense of how quickly the numbers get large in our iterative process.

That's how Julia Sets are made. There is one Julia Set for each complex number $a + bi$, and we call it the $(a + bi)$-Julia Set. We just fix $a + bi$ and then ask which complex numbers in the plane, when viewed as starting seeds for the squaring and then adding $a + bi$ iteration process, result in lists of numbers that get larger and larger without bound and which starting seeds result in lists of numbers that do not get arbitrarily large. The interface between these two collections is the $(a + bi)$-Julia Set. One fascinating feature of Julia Sets is that they reveal complicated structure at any magnification. They truly are infinitely detailed. Here is a Julia Set with a magnified area to show the fractal-like detail.

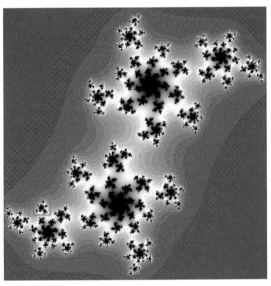

The $(-.11031 + .67037i)$—Julia Set.

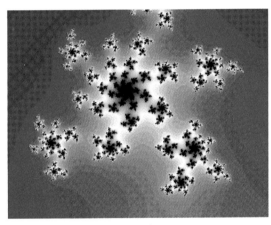

An enlarged view of the tiny red square.

## Mandelbrot Set

After being immersed in the (seemingly infinite) details of the construction of each Julia Set, it is time to step back and look at the whole world of all $(a + bi)$-Julia Sets at once. We are now ready to construct the Mandelbrot Set. We begin by making a simple observation about each preceding filled-in Julia Set: Either the filled-in region is just one piece, or it's made up of lots of smaller pieces. This basic distinction is the key to understanding the Mandelbrot Set. The Mandelbrot Set is an object that captures information about the collection of *all* $(a + bi)$-Julia Sets.

For each point $a + bi$ in the complex plane, we can go off and draw the $(a + bi)$-Julia Set. In other words, we can go off and draw the Julia Set in which the number we add at each stage after squaring is $a + bi$.

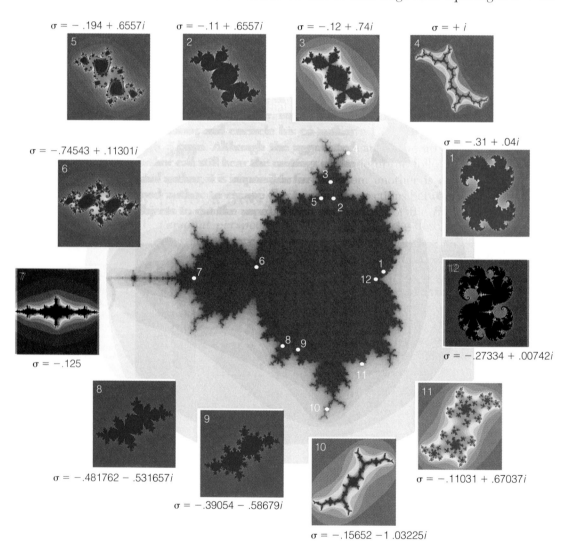

472 ◆ CHAOS AND FRACTALS

That filled-in $(a + bi)$-Julia Set will either be one piece or more than one piece. If the filled-in $(a + bi)$-Julia Set is connected, then we say that the point $a + bi$ is in the Mandelbrot Set. So the Mandelbrot Set is the collection of all complex numbers $a + bi$ with the property that the filled-in $(a + bi)$-Julia Set is just one piece.

So, for each point in and near the Mandelbrot Set, we can associate that point with a certain $(a + bi)$-Julia Set. If the point is in the Mandelbrot Set, then the filled-in Julia Set associated with it will be in one piece. If the point is outside the Mandelbrot Set, then the associated filled-in Julia Set will be made of many pieces. In the picture, we select various points in and around the Mandelbrot Set and show inlaid pictures of the associated filled-in Julia Sets.

Here are certain enlarged regions of the Mandelbrot Set to highlight its amazing fractal-like structure. Since there are infinitely many Julia Sets and both the Julia and Mandelbrot Sets are infinitely detailed at increasing magnifications, it is heartening to know that we can never run out of glorious images for calendars, posters, and screen savers.

## *P. S. (Processes besides Squaring)*

We close this section by noting that there are other iterative processes we could have used besides squaring and then adding a fixed number. If

we used other iterative processes, repeated forever, and then asked if the starting number leads to a list of numbers getting arbitrarily large, or not, then we can construct other types of Julia Sets. We close this chapter (see the last page of Section 6.6) with a Julia Set constructed in the same manner we described, except with more elaborate iterative processes. Enjoy its wondrous beauty.

## A LOOK BACK

We are able to generate truly imaginative images using the geometry and arithmetic of complex numbers. Every complex number leads to a different Julia Set. The $(a + bi)$-Julia Set captures those points in the complex plane that do not head off to infinity under the iterative process of repeated squaring and adding $a + bi$. The Mandelbrot Set includes all complex points whose Julia Sets are connected—that is, are in one piece. The Julia and Mandelbrot Sets are the icons of fractal art, and they arise from iterative processes of complex numbers.

Complex numbers arise as an abstract idea to create solutions to equations that have no real-number solutions. By conceiving of complex numbers geometrically as points in the plane, we gain new insight about their relationships. Visualization is a powerful tool for detecting patterns and suggesting relationships. The Julia Set began as an abstract idea about complex numbers, but, after seeing actual images of them, we discover new worlds of interest and new questions to ponder.

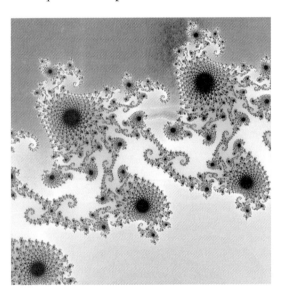

Finding a variety of ways to represent the same idea adds depth to our understanding. Visualizing ideas through images is a powerful way of noticing previously unseen patterns and relationships. By expressing the same idea in different ways, we not only see new sides of the idea, but we can explore the connections among the different views. Those connections create new insights of their own.

Visualize the same idea in different ways to see relationships.

If something is an anomaly in your world view, expand your world view to make it fit.

Understanding is rock. Memory is sand. Build on rock.

# MINDSCAPES   Invitations to Further Thought*

## I. SOLIDIFYING IDEAS

1. **Arithmetic.** Compute the following sums and products of complex numbers: $(2 + 3i) + (3 - 7i)$; $(-2 + 5i) + (4 + 6i)$; $(1 - 2i)(2 + 3i)$; $(-i)(4 - 2i)$; $(3 - 2i)^2$.

2. **More arithmetic (H).** Compute the following sums and products of complex numbers: $(1/2 + 3i) + (5/2 - i)$; $(-1 + i) + (3 + 2i)$; $(1 - \sqrt{2}i)(3 + \sqrt{2}i)$; $(3 - 2i)(3 + 2i)$; $(2 - \sqrt{5}i)^2$.

3. **Quick draw.** Two complex numbers are marked and labeled on each axis, as shown. Geometrically compute their sums using the parallelogram rule. Geometrically compute their products by adding the angles to the two points measuring counterclockwise from the positive $x$-axis to find the direction and then multiplying their lengths to get the distance out from the origin. Check your answers by adding and multiplying the complex numbers algebraically.

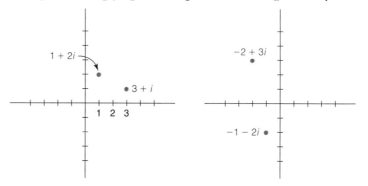

4. **Quick draw II.** Two complex numbers are marked and labeled on each axis, as shown. Geometrically compute their sums and products. Check your answers by adding and multiplying the complex numbers algebraically.

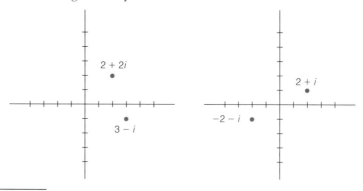

---

*In the Mindscapes section, exercises marked (H) have hints for solutions at the back of the book. Exercises marked (S) have solutions.

5. **Be square** (H). Let's get in the swing of squaring the imagination by squaring each of the following complex numbers: 1/2, 2 + 2i, −1 + i, 2 + i. First compute the squares just using algebra. Then, for each one, draw it on the complex plane and measure its angle from the x-axis and its distance from the origin. Finally measure the angle from the x-axis and the distance from the origin for the square. Did the angle double and distance square? Yes? Good. No? Oops!

6. **Squarer.** For each of the complex numbers, 3 + 4i and 3/7 − (4/7)i, find its square geometrically. Check your answers by computing the squares just using algebra.

7–10. **Ill iterate I.** Feeling queasy? Then compute the first three iterates of $z^2 + i$ for each of the following starting complex numbers z: 0 + 0i, i, 2 + 3i, and 1 − i. Graph the points and their iterates on the same graph, and color green those points that seem to you would be headed toward infinity if you continued the iteration process (I.7:S).

11–14. **Ill iterate II.** Compute the first three iterates of $z^2 + .27$ for each of the following starting complex numbers z: 0 + 0i, .5 +.5i, 2 + 3i, i. Graph those points and their iterates on the same graph, and color green those points that seem to you would be headed toward infinity if you continued the iteration process (I.13:S).

15–18. **Ill iterate III.** Compute the first three iterates of $z^2 + (.11 - .67i)$ for each of the following starting complex numbers z: 0 + 0i, .5 +.5i, 1 + 0i, 0 +.5i. Graph those points and their iterates on the same graph, and color green those points that seem to you would be headed toward infinity if you continued the iteration process.

19. **Orange Julias.** Pictured are some Julia Sets surrounded by contour drawings. Each contour line is numbered. The number represents how many iterations of the process of squaring and adding a + bi are required before a point on that line is of distance 2 or more from the origin. Color the regions between the contour lines different colors to show how many iterations are required to have the various points outside the Julia

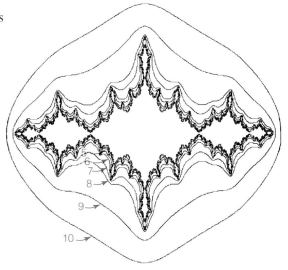

Set exceed distance 2 from the origin. Choose your colors to create the most attractive image. Recall that The University of Texas at Austin's color is burnt orange, so use burnt orange prominently.

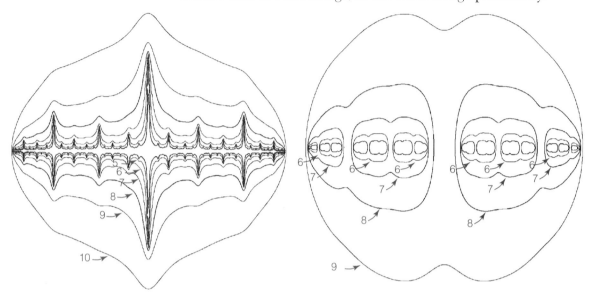

20. *Julia Webbed.* Visit our Web site at **http://www.heartofmath.com** or your CD-ROM in your kit. There, plug in some random complex numbers, and create their associated Julia Sets. Print out your favorites and hang them in your dorm room as psychedelic posters.

## II. CREATING NEW IDEAS

**1–3.** *Great escape?* For each of the complex numbers $(-1 + i)$, $(0 + 0i)$, $(.25 + .25i)$, guess whether the iterates of the number escape to infinity or not under the transformation $z^2 + (-.2 + .65i)$.

**4–5.** *Mandelbrot or not?* (H). Here are little Julia Sets for each of several complex numbers. Which of these complex numbers should be in the Mandelbrot Set? Mark the points on the picture of the Mandelbrot Set here, and confirm your answers.

**6–9.** *Zero in* (S). For each picture of the Julia Sets in II.4–5, note whether the Julia Set contains or does not contain the origin. Make a conjecture about the relationship between the origin being in the Julia Set associated with $a + bi$ and that Julia Set being connected.

**10.** *Mandelbrot origins.* Using your insights from the preceding Zero In question, make a conjecture about an alternative description of the Mandelbrot Set that refers to the origin.

Figures for II.4

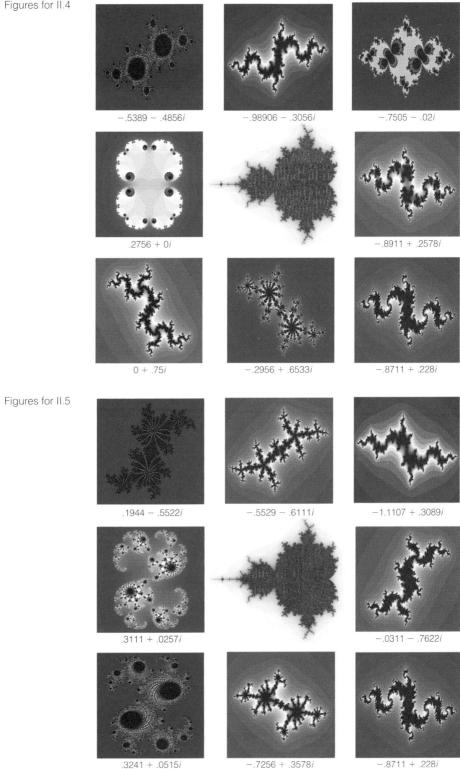

Figures for II.5

6.4 / THE MYSTERIOUS ART OF IMAGINARY FRACTALS   ◆   479

## III. FURTHER CHALLENGES

1. ***The great escape*** (**H**). Consider $z^2 + a + bi$ where $a + bi$ is less than one unit from the origin. Suppose $z$ is a complex number that lies more than two units from the origin. Show that iterates of $z$ will go off to infinity.

2. ***Bounded Julia.*** Why is no Julia Set infinitely big?

3. ***Prisoner.*** Find a complex number $a + bi$ that is fixed under the transformation $z^2 + (1 + i)$—that is, for which $(a + bi)^2 + 1 + i = a + bi$.

4. ***Always a prisoner.*** For any complex number $c + di$, find a complex number $a + bi$ that is fixed under the transformation $z^2 + (c + di)$—that is, for which $(a + bi)^2 + c + di = a + bi$. This question establishes that every Julia Set has some points in it.

5. ***Mandelbrot connections.*** Can every two points on the Mandelbrot Set that are close to each other be connected by a fairly small squiggly path? Look at pictures of the Mandelbrot Set and make a guess. (*Hint*: Don't be discouraged if you are unable to settle this question, since it is an unsolved problem. No one in the world knows whether the answer is yes or no.)

## IV. IN YOUR OWN WORDS

1. ***Personal perspectives.*** Write a short essay describing the most interesting or surprising discovery you made in exploring this section's material. If any material seemed puzzling or even unbelievable, address that as well. Explain why you chose the topics you did. Finally, comment on the aesthetics of the mathematics and ideas in this section.

2. ***With a group of folks.*** In a small group, discuss what the Julia Set associated with the complex number $a + bi$ is. After your discussion, write a brief narrative describing which points are in the Mandelbrot Set.

3. ***Creative writing.*** Write an imaginative story (it can be comical, dramatic, whatever you like) that involves or evokes the ideas of this section.

4. ***Power beyond the mathematics.*** Provide several real-life situations—ideally, from your own experience—for which some of the strategies of thought presented in this section would provide effective methods for approaching and resolving them.

# 6.5 Predetermined Chaos:
## How Repeated Simple Processes Result in Utter Chaos

*Chaos umpire sits And by decision more embroils the fray by which he reigns: next him high arbiter Chance governs all.*

—John Milton

We all know that the world is unpredictable and that we must expect the unexpected from time to time. But, somewhere deep in our minds, we might wonder whether the world is fundamentally orderly. Although we cannot find the key that organizes our surroundings and their meanings, perhaps a better theory, a better computer, or a better brain might do the trick. If we just had enough information, maybe we could predict the weather and the stock prices. Such predictions would keep us both dry and rich. We have made tremendous strides over the centuries in describing laws of nature. The accoutrements of modern life are a legacy of that insight. But how far can we expect these insights to take us? Are there limits to our ability to understand nature?

We now will discover another side of scientific progress—namely, its limitations. Many phenomena of the world are the result of repeatedly applied interactions at all scales—molecules jostling one another, animals reproducing, planets orbiting. In each case, inevitable minute inaccuracies in measuring the position, speed, or the laws of interaction are dramatically magnified over time, and our uncertainty about the future grows. It may not be surprising that there will be limits to our ability to accurately predict future outcomes due to our vagueness about the current conditions. What is shocking, however, is that uncertainty about the future grows even when there is *no* vagueness at all. Chaos may be a more fundamental feature of our world than we would prefer to believe.

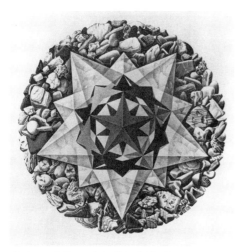

*Order and Chaos* (1950) by M.C. Escher.

Our journey through chaos will illustrate that direct experience is the best teacher. Ideas become real when we experience them first hand, particularly if the experience is surprising. The feeling of surprise tells us that our understanding was somehow not complete and invites us to seek a deeper understanding. Experience, then discomfort or surprise, then analysis is a firm path toward understanding.

## An Experiment—In Search of a Sign

*Just do it. Make the effort to get experience.*

◆ ◆ ◆ ◆ ◆ ◆ ◆

We want you to experience something pretty surprising... but you'll need a calculator. So, before reading any more, put down this book (and no one will get hurt), find a scientific calculator that has a SIN (sine) key, and then pick up this book and continue reading.

With the help of the calculator, we're about to create a list of numbers. We're going to be relying heavily on the SIN (sine) key, so it will be the "key" player in generating our list of numbers. To begin, set the SIN key for angles in degrees (rather than radians). To check that your calculator is set properly, enter the number 180 and press the SIN key. If you get an answer of 0, then you're all set (perfectly). If you see something like $-0.801152\ldots$, then your calculator is set on radians and you need to change it to degrees.

We're now ready to compute away. To record the results, make two columns with the numbers from 1 to 25 in the first column. Start the experiment by writing down a random decimal number between 0 and 1 and putting it in the second column next to 1. We are now ready to produce the remaining numbers in the second column by following these steps:

1. Type in your random number on the calculator.
2. Multiply it by 180.
3. Hit the SIN key and write down your answer in the second column. Keep all the decimal places.
4. Go back to step 2 then step 3. Repeat, repeat, repeat. (That SIN key is getting lots of *hits* today.) Continue to record your results until you have 25 numbers on your list.

To illustrate this procedure (and to prove to you we actually did it ourselves), we selected the first number (the random one) to be 0.287. We multiplied it by 180 and then pressed the SIN key. The answer was 0.78434349, and this was our next number. We then multiplied it by 180 and hit the SIN key to get our third number. Here is our list, but we hope that you made up your own with your own first random number.

    1    0.287

    2    0.78434349

    3    0.62685095

| | |
|---|---|
| 4 | 0.92163865 |
| 5 | 0.24370037 |
| 6 | 0.69297498 |
| 7 | 0.82179119 |
| 8 | 0.53106712 |
| 9 | 0.99524087 |
| 10 | 0.01495068 |
| 11 | 0.04695167 |
| 12 | 0.14696871 |
| 13 | 0.44548493 |
| 14 | 0.98537011 |
| 15 | 0.04594498 |
| 16 | 0.14383972 |
| 17 | 0.43666283 |
| 18 | 0.98026879 |
| 19 | 0.06194772 |
| 20 | 0.19338834 |
| 21 | 0.57085546 |
| 22 | 0.97532697 |
| 23 | 0.07743501 |
| 24 | 0.24087690 |
| 25 | 0.68655271 |

Pretty exciting, huh? Well, okay, so it's not so interesting . . . not yet, anyway. Now that you've made up your list, let's make up the exact list yet again, but this time something scary will happen!

## *Variation on a SIN Theme*

Start by entering exactly the same first decimal number as before and repeat the process for the first five steps. The moment you record your fifth number, suppose your calculator makes a slight buzzing sound followed by the appearance of a cryptic symbol on its display and then goes blank. It turns out that your batteries have died. (To simulate this, shut off your calculator now.) Suppose you then rush out and replace the batteries. Well, now you've got to finish the list you started. But hey, why bother starting from the beginning? Instead, you decide to save some time and start with the fifth number (the last one you wrote down, before your calculator fainted). In fact, why bother typing in all those digits from that

**TABLE 6.5-1**  SIN Experiment Repeated with Rounded Fifth Value

| Original Time | | This Time | |
| --- | --- | --- | --- |
| Step Number | Value | Step Number | Value |
| 1 | 0.287 | 1 | 0.287 |
| 2 | 0.78434349 | 2 | 0.78434349 |
| 3 | 0.62685095 | 3 | 0.62685095 |
| 4 | 0.92163865 | 4 | 0.92163865 |
| 5 | 0.24370037 | 5 | **0.243700** |
| 6 | 0.69297498 | 6 | 0.69297412 |
| 7 | 0.82179119 | 7 | 0.82179273 |
| 8 | 0.53106712 | 8 | 0.53106303 |
| 9 | 0.99524087 | 9 | 0.99524212 |
| 10 | 0.01495068 | 10 | 0.01494674 |
| 11 | 0.04695167 | 11 | 0.04693932 |
| 12 | 0.14696871 | 12 | 0.14693033 |
| 13 | 0.44548493 | 13 | 0.44537699 |
| 14 | 0.98537011 | 14 | 0.98531226 |
| 15 | 0.04594498 | 15 | 0.04612653 |
| 16 | 0.14383972 | 16 | 0.14440412 |
| 17 | 0.43666283 | 17 | 0.43825728 |
| 18 | 0.98026879 | 18 | 0.98124664 |
| 19 | 0.06194772 | 19 | 0.05888135 |
| 20 | 0.19338834 | 20 | 0.18392805 |
| 21 | 0.57085546 | 21 | 0.54620504 |
| 22 | 0.97532697 | 22 | 0.98948314 |
| 23 | 0.07743501 | 23 | 0.03303366 |
| 24 | 0.24087690 | 24 | 0.10359211 |
| 25 | 0.68655271 | 25 | 0.31972972 |

number? Instead, type in the first six digits after the decimal point from the number and finish the list.

Try that scenario now. That is, record the first four numbers as they appear in your previous list and then for the fifth number, just enter the first six digits after the decimal point. Our fifth number is boldface in Table 6.5-1. Now, from that point, using that truncated number as the fifth number, complete the list and record your results. Now compare your new list to your original list. Go ahead and finish your list before looking at our results.

Table 6.5-1 shows what we got.

Look at the original 25th number and compare it to the 25th number computed using the slightly truncated version of the fifth number. Why is there such an enormous difference between these two last numbers on our lists given that there was only an extremely tiny difference in our fifth numbers? One short answer is that sometimes very slight differences in initial conditions can have an enormous and dramatic effect on the final outcomes. A shorter answer is *chaos*!

## *Dueling Calculators*

Here is a similar experiment that has even more surprising results. We borrowed a calculator from a student. The precision of the borrowed calculator was 8 digits, whereas ours was 9. We entered the exact same decimal number between 0 and 1 on each calculator and began the previous process of multiplying by 180, hitting the SIN key, recording our answer, and repeating. Now, since we started with the same number, the answers should be the same, right? We repeated the process about 35 times. What do we observe? Why? Once again: *chaos!*

We started with the random number 0.7391. Table 6.5-2 shows what we got with successive trials.

**TABLE 6.5-2** Experiment Using Dueling Calculators

| Round | Our Calculator | The Borrowed Calculator |
|---|---|---|
| 1 | 0.7391 | 0.7391 |
| 2 | 0.73090122 | 0.7309012 |
| 3 | 0.74823571 | 0.7482357 |
| 4 | 0.71101516 | 0.7110152 |
| 5 | 0.78819629 | 0.7881963 |
| 6 | 0.61737461 | 0.6173746 |
| 7 | 0.93278107 | 0.9327811 |
| 8 | 0.20960844 | 0.2096084 |
| 9 | 0.61193460 | 0.6119345 |
| 10 | 0.93880464 | 0.9388047 |
| 11 | 0.19106878 | 0.1910686 |
| 12 | 0.56485729 | 0.5648568 |
| 13 | 0.97931362 | 0.9793139 |
| 14 | 0.06494242 | 0.0649414 |
| 15 | 0.20261016 | 0.2026071 |
| 16 | 0.59439941 | 0.5943916 |
| 17 | 0.95634610 | 0.9563533 |
| 18 | 0.13671325 | 0.1366909 |
| 19 | 0.41641387 | 0.4163499 |
| 20 | 0.96571997 | 0.9656678 |
| 21 | 0.10748581 | 0.1076489 |
| 22 | 0.33129584 | 0.3317792 |
| 23 | 0.86280720 | 0.8635739 |
| 24 | 0.41778307 | 0.4155936 |
| 25 | 0.96682764 | 0.9650479 |
| 26 | 0.10402551 | 0.1095849 |
| 27 | 0.32101951 | 0.3375106 |
| 28 | 0.84603979 | 0.8725123 |
| 29 | 0.46504041 | 0.3898922 |
| 30 | 0.99397488 | 0.9407660 |
| 31 | 0.01892734 | 0.1850170 |
| 32 | 0.05942697 | 0.5490674 |
| 33 | 0.18561266 | 0.9881424 |
| 34 | 0.55063050 | 0.0372430 |
| 35 | 0.98737654 | 0.1167356 |

Huh? What could be more precise and accurate than the results we get from a calculator? Surely the answer we get is completely determined by what we punch in, and, indeed, it is. However, even with calculators we find that surprising developments can occur. As you can see, repeating a procedure like multiplying by 180 and then pressing the SIN key gives grossly different answers after relatively few repetitions.

Suppose we are trying to predict the future of the population, the stock market, or the course of an epidemic or the weather. Our method for prediction may well involve an idea of how the system will appear after one step—a generation, a day, a year, or a second. Then the answer to that one-step prediction is fed back into our predictive engine, and the process is reapplied to generate a prediction for the state of the system after two steps. And so on. Based on our experience with the SIN key, our confidence in the state of a system after 35 steps should be extremely shaky. We are dealing with some strange observations, so let's try to understand them.

These examples illustrate how tiny differences result in completely different outcomes after several repetitions of a process. This sensitivity is at the heart of mathematical chaos, a theme with profound implications for how well we can ever hope to know and understand the future.

## *Predicting Future Populations*

Let's try another example, this one concerning population growth. Recall that, by making sensible assumptions about how fast populations will grow, we were led to a simple mathematical model for population growth. In section 6.2, the Verhulst Model, sometimes referred to as the *logistic equation*, captured the reasonable idea that populations will increase in proportion to how much unused capacity exists in the environment. This assumption led Verhulst to the following relationship between one year's population and the next year's population. In this relationship $P_n$ stands for the fraction of the sustainable capacity of the environment at year $n$—that is, the population density at year $n$. The Verhulst Model contains a constant that is selected to fit the specific environment being modeled. In the example here, we use the constant 3.

$$P_{n+1} = P_n + 3P_n(1 - P_n).$$

Let's see what pattern of populations this model predicts from year to year. As an example, we started with the decimal number 0.058 for $P_0$ and computed future years' populations; that is, we plugged in 0.058 for $P_n$ in the preceding equation and computed $P_{n+1}$ (in this case, $P_1$). We did this calculation on two different calculators to see what might happen. Table 6.5-3 shows what we found.

One calculator is predicting that the population density at year 45 is 1.3159026, and the other calculator using the exact same formula is predicting that the population density at year 45 is radically different—specifically, 0.1293406901. So, if the maximum sustainable population of the pond is 10,000 fish and we start with 580 fish, one calculator tells us we should expect 13,159 fish after 45 years (overpopulation), and the other calculator tells us we should expect only 1,293 fish. So what's the answer? After 45 years do we have a fish

**TABLE 6.5-3** Population Predictions on Two Different Calculators

| n | $P_n$ (calculator 1) | $P_n$ (calculator 2) | n | $P_n$ (calculator 1) | $P_n$ (calculator 2) |
|---|---|---|---|---|---|
| 1 | 0.22190800 | 0.2219080000 | 24 | 1.2938053 | 1.294691684 |
| 2 | 0.73990251 | 0.7399025186 | 25 | 0.15342458 | 0.1500870658 |
| 3 | 1.3172428 | 1.317242863 | 26 | 0.54308101 | 0.5327698814 |
| 4 | 0.063585171 | 0.0635851704 | 27 | 1.2875130 | 1.279548286 |
| 5 | 0.24221146 | 0.2422114600 | 28 | 0.17698247 | 0.2064616951 |
| 6 | 0.79284667 | 0.7928466660 | 29 | 0.61396151 | 0.6979674858 |
| 7 | 1.2855691 | 1.285569156 | 30 | 1.3249998 | 1.330394109 |
| 8 | 0.18421248 | 0.1842124572 | 31 | 0.033125663 | 0.0117309782 |
| 9 | 0.63504722 | 0.6350471407 | 32 | 0.12921072 | 0.0465110654 |
| 10 | 1.3303339 | 1.330333950 | 33 | 0.46675666 | 0.1795544243 |
| 11 | 0.011970483 | 0.0119705441 | 34 | 1.2134413 | 0.6214983234 |
| 12 | 0.047452057 | 0.0474522947 | 35 | 0.43644580 | 1.327212795 |
| 13 | 0.18305313 | 0.1830540180 | 36 | 1.1743283 | 0.0243697677 |
| 14 | 0.63168719 | 0.6316897517 | 37 | 0.56017204 | 0.0956974144 |
| 15 | 1.3296626 | 1.329663179 | 38 | 1.2993100 | 0.3553156723 |
| 16 | 0.014642343 | 0.0146402048 | 39 | 0.13262049 | 1.042515008 |
| 17 | 0.057926180 | 0.0579178125 | 40 | 0.47771737 | 0.9095474057 |
| 18 | 0.22163839 | 0.2216078311 | 41 | 1.2262278 | 1.156360173 |
| 19 | 0.73918284 | 0.7391012320 | 42 | 0.39400724 | 0.6139341425 |
| 20 | 1.3175575 | 1.317593034 | 43 | 1.1103038 | 1.324991176 |
| 21 | 0.062356531 | 0.0622179240 | 44 | 0.74289148 | 0.0331598544 |
| 22 | 0.23776111 | 0.2372584859 | 45 | 1.3159026 | 0.1293406901 |
| 23 | 0.78145341 | 0.7801591763 | | | |

population that has exploded to 22 times our initial size or one that has merely doubled? We may well ask: Is one of these calculators broken? The answer is no. The enormous difference in the results arises from how many digits the two calculators carry before rounding off. The number of fish expected to be alive in a pond or the population of people on Earth might well be a crucial piece of knowledge for making policy decisions. Yet here we see that even a completely deterministic, simplistic model of population growth is susceptible to the problem of hypersensitivity. The sensitivity of this model to slight changes along the way makes it completely unreliable for predicting the future after 45 steps.

These radical differences exist even when we are using the exact same mathematical model. The sensitivities of the mathematical model suggest that real populations may themselves have greater natural variations than we might intuitively have first thought. Once again, these examples indicate that, when we look at reality, we might find instances of unexpected sensitivity. Of course, a mathematical model is itself only an approximation of reality. Thus, we would expect the actual populations of fish to differ from the predicted numbers not only because of round-off problems but also because of inadequacies of the model. It is safe to say that the distant future is difficult to predict.

## *Predicting Planetary Positions*

Using our knowledge of gravity and the masses of the planets, as well as their current locations and current velocities, we can predict where the planets will be in a year, and our answers will be almost correct. However, if we take that prediction, use the same method again for predicting locations of the planets after two years, we will be close to correct again, but slight differences in our estimates of the masses of the planets and their locations and velocities will make the two-year prediction a little less certain than the one-year prediction. After hundreds, thousands, or millions of years, the predictions will be completely different depending on whether we used one value for the mass of a planet or another value. The future locations of the planets in the solar system are also sensitive to initial conditions to such an extent that astronomers and mathematicians are currently debating whether the solar system is stable or whether after some time the planets might not attain increasingly eccentric orbits and literally fly apart. Although it's an ominous thought, don't count on that as a means of getting out of the final exam.

All these examples are leading us to appreciate that even determined and theoretically predictable mathematical models or physical systems can be so sensitive that we cannot effectively use them to predict the future. *Deterministic chaos* refers to the idea that wildly different futures can result from beginnings that are only minutely different, even in totally deterministic systems.

## *The Dawn of Chaos*

> Luck is a powerful means of making progress. Keep an eye out for it.

Edward N. Lorenz, a meteorologist at the Massachusetts Institute of Technology, was the first to notice this idea of chaos—not in the weather but in systems of equations that describe the weather. In making his seminal discoveries about chaos, Lorenz used one of the most effective methods of investigation in history: discovery by accident.

Luck is a powerful means of making progress. It's something we all want to keep an eye out for. However, luck is not enough. Probably most of us see potentially significant accidents around us all the time, but, unlike Lorenz, we are not prepared to interpret the importance of them.

*Chance favors the prepared mind.* —Louis Pasteur

In the 1960s Lorenz had a primitive computer that he had programmed with equations designed to simulate global weather conditions. One day the program stopped, so he restarted it at an intermediate point. He wanted to save himself a little time, so, instead of copying in their entirety the decimal numbers that were describing the state of his weather simulation, he rounded off the parameter values to two or three decimal points, thinking that this approximation would give nearly the same answers. (Does this story sound familiar?) He noticed that the answers he got were radically different from those he had gotten when he ran the simulation without the interruption. He investigated and discovered that the future weather predictions came out entirely differently when he rounded the parameters to two decimal places rather than using three or four. Just a tiny change in those parameters made a huge difference in his model's prediction of future weather. His insight arising from his accidental experiment of rounding the parameters led him to crystallize the idea of sensitivity to initial conditions in many mathematical models and in nature as well.

Here is the Lorenz attractor associated with weather prediction models whose attractive double spiral ushered in the era of chaos and fractals.

Lorenz attractor

## "Chaos"—The Word

Some words have related yet significantly different meanings. The dictionary defines *chaos* as *state of utter confusion and disorder; a total lack of organization or order.* In 1975 James A. Yorke, from the University of Maryland, and his student Tien-Lien Li wrote a paper in the *American Mathematical Monthly* titled "Period three implies chaos." They thereby applied this word to refer to a scientific concept whose meaning is related, but different. The chaos of mathematics, physics, and biology is not actually utterly disordered. In fact, scientific and mathematical chaos refers to systems, like the simple equations we have seen, that are ultimately completely deterministic and whose future states depend in a fixed and describable way on their current state. There are no uncertainties, no randomness, and no unknowns involved. However, as we have seen with our examples, these completely deterministic systems nevertheless display behavior that is surprisingly chaotic in appearance. They are chaotic in the sense that tiny variations in where we start or how many decimal digits we use in our calculations make an enormous difference in the final outcome.

We never know the current state of any physical system exactly, because there are always errors of measurement. But even when no measurement errors are involved, such as in calculating sines repeatedly, tiny errors due to rounding soon cause tremendous differences in the outcomes. The weather, populations of animals, and prices of goods are real-world phenomena for which we can develop mathematical models that try to predict the future. Unfortunately, these models are susceptible to the problem that slight errors in the beginning data and slight errors along the way propagate and expand. After a relatively short period of time, the models predict grossly erroneous outcomes.

## *Butterflies and Tornadoes*

Because it was a meteorologist who first recognized the phenomenon of the dramatic sensitivity to initial conditions that inspired the mathematical notion of chaos, the parable of the *Butterfly from Brazil* has become an icon of chaos. Thus, we cannot resist giving our own rendition of the story of this famous, beautiful insect and its unfortunate effects on a girl and her dog in Kansas. Please read the boxed story now.

We might consider several responses to this classic butterfly tale—for example, "Can't that butterfly be found and stopped?" Although this question sounds silly, let's rephrase it in a more plausible way. Could we understand weather patterns so accurately that during a major drought we could perhaps set off a bomb over the ocean to cause a crop-saving rain on the drought area two weeks later? During a war, could we set off a bomb over the ocean to cause a hurricane to wipe out the enemies' principal towns two weeks later? Edward Teller, sometimes referred to as the father of the atomic bomb, once said that the control of the weather would make atomic bombs look like children's toys. Is control of the weather more appropriately relegated to science fiction, or is it within the reach of future science fact? Today no one knows. We do know that little changes make dramatic differences, but we are unsure which changes lead to which differences.

Keep alert for significance even in ridiculous ideas.

◆ ◆ ◆ ◆ ◆ ◆

These insights reveal that our potential to predict the future is severely limited. The Brazilian butterfly's wing flap is an example of a tiny change in current conditions that leads to massive changes as a consequence. If we had millions of temperature, pressure, and wind meters spread around the entire atmosphere, we could have a good sense of the weather now and also a pretty good way to predict the weather a few minutes or hours from now. But we would have some small uncertainty in the details of our prediction even a few minutes or hours later. As time passes, our uncertainties grow until quite soon our uncertainties outweigh our predictions. As a practical matter, since small uncertainties in our knowledge of the weather always exist, we cannot hope to ever have

## THE BUTTERFLY FROM BRAZIL

On a sultry summer morning in the Amazon basin, a beautiful butterfly perched on the petal of an exotic purple flower. The still air and a lack of anything good on cable made the butterfly less energetic than was customary. The vision of her on that flower was indeed beautiful.  Her eyelids, if butterflies actually had eyelids, drooped languidly over her eyes, and her mind, if butterflies actually had minds, slipped lazily from one inconsequential thought to another. As her thoughts drifted, she saw out of the corner of her eye what she thought to be the shadow of a hungry bird swooping toward her. (In fact she had inadvertently been startled by a floating leaf.)

As graceful as a ballerina, she sprang into action. As she gently pushed off the petal, she flapped her wings and altered the course of tiny micrograms of air surrounding her slender body. This minute change had enormous consequences. Air masses have some coherence, and they move somewhat in a pack. If air masses meet, a tiny difference can result in one mass sliding under another or over it—like the leading edge of a wedge. As the masses slide, they push neighboring air masses. The butterfly's flapping caused a small air mass to rise instead of fall, which in turn caused a bigger air mass to change its course, which made an even bigger air mass alter its course, and so on. The unpredictable event of a slight change in the air movement caused by the butterfly's flap led to other unpredictable events, which in their turn altered the patterns of increasingly larger bodies of air until two weeks later Dorothy was knocked unconscious when she was hit in the head by her flying dog Toto during a tornado in Kansas.

If that butterfly had had eyelids, perhaps she would not have been startled by the harmless leaf and would have remained still for several moments longer and would not have altered the course of those particular micrograms of air. Those molecules would have traveled a different path, other air masses would have been altered instead, and the good people of Kansas would have seen a beautiful clear day instead of a devastating tornado raining cats and dogs (in particular, Toto). This concludes our version of the parable of the *Butterfly from Brazil*.

sufficient knowledge of the current state of the weather conditions at any one time to predict successfully the weather in several weeks.

*A LOOK BACK*

Even when we use reasonably accurate mathematical models and reasonably accurate data, accurate long-term predictions become impossible due to the overabundance of accumulated error and uncertainty. Rounding off numbers in calculators as far out as the eighth or ninth decimal place leads to huge errors after only a handful of repeated applications of completely deterministic calculations. There is only one certainty here: Accumulated errors and uncertainty will always exist, and sadly, prevail. In a single word, *chaos!*

Performing experiments with calculators and computing future values of populations following the Verhulst Model give us direct experience with the idea of deterministic chaos. From these examples, we extrapolate the idea of systems that are hypersensitive to initial conditions and to accumulated error.

Getting direct experience makes ideas much more real and immediate. As we observe the world, we sometimes notice unusual events. Often we simply ignore them, but, by exploring the unusual, we can find whole new worlds. Today's unusual or extraordinary anomaly may well point the way toward tomorrow's main issue.

*God has put a secret art into the forces of Nature so as to enable it to fashion itself out of chaos into a perfect world system.* —Immanuel Kant

Make the effort to gain experience—*Just do it.*

Keep your eyes open for unusual occurrences.

Seek explanations for anomalies.

◆ ◆ ◆ ◆ ◆ ◆ ◆ ◆ ◆ ◆ ◆ ◆ ◆ ◆ ◆ ◆ ◆ ◆

# MINDSCAPES  Invitations to Further Thought*

### I. SOLIDIFYING IDEAS

**1–13. The cobweb tent.** Take a square in which the diagonal is drawn and in which an inverted V is drawn. Start at any point on the diagonal. From there, everything else is determined. Go vertically up or down as needed to head toward the V. When you hit the V, go horizontally right or left until you hit the diagonal. From there repeat the pattern going vertically until you hit the V and then horizontally until you hit the diagonal. Repeat.

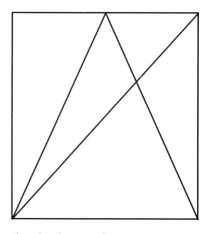

Following this pattern creates the cobweb plot you have seen before. Look at the following examples. These cobweb plots are the result of a repeated process, and they illustrate many of the ideas from this section. Questions 1–13 all refer to this process.

---

*In the Mindscapes section, exercises marked (H) have hints for solutions at the back of the book. Exercises marked (S) have solutions.

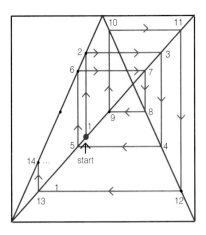
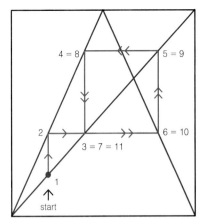

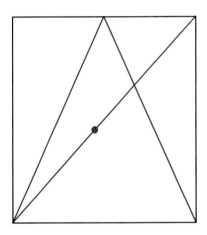

1. **Go.** Start at the point marked on the diagonal, and draw the first four iterates of the cobweb plot. Use a straight edge and be as careful as possible to keep your lines vertical and horizontal.

2. **Staircase.** Start at the point marked on the diagonal, and draw the first four iterates of the cobweb plot. Use a straight edge and be as careful as possible to keep your lines vertical and horizontal.

3. **How far?** Start at the point marked on the diagonal, and draw the cobweb plot. How many steps are in the cobweb plot before the point quits moving?

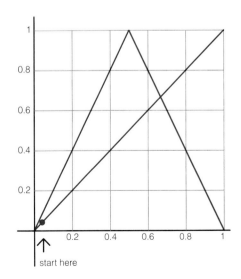
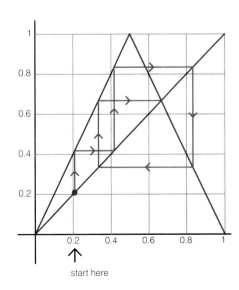

4. **Points that quit (H).** Find three different points that move at least three times each and then are fixed.

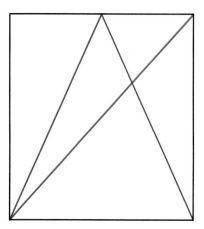

5. **Ups and downs.** Start at the point marked on the diagonal. Draw the cobweb plot until the path goes up to near the top of the diagonal and then down to near the bottom and repeats. This oscillation could go on forever.

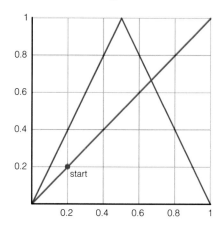

6. **Don't move (S).** Find the point on the diagonal other than 0 that does not move at all in its cobweb plot.

7. **Two step (S).** Find a point on the diagonal that goes up to the curve, over to the diagonal, down to the curve, and returns to the starting point. This point follows a repeating path that hits only two points of the diagonal and only two points of the curve.

8. **Three step.** Find a starting point on the diagonal where the path hits the diagonal at three points. The path starts by going up to the curve, then right over to the diagonal, and down to the curve. It then goes left to the diagonal, up to the curve, and over to the right, at which point it has returned to its original position. So, this point repeats after three cycles; it has period three.

9. **Four step.** Find a starting point on the diagonal where the path hits the diagonal at exactly four points. So, this point repeats after four cycles; it has period four. (*Hint:* You might start near the lower left and create a sort of stair step. Try sliding the starting point and see what happens.)

10. **Grow up.** What part of the diagonal is covered if you start with all the points on the diagonal above the interval $[0, .1]$ and apply the process once to each point?

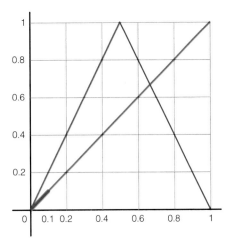

11. **Grow up again.** What part of the diagonal is covered if you start with all the points on the diagonal above the interval $[0, .1]$ and apply the process twice?

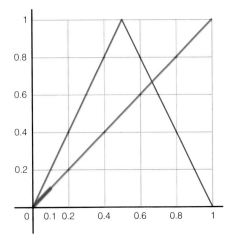

12. **A different patch (H).** Suppose you start with all the points on the diagonal above the interval $[0, .01]$. How many steps are required before the future positions of those points cover up the whole diagonal?

13. **Target practice.** Suppose we give you a target interval—for example, all the points on the diagonal above [.31, .32]. Why must there be a point on the diagonal above the interval [.0, .01] that hits the target interval after some number of steps?

14. **Too high.** Consider a tent function that got too high and sticks out over the top of the unit square. Let's play the cobweb game and see which points stay within the unit square. Any point between 1/3 and 2/3 of the diagonal goes out immediately. Show that the point (1/6, 1/6), which lies on the diagonal above the interval [1/9, 2/9], leaves the square. How many steps are required? Likewise, show that the point (5/6, 5/6), which lies on the diagonal above the interval [7/9, 8/9], also leaves the square.

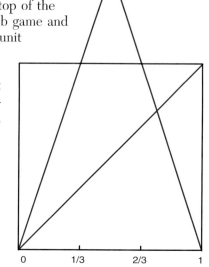

15. **More gone.** In the Too High tent function, show that the point (1/18, 1/18), which is on the diagonal above the interval [1/27, 2/27], leaves the square eventually. How many steps are required before it leaves?

16. **Too short.** Consider a short tent function. Will any point on the diagonal above the interval [1/3, 2/3] ever hit the diagonal at a point above the interval [0, 1/4]? Why not?

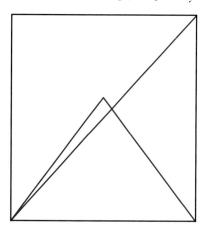

17. **Where to?** Using the logistic transformation $y = 3.5x(1 - x)$, calculate the first 30 values starting from .437 and from .438. Do the results stay fairly close to each other, or do they become quite different?

18. **Calculator slips.** Using the logistic transformation $y = 4x(1 - x)$, calculate the first 30 values starting from .437 and from .438. Do the results stay fairly close to each other, or do they become quite different?

19. **Just missed.** Find several examples in everyday life of tiny differences that led to huge differences in your own future—good, bad, or just different.

20. **Take stock.** Pick a stock. Some Web sites will display graphs about stocks. Look at the graph of the highs of that stock during each of the last 100 days. Next look at the graph of the highs of that stock during each of the last 100 weeks. Finally, look at the graph of the highs of that stock during each of the last 100 months. To what extent are these graphs similar? Do they seem to describe some chaotic features?

## II. CREATING NEW IDEAS

1. **Repulsive.** If you take the number 1 and square it, then square the answer, then square the answer again, of course, you will get 1 again and again. So, 1 is a fixed point under that process. Show that every point near 1 gets increasingly distant from 1 under repeated squaring. We call 1 a *repeller*, since nearby points are repelled from it.

2. **Attractive.** Again consider the process of repeated squaring. If you take the number 0 and square it, you get 0 again. So, 0 is a fixed point under that process. Show that every point near 0 gets increasingly close to 0 under repeated squaring. We call 0 an *attractor*, since nearby points are attracted toward it.

3. **Sierpinski attractor.** Remember how the Sierpinski Triangle was created using a collage-making process. Consider the pictured image that is similar to it. Sketch the result of applying the Sierpinski collage-making instructions to this picture. Did the new image become more like the Sierpinski Triangle? If it did, then we would call the Sierpinski Triangle an attractor under the collage-making transformation.

4. **Two step.** Consider the transformation that takes any point $x$ in the interval $[0, 1]$ to $(1 + \sqrt{5})x(1 - x)$. Compute the future values of 1/2 under repeated applications of this process.

5. **Periodic attraction (S).** Again consider the transformation that takes any point $x$ in the interval $[0, 1]$ to $(1 + \sqrt{5})x(1 - x)$. As you saw in the previous question, the number 1/2 is a periodic point of period 2, meaning that it returns to itself after the process has been

repeated twice. Compute 10 future values of the point .48 under repeated applications of this process. Do these points get closer to the repeated values of 1/2, or do they get farther away from the repeated values of 1/2?

6. ***Periodic attraction.*** Again consider the transformation that takes any point $x$ in the interval $[0, 1]$ to $(1 + \sqrt{5})x(1 - x)$. Compute 10 future values of the point .8 under repeated applications of this process. Do these points get closer to the repeated values of 1/2, or do they get farther away from the repeated values of 1/2?

7. ***Four-peat* (H).** Consider the equation $y = 3.5x(1 - x)$. The point .5009 . . . is periodic. Compute future values until you find the period.

8. ***Nearly fourly.*** Consider the equation $y = 3.5x(1 - x)$. The point .36 is not periodic. Compute 12 future values under repeated applications of this equation. To what are they tending?

9. ***Tent attraction?*** Consider the point .4, which is of period 2 in the cobweb tent. Sketch the cobweb plot of the point .38. Do these values seem to be tending toward the periodic point's values or not?

10. ***Becoming periodic.*** The point .4 is of period 2 in the cobweb tent transformation. Find a point that takes two steps first and then becomes of period 2.

### III. FURTHER CHALLENGES

1. ***The Earth moved.*** Consider a transformation of the sphere that consists of moving each point by shifting its longitude by 30° to the west and by increasing its latitude by squeezing points upward away from the south pole and toward the north pole. This process keeps the north and south poles fixed and spirals points upward toward the north pole and away from the south pole. Describe the movement of the equator under the first two iterations of this process. What points are fixed under this transformation?

2. ***Poles apart.*** Consider the same transformation as in the previous question. For each fixed point, state whether points within 10° in latitude and longitude from each of the fixed points get increasingly close to that fixed point, making it an attractor, or increasingly far from that fixed point, making it a repeller.

3–5. ***Logistic cobwebs.*** Let's explore the relatively simple looking transformation $y = rx(1 - x)$ in the following three questions. A *fixed point* is one where $x = rx(1 - x)$.

**3.** $r = 2$. For $y = 2x(1 - x)$, what points remain fixed? Draw the first 10 steps of the cobweb plot starting at the point marked. How does it relate to the fixed point?

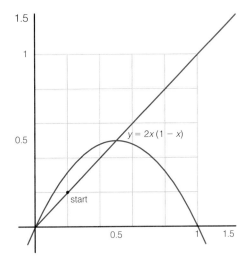

**4.** $r = 4$ (**H**). For $y = 4x(1 - x)$, sketch a cobweb pattern that indicates that there must be periodic points of period 3. (*Hint:* Start fairly near the lower left part of the diagonal and create a stair-step pattern that ends near the top, thus pushing you back to where you started. Slide your starting point if necessary.)

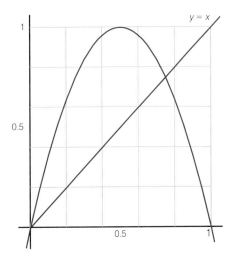

5. **$r = 4$.** For $y = 4x(1 - x)$, start at two random points near your periodic point of period 3 and compute the first 10 values under repeated application of the transformation. Do these values seem to be converging toward the plot of the period 3 point or not?

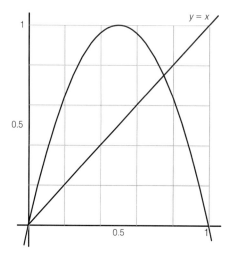

## IV. IN YOUR OWN WORDS

1. **Personal perspectives.** Write a short essay describing the most interesting or surprising discovery you made in exploring this section's material. If any material seemed puzzling or even unbelievable, address that as well. Explain why you chose the topics you did. Finally, comment on the aesthetics of the mathematics and ideas in this section.

2. **With a group of folks.** In a small group, discuss the concept of mathematical chaos. After your discussion, write a brief narrative describing mathematical and natural systems that exhibit chaos.

3. **Creative writing.** Write an imaginative story (it can be comical, dramatic, whatever you like) that involves or evokes the ideas of this section.

4. **Power beyond the mathematics.** Provide several real-life situations—ideally, from your own experience—for which some of the strategies of thought presented in this section would provide effective methods for approaching and resolving them.

# 6.6 Between Dimensions:

## Can the Dimensions of Fractals Fall Through the Cracks?

*Without dimension, where length,*
  *breadth and height,*
*And time and place are lost;*
  *where eldest night*
*And chaos, ancestors of nature, hold*
*Eternal anarchy.*

—John Milton

What is dimension? In our look at geometry in Chapter 4, we developed an intuitive idea of the dimension of space as measuring how many degrees of freedom we have. But this intuitive meaning of dimension seems a bit limiting and difficult to apply if we consider objects more intricate than the line, the plane, and so on. Can we sensibly associate a dimension with any object, or is dimension an idea that applies only to rather regularly shaped things?

To explore this question we begin, as usual, by examining carefully the simple and familiar with an eye toward understanding the more complex. In regular objects we see patterns associated with dimension that suggest how to extend the idea of dimension to objects that are irregular.

Are these swiss cheeselike Platonic solid fractals 3-dimensional?

501

Start easy.

## Easy Does It

We begin by looking at the familiar. Here are some warm-up questions to get you in the mood for what is ahead. Try to answer all of them before reading on.

- How many dimensions does a line have?
- Can you give an example of an object that is 2-dimensional?
- The cube is ???-dimensional. What does ??? equal?

You will not be shocked by the answers we are about to give to these questions. The line is 1-dimensional. One of the simplest examples of a 2-dimensional object is the filled-in square. The cube is 3-dimensional. Okay, not too tricky. It seems as though we have some reasonable notion of dimensionality: If the object just has length but no thickness or height, then it is 1-dimensional; if it has length and thickness but no height, then it's 2-dimensional; if it has length, thickness, and height, then it's 3-dimensional. That's that.

Well, that *is* that as long as we want to look only at simple shapes. Of course, the coastline of Norway isn't so simple, and yet we do like to look at it. Suppose we consider (as we tend to do in this chapter) objects of infinite intricacy. For example, what is the dimension of the Koch Curve? On the one hand, since it's basically made out of 1-dimensional line segments, it just has length and so it's reasonable to think that it's 1-dimensional. On the other hand, it has infinitely many bends and corners. If we look at it, it's so incredibly jagged that we might call it fuzzy—like mold on stale bread—and it seems to have some slight thickness. Thus, perhaps it is 2-dimensional. But wait: It cannot be 2-dimensional since it's just lines, yet it cannot be just 1-dimensional since its infinity fuzziness seems to take up a bit of area. What is the dimension of the Koch Curve?

Our conflicting thoughts are leading us to a strange possibility: Perhaps the dimension of the Koch Curve is some number that is bigger than 1, but less than 2. Huh? Does it make sense to have a dimension between 1 and 2? Answer: We need to pin down our idea of dimension. Let's examine simple things carefully and search for patterns.

## What Is Dimension?

A straight line is 1-dimensional. A filled-in square is 2-dimensional. A solid cube is 3-dimensional. Somehow we intuitively know these facts. The problem is that our intuition fails us when we look at the infinitely intricate. Thus, let's try to devise, without appealing to our intuition, a means of determining the dimensions of the line, the square, and the cube—dimensions we already know. If we could accomplish this task, then perhaps we could use the method to compute the dimensions of more exotic objects.

Let's begin with the filled-in square. Suppose the length of each side is 1. How many copies of this square can be assembled to produce a larger square? Well, four will do, or nine would work. If we use four squares to build a big square, we notice that the new square is twice as long as the original square. So, to make a square two times as large as the original square (that is, if we set scaling on the copier machine to print twice as large), we need four of the original-size squares to make the larger square.

What if we consider a cube having each side equal to 1? How many copies of this cube could be used to build a cube whose sides are twice as long? The

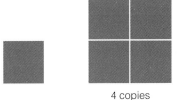

4 copies

answer is that we require eight cubes to build a cube with edges two times as large as the first cube. So, if we had a 3-dimensional photocopier, we would scale things by two to go from the first cube whose sides are 1 to the larger cube whose sides are 2.

Let's also consider a straight-line segment having length 1. Of course, two copies of this line produce a line that is twice the size of the original line. So, in this case, if we set the copying machine to make an image twice the size, we could construct that enlarged image with just two copies of the original segment.

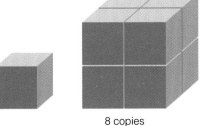

8 copies

2 copies — twice as long as first segment

## In Search of a Pattern

Let's organize our findings in a chart.

| Original Object | Dimension of the Object | Scaling Factor to Make a Larger Copy | Number of Copies Needed to Build the Larger Copy |
|---|---|---|---|
| line | 1 | 2 | $2 = 2^1$ |
| square | 2 | 2 | $4 = 2^2$ |
| cube | 3 | 2 | $8 = 2^3$ |

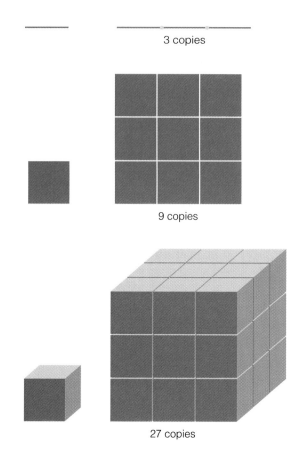

3 copies

9 copies

27 copies

Look for patterns.

♦ ♦ ♦ ♦ ♦ ♦

Do you see a pattern? It turns out that, if we write the number of copies needed to build the next larger copy of the original object as the scaling factor raised to some power, then that exponent is equal to the dimension of the object.

Now that we see a pattern, let's confirm our theory by scaling by a factor of 3 instead of 2 to see whether the pattern continues. Here is the chart we construct for a scaling factor of 3.

| Original Object | Dimension of the Object | Scaling Factor to Make a Larger Copy | Number of Copies Needed to Build the Larger Copy |
|---|---|---|---|
| line | 1 | 3 | $3 = 3^1$ |
| square | 2 | 3 | $9 = 3^2$ |
| cube | 3 | 3 | $27 = 3^3$ |

So, our hypothesis about the relationship between the scaling factor and the number of copies needed to construct a scaled-up version is confirmed. Raising the scaling factor to the power of the dimension gives us the number of copies needed to construct that larger-scale version of the object.

## Testing Our Conjecture

Let's test this hypothesis with some other objects and see if we can use this procedure to compute their dimensions.

Let's consider a filled-in equilateral triangle with all sides of length 5. How many copies of this triangle are needed to make an equilateral triangle with sides of length 10?

4 copies

We need four copies to build an equilateral triangle with sides twice as long. If we take the number of copies we need and write that number as the scaling factor raised to some power, then we see $4 = 2^2$. We need nine copies to build an equilateral triangle with sides three times as long. Once again, if we take the number of copies we need and write that number as the scaling factor raised to some power, then we see $9 = 3^2$. So, we see that the exponent we need is 2. Since we believe that a filled-in triangle is 2-dimensional, we see once again that the exponent is telling us the dimension.

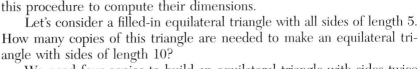
9 copies

Using this method, figure out the dimension of a filled-in rectangle whose base is 2 and height is 1. First, we must figure out how many copies of this rectangle are needed to build a similar rectangle that is larger. Again, we need 4 copies to construct a rectangle with sides twice as long. Thus, if we write the number of copies we need as the scaling factor raised to some power, then we again see $4 = 2^2$. So, the dimension of the rectangle is again the power 2. Since we have observed a relationship between dimension and this exponent, let's crystallize these observations into an official definition of dimension.

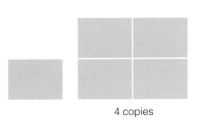
4 copies

## Dimension Finally Defined

*Let examples lead to definitions.*

❖ ❖ ❖ ❖ ❖ ❖

Suppose we have an object for which $N$ copies can be assembled to construct a larger version of it that has been scaled by a factor of $S$. We now define the *dimension*, which we call $d$, of the object to be the power to which we have to raise $S$ to have it equal $N$. That is, $d$ is the number such that

$$S^d = N.$$

We already know that this definition of dimension gives us the correct answers in the cases of the line segment, the square, the cube, the filled-in triangle, and the filled-in rectangle, so it is a plausible distillation of our intuition. We now want to apply our clear understanding of the known to analyze the unknown.

## From the Ordinary to the Exotic

We have worked hard to discover a precise definition of dimension. Now let's test our wings and compute the dimensions of more interesting and intricate objects. We begin with the Koch Curve.

## Example 1—Koch Curve Dimension

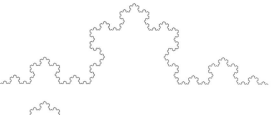

Recall that the Koch Curve is the curve we get by iterating a process infinitely often. We begin with a line segment and then cut it into three equal pieces. We then replace the middle piece with an inverted V made of two segments, each having the same length as the middle piece we removed. We now repeat this procedure with each of the four smaller line segments. We continue this process, creating an object of infinite intricacy and self-similarity.

What is its dimension? This question started our entire discussion. Our first, strange guess was that it would be some number between 1 and 2. Think with us now as we compute the dimension of a Koch Curve.

How do we compute dimension? We must first ask how many copies of the Koch Curve are needed to build a larger version of the Koch Curve. How many would you guess? One answer is 4. Imagine gluing four copies together to produce a big version of the Koch Curve.

Now we come to a more challenging question: How much do we have to scale the original Koch Curve to enlarge it to the bigger one we just built? It is worthwhile to think about this question for a bit.

The answer is that we scale by 3. Notice that, if the original Koch Curve had a horizontal length of 1, then our enlarged version would have a horizontal length of 3. Thus we see that the scaling factor is 3. So, what is the dimension $d$? According to our definition of dimension, the dimension $d$ of the Koch Curve is the power such that

$$3^d = 4.$$

Finding the numerical value of $d$ involves logarithms, which we will not review here. As a practical matter, we just need a calculator. To solve for $d$, use a calculator that has an LN or LOG key. Specifically, $d$ is ln 4/ln 3 or, equivalently, log 4/log 3. If we enter this number into a calculator we get this result:

$$d \approx \frac{0.60205}{0.47712} = 1.26185\ldots$$

On your calculator check that 3 raised to the 1.26185 . . . power equals 4. So, the dimension of the Koch Curve is 1.26185. . . . It is indeed bigger than 1-dimensional and yet smaller than 2-dimensional. Notice how the infinite intricacy of the Koch Curve creates an interesting fuzziness that increases the dimension ever so slightly.

## Example 2—Sierpinski Dimension

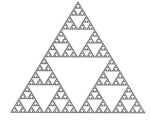

Let's now return to the Sierpinski Triangle. What is its dimension? Notice that three copies of the Sierpinski Triangle can be assembled to create a larger version and that the larger version is twice the size of the original one (the scaling factor in this case is 2). Thus, the dimension $d$ of the Sierpinski Triangle is the number such that

$$2^d = 3$$

We can compute this dimension using a calculator by calculating

$$\ln 3 / \ln 2$$

or, equivalently,

$$\frac{\log 3}{\log 2}$$

and we see that

$$d \approx 0.47714/0.30103 = 1.58496\ldots$$

So, again we see that the dimension of this fractal is greater than 1 but smaller than 2.

These two examples give us the tools by which we can compute the dimensions of many fractals that were constructed using the collage-making procedure—as long as the shrinking factors can be determined. We can use a calculator to find that

$$d = \frac{\ln N}{\ln S}$$

or, equivalently,

$$\frac{\log N}{\log S}$$

where, again, $N$ is the number of copies required to make a larger version of the original object and $S$ is the scaling factor required to enlarge the original object to the size of the larger version.

**A LOOK BACK**

We can take the idea of dimension and extend it to encompass and describe many of the interesting fractal images we constructed. Surprisingly, the dimensions of fractal objects are not restricted to the numbers 1, 2 or 3 but instead fall somewhere between these dimensions. Perhaps this fact is not so shocking, because a fractal is cloudlike: It is difficult to see where it begins and where it ends. Fractal dimension captures and in some sense measures the beautiful cloudlike essence of fractals.

Our strategy for developing the idea of fractal dimension is to start with observations of familiar objects. For familiar, regular objects like cubes and triangles, we find a relationship connecting a scaling factor with the number of copies needed to produce similar objects at a larger scale. That relationship allows us to give a more encompassing definition of dimension. Our extended idea of dimension applies to more complicated objects like fractals.

Starting with simple and familiar cases and understanding them deeply allows us to reach beyond what we know now. Patterns among familiar objects and ideas can then be fit on unfamiliar terrain to show features not previously evident. By defining carefully an idea that we cull from common experience, we have a beacon for exploring the unknown.

Start with the simple and familiar.

Look for patterns.

Apply patterns to new settings.

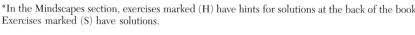

# MINDSCAPES  Invitations to Further Thought*

## I. SOLIDIFYING IDEAS

1. **Stay inbounds.** Give two consecutive integers that bound the fractal dimension of (1) the Mandelbrot Set, (2) the Menger Sponge, (3) the Cantor Set.

2. **Regular things (H).** Find the fractal dimension of these two objects using the definition of fractal dimension from the section.

3. **More regular things.** Find the fractal dimension of a parallelogram.

4. **Any right triangle.** Take any right triangle. It can be broken into four similar triangles, each one having edges that are 1/2 the length of their corresponding edges in the big triangle. Use this fact to show that, using the fractal dimension definition, any filled-in right triangle has dimension 2.

5. **Sierpinski carpet (S).** Compute the fractal dimension of the purple part of the Sierpinski carpet—that is, the fractal constructed in Chapter 1, "A Tight Weave" (p. 10).

---

*In the Mindscapes section, exercises marked (H) have hints for solutions at the back of the book. Exercises marked (S) have solutions.

6. **Koch Stool.** Compute the fractal dimension of the Koch Stool described in section 6.3. Mindscapes (I.16).

7. **Cantor Set (H).** The Cantor set was constructed by taking a unit segment, making two copies each of length 1/3, and putting those copies at the two ends of the unit segment. Repeating this collage-making process infinitely often results in the Cantor Set. What is the dimension of the Cantor Set?

8. **Cantor reduced.** Suppose you take a unit interval, make two copies each shrunk to 1/4 its length, and position them at the ends of the unit interval. Repeating this process infinitely results in a fractal similar to the standard Cantor Set. What is its fractal dimension?

9. **Long Koch.** The first stage in the construction of the Koch Curve is a line segment of length 1. In the second stage that segment is replaced by four segments, each of length 1/3. So, the length at the second stage is 4/3. What is the length of the third stage? What is the length of the fourth stage? What is the length of the $n$th stage in the Koch construction? What would you say is the length of the final Koch Curve?

10. **Plus.** This fractal plus sign is self-similar in that, if it is reduced by 1/3, four reduced copies can be put together to create it again. What is its fractal dimension?

The first few steps in the fractal plus sign.

## II. CREATING NEW IDEAS

1. **Tinier triangles (S).** Suppose you make something similar to the Sierpinski Triangle, but this time you make three copies each reduced so that the sides are only 1/3 of the original's sides' lengths, and then you position the reduced triangles at the three corners. Notice that this process has the same result as removing a hexagon from the triangle, as we did before. Continue this collage-making process infinitely. What is the dimension of the resulting collage?

2. ***Menger Sponge.*** Compute the fractal dimension of the Menger Sponge.

3. ***Thinning.*** Take a square. Make two copies, each reduced to half size in length and width, and position them in diagonal corners. Repeat the process infinitely often to create a fractal. What is the fractal?

4. ***Not much.*** What is the fractal dimension of the fractal in the previous question?

5. ***Koched*** (H). Create a Koch-like curve with fractal dimension $\ln 5/\ln 4$.

### III. FURTHER CHALLENGES

1. ***Find a fractal.*** Describe a fractal having dimension $\ln 8/\ln 3$.
2. ***Find a 1.5 fractal.*** Describe a fractal having dimension 1.5.

### IV. IN YOUR OWN WORDS

1. ***Personal perspectives.*** Write a short essay describing the most interesting or surprising discovery you made in exploring this section's material. If any material seemed puzzling or even unbelievable, address that as well. Explain why you chose the topics you did. Finally, comment on the aesthetics of the mathematics and ideas in this section.

2. ***With a group of folks.*** In a small group, discuss the meaning of fractal dimension. After your discussion, write a brief narrative describing the fractal dimension of the Koch Curve in your own words.

3. ***Creative writing.*** Write an imaginative story (it can be comical, dramatic, whatever you like) that involves or evokes the ideas of this section.

4. ***Power beyond the mathematics.*** Provide several real-life situations—ideally, from your own experience—for which some of the strategies of thought presented in this section would provide effective methods for approaching and resolving them.

# CHAPTER SEVEN

# Risky Business

*But to us, probability is the very guide of life.*
—Bishop Joseph Butler

### 7.1

**Chance Surprises**

### 7.2

**Predicting the Future in an Uncertain World**

### 7.3

**Random Thoughts**

### 7.4

**Down for the Count**

Many, if not most, significant events in our lives arise from coincidence, randomness, and uncertainty. We meet friends and loved ones, we find intriguing opportunities, we fall into a profession or lifestyle. At a deeper level, the most basic interactions of molecules and subatomic particles are described in terms of probabilities and statistics. Nothing is more fundamental than chance. However, the uncertain and the unknown are not forbidding territories into which we dare not tread. They too can be organized and understood.

Probability and statistics enable us to better understand our world. They move us from a vague sense of disordered randomness to a sense of measured proportion. They are the mathematical foundations of common sense, wisdom, and good judgment. They weigh our expectations and give us a refined sense of valuing the unknown. Perhaps, too, they let us view our world more truly as it is—a place where the totality follows rules of the aggregate while leaving individuals to their wild variation and unbridled possibilities.

*Measure what is measurable, and make measurable what is not so.*
—Galileo Galilei

We develop a measure of likelihood by looking at situations in which the future is uncertain but the possible outcomes are definite and easily described. Gambling games provide concrete and clear illustrations in that dice and coins teach us how to measure likelihood. We learn to count possible outcomes and determine which are of use in various different scenarios. We apply the principles we develop to measure the value of future possibilities, thus allowing us to weigh decisions involving the unknown future. Frequently we need to extrapolate the probable future from evidence from the past. This need presents us with the challenge of collecting and meaningfully interpreting data. By examining errors in such processes, perhaps we can avoid some of the many pitfalls associated with amassing and interpreting a sea of data.

Surprises lead the way in the study of the uncertain and the unknown. But we progress by looking at concrete examples, by doing experiments to ground our theory in experience, and by looking at fallacies. Simple, clear cases let us develop principles that we can apply widely. So, even the uncertain and unknown are best understood by starting with the simple and the familiar.

◆ ◆ ◆ ◆ ◆ ◆ ◆ ◆ ◆

**7.5**

*Great Expectations*

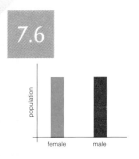

**7.6**

*What the Average American Has*

**7.7**

*Navigating Through a Sea of Data*

# 7.1 Chance Surprises:
## Some Scenarios Involving Chance That Confound Our Intuition

*Chance, too, which seems to rush along with slack reins, is bridled and governed by law.*

—Boethius

Many surprises lurk in the world of chance. We guess wrong because our intuition is untrained or mistrained. Each surprise is an enticement for us to find structure among the forces of the uncertain and unknown. In the following sections, we develop methods for analyzing chance scenarios. Let's begin by resolving the experiments below.

### Lincoln on Edge

First, a thought experiment: Imagine taking a collection of pennies (at least 20) and carefully balancing each on its edge on a table. Envision this spectacle in your mind's eye—it's something one does not see every day. Now suppose that an aggressive person comes along, sees the pennies delicately balanced on their edges, and proceeds to slam one hand on the tabletop. Not only does this act of violence cause a loud BANG, but it also causes a sufficient vibration on the table to bring all the pennies crashing down on their sides. In the aftermath of this event, we see some pennies resting in peace on their heads and others resting on their tails.

- *What we expect.* Naturally, we expect to see about the same number of heads as tails.
- *Surprise.* Significantly more than half will land heads up. Why?

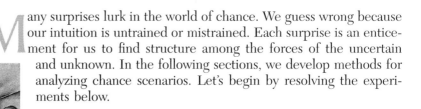

*The Dream* (1921) by Max Beckmann. Analyzing what we see sometimes leads to surprising, counter-intuitive results.

## Dizzy Lincoln

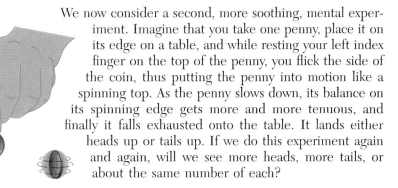

We now consider a second, more soothing, mental experiment. Imagine that you take one penny, place it on its edge on a table, and while resting your left index finger on the top of the penny, you flick the side of the coin, thus putting the penny into motion like a spinning top. As the penny slows down, its balance on its spinning edge gets more and more tenuous, and finally it falls exhausted onto the table. It lands either heads up or tails up. If we do this experiment again and again, will we see more heads, more tails, or about the same number of each?

- *What we expect.* Naturally, we expect to see about the same number of heads as tails.
- *Surprise.* Significantly more than half will land tails up. Why?

*Try it.* Now make it physical. First, balance at least 30 pennies on their edges on a table. (This delicate step is infinitely easier to say than to do—you must be patient, but we promise it will be worth it.) Then slam your hand

on the table, causing the pennies to topple over. How many heads do you see? How many tails? Try this experiment a few times, and record the outcomes. Next, spin a number of pennies on their edges and see how they land. Repeat a number of times, and again record your results. You will notice that your experimental results will tend to confirm the surprising reality rather than our initial guess.

Why is there such a dramatic difference between our guesses and the outcomes of the physical experiments? Try to think of various reasonable explanations for these phenomena. Discuss them with others and see if you can reach a consensus. There *is* an explanation.

## Let's Make a Deal

Revisit the Let's Make a Deal scenario from Chapter 1 (Section 1, story 7). The contestant selects one of three doors. Another door is opened to reveal a mule rather than the Cadillac. The contestant can stick to the original guess or switch.

- *What we expect.* Probably we expect that switching or sticking makes no difference. We might think that in either case the chance of getting the Cadillac is 1 out of 2 since there are two closed doors left.

- *Surprise.* Switching gives the contestant a 2 of 3 chance of winning the car. Why? The explanation is in Chapter 1 (Section 3, story 7).

*Try it.* Here is an experiment that verifies the Let's Make a Deal probability. From a deck of regular playing cards, remove three—a king and two aces. The king represents the Cadillac, and the aces represent the mules. Have a friend act as the dealer. The dealer shuffles these three cards and places them face down on a table, side by side, without looking at them. Once the cards are on the table, the dealer peeks under each card so that the location of the king is known to the dealer but not to you. Point to a card. The dealer then turns over one of the other two cards to reveal one of the aces (the mules). You now have the chance to switch cards. Stick with your original guess and see what happens. Have the dealer shuffle the cards again and repeat the exact same scenario—don't switch—at least 20 times and see what fraction of the time you end up with the king (the car). Next do the same experiment, but try switching and again record how often you find the king. After repeating this experiment several times, you will discover that about ⅓ of the time you find the king if you stick to your original guess, and about ⅔ of the time you will find the king if you switch.

## *Reunion Scene—Take One*

Suppose we return to our 25th college reunion and one of our old classmates tells us, "I have two children: *The older* is named Jonathan and . . ." At that point she chokes on an hors d'oeuvre and collapses.

## *Reunion Scene—Take Two*

Suppose we return to our 25th college reunion and one of our old classmates tells us, "I have two children: *One is* named Jonathan and . . ." At that point she chokes on an hors d'oeuvre and collapses.

Naturally, we have some concern for the respiratory challenges of our former classmate; however, we probably would be more consumed with the following burning question: Assuming that Jonathan is a boy, what is the chance that our classmate has two boys? Is the chance the same in both scenarios?

?
or
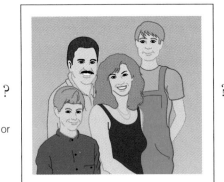
?

- *What we expect.* We probably expect that in both cases there is a 50-50 chance that she has two boys.
- *Surprise.* In Take One, the chance is exactly one half (as expected); however, in Take Two the chance is only 1 in 3. Why?

*Try it.* You could simulate this situation as follows: Take a deck of cards. Think of the boys as the black cards and the girls as the red cards. Shuffle the deck and remove cards from the top of the deck in pairs. First look at all the pairs whose first card is black. What fraction of those have both cards black? Now start over. Shuffle the cards and take them off in pairs; however, this time look at all pairs that contain at least one black card. What proportion of those have both cards black? Does this experiment tend to confirm our original intuition or the surprising result?

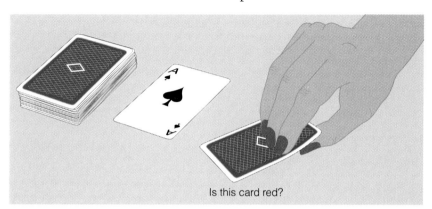

Is this card red?

## The Birthday Question

How many people are needed in a room so that the probability that there are two people whose birthdays are the exact same day is roughly 1/2?

- *What we expect.* We might expect to need about 183 people. If we had 367 people in the room, then we would be guaranteed to have at least two people who have the same birthday since 367 people can't all have different birthdays. Therefore, if we want the chances of a matched pair of birthdays to be 50-50, then it seems we would need about 183 people in the room (about half of the 367 people) for the chance of finding a shared birthday to be roughly 50%.
- *Surprise.* In a room containing only 23 people, the chance of two people having the same birthday is just *over* 50%. That is, in a random gathering of 23 people we will more often than not find a pair of people with the same birthday. In a room with 183 people the chance of finding a pair of people with the same birthday is over 99.999999%. Why?

*Try it.* The next time you are in a room with 40 people or so, ask them for their birthdays and see whether you find a common birthday.

**A LOOK AHEAD**

Matters of chance do have satisfying explanations. However, we need to develop a sense of measuring uncertain events so that the experiences associated with the types of surprises discussed in this section come to appear natural and expected.

*But to us, probability is the very guide of life.*
—Bishop Joseph Butler

Thinking about situations that jar our intuition can lead to new and important insights.

◆ ◆ ◆ ◆ ◆ ◆ ◆ ◆ ◆ ◆ ◆ ◆ ◆ ◆ ◆ ◆ ◆ ◆

# MINDSCAPES  Invitations to Further Thought

### I. SOLIDIFYING IDEAS

1. *Edgy Lincoln.* Balance pennies on their edges on a table. Then slam your hand on the table, causing the pennies to topple over. Perform this experiment until 100 pennies have fallen. Record how many pennies landed heads up and how many landed tails up. What percentage of the pennies landed heads up?

2. *Spinning Lincoln.* Spin pennies on their edges 100 times and see how they land. Record your results. What percentage of the pennies landed heads up?

3. *Flipping Lincoln.* Flip a penny 100 times and record how many pennies land heads up and tails up. What percentage of the pennies landed heads up?

4. *A card deal stick.* From a deck of regular playing cards, remove three—a king and two aces. Have a friend act as the dealer. The dealer shuffles these three cards and places them face down on a table, side by side, without looking at them. Once the cards are on the table, the dealer peeks under each card so that the location of the king is known to the dealer but not to you. Point to a card. The dealer then turns over one of the other two cards to reveal one of the aces. Stick with your original guess, turn over that card, and record whether you have chosen the king. Have the dealer scramble the cards again and repeat the exact same scenario—again, don't switch—and record the result again. Repeat this experiment 50 times (you can get very quick at it). What percentage of the time did you choose the king?

---

*In the Mindscapes section, exercises marked (H) have hints for solutions at the back of the book. Exercises marked (S) have solutions.

5. ***A card deal switch.*** From a deck of regular playing cards, remove three—a king and two aces. Have a friend act as the dealer. The dealer shuffles these three cards and places them face down on a table, side by side, without looking at them. Once the cards are on the table, the dealer peeks under each card so that the location of the king is known to the dealer but not to you. Point to a card. The dealer then turns over one of the other two cards to reveal one of the aces. Now switch your guess to the other face-down card. Turn over that card and record whether you have chosen the king. Have the dealer shuffle the cards again and repeat the exact same scenario—that is, switch your guess each time after the dealer turns over an ace—and record the result again. Repeat this experiment 50 times (you can get very quick at it). What percentage of the time did you choose the king?

6. ***A card reunion—black first*** (S). Take a shuffled deck of cards and remove cards from the top of the deck in pairs. For each pair where the first card is black, record whether the second pair is red or black. After you have done 10 pairs, reshuffle the deck and repeat until you have recorded 50 cases where the first card of the pair was black. What percentage of those pairs had both cards black?

7. ***A card reunion*** (H). Take a shuffled deck of cards and remove cards from the top of the deck in pairs. For each pair where at least one of the cards is black, record whether both cards are black or whether one is black and one is red. After you have done 10 pairs, reshuffle the deck and repeat until you have recorded 50 cases where at least one card of the pair was black. What percentage of those pairs had both cards black?

8. ***Birthday bash.*** The next time you are in a room with 40 people or so, ask them for their birthdays and see whether you find a common birthday.

9. ***Presidential birthdays.*** Have there been two presidents of the United States who had the same birthday?

10. ***Vice-presidential birthdays.*** Have there been two vice-presidents of the United States who had the same birthday?

## II. IN YOUR OWN WORDS

1. ***Personal perspectives.*** Write a short essay describing the most interesting or surprising discovery you made in exploring this section's material. If any material seemed puzzling or even unbelievable, address that as well. Explain why you chose the topics you did. Finally, comment on the aesthetics of the mathematics and ideas in this section.

2. ***With a group of folks.*** In a small group, discuss the surprises involving chance found in this section. After your discussion, write a brief narrative describing the surprising features in your own words.

## 7.2 Predicting the Future in an Uncertain World:

### How to Measure Uncertainty Through the Idea of Probability

*To be, or not to be . . .*

—Shakespeare

*Understand simple things deeply.*

What will be? How can we cope with the unknowable—the uncertain future or unpredictable present? Some seek insight from tea leaves, the stars, or the entrails of sheep. Some gaze deeply into the translucent beauty of a crystal ball. Let's not. Instead, let's gaze deeply into the powerful world of transcendent ideas and take our vague view of the future and give it some structure. Specifically, let's construct a means to measure the possibilities for a future we do not know. Quantifying the likelihoods of various uncertain possibilities is an impressively grand idea. How can we sensibly measure what we admit we do not know?

We adopt a strategy used in previous investigations. We have already confronted numerous mysteries, including infinity and the fourth dimension. We uncovered their secrets by first understanding basic ideas deeply. Clarifying fundamental ideas enabled us to effectively develop precise notions and led us to new discoveries. Now we wish to delve into the uncertain and the unknown, so we seek examples where we have an intuitive idea of how to measure the likelihood of a future event. We look at those examples with the goal of finding patterns and techniques that can be applied more broadly. A careful examination of our intuition often leads to new insights and discoveries.

*Le Tricheur à l'As de Carreau* (1635) by Georges de la Tour. (The cheater with the ace of diamonds.) *Watch those hands!*

## Likelihood in Everyday Life

The notion of likelihood is a major component of our everyday lives. How likely is it that a certain scenario will actually happen? What are the chances? As we will continue to see, often the answers to such everyday questions are surprising and counterintuitive.

Tomorrow it will either snow or not snow. Does this fact imply that there is a 50-50 chance of snow? If we are reading this book in Hawaii in June, then we would not expect it to snow tomorrow. If, however, we are reading this book in Buffalo, New York, in June, then the answer is not so clear. The point is, there is certainly a *chance* of snow, but is it *likely* to snow?

An amazing number of our actions and decisions are based on an intuitive sense of likelihood. In fact, "likelihood" often provides the foundation for what we think of as "common sense." Here is just a sample of some everyday issues and questions involving likelihoods, risks, and chances.

- Do you go to the dining hall for lunch the moment classes let out, or do you wait because you expect there to be long lines?
- While walking home at night, do you take the shortcut through the dark alley, or do you walk around the well-lit block? Why?
- You are driving on a four-lane highway, and you pass a car. You assume that car will continue to stay in its lane. Would you pass someone who was swerving in and out of his lane?
- Why do so many sexually active people practice safe sex?
- There is no nearby parking. Do you park illegally in front of a store to run in for 5 minutes and take the chance of getting a ticket? What about for 15 minutes? An hour?

Knowing how the future will unfold would be valuable: We'd know what number will come up on the roulette wheel, which stock will skyrocket, and which numbered Ping-Pong balls will bubble out of the Lotto machine. We are constantly attempting to predict what will happen in our lives and act accordingly. To develop a measurement of likelihood, let's find some concrete, simple situations in which the future, though uncertain, presents clear, quantifiable alternatives.

## On a Roll—The Measure of Likelihood

Games of chance provide basic examples where the measurements of likelihood are reasonably clear. Therefore, let's measure uncertainties in the high-rolling domain of dice. Suppose we have an ordinary die with sides numbered from 1 to 6, and suppose it is a *fair die*, which means that no particular side is more likely to be rolled than any other. If we roll the die, what is the probability of rolling a 4? In other words, what number would you associate with the likelihood of a 4 coming up if you roll a die one time?

What are the chances?

Probably you came up with 1/6. Why is 1/6 the probability of rolling a 4? Well, there are six possible outcomes of rolling a die. We could roll a 1, or a 2, or a 3, or a 4, or a 5, or finally a 6. All these outcomes are equally likely since the die is fair. Exactly one of these outcomes (rolling a 4) is the outcome whose likelihood we are assessing. Thus, there is exactly one way out of the six equally likely possible outcomes to roll a 4, and so there is a 1 in 6 chance of rolling a 4. The number 1/6 captures the idea that rolling a 4 is *one* of the *six* equally likely outcomes that are possible when the die is rolled.

Let's put our intuition to the test and experiment by rolling a die a bunch of times and recording the outcomes. We did some experimenting ourselves with 100 rolls of a die. Here are our results.

| Number Appearing on Die | Times Rolled (out of 100) |
|---|---|
| 1 | 18 |
| 2 | 16 |
| 3 | 20 |
| 4 | 17 |
| 5 | 15 |
| 6 | 14 |

We see that 17 out of our 100 rolls were a 4. Thus, 17/100 (or .17) of the time we saw a 4. This experiment seems to support our thinking that the probability of rolling a 4 is 1/6 or .1666 . . . .

Let's now try to figure out the probability of rolling the die and seeing an even number. As before, there are a total of six possible outcomes, each equally likely. However, now more than one outcome would lead to success (an even number): We could roll a 2, or a 4, or a 6. Thus, there are a total of three different ways of seeing an even number. If we divide the total number of ways of seeing an even number by the total number of possible outcomes, we have 3/6, which equals 1/2, or .5. So the probability of rolling an even number is 1/2. This answer makes sense since half the numbers on the die are even, and, therefore, half the time we would expect to roll an even number.

## *A Measure of Likelihood—Probability*

The concept of dividing the number of successful outcomes by the total number of possible outcomes provides us with a measure of likelihood. Notice that this fraction will always be a number between 0 and 1; where the closer this fraction is to 1, the greater our confidence that the successful outcome will actually occur, and the closer the fraction is to 0, the lower our confidence that the successful outcome will happen. Let's extend this concept of measuring likelihood into a precise definition.

Suppose we perform a certain activity in which there are only finitely many possible outcomes, any one of which is just as likely as any other. We then define the *probability of a particular event* to be the number of different outcomes that would result in that particular event divided by the total number of possible outcomes. Let's say that again. Suppose that a certain activity (say, rolling a die) will result in a total of $T$ possible outcomes, all of which are equally likely to occur (for rolling a die, $T$ would be 6). We now consider a specific event, which we'll call $E$. (For instance, $E$ might be the event of "rolling an even number.") If we know that there are $N$ different outcomes that would result in the event $E$ (in this case $N$ would equal 3, since there are three different outcomes that would result in rolling an even number), then we define the probability of the event $E$ to be the number $N/T$. (In our example, this probability would be 3/6, or just 1/2.) So,

> **PROBABILITY.** *The probability of the event E occurring =*
> $$\frac{N}{T} = \frac{(number\ of\ different\ outcomes\ giving\ E)}{(total\ number\ of\ equally\ likely\ outcomes)}.$$

Notice that $N$ is some number from 0 to $T$. Therefore, the smallest the probability could be is $0/T = 0$, and the largest the probability could be is $T/T = 1$. Observe that the larger the probability of an outcome, the more likely it is that the outcome will occur.

## *Relative Frequency*

As we repeat an experiment again and again, we can keep track of the number of times a particular outcome occurs. We can calculate the *relative frequency* of that particular outcome by taking the number of times that particular outcome occurred and dividing it by the total number of times we repeated the experiment. In other words:

> **RELATIVE FREQUENCY.** *A relative frequency of an outcome =*
> $$\frac{(the\ number\ of\ times\ that\ outcome\ occurred)}{(the\ total\ number\ of\ times\ the\ experiment\ is\ repeated)}.$$

For example, in our first die-rolling experiment, we saw that 17 out of our 100 rolls were a 4. So, the relative frequency of rolling a 4 in this repeated die-rolling experiment is 17/100, which equals .17. The probability of rolling a 4 is equal to $1/6 = .1666\ldots$. Notice how close .17 is to $.1666\ldots$. It seems reasonable that the more times we repeat an experiment and then compute the relative frequency of an outcome, the closer that frequency should get to the actual probability of that outcome. This insight is known as the *Law of Large Numbers*.

## THE BIRTH OF PROBABILITY

In fact, probability started with dice. The French nobleman Antoine Gombauld, the Chevalier de Méré, was a famous 17th-century French gambler. He loved dice games. One of his favorites was betting that a 6 would appear at least once in four consecutive rolls of a die. After some time, Gombauld became bored with this game of chance and devised a new game by scaling up from one die to two dice. In the new game, he bet there would be at least one pair of 6s in 24 consecutive rolls of a pair of dice. He soon noticed that he was not winning as much as before. Bothered by this discovery, in 1654, Gombauld wrote a letter to the French mathematician Blaise Pascal, who in turn mentioned this problem to Pierre de Fermat. The two mathematicians solved the mystery. They computed that the probability that a 6 would appear at least once in four consecutive rolls of a die is equal to .52. Since this probability is slightly greater than .5, over the long run Gombauld would win slightly more often than he would lose. However, the probability of seeing at least one pair of double 6s in 24 consecutive rolls of a pair of dice is equal to .49. (We will verify both of these probabilities ourselves in section 7.4.) Since this probability is slightly less than .5, Gombauld would lose more, on average, than he would win.

This observation by the gambler Gombauld and the answer given by Fermat and Pascal led to the birth of the study of probability. You may be amused by Pascal's view of humanity. In a letter to Fermat referring to Gombauld, Pascal wrote:

> *He is very intelligent but he is not a mathematician: this as you know is a great defect.*

**LAW OF LARGE NUMBERS.** *If an experiment is repeated a large number of times, then the relative frequency of a particular outcome will tend to be close to the probability of that particular outcome.*

### *Analyzing Our Chance Surprises*

Armed with the ideas of probability and relative frequency, let's take another look at the chance surprises from the previous section as well as some additional dicey issues.

## Lincoln on Edge

In the previous section, we described two experiments with pennies whose results surprised us. Balancing pennies on edge and then banging the table usually results in more heads than tails showing up. Spinning pennies usually results in more tails.

It might be that the edge of a penny is not perpendicular to its faces. This slight beveling allows the penny to drop easily out of the mold when it is cast. Although it is difficult to see the beveling, we might try to measure it if we take many pennies and first line them up—edge to edge—all heads on a table and measure how long the line is. Now line them up again, but alternate head, tail, head, tail, and so on. Measure now. This number should be less than the first one, since the alternating bevels make the pennies fit more snugly together. This beveling is the key to the unexpected probabilities of falling pennies.

To get a reasonable measure of the probability of pennies on edge falling heads up, we could do many experiments and record the results. The ratio of heads-up pennies to total trials—that is, the relative frequency—will approach the probability. In this case, there is no theoretical or logical argument that would lead us to a specific probability other than what we get from experiments. Once we suspect the edge is beveled and that the guess of equal probability is no longer valid, we must rely on experimental data to determine the probability.

## All Boys?

The two Reunion scenarios from the last section asked what we can deduce from a mother's slightly different sentence fragments: (1) "I have two children: The older is named Jonathan and . . ." versus (2) "I have two children: One is named Jonathan and . . ." In each case, what is the probability that the speaker has two boys? Our analysis must begin with a careful listing of the possibilities.

A person with two children may have had first a boy then a girl, first a girl then a boy, first a boy then a boy, or first a girl then a girl. These are the four equally likely ways to have two children. Enumerating these possibilities helps us analyze the scenarios, whereas if we rely on vague intuition, we could easily be led astray.

In (1), we know that her older child is a boy. So, either she had first a boy then a girl or first a boy and then another boy. Thus, the probability of her having two sons is 1/2. However, in (2), we know that she has at least one boy, but we do not know whether Jonathan is the older or the younger child. So the three possibilities are first boy then girl, first girl then boy, or two boys. In only one of these three equally likely possibilities would she have two sons. Therefore, the probability of her having two boys is only one in three, or 1/3. The number of equally likely possibilities (3) divided into the number of those we are interested in (1) gives the probability.

## More Dicey Issues

Given the genesis of probability, it seems only fitting that we roll some more dice. Suppose we now roll *two* fair dice. What is the probability of rolling "snake eyes" (rolling a sum of 2)?

Well, there's only one way of getting the numbers on the two dice to add up to 2: Each die must be showing a 1. So, there is only one outcome to yield "snake eyes." We now need to figure out the total number of possible outcomes. A reasonable guess is 11, since, when we roll two dice, we see a 2, 3, 4, 5, 6, 7, 8, 9, 10, 11, or 12. The trouble with this guess is that not all these outcomes are equally likely. For example, we have already seen that there is only one way to roll a 2: Each die must be showing a 1. However, there are two different ways of rolling a 3: The first die could be a 1 and the second die could be a 2; *or* the first die could be a 2 and the second die could be a 1. These two possibilities are different outcomes. To get a clear picture of all the possible outcomes, it is better to color the dice different colors so we can distinguish one from another. The table at the top of the facing page illustrates all the possible outcomes of rolling two dice, a red one and a green one. Notice that there are a lot more than 11 different outcomes.

Notice that there are 36 possible, equally likely, outcomes. Since only one of them produces "snake eyes," we conclude that the probability of rolling "snake eyes" is 1/36, which equals .0277 . . . . This probability is pretty small, so we would not expect to see "snake eyes" too often. What is the probability of rolling a total of 4? What is the probability of rolling a total of 13? What is the probability of rolling a total of 7? Figure out these probabilities using the table.

## The Probability of Success Versus Failure

Let's think about the probability of an event *not* happening. What is the probability of rolling anything other than 7? There are 30 outcomes that do not give 7. Thus, the probability of not rolling 7 is 30/36, which equals 5/6 = .8333 . . . . Therefore, it is likely that we will not roll a 7. How does this answer relate to the probability of rolling a 7? Do you notice an interesting connection between these two probabilities? The probability of rolling a 7 is 6/36 = 1/6 = .1666 . . . . When we add the probability of rolling a 7 to the probability of not rolling a 7, we get exactly 1. Take a few moments to extend this observation into a general principle. Once you have formulated a specific idea of the relationship between the probability that an event will happen and the probability that an event will not happen, continue reading.

When we roll a pair of dice there are 36 equally likely outcomes. Six of these outcomes add up to 7, and the other 30 outcomes add up to something other than 7. Thus, the 36 total outcomes can be divided into the successes (6 outcomes) and the nonsuccesses (30 outcomes). So, the probability of getting a 7 (6/36) plus the probability of not getting a 7 (30/36)

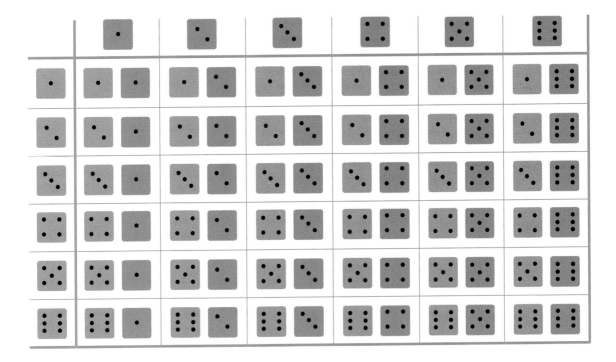

must add up to 1 (36/36). This insight lets us find the probability of an event if we know the probability that the event *doesn't* happen. The probability that an event $E$ does happen is equal to 1 minus the probability that the event $E$ does not happen. This relationship may be stated as:

> **IT EITHER HAPPENS OR IT DOESN'T.** *The probability that the event E does happen = 1 − the probability that the event E does not happen.*

So, the probability of something happening can be found easily if we know the probability that the thing does not happen. Often, as we will discover, computing the probability that something doesn't happen is easier than finding the probability that it does happen.

What is the probability of rolling a pair of dice and seeing a number that is greater than 3? We could count all the entries in the chart where the sum exceeds 3, or we could count the opposite outcomes: Those rolls whose sum is less than or equal to 3. There are only three such outcomes: 1 and 1; 1 and 2; 2 and 1. Therefore, it is easy to see that the probability of *not* rolling a number that *is* greater that 3 is 3/36, and so, by our previous observation, the probability of rolling a number that *is* greater than 3 equals 1 − 3/36, which equals 33/36 = 11/12 = .91666 . . . . This high probability indicates that it is very likely that we will roll a number greater than 3. This example also illustrates the power of looking at a problem in a different way. Looking at a situation in an alternative way may lead to an easy solution.

## When Must a Pair Share a Birthday Cake?

We now return to the seemingly straightforward, harmless question posed in the previous section.

> **THE BIRTHDAY QUESTION.** *How many people are needed in a room so that the probability that there are two people whose birthdays are the exact same day is roughly 1/2?*

Let's make the reasonable assumption that it is equally likely to be born on one day than any other—no one day in particular is more popular for celebrating birthdays. That is, the probability that someone is born on any given day, say December 9, is 1/365, since there are 365 days in the year (let's pretend we never leap) that are all equally likely candidates for one's birthday, and exactly one of them is December 9.

## Starting with Small Crowds

Let's first try the warm-up exercise of explicitly finding the probability that two people share the same birthday. The first person has some birthday, so the question could be rephrased as, What is the probability that the second person has the same birthday as the first person? Well, there are 365 possible days for a birthday, and to have the same birthday, the second person must have the one birthday that the first person has. So, the probability of that happening is 1/365. To check our reasoning, let's again compute the probability, but this time let's use a chart as we did for the dice-rolling experiment.

How many different possible pairs of dates are there for the birthdays of two people? Well, there are 365 possibilities for the first person and 365 possibilities for the second person. We could make a huge chart similar to the two-dice chart, only this one would have 365 rows and 365 columns. Making the chart is hard, but figuring out how many outcomes are represented in the chart is easy: $365 \times 365 = 133{,}225$. Which entries on that chart correspond to the two people having the same birthday? The entries along the diagonal starting at the upper-left corner have both dates the same. There are 365 entries on that diagonal, so the probability of the two people having the same birthday is $365/(365 \times 365)$, which again, happily, is 1/365.

*Looking at a situation in an alternative way may lead to an easy solution.*

## Different Birthdays

It will be useful to consider another method of computing the probability of two people having the same birth date. This time, let's first find the probability of the opposite outcome—the probability that they do *not* share the same birthday. It is this strategy that will enable us to solve the original birthday question.

As before, there are 365 × 365 different pairs of birth dates. How many of these 133,225 possible outcomes produce a pair of different birthdays? There are 365 possible and allowable birth dates for the first person. However, once that person's birthday is known, the second person must avoid that particular date like the plague, thus leaving only 364 possible birth dates to ensure that the two people do not share birthdays. So, for every one of the 365 possible dates for the first person, we have 364 possible dates for the second person (all the dates except the first person's date). This gives a total number of 365 × 364 = 132,860 pairs of dates where the two dates are different. Therefore, the probability of two people having different birthdays is

$$\frac{365 \times 364}{365 \times 365} = \frac{364}{365} = .9972\ldots.$$

So, using our previous relation between an event happening and it not happening, we conclude that the probability that two people have the same birthday is equal to

$$1 - \frac{364}{365} = \frac{1}{365} = .00273\ldots,$$

once again confirming our previous computations.

## A Few More People

So, the probability that two people in a room share a common birthday is extremely low. What if we had three people in the room? What is the probability that two or more of the three share the same birthday? We examine this case while looking to develop a pattern that will work for larger numbers of people.

*Look for patterns.*

Let's again consider the opposite outcome: All three people have different birthdays. Let's first ask, How many possible triples of dates are there for the birthdays of any three people?

There are 365 possibilities for the first person, 365 possibilities for the second person, and 365 possibilities for the third person. Therefore, there must be

$$365 \times 365 \times 365 = 48,627,125$$

possible triples of dates.

How many of these triples have the property that all three dates are different? The first person can have any date, so there are 365 possibilities. The second person can have any date, except for the date of the first person, so the second person has 364 possible dates. The third person has to avoid the dates of both the first and the second person, which leaves 363 possible dates. How many in all? Well, for each of the 365 possible dates for the first person, we have any one of 364 possible dates for the second person, and for each of those combinations, we have 363 possible dates for the third person. Thus, there would be

$$365 \times 364 \times 363 = 48{,}228{,}180$$

possible triple dates where none of the three dates are the same.

So, the probability of having three people with three different birthdays is equal to

$$\frac{(365 \times 364 \times 363)}{(365 \times 365 \times 365)} = .9917\ldots.$$

Therefore, the probability of the opposite (having at least two people out of three share the same birthday) is

$$1 - \frac{(365 \times 364 \times 363)}{(365 \times 365 \times 365)} = 1 - .9917\ldots = .0082\ldots.$$

Although this probability is still extremely small and nowhere near the .5 probability that we seek, we do notice that having a birthday match with three people is about *four times* as likely as with two people. Can you now figure out the probability of a match if we have four people in the room? Try it.

The previous reasoning can be used to show that the probability of having a pair of matched birthdays among four people is equal to

$$1 - \frac{(365 \times 364 \times 363 \times 362)}{(365 \times 365 \times 365 \times 365)} = .01635\ldots,$$

which, although still nowhere near 1/2, is almost twice as large as the probability of finding a match among three people. We can continue to compute the probabilities in this manner. For example, a match among five people would have a probability of

$$1 - \frac{(365 \times 364 \times 363 \times 362 \times 361)}{(365 \times 365 \times 365 \times 365 \times 365)} = .0271\ldots.$$

## *We've Got the Pattern*

We now see the pattern. If we continue for various numbers of people, we could produce the following chart.

| Number of People in the Room | Probability of Two Sharing the Same Birthday |
|---|---|
| 5 | .027 . . . |
| 10 | .116 . . . |
| 15 | .252 . . . |
| 20 | .411 . . . |
| 25 | .568 . . . |
| 30 | .706 . . . |
| 40 | .891 . . . |
| 50 | .970 . . . |
| 60 | .994 . . . |
| 70 | .9991 . . . |
| 80 | .99991 . . . |
| 90 | .999993 . . . |

It is truly surprising how quickly the probability heads toward 1. With only 50 people it is almost a sure thing that there will be a match. With 90 people we are essentially 100% confident of a match; yet 90 is a far cry from 366 people, which guarantees a match for sure. We also have an answer to our Birthday Question: The probability of a birthday match with 23 people is .5072 . . . .

## *Retraining Our Intuition*

If our intuition leads us astray, we need to look at the situation in different ways until not only our reason but also our intuition believes it.

Why is the actual answer of 23 people so much lower than we first guessed? In this case, intuition does not correspond to reality. Before thinking about the Birthday Question, our intuition was probably influenced by some simple yet wrong reasoning. We might have reasoned that, since 366 people are required to guarantee that two people will have the same birthday, then 183 people will be needed to give a .5 probability. Now we see that such reasoning is far from correct. Somehow, to make this birthday principle real to us, we must retrain our intuition.

A helpful technique in retraining the intuition is literally to try the birthday experiment in several gatherings, as we have done, and see that in fact pairs of people will have the same birthday. Another approach is to examine situations similar to the birthday question and discover the answer in the new setting. We could try analogous experiments in other settings, such as with cards, to experience the underlying principles at work. In the Mindscapes we invite you to try several.

The surprising answer to the birthday question illustrates the power of putting quantities together carefully. It also demonstrates that, when dealing with many small events simultaneously, our intuition has difficulty seeing the true story. In the Mindscape called Cool Dice, you will have an opportunity to discover another cool counterintuitive fact that you can share with your friends (and use to get rich). The four funky dice are included in your kit.

Our entire discussion of probability is implicitly based on a concept of *randomness*. What is randomness? Does that mean unpredictable? We will either visit the notion of randomness in the next section . . . or not.

## A LOOK BACK

Probability provides us with a quantitative method to analyze the uncertain and the unknown. It is a measure of the likelihood of an event, such as rolling a 7 with two dice. For an activity (like rolling two dice) with only finitely many equally likely outcomes, the probability of a particular event is the number of different outcomes that would result in that particular event divided by the total number of possible outcomes. So the probability of rolling 7 is $6/36 = 1/6$. Using this basic definition and careful analysis, we can understand many probabilistic situations, some leading to surprising results. Perhaps the most famous and surprising example of unexpected probability is the Birthday Question, whose answer is that, in a group of 23 people, it is slightly more likely than not that two of them have the same birth date.

We develop ideas about probability by starting with familiar situations in which the probabilities are intuitively clear—for example, the simple cases of rolling dice. From those examples we extrapolate the basic idea of probability. We then formulate a specific definition. Finally, we explore consequences of that definition and discover some surprising results.

Carefully analyzing simple and familiar events opens the door for us to understand more complex and puzzling situations. We can more easily see patterns and develop insights from simple and clear examples than we can from complex or muddy examples. So, focusing on the simple and familiar allows us to concentrate on uncovering the essential principles.

Analyze simple things deeply.

Deduce general principles.

Apply them to more complex settings.

# MINDSCAPES   Invitations to Further Thought

## I. SOLIDIFYING IDEAS

1. **Lincoln takes a hit.** Using your data from your banging and spinning experiments with pennies, deduce the direction of the beveling. Why do you get different outcomes when you bang versus when you spin?

2. **Giving orders.** Order the following events in terms of likelihood. Start with the least likely and end with the most likely event.

   - You randomly pick an ace from a regular deck of 52 playing cards.
   - There is a full moon at night.
   - You roll a die and you see a 6.
   - A politician fulfills all campaign promises.
   - You randomly pick the queen of hearts from a regular deck of 52 playing cards.
   - Someone flies safely from Chicago to New York City (their luggage may or may not have been so lucky).
   - You randomly pick a black card from a regular deck of 52 playing cards.

3. **Two heads are better.** Simultaneously flip a dime and a quarter. If you see two tails, ignore that flip. If you see at least one head, record whether you see one or two heads. Repeat this experiment so you have 20 flips having at least one head. Calculate the number of double heads divided by 20. How close is this answer to the computed probability of having two boys in the second Reunion scenario (p. 525) from this section?

4. **Tacky probabilities.** Before doing the following experiment, think a bit and then guess what you think the probability will be and write down your guess. Take five identical-looking, standard thumb tacks. Cup them in your hands, shake them, and then toss them slightly upward and let them fall onto a smooth, tiled floor. Count how many of the tacks land completely on their flat side and how many land resting against their points. Repeat this experiment ten times and use your data to estimate the probability of tossing a thumb tack and having it land resting against its point.

tack landing flat

tack landing against the point

---

*In the Mindscapes section, exercises marked (H) have hints for solutions at the back of the book. Exercises marked (S) have solutions.

5. **BURGER AND STARBIRD.** Suppose you randomly select a letter from BURGER AND STARBIRD. You could imagine writing these letters on Ping-Pong balls—one letter per ball—then putting them all in a barrel and pulling one out. What is the probability of pulling out an R? What is the probability of pulling out a B? What is the probability of pulling out a letter appearing in the first half of the alphabet? What is the probability of pulling out a vowel?

6. **Monty Hall.** Read and rework the Monty Hall scenario from the Fun and Games chapter. Work through the solution. Then find a friend and simulate the Monty Hall situation keeping track of the outcomes under the two possible strategies—the switch strategy and the stick strategy. Perform the experiment perhaps 40 times and record the results. Do the experimental data accord with the analysis of the probabilities?

7. **7 or 11 (S).** What is the probability of rolling two fair dice and seeing a 7 or an 11?

8. **D and D.** You simultaneously flip a dime and roll a die. Make a table of all the possible outcomes. What is the probability of seeing Roosevelt and a 4? Suppose now that someone else flipped and rolled, did not show you the result, but reported that the die shows a 2. What is the probability that the dime is showing tails? Justify your answer.

9. **The top ten.** Suppose you have 10 marbles. They are each marked with one number: 1, 2, 3, 4, 5, 6, 7, 8, 9, 10. They are placed in a jar, and you reach in and pick one out. What is the probability that the number you pick has a factor of 3? What is the probability that the number you pick is a prime number? What is the probability that the number you pick is even? What is the probability that the number you pick is evenly divisible by 13?

10. **One five and dime (H).** Someone simultaneously flips a penny, a nickel, and a dime. Make a list of all the possible outcomes. What is the probability of seeing three presidents? What is the probability of seeing exactly two presidents? Suppose now that you did not see the outcome, but you are told that a president is showing. What now is the probability of seeing three presidents? Suppose instead that you are told that Lincoln is showing. What now is the probability of seeing three presidents? Why is there a difference in your answers?

11. **Five flip.** Someone flips five coins without your seeing the outcome. The person reports that no tails are showing. What is the probability that the person flipped five heads?

12. **Flipped out.** We take a coin and flip it 10,000,000 times (okay, we have a lot of time on our hands). We notice that 6,010,375 times it landed on heads. What do you suspect about the coin?

13. **Spinning wheel.** A roulette wheel has 36 spaces marked from 1 to 36 with half marked red and half marked black, together with two

green spaces marked 0 and 00. What is the probability of having the little ball land on 13? What is the probability of red?

14. **December 9.** Choose two people at random. What is the probability that they were both born on December 9?

15. **High roller** (H). What is the probability of rolling two fair dice and having the sum exceed 4?

16. **Double dice.** You roll two fair dice. What is the probability that you will see a double (1, 1 or 2, 2 or 3, 3, and so on)?

17. **Silly puzzle.** After a professor explains the birthday problem to her class of 20, she points out that the probability of having a birthday match in the class is around .4. A student raises her hand and states that she is certain that there will be a birthday match. She knows no one's birthday except her own. Explain why she was able to correctly state this fact.

18. **Just do it.** Find groups of roughly 35 people together (in a class, dorm, or dining hall) and have each person in turn shout out his or her birth date. Is there more than one pair of matches? Record your results.

19. **No matches** (S). Suppose 40 people are in a room. What is the probability that no two people share the same birthday?

20. **Spinner winner.** You spin the spinner shown. The spinner is equally likely to stop at any particular place. You win if you land on a space that is 6 or higher and lose otherwise. What is the probability of winning?

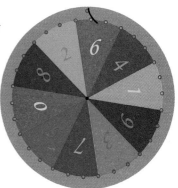

## II. CREATING NEW IDEAS

1. **Flip side** (S). Someone flips three coins behind a screen. The person says, "I flipped at least two heads." What is the probability that the flipper flipped three heads?

2. **Other flip side.** Someone flips three coins behind a screen. The person says, "I didn't flip all tails." What is the probability that the flipper flipped all three heads?

3. **Blackjack.** From a regular deck of 52 playing cards, you turn over a 5 and then a 6. What is the probability that the next card you turn over will be a face card?

4. **Be rational.** Suppose someone has randomly generated two natural numbers and used them to make a fraction. Reduce the fraction to lowest terms. Is there a 0.5 probability that both the numerator and the denominator are odd numbers? Why or why not?

5. **Well red (H).** Someone shows you three cards. One is red on both sides, another is blue on both sides, and the last is red on one side and blue on the other. They are shuffled and you are then shown one side of one card. You see red. What is the probability that the other side is blue? Is it .5? Explain.

6. **Regular dice.** In Dungeons and Dragons, there are dice in the shape of each of the regular solids (see section 4.5). The faces are always numbered 1 through whatever the number of faces there are. You shake all five dice. What is your probability of throwing a total of 6?

7. **Take your seat.** You decide to fly to California on EconoJet Airlines. You are randomly assigned a seat. Seats are numbered by row from 1 to 40 and in each row by A, B, C, or D (amazingly, there is only one window seat in each row). The plane is boarded starting from the rear in groups of 10 rows at a time. What is the probability that you will be in the first group to board the plane? What is the probability that you get a window seat?

8. **Eight flips.** What is the probability of flipping a half dollar eight times and seeing at least one head?

9. **Lottery (S).** The lottery in an extremely small state consists of picking two different numbers from 1 to 10. Ten numbered Ping-Pong balls are dropped in a fish bowl, and two are selected. Suppose you bet on 2 and 9. What is the probability that you match at least one number? What is the probability you match both numbers?

10. **Making the grade.** What is wrong with the following statement? "The way I figure, the probability I get a 4.0 average this term is .2. The probability I get below a 4.0 average this term is .9." Explain. Given the statement, what is your guess as to the person's actual G.P.A.?

### III. FURTHER CHALLENGES

1. **Cool dice.** Put together the four dice from your kit (see the kit instructions). Show that, no matter which die someone picks, there is always another die such that the probability of rolling a higher number on the second die is greater than .5. That is, there is no best die. Order the dice A, B, C, D such that B beats A, C beats B, D beats C, and A beats D. This collection is similar to four sporting teams whereby B generally beats A, C generally beats B, and D generally beats C, but, because D has some weakness that one of A's strengths can take advantage of, A generally beats D. Play this dice game with several friends.

2. **Don't squeeze.** Five shoppers buy Charmin toilet paper. One Charmin out of 10 in this batch is defective—it's unsqueezable. You want to save everyone from this catastrophe, so you stop them at the door and ask to squeeze their Charmin. After squeezing five

rolls, what is the probability that you have located one or more defective Charmins?

3. **Birthday cards.** Take a regular deck of 52 playing cards, pick a card at random, record it, and then put it back in the deck. Shuffle the cards and pick another card at random and record it, put it back, and so on. How many cards do you draw before you pick the same card for the second time? Do this experiment several times. Calculate the probability of choosing 10 times and seeing 10 different cards.

4. **Too many boys.** Long, long ago and far, far away, an emperor believed that there were too, too many males and not enough females. To correct this wrong, the emperor decreed that, as soon as a woman gave birth to a male child, she was not permitted to have any more children. If the woman gave birth to a female, she was allowed to continue to have children. What was the result of this decree? After the decree, what fraction of the babies will be male? Carefully explain your answer.

5. **Three paradox (H).** The probability of tossing three coins and seeing all three the same is 1/4. What's wrong with the following dubious reasoning? When we toss three coins, we know for a fact that two of the coins will be the same, thus we only have to get the third one to match. Thus the probability is 1/2.

## IV. IN YOUR OWN WORDS

1. **Personal perspectives.** Write a short essay describing the most interesting or surprising discovery you made in exploring this section's material. If any material seemed puzzling or even unbelievable, address that as well. Explain why you chose the topics you did. Finally, comment on the aesthetics of the mathematics and ideas in this section.

2. **With a group of folks.** In a small group, discuss and work through the details involved in the answer to the Birthday Question. After your discussion, write a brief narrative describing the reasoning in your own words.

3. **Creative writing.** Write an imaginative story (it can be comical, dramatic, whatever you like) that involves or evokes the ideas of this section.

4. **Power beyond the mathematics.** Provide several real-life situations—ideally, from your own experience—for which some of the strategies of thought presented in this section would provide effective methods for approaching and resolving them.

# 7.3 Random Thoughts:
## Are Coincidences as Truly Amazing as They First Appear?

*How dare we speak of the laws of chance? Is not chance the antithesis of all law?*

—Bertrand Russell

Coincidences are so striking because any particular one is extremely improbable. However, what is even more improbable is that no coincidence will occur. We saw in the birthday scenario that finding two people having the same birthday in a room of 50 is extremely likely, even though the probability of any particular two people having the same birthday is extremely low. If you were one of a pair of people in that room with the same birthday as someone else, you would feel that a surprising coincidence had occurred—as indeed it had. But almost certainly some pair of people in the room would experience that coincidence. Let's now delve into the mysterious world of coincidences.

Coincidences and random happenings easily befuddle our intuition. To expose them for what they are, we must describe them clearly and analyze them quantitatively. Looking at simplified situations will help us to understand whence misleading impressions arise. As usual, we start with concrete examples.

During the great Sammy Sosa–Mark McGwire home-run race of 1998, Mark McGwire tied a home-run record of 61 home runs on his own father's 61st birthday. *What an amazing coincidence!*

### A Deadly Coincidence

We began working on a first draft of this section during a two-week period in late June and early July of 1997. During that time five famous people died—television celebrity Brian Keith (June 24), deep-sea diver Jacques Cousteau (June 25), actor Robert Mitchum (July 1), actor Jimmy Stewart (July 2), and news commentator Charles Kuralt (July 4). As we were writing about randomness, we began to ponder: What are the chances of five famous people dying during those two weeks? Isn't it strange that during the two weeks we were writing a section about coincidence that such a public coincidence actually occurred?

Having noticed this interesting phenomenon, we decided to analyze it and attempt to understand it. Is it strange or not that five famous people died during a two-week period? We know we should be sad, but should we be surprised? Contemplating this question brings up several of the main ideas about randomness, probability, and coincidence. The first idea is the meaning of *strange*. Presumably we mean that the event had a low probability of occurring. But to associate a probability with the event of five famous people dying, we are obliged to specify the total collection of possible occurrences with which the death of the five is to be compared. One possibility is to consider all deaths during that period and ask how likely it is that five of them would be famous.

At least 52 million people died in 1997, which on average is one million deaths per week around the world. Thus, our question might be rephrased in a provocative way: Among the roughly two million people on Earth who die during any two-week period, what is the probability that five of them will be famous? Already this phrasing of the question makes the fact of five famous deaths a little less surprising. But perhaps these five deaths would still be surprising if the total number of famous people is extremely low. How many famous people are there on earth?

We just made some conservative guesses of the number of famous people in different categories: 300 singers; 600 actors; 600 sports figures; 1,200 leaders; 150 scientists; 200 business people; 200 artists and writers; and 750 miscellaneous people. So, let's say there are 4,000 famous people in the world. Nearly all are famous for less than 40 years of their lives, so we estimate that at least 100 must die each year, which amounts to approximately two per week. Therefore, as a matter of fact, our five deaths are pretty much right on target for the average number of deaths of famous people for a two-week period. To check our figures, we looked in a world almanac that listed famous people who died between October 1995 and October 1996. The list consisted of 105 names, so our estimates appear to be reasonably accurate.

Actually, the coincidence of five famous deaths should probably be put in the context of all possible remarkable coincidences that might have happened during that two-week period. We didn't set out to look for a coincidence involving deaths of celebrities. The huge collection of all conceivable coincidences diminishes the significance of the five-death coincidence even further. When looked at in this way, many coincidences lose some of their luster.

Let's now see if we can come to understand the presence of coincidence in our own lives.

## Personal Coincidence

Think of the most amazing coincidences in your life. Perhaps you were walking in the airport in Chicago and ran into an old friend you hadn't seen in 10 years. Perhaps you thought about a car crash, and the very next day, a relative had an automobile accident. Perhaps you and a college classmate independently decided to open a Starbucks rather than go to grad school in math. Perhaps your birthday is the same as a string of numbers in your social security number. Some remarkable coincidences have occurred in your life.

How unlikely are any of these events? The answer is that each one of them is extremely unlikely. However, let's look at the same situations from a different point of view. Let's consider the probability that you will *avoid* remarkable coincidences. Often, to better understand a possibility, it is valuable to consider the opposite possibility.

Each day, suppose you wake up in the morning and think of an event that has a one-in-a-thousand chance of happening that day. Let's compute the probability of not one of those coincidences coming to pass during a year. The first day, you have a 999/1000 probability of not experiencing that coincidence (pretty likely it will not happen). Using the ideas developed in the previous section, we see that your chance of missing rare coincidences both the first and second days is 999/1000 multiplied by 999/1000. Your chance of missing out for any number of days is simply 999/1000 raised to the power of the number of days. Using a calculator and taking 999/1000 to the 365 power, we see that missing out every day for a year has a probability of .69. So, your chance of experiencing one of your one-in-a-thousand coincidences during one year is .31—nearly 1/3 (one in three chances—not that unlikely). During a three-year period, your chance of missing every single day is 999/1000 raised to the 1,095, a mere .33. In other words, the probability that during a three-year period at least once your one-in-a-thousand event will occur is a whopping 2/3 (two in three chances).

The probability that such a one-in-a-thousand event will happen at least once in 10 years is .97, and after 20 years the probability that at least one such unlikely event will happen to you is .9993. In other words, even if we select, in advance, each morning the one-in-a-thousand coincidence that we would count for that day, each of us is almost certain to experience that coincidence from time to time. Of course, in practice we would take note of a one-in-a-thousand coincidence even if it did not happen to be our particular coincidence *du jour*. Therefore, we see that it would be truly remarkable if we never experienced such a coincidence.

Moral: *Coincidence Happens*.

## How to Get Rich Quick as a Stock Whiz

Predicting the future is a feat full of folly. One method of beating the odds is to make many predictions and then declare success if a few of them

> To better understand a possibility, consider the opposite possibility.

Another up and down day on the market.

materialize (and hope that people will just forget the incorrect ones). Another is to cover all bets. Let's take a look at how we can predict the future in the stock market with impressive accuracy, from some people's point of view.

Let's take a list of 1,024 investors and send them a letter on Monday. To 512 investors we write, "IBM stock will go up next week"; to the other 512 we write, "IBM stock will go down next week." The following week, we send 512 letters to the group for whom we were correct and write to 256 of them, "IBM stock will go up next week," and to the other 256 we write, "IBM stock will go down next week." At the end of that week 256 people will begin to pay attention to our ability to predict the future. We continue the pattern. After nine weeks, two people will have seen us predict the future nine times in a row. Now we ask them to send us a large check requesting our next week's prediction. We then send an "up" letter to one, a "down" letter to the other, and, to assuage our conscience, refund the payment for the person whom we misled.

In real life we presume that such a scam is illegal, but it does happen inadvertently in another way. There are thousands of people who predict stock market activity. Some are correct sometimes. Suppose these stock analysts were literally flipping coins to make their predictions. Still, we would expect that someone would have a good track record if enough of them were flipping coins. The moral of this story is beware of investment counsel that says, "This expert correctly predicted the big crash of 1987." That person may well have done so, and that person no doubt believes that the prediction was not based on randomness but was, instead, based on special insight. How are we going to determine whether the truth lies in randomness among a lot of predictors who really don't have any special insight or in the incredible instincts of a few special people? This question is essentially impossible to decide. Or is it? We leave these conundrums for you but do caution you not to take advice blindly.

## Hey, Hey, We're the Monkeys

As we have said, although it is rare to see a particular coincidence within a random event, it is even more improbable to *never* see a coincidence. We further illustrate this idea with perhaps the most famous example of randomness.

Suppose we have a very large number of monkeys, each banging away randomly on his or her own word processor. Their typing is completely random. If we let them type indefinitely, would one of them at some point randomly type out Shakespeare's entire *Hamlet*?

The answer is yes. If we have enough monkeys typing long enough, we are bound to get *Hamlet*. Why? Because if we perform a random event enough times, we would expect to see any possible outcome, no matter

*Don't be shocked if the improbable occurs from time to time.*

✦ ✦ ✦ ✦ ✦ ✦

how unlikely it may be. This result is known as the Infinite Monkey Theorem or, as we like to refer to it, *Hamlet Happens*.

> **THE INFINITE MONKEY THEOREM.** *If we put an army of monkeys at word processors, eventually one will bash out the script for* Hamlet.

This observation was first made by the astronomer Sir Arthur Eddington in 1929 while describing some features of the second law of thermodynamics. He wrote: "If I let my fingers wander idly over the keys of a typewriter it might happen that my screed made an intelligible sentence. If an army of monkeys were strumming on typewriters they might write all the books in the British Museum. The chance of their doing so is decidedly more favourable than the chance of the molecules returning to one half of the vessel."

Although *Hamlet* does happen, it does not happen often. Suppose 1,000,000 monkeys randomly typed at standard 48-key computer keyboards, and each typed one character per second. We would expect to wait more than

1,000,000,000,000,000,000,000,000,000,000,000,000,000,000,000,000

years before they typed "To be or not to be. That is the question."

There are many literary references to this random fact about monkeys and Shakespeare. One is from Douglas Adams's book, *The Hitchhiker's Guide to the Galaxy*: "'Ford!' he said, 'there's an infinite number of monkeys outside who want to talk to us about this script for *Hamlet* they've worked out.'"

Our favorite story in this direction is a 1950s television sketch with Steve Allen, who was the first host of the *Tonight Show*. He portrayed a reporter reporting on a scientific experiment. In the scene we see Allen talking into a microphone, and beside him are a bunch of monkeys typing on typewriters. His words were roughly: "Scientists have claimed that if you have enough monkeys typing randomly for enough time, one of them will eventually produce *Hamlet*. To test this theory, we have brought in a pack of monkeys and have been letting them type away for 72 days now. Let's see how they're doing. [He reaches over and pulls out the paper from one of the typewriters and reads.] 'To be, or not to be, that is the zinfenblatte.' Well, I guess we're not quite there yet—back to the studio." Besides being a classic routine, it does illustrate another important point. If we let those monkeys type away, we will not only see *Hamlet*, but we will see any possible variation of it. Basically, if we have enough monkeys and enough time, we could generate every book ever written—although for this book only two monkeys were required.

This monkey business has more than just mere entertainment value. Recently, in 1993, George Marsaglia and Arif Zaman from Florida State University used this monkey fact to detect errors in computer programs that generate random numbers. Their basic strategy was to convert the numbers to letters and, roughly speaking, determine how likely it would be for the random-number generator to type a particular phrase like "To be or not to be." Building a computer program to generate random numbers is a challenging task. However, as we've seen with the monkeys, within the purely random we occasionally will see familiar patterns. As one may expect, we might have to wait a long, long time to stumble across a desired pattern. Thus, if you do attempt the monkey experiment, don't be discouraged if you have them type away for 500 years and they only produce *Macbeth*.

## *From Needle Droppings to an Approximation of $\pi$*

Do random events ever lead to concrete results? Seems unlikely—after all, they're random. Let's consider the following random experiment. Suppose we have a sheet of lined notebook paper and a needle whose length is equal to the distance between consecutive lines on the paper. We now randomly drop the needle onto the paper. We notice that sometimes the needle crosses one of the lines, and sometimes the needle does not cross any line. What is the probability that the needle will cross a line?

Some needles cross lines—others do not.

This question was first raised and then answered by the 18th-century French scientist Georges Louis Leclerc Comte de Buffon (if you drop his name on the paper, it will definitely hit a line). The surprising answer is that the probability of the needle hitting a line can be computed and is exactly equal to $2/\pi$. Using this fact, Buffon was able to give estimates for $\pi$ by, we kid you not, throwing French bread sticks over his shoulders numerous times onto a tiled floor and counting the number of times the bread sticks crossed the lines between the tiles. Although we are told not to play with food, Buffon's food tossing actually gave birth to an entire realm of mathematics now known as *geometric probability*. Hundreds of years after Buffon tossed his bread sticks, atomic scientists discovered that a similar needle-dropping model seems to accurately predict the chances that a neutron produced by the fission of an atomic nucleus would either be stopped or deflected by another nucleus near it—so, even nature appears to drop needles.

Since the probability of the needle hitting the line is equal to $2/\pi$, we can get a good approximation to $\pi$ by just dropping a needle onto paper many, many times. Here's how: First drop the needle a good number of times and keep track of the number of line hits and the number of total droppings. We know that the relative frequency of line hits, which equals the number of line hits divided by the total number of drops, will be approximately the probability of hitting the line. But the probability of hitting the line is $2/\pi$. Therefore, we can solve for $\pi$ and see that

$$\pi \approx 2 \times \frac{\text{(total number of drops)}}{\text{(the number of line hits)}}.$$

In 1864, by making 1,100 droppings, the English scientist Fox repeated this experiment and used his findings to give an estimate for $\pi$ of 3.1419. Today we can visit a number of Web sites and watch computers simulate this experiment with thousands of virtual droppings. The occurrence of $\pi$ in the needle-dropping experiment shows that even randomness has a rich and precise structure. Our next random journey could be an opening scene from the classic 1960s TV series, *Mission Impossible*.

## *Random Journeys—Mission Impossible?*

"Good morning, Mr. Phelps. The road you are standing on runs east and west and goes on forever in both directions. You are holding a penny. Somewhere (although you don't know where) along the side of the road is a small tape recorder that will self-destruct in five seconds after you play the message (thus it has an extremely limited manufacturer's warranty). Your mission, should you decide to accept it, is to find the tape recorder by walking one block at a time as follows: You flip the penny. If the coin comes up heads, then you walk one block east, if it comes up tails, you walk one block west. You repeat this process indefinitely until you find the recorder. Question: Is this mission possible? As always, if you or any of your I. M. Force is caught or killed (or dies of old age), the Secretary will disavow any knowledge of your actions. Good luck, Jim."

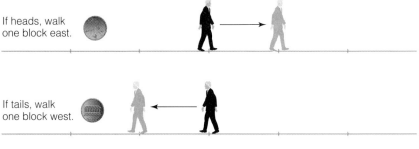

If heads, walk one block east.

If tails, walk one block west.

. . . Repeat!

The preceding walking scenario is an illustration of a *random walk*—a walk whereby the direction we move at any particular stage is selected at random. In this case we are walking one block at a time, and it is equally likely that we walk east or west. For example, suppose our flipping gave a sequence of H H T H H T T T H T H H T T and we started at the marked spot. Draw what our path would look like below.

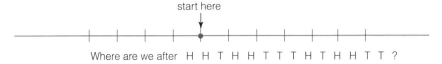

Although we have no idea how far away or even in which direction the poorly made recorder resides, the surprising fact is that the probability of finding the recorder in this manner is equal to 1—we will find it with (probabilistic) certainty. Why? If we flipped the coin indefinitely, we should not be shocked to see long, long, long strings of heads. That string would result in a long, long, long journey eastward. This observation illustrates that it is possible and even reasonable for us to migrate far from our starting position (and similarly, eventually return back home). Although this informal observation certainly does not *prove* that the probability is, in fact, equal to 1, it does make the result plausible.

If we walk randomly on a grid in the plane where we move east, west, north, or south at each step, then the probability of finding the hidden recorder remains equal to 1, but we won't try to justify that fact here. Surprisingly, however, if we consider walking randomly in *three* dimensions (we now have wings and can move up and down), then the probability that we will ever return back to where we start is only about .699 . . . , and we are not at all certain of finding a hidden tape. Why the change in likelihood? Well, as we add dimensions (and thus new directions to randomly journey), we move in a space with greater degrees of freedom. When we have three or more dimensions, it turns out that it is possible for us to get lost in space and never return home.

Random walks actually occur in nature. The path of a liquid or gas molecule is determined by its knocking around and bumping into nearby molecules. This path is an example of a random walk, and such movement is known as *Brownian motion*. The theory of random walks has also been used to study and analyze other phenomena including the behavior of the stock market.

## A LOOK BACK

Coincidences and random behavior do occur and often with predictable frequency. A bit of careful thought reveals that coincidences are not as shocking as they may first appear. One of the most famous illustrations of randomness is the scenario of monkeys randomly typing Hamlet—*Hamlet Happens*. Other examples, such as Buffon's needle, show how random behavior can be used to estimate numbers, such as $\pi$. The theory of random walks is filled with counterintuitive and surprising outcomes and appears in Brownian motion and the stock market.

We can apply the principles of probability to understand coincidences and random behavior. The basic definition of probability and how to compute it allows us to gauge more meaningfully how rare or common a seemingly unlikely event really is. We can estimate the probability of no coincidences occurring. This opposite view helps us to understand how likely coincidences really are.

If an experience or an idea seems surprising or vague, often we do not fully understand it. We can understand it more deeply by looking at the opposite point of view or by analyzing it using familiar principles. Often such analyses not only solve the mystery but also lead to deeper insights into related issues.

*Don't be shocked if the improbable occurs from time to time.*

*Apply reason to understand the mysterious or the unknown.*

*Consider all points of view.*

# MINDSCAPES  Invitations to Further Thought*

### I. SOLIDIFYING IDEAS

1. **Pick a number.** Pick a number: 1 2 3 4. Write it down. In a class of 50 students, how many would you guess have selected the same number as you did?

2. **Personal coincidences.** List three coincidences that you experienced in your life.

3. **No way.** It is the last Sunday of spring break and you are flying back to school. You have a connecting flight in Chicago. In the gate area, you see a bunch of friends from school. Are you shocked? Discuss this coincidence.

4. **Enquiring minds.** For a previous year find the end-of-the-year issue of the *National Enquirer* or a similar fine publication and find all the predictions made by the psychics. How many of them actually happened? Based on your investigation, can psychics predict the future?

5. **Milestones.** Take a look at the obituary sections of two consecutive weeks of either *Time* or *Newsweek* magazine. How many famous people died? Use your answer to make your own estimate of the number of famous people who die in one year.

6. **Local mortality.** By looking at one issue of a local daily paper, estimate how many people in your city will die during the next two weeks.

---

*In the Mindscapes section, exercises marked (H) have hints for solutions at the back of the book. Exercises marked (S) have solutions.

7. ***Unlucky numbers.*** Suppose you randomly picked 1,000 people from the telephone book. What would you estimate the probability to be that one of them will die within the next year? Justify your estimate.

8. ***A bad block*** (S). Suppose 1,054 people died in Datasville last year. Why must there be a two-week period during which 40 of those people died?

9. ***Eerie.*** What is the probability that two celebrities who died in the exact same year actually have the exact same birthday?

10. ***Murphy's Law.*** *If something can go wrong, it will.* Given our discussion of randomness, do you agree or disagree with this law? Using examples, try to place it in the context of our discussion of coincidence.

11. ***A striking deal.*** Get two decks of ordinary playing cards, give one deck to a friend and have the person arrange it in any order, either random or planned. Bet the person that you will be able to take the other deck and place it in an order so that, if you both turn over the cards one at a time, at least once both cards will come up exactly the same. Shuffle your deck randomly. Now turn one card from each deck over and continue until there is a match. If there is a match, you win, otherwise you lose. Play the game six or more times and record the results. Given your data, what would you guess to be the probability that you win?

12. ***Drop the needle.*** Try the Buffon's needle experiment by dropping a needle 50 times and recording your results. Use your numbers to give an approximation for $\pi$. Search the Web for a site that will allow you to simulate the Buffon's needle experiment and try the computer simulation. Record the results and the associated approximations to $\pi$.

13. ***IBM again*** (H). Suppose we try the IBM stock-prediction scheme of sending half (or as close to half as possible) of the group the "up" message and the other half the "down" message, but this time we start with only 600 people. How many weeks could we go until we are down to just two people seeing a perfect track record?

14. ***The dart index.*** Take a page of stock quotes from a newspaper. Mount it on a piece of cardboard and throw 10 darts randomly at the page. Record the 10 stocks you hit. Now get a newspaper that is exactly six months old and look up the prices of those 10 stocks. Suppose you bought 100 shares of each of those 10 stocks six months ago. How would you do if you sold them all today?

15. ***Random walks.*** Using a piece of graph paper, a penny, and a dime, embark on a random walk on the grid. Flip both coins. If the penny lands heads up, you move one unit to the right, if it lands tails up, you move one unit to the left. If the dime is heads, you move one unit up, and if it is tails you move one unit down. Take 50 steps and mark your trail on the graph paper.

16. *Random guesses* (S). A multiple-choice test has 100 questions; each question has four possible answers from which to choose. Each question is worth one point, and there are no points taken off if the answer is incorrect. Someone decides to take the test by picking answers randomly. What is the probability that this person gets 100%? Suppose now that the person actually reads the questions, is correctly able to eliminate two of the incorrect choices, and then guesses randomly from the other two choices. What is the probability that the person gets 100%?

17. *Random dates.* There is a room filled with exotic people, any one of whom you would be happy to ask out on a date. Suppose that the probability that any particular person agrees to go on a date with you is .5. What is the probability that the first person you ask says yes? What is the probability that the first five people you ask say no?

18. *Random phones* (H). Suppose you roll a 10-sided die with sides marked from 0 to 9. If you kept rolling and recorded the outcomes, what do you think the probability is that at some point you will see seven digits in a row that make up your telephone number? Explain your thinking.

19. *All sixes.* Suppose you roll a die repeatedly, forever. Is it possible that you would roll only 6's? Is it likely? What is the probability of rolling only 6's?

20. *Pick a number, revisited.* In the first question, did you pick 3? Are you impressed? What is the probability that we correctly guessed your answer?

## II. CREATING NEW IDEAS

1. *Good start* (H). Suppose the monkey is typing using only the 26 letter keys. What is the probability that the monkey will type "cat" right off the bat?

2. *Even moves.* Suppose you embark on a random walk on the real number line. Show that, if you return back to your starting point, you must have made an even number of coin flips.

3. *Playing the numbers.* Here is a numbers game. You choose a number from 000 to 999 each morning and compare it to the last three digits of the official attendance figures at the nearest racetrack. What is the probability of guessing the correct number at least once if you make a guess each day for three years? (*Hint*: Consider a related example from this section.)

4. *Random results.* Someone looks at a list of 10,000 numbers, each from 1 to 100, generated by a random-number generator and states that the program must not work correctly because there is a string of 17 numbers starting with the 2,713th number that reads 1, 2, 3, 4, 5, 6, 7, 8, 9, 10, 11, 12, 13, 14, 15, 16, 17. How do you respond to this conclusion?

5. **Monkey names.** Suppose we have a monkey typing on a word processor that has only 26 keys (no spaces, numbers, punctuation, and so on). The monkey types randomly for a long, long time. What is more likely to be seen: MICHAELSTARBIRD or EDWARDBURGER? Explain your answer.

6. **The streak.** Suppose you flip a fair coin 10 times on two different occasions. One time you see 10 heads, the other time you see H H T H H H T T H T. Is either one of these outcomes more likely than the other? Which one is random? Explain.

7. **Girl, Girl, . . . (S).** A couple has eight children. Suppose that the probability of having a girl is .5. What is the probability of producing eight girls? How does that answer compare with the probability of producing boy, girl, boy, girl, boy, girl, boy, girl? Explain. (*Note:* This does happen.)

What an amazing coincidence. . . . All the daughters have the same outfits!

8. **One mistake is okay.** Suppose we try the IBM stock-prediction scheme with only 128 people. Now, however, we are allowed one mistake. That is, if we send a batch of letters out saying that IBM stock will go down next week, and it actually goes up, we can keep sending letters to this group as long as we never make another mistake. After five weeks, how many people will have seen you make at most one wrong prediction?

9. **Picking and matching.** You and a friend individually and secretly pick a number from 1 2 3 4. What is the probability that you both picked 3? What is the probability that you both picked the same number?

10. **Picking and matching.** You and a friend individually and secretly pick a number from 1 2 3 4. What is the probability that you both picked 3? What is the probability that you both picked the same number? Why is this question here? Think about what this section is about.

## III. FURTHER CHALLENGES

1. ***Death row*** **(H).** You may have noticed that two pairs of the celebrity deaths we mentioned occurred one day after the other (Brian Keith on June 24 and Jacques Cousteau the next day; Robert Mitchum on July 1 and Jimmy Stewart the next day). Suppose that two celebrities die one week (Monday through Sunday). What is the probability that they die on the exact same day? What is the probability that they die one day after the other in that same week?

2. ***Striking again.*** Consider the Striking Deal game described in question I.11. Compute the actual probability that you will win. (*Hint*: It might be easier to first compute the probability that your opponent will win.)

3. ***Random returns.*** Suppose we journey on a random walk on the line. Show that, with probability 1, we walk past every point on the street *arbitrarily* often. (*Hint*: Once you land on a particular point, imagine that you are starting a random walk from scratch.)

4. ***Random natural.*** Suppose you have a 10-sided die with sides labeled from 0 to 9. You roll it 50 times and record the digits to create a 50-digit natural number. What is the probability that the digit 9 occurs at least once in your random 50-digit number? What do you conclude about the digits of very large random natural numbers?

5. ***Ace of spades.*** You randomly shuffle a deck of cards and then look at the first card. If it is the ace of spades, you win; if not, then you lose. What is the probability that you will win after 36 tries?

## IV. IN YOUR OWN WORDS

1. ***Personal perspectives.*** Write a short essay describing the most interesting or surprising discovery you made in exploring this section's material. If any material seemed puzzling or even unbelievable, address that as well. Explain why you chose the topics you did. Finally, comment on the aesthetics of the mathematics and ideas in this section.

2. ***With a group of folks.*** In a small group, discuss the ideas of coincidence and the Infinite Monkey Theorem. After your discussion, write a brief narrative describing these ideas in your own words.

3. ***Creative writing.*** Write an imaginative story (it can be comical, dramatic, whatever you like) that involves or evokes the ideas of this section.

4. ***Power beyond the mathematics.*** Provide several real-life situations—ideally, from your own experience—for which some of the strategies of thought presented in this section would provide effective methods for approaching and resolving them.

# 7.4 Down for the Count:
## Systematically Counting All Possible Outcomes

*No priest or soothsayer that ever lived could hold his own against old probabilities.*

—Oliver Wendell Holmes

Did you win?

Let's count. To determine probabilities, we often have to do some serious counting. Since probability is a fraction of the number of favorable outcomes divided by the total number of all possible outcomes, we must count how many outcomes are involved, and that's not always so easy. Perhaps you are thinking that you learned how to count in kindergarten, so you'll just skip this section. It's true that, when we were children, we learned to count one at a time, and that is a simple task for small collections of objects. But when we count big, complicated collections, such as how many lottery outcomes are possible, we find ourselves perplexed and prone to error.

In principle, counting a collection of objects is easy. All we must do is

- count every object, and
- not count the same object more than once.

These two rules sound so easy and obvious that we might think that counting is a piece of cake, but let's list a few things to count that might convince us that counting is not, in fact, a dessert item.

How many lock combinations are possible in a standard padlock? How many passwords are possible for your e-mail account? How many different poker hands are there? Experienced counters tend to group different counting scenarios into various categories,

Darn!

but, unless we are intending to become professional counters (known as *combinatorists*), dividing counting problems into a taxonomy of types can be a perilous enterprise. The perils center around the possibility of applying an incorrect counting method to a particular situation. Instead, let's explore some principles of analyzing questions on counting so that we can correctly analyze whatever case is at hand.

### A Truly Merry Festival

In Florida, gambling is allowed on jai alai games. Jai alai (meaning *merry festival* in the Basque language) is the world's fastest ball game. The players use scoops (called *cestas*) to catch and throw the ball in one motion. During the day, six games are played in a type of round-robin play with eight players. One can make a "super 6" bet for $2.00 stating who will win each of the six contests. Of course, the chance of anyone winning such a bet is rather small since one has to be correct on all six games. Thus, this bet is run like the lottery in that money that is bet is carried over into the jackpot day after day until there is finally a winner.

On March 2, 1988, the pool of money at West Palm Beach Jai Alai reached an enormous amount, so one wealthy gambler decided not to gamble. He simply bet on every possible outcome, thus assuring himself of the prize money. Of course, there was the slight danger that some other person would also happen to win that day, in which case he would have to share his bounty.

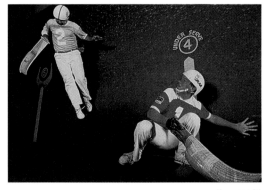

I've got it!

Our questions are as follows.

- If each bet costs $2.00, how big would the prize need to be to make it worthwhile to place every possible bet?
- How many bets would we have to make to be absolutely guaranteed of winning?

These questions present us with our first real counting issue. We have to pick the exact winner from each of the six contests. Naturally, we will have to guess all eight players as potential winners of the first game, because any one of them might win. So, for each of those we must choose all eight of the players of the second game. For each of those 64 patterns of potential winners in the first two games, we will have to choose all eight potential winners of the third game. So, there are $8 \times 8 \times 8$ potential sets of winners from the first three games. We see the pattern and conclude that there are $8^6$ different possible sets of winners for the six games, for a total of 262,144 possible bets costing a total of $524,288.

If the prize exceeds that amount, then we are sure to win. In the case of the West Palm Beach game in March of 1988, the payoff was $988,326. Unfortunately, if someone else also wins, the prize must be shared. This possibility of having to share the prize is what prevents wealthy people from actually buying every combination in lottery games whenever the prize exceeds the number of combinations. Despite this nagging possibility, this method of making money by covering all the possibilities has been used throughout history. In 1729 the French writer François Voltaire used such a method to win the Parisian city lottery, and more recently, in 1992, a group of Australian investors essentially cornered the market on Virginia lottery tickets, winning about twenty-seven million dollars.

## *Lottery*

Let's dream about hitting the jackpot in the lottery and spending the remainder of our lives basking in the lap of luxury. In a typical lottery, we pay a dollar and choose six different numbers, each from 1 to 50; the order of the six numbers does not matter. If we guess all six, we win the big jackpot. How high would the lottery prize have to get before it would be worth our while to buy a ticket for each possible outcome, thus assuring ourselves the title of "winner"? Or, equivalently, how many collections of six different numbers from 1 to 50 are there?

Choosing six numbers from 50 is far too large a task to think about yet, so let's first examine a simpler task. Remember: When the going gets tough, the smart stop going and instead do something easy.

Rather than choosing six numbers from 50, let's figure out how many ways there are of choosing two numbers from 5. We could first list all pairs of numbers in all possible orders by writing down the pairs whose first number is 1, then 2, then 3, and so on.

*When faced with a difficult challenge, it's wise to begin by considering related challenges that we can actually meet.*

❖ ❖ ❖ ❖ ❖ ❖

| 1,2 | 1,3 | 1,4 | 1,5 |
| 2,1 | 2,3 | 2,4 | 2,5 |
| 3,1 | 3,2 | 3,4 | 3,5 |
| 4,1 | 4,2 | 4,3 | 4,5 |
| 5,1 | 5,2 | 5,3 | 5,4 |

It is easy to see how many of these ordered pairs we have, since they form a rectangle. There are five rows with four in each row for a total of 20.

Notice that every pair of numbers—for example, 2,4—occurs twice, once as 2,4 and once as 4,2. So, we have systematically counted each pair of numbers twice instead of once. To get the actual number of pairs, we need to take our 20 ordered pairs and divide by 2 (the number of times each pair appears) to give the number of unordered pairs, namely, 10. We can now use a similar type of analysis to figure out how many different ways we can select six different numbers, each from 1 to 50: We first count all *ordered* collections of six numbers and then divide by how many times each unordered collection was counted.

The number of different ways of selecting six different numbers from 1 to 50 is

$$\frac{50 \times 49 \times 48 \times 47 \times 46 \times 45}{6 \times 5 \times 4 \times 3 \times 2 \times 1},$$

which equals 15,890,700.

Why? We can choose any of the 50 numbers as the first number. For each first choice, we can choose any of the remaining 49 numbers second. So, we have $50 \times 49$ ways to choose the first two numbers. Likewise, we have $50 \times 49 \times 48 \times 47 \times 46 \times 45$ ways of choosing six numbers in specific orders.

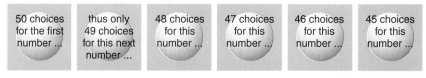

So, the total number of possible ways of choosing six numbers from 50 if order matters =
50 × 49 × 48 × 47 × 46 × 45.

Since we counted the number of ways of selecting six numbers where order matters, we counted

2, 5, 4, 16, 27, 45;

5, 2, 4, 16, 27, 45;

2, 4, 5, 45, 27, 16; and so on,

separately since their orderings are different even though each ordering involves the same collection of six numbers; therefore, we did some major overcounting. How many ways can we order these six numbers?

There are $6 \times 5 \times 4 \times 3 \times 2 \times 1$ ways to order those six numbers: Any of the six could be the first number, so there are six different possibilities for the first number. Once the first number is determined, any of the remaining five could be the second. Thus, for each choice of first number, there are five different possible choices for the second number. Similarly, there are four different choices for the third number, and so forth. Thus we have $6 \times 5 \times 4 \times 3 \times 2 \times 1$ many different orderings of each group of six numbers. So, since each group of six numbers is counted exactly $6 \times 5 \times 4 \times 3 \times 2 \times 1$ times in our count of $50 \times 49 \times 48 \times 47 \times 46 \times 45$, we simply divide

$50 \times 49 \times 48 \times 47 \times 46 \times 45$ by $6 \times 5 \times 4 \times 3 \times 2 \times 1$

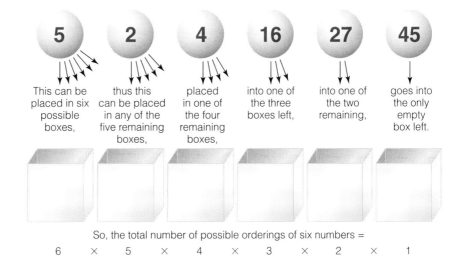

to arrive at the number of different six number groupings selected from 1 to 50.

## Generalized Counting

We can extend the calculations we just made to a more general situation. Suppose we have a total of $T$ objects and we wish to select $S$ of them. How many different ways can we do that? We see that the answer is that there are

$$\frac{T \times (T-1) \times (T-2) \times \ldots \times (T-(S-1))}{(S \times (S-1) \times (S-2) \times \ldots \times 2 \times 1)}$$

different ways of selecting $S$ objects from $T$ things where the order doesn't matter.

Calculators and computers have no difficulty counting these unwieldy collections. If from a set of 50 numbers we want to know how many different ways there are of choosing 6 where order does not matter, in real life we simply find the combination key on our calculator, enter the numbers 50 and 6, and we have the answer. The trick is to understand thoroughly when we are seeking the number of unordered sets of 6 from a set of 50 and when we really want something else. In this case, the calculator would report 15,890,700, and so, if the Lotto jackpot exceeds $15,890,700, we will make money if we buy every single possible combination of six numbers—assuming that we don't share the prize.

## Dealing with Cards

Some of the most challenging counting questions occur in dealing with playing cards. Let's use a standard 52-card deck of playing cards and consider, for example, five-card poker hands. First of all, how many different five-card hands could we be dealt? The answer is the number of ways of picking five things from that group of 52 things. We just figured out how to compute such a count (notice that order does not matter). The answer is (52 × 51 × 50 × 49 × 48)/(5 × 4 × 3 × 2 × 1), which is 2,598,960. Among all those hands, we may now ask how many are of the various poker types, such as four-of-a-kind, flush, straight, full house, and so forth. Many of these hands are not so easy to count—especially if we don't know how to play poker.

To count how many four-of-a-kind hands there are, let's first ask: How many hands have four aces? This question is not too difficult, because after we have been told that there are four aces, there can be only one additional (non-ace) card in the five-card hand. That card could be any one of the remaining 48 cards (52 minus the four aces). So, there are 48 poker hands that have four aces. There are, of course, the same number of poker hands that contain four kings, or four queens, and so on. So, the total number of hands that contain four-of-a-kind is 48 × 13. If we are dealt a five-card hand, the probability of getting four-of-a-kind is (48 × 13)/2,598,960, or .00024—not likely.

Such a hand is rare indeed. Let's now see how it compares to a straight flush. A straight flush is a run of five consecutive cards all of the same suit. For example, 5, 6, 7, 8, 9 all of diamonds and 8, 9, 10, J, Q all of spades are both straight flushes. How many possible straight flushes are there? The answer requires some careful counting. Let's start with one suit, say spades. How many straight flushes are there in spades? Well, we could have the 2, 3, 4, 5, 6 or the 3, 4, 5, 6, 7, and so forth until 10, J, Q, K, A. (We assume that the ace can be used only as the highest card.) That is a total of nine straight flushes in spades—the same for hearts, diamonds, or clubs. So the total number of straight flushes possible is 9 × 4. Therefore, a straight flush is rarer than four-of-a-kind. The probability of being dealt a straight flush is (9 × 4)/2,598,960, or .000014—which makes getting four-of-a-kind look pretty easy. Counting the number of hands of each type is the method used to determine which hands beat which other hands. So, the counting we have done here shows why a straight flush beats four-of-a-kind.

## "Or" and/or "And"

Suppose we have a die and a coin. We are going to throw them both. Here are three questions.

- What is the probability we will roll a 6 on the die **and** flip a heads on the coin?
- What is the probability we will roll a 6 on the die **or** flip a heads on the coin?
- Which of these two outcomes is more likely?

The first two questions are similar and bring up a significant feature of counting. As always, when we want to compute a probability, we must know how many total outcomes are possible, and then we need to know how many of those are the outcomes we desire. In this case, for every one of the six equally likely outcomes from rolling the die, there are two possibilities for the coin. Therefore, there are a total of 12 outcomes altogether.

|   | 1 | 2 | 3 | 4 | 5 | 6 |
|---|---|---|---|---|---|---|
| H | H, 1 | H, 2 | H, 3 | H, 4 | H, 5 | H, 6 |
| T | T, 1 | T, 2 | T, 3 | T, 4 | T, 5 | T, 6 |

Now let's count how many outcomes have a 6 on the die **and** a heads up on the coin. Well, only one. So, the probability of rolling a 6 on the die and flipping a heads on the coin is 1/12. Notice that the probability of rolling a 6, 1/6, multiplied by the probability of flipping a heads, 1/2, yields the answer of 1/12. If we wish to compute the probability that one event happens **and** simultaneously another unrelated event also happens, we need only multiply the individual probabilities together. Why multiply? Because the number of outcomes of the two events is obtained by multiplying the number of outcomes for the first times the number of outcomes for the second.

|   | 1 | 2 | 3 | 4 | 5 | 6 |
|---|---|---|---|---|---|---|
| H | H, 1 | H, 2 | H, 3 | H, 4 | H, 5 | **H, 6** |
| T | T, 1 | T, 2 | T, 3 | T, 4 | T, 5 | T, 6 |

Heads **and** 6: 1 out of 12

**PROBABILITY OF TWO EVENTS BOTH OCCURRING.** *The probability of one thing happening **and** some unrelated thing happening is equal to the **product** of the individual probabilities.*

Now let's count how many of the outcomes have a 6 on the die **or** a heads on the coin. Well, we could have a 6 and either a heads or a tails on the coin. So, that's two ways. We could also have a heads on the coin and *any* number 1, 2, 3, 4, 5, or 6 on the die. That sounds like 6 more, but we must avoid the double counting of a 6 and a heads. The total of number of outcomes that have a 6 **or** a heads is 7. So, the probability of getting a 6 **or** a heads is 7/12.

|   | 1 | 2 | 3 | 4 | 5 | 6 |
|---|---|---|---|---|---|---|
| H | **H, 1** | **H, 2** | **H, 3** | **H, 4** | **H, 5** | **H, 6** |
| T | T, 1 | T, 2 | T, 3 | T, 4 | T, 5 | **T, 6** |

Heads **or** 6: 7 out of 12

Frequently in computing the probability of getting something **or** something else, it is more convenient to compute the probability of the opposite scenario. In the preceding case, for example, we were asked to consider the possibility of getting a 6 or a heads. The alternative is that we do not roll a 6 **and** we do not flip a heads. This question is easier, because we just saw how to find the probabilities of two such events that happen simultaneously: We multiply the individual probabilities. What is the probability of not rolling a 6 on the die? There are five out of the six possibilities that lead to success, so the probability is 5/6. What is the probability of not flipping a heads? We know that is 1/2. So, the probability of avoiding both a 6 **and** a heads is the product (5/6) × (1/2), which equals 5/12. We can now deduce that the probability of getting a 6 **or** a heads is 1 − 5/12, which equals 7/12, as we saw before. So, the opposite of an event where *something happens* **or** *something else happens* is that *the first does not happen* **and** *the second does not happen*. With careful counting and multiplying probabilities, we can now find probabilities of several events happening together. Let's see these ideas in action.

## Dicey Issues

Recall that the seminal counting questions posed by our French nobleman concerned the frequency with which a six would appear in four rolls of a die. Let's count how many different ways a die can be thrown four times and then count how many of those outcomes contain a six. These two numbers allow us to compute the probability of throwing a six in four rolls of a die. Let's count.

1. Any of the six numbers could come up first.
2. For each of the six possible first numbers, any of the six numbers could come up second.
3. For each of the 36 possible first and second rolls, any of the six numbers could come up third.
4. For each of the 36 × 6 = 216 possible first, second, and third rolls, any of the six numbers could come up fourth.

So, altogether there are

$6^4 = 216 \times 6 = 1{,}296$

equally likely possible outcomes of rolling a die four times.

How many of those contain a 6? Well, the 6 could be in the first, **or** the second, **or** the third, **or** the fourth place. But beware: We must not double count the outcomes of rolling two 6's in a row. We might also roll three 6's, and so on, so we have to be careful not to double, triple, and whatever count those outcomes involving 6's. Looking at the counting question in this way is sufficiently complicated and perplexing that it is best to try to think of an alternative approach.

*If the going gets tough, do something else.*

In this case, a little thought saves a great deal of toil. Instead of thinking about what's there, let's think about what's not. Often it is best to count what's missing rather than what's present. In this case, we want to know how many of the 1,296 outcomes include a 6. Why don't we think about the alternative? Namely, how many outcomes do not contain *any* 6's? That is, let's count how many outcomes involve only the numbers 1 through 5. Well, that is pretty easy once we realize that we just answered that same kind of question. There are five possibilities for the first roll, five for the next, five for the third, and five for the fourth. So, there are $5 \times 5 \times 5 \times 5$ outcomes altogether that involve only the numbers 1 to 5. That is a grand total of 625.

Therefore, of the 1,296 possible outcomes of rolling a die four times, 625 do not involve a 6, and therefore the rest ($1,296 - 625 = 671$) must involve a 6 somewhere. Therefore, the probability of rolling at least one 6 among the four rolls is 671/1296, which equals .5177 . . . . Since this probability is greater than one half, a person betting on rolling a 6 in four rolls will tend to win slightly more than lose.

## *Dicier Issues*

When the French gambler tried his game of rolling pairs of dice 24 times and asked whether there would be a pair of 6's, his winning streak evaporated. Why? He may have thought that, since 24 is 2/3 of 36 (the number of possible outcomes when throwing a pair of dice) and 4 is 2/3 of 6 (the number of possible outcomes when throwing one die), his winning percentage should be the same. Let's see why his winning streak came to a screeching halt.

Our whole goal is to carefully compute the number of possible outcomes of rolling two dice 24 times and seeing how many equally likely outcomes there are. With any one roll there are 36 outcomes. So, let's get rid of the red herring of two dice and instead concentrate on the fact that one of 36 outcomes are possible in each of the 24 repetitions. By the same reasoning as before, there are then a total of 36 to the 24th power outcomes, or about $2.245 \times 10^{37}$—which is one heck of a big number. We now need to know how often a particular outcome (a pair of 6's) from the 36 possible outcomes will occur at least once among the 24 trials. Since counting such a thing is difficult, we will instead count the number of outcomes that avoid a pair of 6's. That is to say, in how many ways can each of the 24 rolls result in one of the 35 outcomes other than a pair of 6's? Again, that number is just 35 to the 24th power, which is about $1.142 \times 10^{37}$. So, the probability of **not** getting a pair of 6's in 24 rolls is approximately

$$1.142 \times 10^{37}$$

divided by

$$2.245 \times 10^{37}.$$

That fraction equals

$$\frac{1142}{2245},$$

which is

$$.508\ldots,$$

and so the probability of actually rolling a pair of 6's is

$$1 - .508\ldots = .491\ldots.$$

Thus, we are slightly more apt **not** to roll a pair of 6's during 24 rolls than we are to roll a pair of 6's, and our French *bon homme*, sadly, is a loser.

## *From Dice to DNA*

With cap pulled down and gloves pulled up, the murderer steals into the enclosed garden and confronts his victims. There is a blood-curdling scream as the murderer slashes the throat of his first victim. In a desperate and vain attempt to save his life, the second victim wrestles with the murderer, inflicting a slight wound while receiving his own mortal cut. A few drops of the murderer's blood fall from the wound onto the carnage as he flees the gruesome scene.

Soon the police arrive and take blood samples from the scene. Most of the blood belongs to the victims, but a small portion belongs to the murderer. Those drops of blood might prove the culprit guilty beyond any reasonable doubt, or they might not. Let's investigate.

DNA is the genetic material contained in each of our cells. All the cells of an individual have identical DNA; however, the DNA from one person differs from the DNA of another. DNA is composed of genes, each of which can be present in one of several forms, known as *alleles*. Human DNA has tens of thousands, of genes, and each gene has several possible forms in which it might appear. By looking at the distinct alleles of the various genes, we will see how DNA can incriminate or exonerate a defendant.

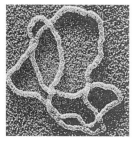

A strand of DNA, but whose is it?

Suppose we examine 20 genes from a DNA sample, and each of these genes can occur as two different alleles. Let's further suppose that those two alleles are roughly equally present among the human population and that the choices of allele of these 20 genes are independent of one another. In other words, having one form of one gene does not make it more or less likely to have any particular form of any of the other genes. Let's try now to compute the probability that the 20 genes from a random DNA sample would have the same alleles as those in the murderer's DNA.

Each of the 20 genes can be present as one of two alleles. So, the first gene could be in the first or second form. For each of those possibilities, the second gene could be in its first or second form. So, there are four possible patterns for the first two genes. The third gene could also appear as either of two alleles. Thus, we observe that for 20 genes, there will be $2^{20}$—which equals 1,048,576—many possible patterns for those 20 genes. So, the chance of another person having all 20 of those genes is 1 in 1,048,576.

DNA evidence is potent. How could a lawyer refute such incriminating evidence? The only possible way to refute such evidence is to recognize that the test is only as valid as the accuracy with which it is performed. If samples are switched, if the defendant's blood contaminates the blood from the murder scene, or if the samples are mislabeled, then the reported match may be completely unreliable. Thus, it could be argued that police and lab error, in a sense, wipe out the accuracy of DNA tests.

## *Counting Tigers*

We have encountered some pretty scary counting, but perhaps none as dangerous as counting tigers in the wild. Suppose someone forced us to go to the wilderness and count the tigers. How would we begin? We might ask the tigers to line up neatly for us to count, but a better method would be to think about probability.

Before facing the tigers, let's think about a related, less threatening counting problem. Suppose we have a large jar of jawbreakers and we want to know approximately how many there are. We do not have time to count them all, so how can we estimate the number? Even this problem seems too hard and, to some, perhaps too threatening, so let's make it easier still. Suppose we were told that in the jar of jawbreakers, exactly 50 of them were green and the green ones are all mixed in with others. Could we use that information to help us count the total number?

Let's reach in and start taking the jawbreakers out of the jar. Suppose we take out 40 jawbreakers, look at them, and see that 20 are green. What could we conclude about the total number of jawbreakers in the jar? There are 50 green jawbreakers in the jar. When we randomly selected 40, half were green. If that proportion holds up, then roughly half of the jawbreakers are green. Since there are 50 green jawbreakers, there must be a total of roughly 100 jawbreakers in the jar.

Now back to the jungle. Remember that we must count the tigers. Unfortunately, we do not know that 50 are green. In fact, none are green—they all are striped. So what will we do? We'll create a situation that we understand. We now understand how to estimate an entire population if we know how many are green. So let's see whether we can make some number of tigers green. They don't start green, so we'll have to capture them first. Suppose we capture some number of tigers, say 50, and mark each one by putting a tag on its ear (probably green) indicating it was one of the captured tigers. Next we let them all go, wait a few months, then recapture some number of tigers, and see what proportion of them are tagged. Suppose we capture 30 tigers and 10 are tagged. Assuming that we have picked a random collection of tigers, we can guess that the tagged tigers represent about a third of the total tiger population. Since there are 50 tagged tigers, we deduce that there must be approximately 150 total tigers in the population.

Of course, there are several assumptions in this process that need to be considered. For example, are we really capturing a random set of the tigers? Perhaps we are capturing only the dumb ones. In which case, all we can conclude is that the population of tigers dumb enough to be captured is about 150—an important piece of information nonetheless.

## A LOOK BACK

Counting can be hard. Careful counting can lead one to better understand probabilities and perhaps even to break the bank at Monte Carlo. In fact, if a casino catches an individual *successfully* counting cards at blackjack, that individual is banned from the casino. However, casinos generally encourage counting, because most people who try counting foul up and lose more money than ever. Counting helps us determine our chances for winning the lottery, being dealt a royal flush in poker, having the same genetic profile as a murderer, and counting tigers in the wild.

In all cases, careful counting consists of two challenges: counting every item and not counting anything twice. To deal with difficult counting challenges, we start with simpler versions of the same thing. Instead of choosing six numbers each from 1 to 50 for the state lottery, we start by assuming we are in a very small state where we choose two numbers from 1 to 5. From such small examples, we gain experience and see patterns that allow us to deal with the more difficult cases. Sometimes a useful strategy is to count the opposite of what we want. That is, instead of counting how many rolls of a pair of dice contain a 6 or a 3, we can count the number of rolls that are made only of other numbers. Then we subtract from the total. Systematic organization is helpful for counting.

Complicated questions are often best approached by thinking hard about simple examples of similar situations. From the simpler situations we can more easily deduce patterns and methods that let us deal with the more complicated settings. Systematically organizing our thinking empowers us to deal with big issues using the same ideas that work for small ones.

Look at simple cases.

Find patterns and methods.

Apply the methods systematically to more complex cases.

◆ ◆ ◆ ◆ ◆ ◆ ◆ ◆ ◆ ◆ ◆ ◆ ◆ ◆ ◆ ◆ ◆

# MINDSCAPES  Invitations to Further Thought*

## I. SOLIDIFYING IDEAS

1. *The gym lock.* A lock has a disk with 36 numbers written around its edge. The combination to the lock is made up of three numbers (such as 12-33-07 or 19-19-08). How many different possible combinations are there for this lock? What is the probability of randomly guessing the correct one?

2. *The dorm door.* A dormitory has an electronic lock. To unlock the door, students must enter their unique five-digit secret code into the key pad (made up of the digits from 0 to 9). How many different secret codes are there? Suppose there are 200 students living in the dorm. What is the probability of randomly guessing a code and having the door unlock?

3. *28 cents.* How many different ways can you make 28 cents using current U.S. currency?

4. *82 cents.* How many different ways can you make 82 cents using current U.S. currency?

5. *Number please.* Someone you really wanted to go out on a date with gave you a beeper number. You didn't write it down and remember only the area code and the exchange (the first three digits of the number). How many different numbers are there with that area code and exchange? What is the probability that you randomly pick the right number?

6. *Dealing with jack.* Suppose you deal three cards from a regular deck of 52 cards. What is the probability that they will all be jacks?

7. *MA Lotto* (H). To win the jackpot of the Massachusetts lottery game, you must correctly pick the six numbers selected from the numbers 1 through 36. What is the probability of winning the Massachusetts lottery?

8. *NY Lotto* (H). To win the jackpot of the New York lottery game, you must correctly pick the six numbers selected from the numbers 1 through 40. What is the probability of winning the New York lottery?

9. *OR Lotto.* To win the jackpot of the Oregon lottery game, you must correctly pick the six numbers selected from the numbers 1 through 42. What is the probability of winning the Oregon lottery?

---

*In the Mindscapes section, exercises marked (H) have hints for solutions at the back of the book. Exercises marked (S) have solutions.

10. **Burger King (S).** You take a summer job making hamburgers. The burgers can be made with any of the following: cheese, lettuce, tomato, pickles, onions, mayo, catsup, and mustard. How many different kinds of burgers can you make?

11. **More burgers.** Suppose you are working at a burger place where the burgers can be made with any of the following: cheese, lettuce, tomato, pickles, onions, mayo, catsup, and mustard. You have a picky clientele: They all want the works, but they wish to specify the order of the placement of the items! For example, one person may want burger, cheese, mayo, onion, catsup, pickle, mustard, tomato, lettuce. Someone else may want lettuce, tomato, burger, onion, cheese, mustard, mayo, catsup, pickle. How many different types of burgers-with-the-works are there?

12. **College town.** You go to school in a college town. You know that there are 2,000 students enrolled in the school, but you don't know the population of the town (without students). You walk up and down the main streets of the town, stop people and ask them if they are students or not. You ask 100 people, and 60 of them say they are students. Estimate the nonstudent population of the town.

13. **Car count (S).** You wonder how many cars there are in your area. You call up the Honda dealership and ask how many Hondas they've sold. They report that they've sold a total of 10,000 cars. You then take a lawn chair and camp out on the most traveled roadway in town and count cars. You count 800 cars and note that 250 of them were Hondas. Estimate the number of cars in your area. Do you believe you would have gotten more, less, or the same accuracy if you had called the Rolls-Royce dealership instead of the Honda? Explain.

14. **One die.** You roll a fair die four times. What is the probability that you see at least one 1?

15. **Dressing for success.** You have five t-shirts, ten shorts, and three pairs of underwear, and you wear one of each. How many different ensembles can you put together? *Bonus*: Assuming all these clothes are clean, how long could you go before doing the laundry?

16. **Band stand.** The Drew Aderburg Band is planning a concert tour of six cities. In how many different orders could the band cover the six cities?

17. **Monday's undies.** You are spending the weekend at a friend's parents' house. You need three pairs of underwear and have ten

clean pairs of underwear of different colors. How many different underwear triples can you pull out to impress your friend's folks?

18. ***Counting classes.*** Your institution will offer 200 courses next semester. You will take four. How many different groups of four courses can you select from?

19. ***Cranking tunes.*** Your car stereo can be programmed to hold six radio stations of your choice. There are nine stations you really like. How many different ways can you program your six buttons with different collections of your favorite stations? (A different ordering of the same six stations counts as different programming.)

20. ***The Great Books.*** There are 20 Great Books, from which you know your English prof for next semester will select 10. You are an overachiever and want to figure out how many different combinations of 10 books the prof can pick. How many are there?

## II. CREATING NEW IDEAS

1. ***Morning variety* (S).** You wish to have a different breakfast every morning. Each day you choose exactly three of the following items: eggs, bagel, pancakes, coffee, orange juice. How many days can you go before you must eat a breakfast combination that you have already had?

2. ***Indian poker.*** You and a friend each pick a card from a regular deck of 52 cards but do not look at them. You then both hold your cards up by your foreheads so that the other person can see your card, but you still have not seen your own card. Your friend is holding a 6. What is the probability that your card is higher?

3. ***Crime story.*** Suppose 20 witnesses saw a perpetrator of a crime, and each supplied a piece of information. One witness said the assailant was wearing a certain type of shoe. Another said that the assailant was taller than 6 feet. A third said the assailant had dark hair, and so on. Each piece of information distinguished the assailant only from half the people on Earth. The different pieces were independent in the sense that any combination of the pieces were possible. Suppose we find a suspect who fits all 20 of the pieces of information. What is the probability that this person is the guilty party?

4. ***There's a 4* (H).** Someone you really wanted to go out on a date with gave you a beeper number. You didn't write it down and remembered only the area code, the exchange (the first three digits of the number), and the fact that there is one 4 in the remaining four digits. How many different such numbers are there? What is the probability you randomly pick the right number?

5. ***Making up the test.*** Your math prof says there will a 15-question test on probability. She also reports that the test will be made up of problems from the Mindscapes of this chapter as follows: two

questions each from sections 7.2, 7.3, and 7.7, and three questions each from sections 7.4, 7.5, 7.6 (no questions from the In Your Own Words category). How many different possible exams could she make up? What is the probability that this very question is on your exam?

6. *Moving up.* You have a part-time job in a department with 20 other people. Word comes out that six of the people from that department will be given a raise. If the six people are chosen at random, what is the probability that you will be one of the six?

7. *Counterfeit bills.* You are given 10 one hundred dollar bills and are told that three of them are counterfeit. You randomly pick three of the bills and burn them. What is the probability that you burned the counterfeit bills?

8. *Car care.* A burglar wishes to break into a car that has a security system. There are five buttons (marked 1 through 5), and he knows the code is a three-digit number. How many possible security codes are there? Knowing this number of combinations, the burglar fears that he will enter the wrong number first—thus tripping the siren (which everyone just ignores). So he writes down his first 20 guesses and then enters his next guess. Does this strategy improve his chances of success? Explain.

9. *Coins count.* On your bureau you have a half dollar, quarter, dime, nickel, and penny. How many different totals can be formed using exactly three coins? How about using four coins? How about using five coins?

10. *Math mania.* There are 90 students enrolled in Math 180. There are three different sections of 30 students, all meeting at the same time. How many different ways can the 90 students be placed into the three sections?

### III. FURTHER CHALLENGES

1. *Party on.* You want to throw a party and can invite only 15 people. You want to invite three people from the soccer team (there are 10 people on the tiny team); four people from the orchestra (there are 20 people in the orchestra); and eight people from your math class (there are 30 other students in your class). Assuming no overlap of the groups, how many different invitation lists could you have?

2. *No dice* (H). You roll a pair of dice 24 times. What is the probability of seeing at least one 11?

3. *Three angles.* Draw 10 points on a piece of paper with no three points lying on the same straight line. How many different triangles can you make using these points as vertices?

4. *Four parties.* You want to have a party and you know 10 men and 10 women. Unfortunately your common room can hold only 10

people. You wish to have enough parties so that each man and woman would, at some point, meet at a party. How many parties are necessary to accomplish this task? Show that you need no more than four parties.

5. *Making the cut.* In 1988, the ignition keys for Ford Escorts were made out of a blank key with five cuts, each cut made of one of five different depths. How many different key types were there? In 1988, Ford sold roughly 380,000 Escorts. What is the probability that one Escort key will unlock a random Escort? (This story was reported in the April, 1989 issue of the *Atlantic Monthly*.)

## IV. IN YOUR OWN WORDS

1. *Personal perspectives.* Write a short essay describing the most interesting or surprising discovery you made in exploring this section's material. If any material seemed puzzling or even unbelievable, address that as well. Explain why you chose the topics you did. Finally, comment on the aesthetics of the mathematics and ideas in this section.

2. *With a group of folks.* In a small group, discuss and work through the details involved in counting the five-card poker hands and the French dice games. After your discussion, write a brief narrative describing the methods in your own words.

3. *Creative writing.* Write an imaginative story (it can be comical, dramatic, whatever you like) that involves or evokes the ideas of this section.

4. *Power beyond the mathematics.* Provide several real-life situations—ideally, from your own experience—for which some of the strategies of thought presented in this section would provide effective methods for approaching and resolving them.

# 7.5 Great Expectations:
## Weighing the Unknown Future Through the Notion of Expected Value

*Chance favors only the prepared mind.*

—Louis Pasteur

Often in life we must make decisions whose wisdom or lack of wisdom becomes clear only in the future—buying stock versus investing in bonds, buying term life insurance versus investing in mutual funds, buying this book versus investing the money in lottery tickets. Perhaps after a year has passed, we can measure how sound our decisions were, but at the time we make these decisions, how can we assess their wisdom? Looking into the future, we must determine, as best we can, the value of the various possible actions we may elect to take, even with the understanding that, in the end, they are all gambles.

We have a vague idea of what we seek: a measure of the value of future actions. Our strategy for developing a concept of future value is to begin with concrete examples. By focusing on situations we know, we can extract essential features and then apply them to settings that are not so clear.

*Soldiers Playing at Cards* (1917) by Fernand Léger.

### Roulette for Scarlett

The best illustrations of the value of future events emerge in the smoky confines of casinos. In gambling, we can analyze the consequences of our decisions in dollars and cents. We might visualize a scene from a James Bond film: handsome tuxedo-clad men and beautiful, elegantly dressed women in Monte Carlo or Saint Moritz. But almost all will leave the casino a little less well-dressed than when they arrived, or at least lighter in the pocketbook. Let's head to an American-style roulette table and see how we fare.

The roulette wheel has 38 slots numbered 1–36 plus 0 and 00. Eighteen of the numbers 1 to 36 are red, and 18 are black; the 0 and 00 are green. On a play of roulette, the roulette wheel spins gracefully around with the ball spinning swiftly in the reverse direction. As the ball slows down, it meets the wheel in an abrupt collision. After several moments of bouncing randomly about, the ball finally settles into one of the 38 slots. As players, we bet on where the ball will come to rest. Many bets are allowed, including betting on one particular number or block of numbers, betting that the number will be even or odd, or betting that the number will be red or black. If we bet on red, and a red number comes up, we double our money. Now, doubling our money sounds promising, so let's learn vicariously how wise this investment truly is.

Gambling is addictive for some people, and one striking, though apocryphal, example of compulsive gambling will illustrate the theme of assessing future values of our actions. Ms. Scarlett was attracted to both roulette and red (not Rhett). One day she arrived at the casino, sat down at the roulette table, bought 3,800 one-dollar chips, and proceeded to bet $1 on red for each of the next 3,800 spins of the wheel. As luck would have it, every number on the roulette wheel came up exactly 100 times. How did Ms. Scarlett do? (Remember, the response, "Frankly, my dear, I don't give a damn" has already been used.)

Since 18 numbers are red, during these spins red numbers came up 1,800 times, black numbers came up 1,800 times, and the green numbers (0 and 00) came up 200 times. On each of the 1,800 occasions that red came up, she won a dollar and retrieved her bet. On the other 2,000 spins she lost her dollar bet. Thus she ended up with $3,600. Poor Ms. Scarlett lost $200, and while she may still have her shirt on her back, she will probably leave clad only in that shirt and her green curtains. On the plus side, she may have learned a valuable lesson about the future value of her bets. She realized that over the course of her 3,800 bets, she lost a total of $200—and that money is gone with the wind. So her average loss was $200/3,800 per bet.

## Average Value

This average value per bet is a measure of the expected value of each bet on red, on average. We arrived at this value by following the repeated exploits of our compulsive, and now red-faced, Ms. Scarlett. However, we can understand this value a bit better by looking at it from a different point of view. Each time Ms. Scarlett plunked down her dollar, there was some chance she would win and some chance she would lose. Specifically, the probability of getting a red number is 18/38, whereas the probability of getting something else is 20/38. So, another way of computing the expected value of her red bet is to note that 18/38 of the time she will win a dollar, whereas the other 20/38 of the time she will lose a dollar. So her expected value is

$$\$1\left(\frac{18}{38}\right) + (-\$1)\left(\frac{20}{38}\right) = \frac{-\$2}{38},$$

the same value we found before.

The negative sign indicates that by betting on red, we are losing on average (we're in the red). Thus, we see that our expected value of betting a dollar on red is $-\$2/38$. So, if we played and played this game again and again, overall we would see an average net loss of 5.26 cents per game. Should we play? No way: We lose more than a nickel each time, on average. However, there is a positive side to this negative outcome.

*Whenever possible, measure rather than guess.*

## Expected Value

The important positive side to our story is that it led us to an idea of how to determine the value of playing a game of chance. The *expected value* is the average net gain or loss that we would expect per game if we played the game many times. Consequently, the expected value is the sum of the products of the value of each possible outcome multiplied by the probability of that outcome. In other words, to find the expected value of playing a game, we list all the possible outcomes of the game and the positive or negative value of each possible outcome. We then compute the probability of each outcome and multiply the value of each outcome by its probability and add up all these products. This sum is the value, on average, we can expect to get from playing the game if we play over and over again. If the expected value is positive, then that means, on average, we would profit; if the expected value is negative, then on average, we would experience a loss.

> **COMPUTING EXPECTED VALUE.** *To compute the expected value, we multiply the value of each outcome with its probability of occurring and then add up all those products.*

## Not Everything Is Fair Game

To further illustrate the meaning and usefulness of expected value, let's consider a different game. This game costs a dollar to play. After paying the dollar, we roll a die. If we roll a 1, then we get our dollar back (the game is a draw). If we roll a 2 or a 3, then we win and are given $3. If we roll a 4, a 5, or a 6, then we lose. What is the expected value of this game? Let's try to compute it by carefully using the definition of expected value as a sum of products of values and probabilities.

The probability of rolling a 1 is 1/6, and the value of this event is $0, since we neither made nor lost money (we paid a dollar to play and got it back). The probability of rolling either a 2 or a 3 is 2/6, which equals 1/3, and, for this event, the value is $2 ($3 winning − $1 initial bet). The probability of rolling a 4, a 5, or a 6 is 3/6, which is 1/2, and the value of any of these outcomes is −$1. So, the expected value of this game for us is

$$(\$1.00 - \$1.00) \times \frac{1}{6} + (\$3.00 - \$1.00) \times \frac{1}{3} + (\$0.00 - \$1.00) \times \frac{1}{2}$$

$$= (\$0.00) \times \frac{1}{6} + (\$2.00) \times \frac{1}{3} + (-\$1.00) \times \frac{1}{2}$$

$$= \frac{\$2}{3} - \frac{\$1}{2} = \frac{\$1}{6} = \$.1666\ldots.$$

Therefore, the expected value is 16.7¢, which means that, if we play this game over and over, we would expect to gain an average profit of about 17¢ per game. If we were inclined to gamble, this game is one we might consider playing.

A game is called a *fair game* if the expected value equals zero. That is, if the game is played again and again, we would expect all players to break even. Suppose, in the spirit of sportsmanship, we wish to make the previous die-rolling game fair. Given the rules and payoffs, how much should we pay to play the game for the game to be fair? Should we pay more or less for the game to make it fair? Answer these questions before reading on.

To make the game fair, we have to pay an amount to play that would make the expected value equal to 0. Let's call the price we have to pay to play $P$. Then we can write out the expected value, just as before, but now replacing the $1.00 with $P$ dollars:

$$(\$1.00 - \$P) \times \frac{1}{6} + (\$3.00 - \$P) \times \frac{1}{3} + (\$0.00 - \$P) \times \frac{1}{2}$$

$$= \frac{\$1}{6} - \frac{\$P}{6} + \$1 - \frac{\$P}{3} - \frac{\$P}{2} = \frac{\$7}{6} - \$P.$$

So, the expected value if you pay $P$ dollars is $\$7/6 - \$P$. We can check this answer by substituting $1.00 for $P$ and see if we get the answer we found before. Verify this check for yourself. Now to make the game fair, we must have the expected value equal to zero. So, we set the expected value equal to zero and solve for the price $P$:

$$\frac{\$7}{6} - \$P = 0,$$

so that means

$$\$P = \frac{\$7}{6} = \$1.1666\ldots.$$

Therefore, we should be charged around $1.17 to have the game be a fair game. It does make sense that we should have to pay an additional amount equal to our previous expected winnings to have the game be fair.

## Life Insurance

Games of chance include more realms than just gambling. Insurance policies (life, home, auto, Chihuahua) are powerful but hidden gaming prospects whose expected values are used to determine the price of a policy. Let's examine term life insurance from the point of view of the insurance company. The insurance company keeps elaborate mortality tables that might state that roughly 1% of 60-year-old men will die within the year. The insurance company may then choose to sell a one-year term life policy to a 60-year-old man for $1,000 that would pay $50,000 if the man dies. Since on average 99% of the policy holders will live the whole year and only 1% of the policy holders will die and cash in on their investment, the expected value of this policy to the insurance company is:

$$\$1{,}000(.99) + (\$1{,}000 - \$50{,}000)(.01) = \$500.$$

That is, the insurance company would expect to make $500 profit on average for such a policy. Of course, this computation is overly simplistic, since we have not computed the expenses incurred by the insurance company. Nevertheless, this example does illustrate how computing the expected value is a central consideration for insurance companies in weighing premiums, coverage, and risk.

*Buy their book? Or buy the lottery tickets?*

## Should You Buy Lottery Tickets or Buy This Book?

We are now about to analyze a treacherous issue that few authors wish to even discuss: the expected value of buying and reading their books. Specifically, we are going to attempt to answer the following question. Suppose this book costs $70. Is it better to use the $70 to buy this book and read it, or is it better to spend the $70 on lottery tickets in the hopes of hitting the big one?

We have already seen in the previous section that the probability of winning the lottery is 1/15,890,700 = 0.00000006292 . . . (pretty slim chance). Let's assume that we play only when the payoff is $6,000,000, and let's assume we are the only ticket purchaser, so we don't worry about sharing the bounty if we hit it big. The cost of each ticket is $1, so the expected value of purchasing a lottery ticket is

$$(\$6{,}000{,}000 - \$1) \times \frac{1}{(15{,}890{,}700)} + (\$0 - \$1) \times \frac{(15{,}890{,}699)}{(15{,}890{,}700)}$$

$$= -\$.62 \ldots.$$

Therefore, the expected value of buying 70 lottery tickets is roughly 70 × (−$.62) = −$43.57, so we would expect the value to be −$43.57. Not a great payoff—you lose more than 43 dollars. Now consider this book.

As this is the first printing of this book, we must make estimates as to the effect of the book on readers' lives. It is difficult to estimate the personal satisfaction of discovering some of the most beautiful and deep ideas of humankind. Thus we will limit ourselves to cold, hard cash. We believe that the ideas in this book can empower one to successfully understand and creatively resolve problems, cut through complicated issues, discover the heart of the matter, ask the right questions, and generally think more effectively. These activities are extremely valuable in the workplace, so, armed with such skills, one will be promoted faster, move up quicker, and reap the benefits of merit raises and bonuses.

We now make, in our minds, *extremely* conservative estimates as to the value of carefully reading this book. We believe that, with probability .5, the ideas honed from this book would result in at least an additional $40-per-year increase in salary each year. We believe that, with probability .4, this book would result in an additional $20-per-year increase in salary each year. Finally, although we honestly do not believe it possible, with probability .1 we will say that the reader sees no additional increase in salary resulting from this book.

Let's consider the case of a 25-year-old person who will work for the next 40 years before retiring. A yearly extra increase of $40 per year in salary for 40 years would result in a $40 increase in salary the first year, a total of $80 increased salary the second year, a total of $120 increased salary the third year, and so on. These increases over 40 years would result in total extra income of $32,800. A yearly extra increase of $20 per year in salary for 40 years would result in total extra income of $16,400. Therefore, the expected value of buying this book for $70 (and carefully reading it) would be equal to

$$(\$32{,}800 - \$70) \times .5 + (\$16{,}400 - \$70) \times .4 + (\$0 - \$70) \times .1$$

$$= \$22{,}890.$$

Thus, we believe a conservative figure for the value of buying this book is $22,890. Plainly, the purchase of this book was a financially sound investment.

Perhaps comparing the expected value of the lottery versus this book is a bit ridiculous, but the idea of expected value is not. Measuring the expected future value of events that have not yet occurred is an important way of assessing prospects for the future. Certainly the strategy of giving more weight to more probable future events and less weight to less probable future events seems eminently reasonable. But the future is a tricky realm, and sometimes reasonable ways to analyze it run contrary to other common sense ideas about cause and effect, as we will now see.

## *Newcomb's Paradox*

Throughout history, paradoxes have played important roles in stimulating new ideas. A paradox presents a situation that has two possible interpretations or resolutions. Each view appears irrefutable, and yet the views are diametrically opposed to each other. The philosopher William Newcomb proposed a paradox that is not fully resolved to this very day. Many have found this paradox intriguing to wrestle with and debate. Perhaps you will find, as we have done, that people who discuss this paradox are nearly evenly divided—usually believing that anyone who thinks the opposite way is completely wacko. Such firm convictions make for interesting exchanges and may reveal important insights. Exploring difficult questions often exposes misperceptions. Good luck in resolving Newcomb's Paradox.

Here is the premise of Newcomb's Paradox. A famous psychologist is showing off her predictive powers by giving you the chance to become rich. You will be led into a room in which stands a table holding two boxes, the Thousand Dollar Box and the Zero-or-Million Dollar Box. You must now choose one of two possible courses of action. You may choose to take the contents of both boxes or you may take the contents of the Zero-or-Million Dollar Box only.

 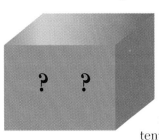

The Thousand Dollar Box has clear glass sides and visibly contains $1,000—*cash*. The other box, the Zero-or-Million Dollar Box, is opaque. When you are led into the room, you will not know the contents of the Zero-or-Million Dollar Box, but the experimenter informs you in advance that in the Zero-or-Million Dollar Box there is nothing at all or there is $1,000,000. The experimenter further informs you of the criterion under which she placed $0 or $1,000,000 in the Zero-or-Million Dollar Box. The criterion is that, if she predicted that you will take the contents of the Zero-or-Million Dollar Box only, then she placed $1,000,000 in the Zero-or-Million Dollar Box. However, if she predicted that you will take the contents of both boxes, then she placed $0 in the Zero-or-Million Dollar Box.

Let us assume that the psychologist is exceptionally good at making her predictions. Specifically, let's suppose that she correctly predicts an individual's decision 90% of the time, meaning that, of those who choose to take the Zero-or-Million Dollar Box only, she correctly predicts their behavior 90% of the time, and, of those who choose to take both boxes, she correctly predicts their behavior 90% of the time. You are led into the room and the door is closed. You are all alone and the two boxes are sealed and affixed to the table. What do you do? We now invite you to pause and ponder this conundrum.

## *Persuasive Opposing Arguments*

When we enter the room, we are faced with two persuasive arguments, leading to two different conclusions. Following are the two arguments.

## *Argument 1—Choose Both Boxes*

The $1,000,000 is either in the Zero-or-Million Dollar Box or it is not there. What we do after we have entered the room does not influence the existence or absence of the $1,000,000. Therefore, we may as well take the contents of both boxes. In either case (that is, if the Zero-or-Million Dollar Box contains $0 or contains $1,000,000), by taking both boxes we will be $1,000 ahead.

## *Argument 2—Choose Only the Zero-or-Million Dollar Box*

If the psychologist predicts that we will take both boxes, she will have put $0 in the Zero-or-Million Dollar Box. Consequently, perhaps it is better to choose only the Zero-or-Million Dollar Box in the hopes that our decision to take the contents of only the Zero-or-Million Dollar Box will somehow have been apparent to the psychologist when she made her prediction. If she is in fact a good predictor, then we take the Zero-or-Million Dollar Box only and hope she predicted our actions correctly. We can formalize the persuasiveness of choosing just the Zero-or-Million Dollar Box by computing the expected value of each possibility.

## *Quantifying Expected Values*

We can quantify the persuasiveness of choosing only the Zero-or-Million Dollar Box by determining the expected value of each possibility. Recall that the psychologist correctly predicts an individual's decision 90% of the time. That is, .9 is the probability that she is accurate in her predictions. So, the expected value of taking the Zero-or-Million Dollar Box only is

$$.9 \times (\$1{,}000{,}000) + .1 \times (\$0) = \$900{,}000.$$

However, the expected value of taking both boxes is

$$.9 \times (\$1{,}000) + .1 \times (\$1{,}001{,}000) = \$101{,}000.$$

## A Paradoxical Situation

Thus we have a paradoxical situation. One plausible analysis (either the $1,000,000 is there or it is not) leads to one conclusion (take both boxes), whereas another plausible analysis (the expected value of choosing only the Zero-or-Million Dollar Box is $900,000 while the expected value of taking both boxes is $101,000) leads to a different conclusion (take the Zero-or-Million Dollar Box only). However, the expected value is the value we would expect on average if we repeated the experiment many, many times. Here we get to play only once. But, suppose we were in line with 100 people in front of us who all do the experiment. The psychologist correctly predicts 90 actions and misjudges a mere 10. Now it's our turn. What would we do?

## Variations

What would *you* do? How can we think about this paradox? If a question is hard, a helpful strategy is to vary the question. For example, we might see how our choices change as we alter the amounts in the two boxes. Does our concept of free will play a role here? Suppose all our actions are predetermined and the psychologist can detect the preexisting conditions for our eventual choices. Or suppose before we are led into the room, we see the experimenter flip a coin and make her decision on the basis of the outcome of that coin toss. Suppose we are taken into the room without ever meeting or even seeing the experimenter. Would that alter our decision?

Many other ideas will come to mind as we discuss this paradox. There is no general consensus about whether Newcomb's Paradox is completely understood. Perhaps this paradox underscores the significance of the expected value perspective. After thinking about the alternatives, get a wealthy friend to try the experiment on you. Good luck and choose carefully.

**A LOOK BACK**

The notion of expected value allows us to determine how much on average we are likely to benefit or lose from a future event. Decisions about insurance and pension plans, horse racing and the lottery, and political policy matters are usefully informed by an expected value point of view. Life decisions we make contrary to the wisdom of their expected value are taken at our peril. Occasionally people do win the lottery, but, if we consistently make decisions whose expected value is negative, over the long run we are likely to lose. Likewise, if we weigh the gains against the losses in the proportionate fashion embodied in the determination of expected value, we will tend to benefit in like proportion. We will discover great value in expected value: It is a powerful idea that can help guide us through our uncertain future.

Our strategy for developing a practical way to measure the value of future possibilities is to start with simple, concrete examples. We envision that uncertain situations are repeated many times to get a proportional sense of the possible results. These thought experiments give us experience with measuring the value of the unknown future, rather than just guessing. By finding a method for determining the average value in a special case, we can use that idea to formulate a definition for *expected value*. Armed with that definition, we apply the concept to analyze several different types of situations.

We gain experience by looking repeatedly at clear and simple cases. Then we can extract specific methods and ideas. As the ideas crystallize, we can form clear definitions and develop broadly applicable methods. These new notions and methods can then be applied to new settings, which, in turn, lead to new insights.

*Even the uncertain can be quantified.*

*When possible, measure rather than guess.*

*Gain experience by looking repeatedly at simple cases.*

❖ ❖ ❖ ❖ ❖ ❖ ❖ ❖ ❖ ❖ ❖ ❖ ❖ ❖ ❖ ❖ ❖ ❖ ❖ ❖

## MINDSCAPES  Invitations to Further Thought*

### I. SOLIDIFYING IDEAS

1. ***Cross on the green*** **(S).** A standard roulette wheel has 38 numbered slots for a small ball to land in: 36 are marked from 1 to 36, with half of those black and half red; two green slots are numbered 0 and 00. An allowable bet is to bet on either red or black. This bet is an even-money bet, which means if you win you receive twice what you bet. Many people think that betting black or red is a fair game. What is the expected value of betting $1,000 on red?

2. ***In the red.*** Given the bet from the previous question, what should the payoff be if the game is to be a fair game?

3. ***Free Lotto.*** For several years in Massachusetts, the lottery commission would mail residents coupons for free Lotto tickets. To win the jackpot in Massachusetts, you have to correctly guess all six numbers drawn from a pool of 36. What is the expected value of the free Lotto ticket if the jackpot is $8,000,000 and there is no splitting of the prize?

4. ***Bank value.*** What is the expected value of putting $100 in a bank for one year if the bank pays 3% compounded annually?

---

*In the Mindscapes section, exercises marked (H) have hints for solutions at the back of the book. Exercises marked (S) have solutions.

5. ***Newcomb your neighbor.*** Explain Newcomb's Paradox to some friends and ask them what they would do and why. Explain the expected value argument. Record your friends' reactions.

6. ***Value of money.*** In Newcomb's Paradox, first suppose that you have no money and have not eaten in two days; next suppose your net worth was $800,000. How would these different scenarios affect your decision?

7. ***Die roll.*** What is the expected value of the outcome (1, 2, 3, 4, 5, or 6) of rolling a fair die?

8. ***Dice roll.*** What is the expected value of the outcome (2, 3, 4, 5, . . . , 12) of rolling two fair dice?

9. ***Fair is foul.*** Someone has a weighted coin that lands heads up with probability 2/3 and tails up with probability 1/3. If the coin comes up heads, you pay $1; if the coin comes up tails, you receive $1.50. What is the expected value of this game? Would you play? Why?

10. ***Foul is fair*** (S). Someone has a weighted coin that lands heads up with probability 2/3 and tails up with probability 1/3. You pay $5 to flip the coin. If the coin comes up heads you lose and receive nothing. For this game to be a fair game, how much would you have to receive if you flip tails?

11. ***Cycle cycle*** (H). You live in an area where the probability that one's bicycle is stolen is .2. You deeply care for your $700 road bike. What is a fair price to pay to insure your bike against theft over the life of your bike? (*Postscript:* On Friday, June 13, 1997, an uninsured mountain bike belonging to one of the authors was stolen out of the garage of the other author. What are the chances?!)

12. ***What's your pleasure?*** You have three options for the evening. (1) You could watch some sitcoms on TV that you are certain to enjoy and that will provide you with a relative pleasure rating of 4. (2) You could go to a movie that is supposed to be good. You would enjoy the movie with probability .5, but, if you enjoy it, it will provide you with a relative pleasure rating of 11. If you don't enjoy it, it will provide a negative rating of −2. (3) You could go on a date. With probability .3 you will experience a pleasure rating of 21, and with probability .7, the date will provide a negative rating of −2. What are the expected values of pleasure for these individual activities? List the activities in order of expected pleasure. What do you do?

13. ***Roulette expectation.*** A standard roulette wheel has 38 numbered spaces for a small ball to land in: 36 are marked from 1 to 36, half black and half red; 0 and 00 are green. If you bet $100 on a particular number and the ball lands on that number, you are paid a whopping $3,600. What is the expected value of betting $100 on red 9?

14. ***Fair wheeling.*** You are at the roulette table and bet $100 on red 9. What payoff should you receive to make the game fair? (See the previous roulette question.)

15. **High rolling (H).** Here is a die game you play against a casino. You roll a fair die. If you roll 1, then the house pays you $25. If you roll 2, the house pays you $5. If you roll 3, you win nothing. If you roll a 4 or a 5, you must pay the house $10, and if you roll a 6, you must pay the house $15. What is the expected value of this game?

16. **Fair rolling.** Suppose you are considering the game described in the previous question. How much would you have to pay, or be paid, to make the game a fair game?

17. **Spinning wheel.** You pay $5 and then pick one of the four spinners below and spin that spinner. Each spinner is balanced, so the spinner is equally likely to land in any direction. Wherever the spinner lands, you receive that much cash. Which spinner would you pick? Justify your answer with an analysis of the expected values of the various spinners.

It costs ONLY $5.00 to play... pick a spinner, *spin* and *win*!!

18. **Dice.** You place a bet and then roll two fair dice. If you roll a 7 or an 11, you receive your bet back (you break even). If you roll a 2, a 3, or a 12, then you lose your bet. If you roll anything else, you receive half of the sum you rolled in dollars. How much should you bet to make this a fair game?

19. **Uncoverable bases.** Show by a specific example how it is possible to lose at the lottery even if the jackpot exceeds the cost of purchasing every possible ticket.

20. **Under the cap.** A national soda company runs a promotional contest. Under the cap of one in a million bottles there is a message saying that you won $1,000,000. Suppose that the two-liter bottle costs $2. What is the expected value of buying a bottle of this soda (assuming, of course, that the soda itself is worthless)?

## II. CREATING NEW IDEAS

1. **From a penny to a million.** Use the data you collected in the penny experiments from section 7.1 to estimate the probability that the penny will land heads up if it is knocked off balance from its edge, and then estimate the probability that the penny will land tails up if it is spun. Now find a 'friend' and make a 'fair' wager. Your friend picks heads or tails. The coin randomly lands heads or tails. If your friend wins, you lose a dollar; if you win, you receive a dollar. The deal is that once your friend calls heads or tails, you will

use either the balancing method or the spin method for landing the penny—whatever is to your advantage. What is the expected value if your friend picks heads? What is the expected value if your friend picks tails? Try this scam (without betting if you wish) with a number of different, soon to be former, friends.

2. **Three coins in a fountain.** You pay someone $5 and then are given three coins to toss in a fountain and see how they land. If you see no heads, then you receive $20. If you see exactly one head, then you receive $5 (the game is a draw), and if you see at least two heads, then you lose. What is the expected value of this game? Is there one single possible outcome whereby you would actually gain or lose the exact amount computed for the expected value? If not, why do we call the expected value, the *expected* value?

3. **Insure (S).** You own a $9,000 car and a $850 mountain bike. The probability that your car will be stolen next year is .02, but the probability that your bike will be snatched is .1. An insurance company offers you theft insurance for your car for $200 and insurance for your bike for $75. What is the expected value of the car insurance? What is the expected value of the bike insurance? (This question pains one of the authors . . . see question I.11 above.)

4. **Get a job (H).** You search for a job. There are three companies that are interested in you, and you will receive at most one offer. The first company has a job open with a salary of $21,000, and the probability of getting an offer there is .5. The second company has a job open paying $32,000, and the probability of getting an offer there is .3. The last company has a job paying $45,000, and the probability of getting an offer there is .1. There is a .1 probability of not landing any job. What is the expected income you will make next year?

5. **Take this job and . . . .** Given the employment scenario described in the previous question, suppose now that the companies do not consult one another about offers, so you may receive more than one offer. Suppose the first company calls you on the phone, offers you the job, and says you have to accept or reject it right there on the spot. You have not heard from either of the other companies yet. Do you accept or reject the offer? What if the call were from the second company? What if it were the third company? Justify your answers using expected value.

6. **Book value.** Refer back to our analysis of the expected value of reading this book (p. 572). Suppose you took the $70 and put it in a bank account guaranteed to pay 10% interest compounded annually for 40 years. Which is the better investment: depositing the money in a bank or buying and reading this book? (*Hint*: You can use the formula $A = P \times (1 + r)^n$, where $A$ is the amount you will have by investing $P$ dollars with an interest rate of $r$ compounded annually for $n$ years. So, in this case, $P = 70$, $r = .10$, and $n = 40$. You may need a calculator.)

7. ***In search of . . . .***  A group of deep-sea divers approaches you with a proposition. They are 60% certain that they know where an ancient shipwreck is; they are also 50% certain that there is a treasure worth about $2,000,000; and finally they are 70% certain that they will be able to get to it. They want you to invest $200,000 in this expedition. If they do indeed find the buried treasure, you receive $1,000,000. What is the expected value of this investment?

8. ***Solid gold.***  There is a 50% chance that the price of gold will go up $25 an ounce, a 20% chance that it will remain the same, and a 30% chance that it will drop to $40. What is the expected value of a purchase of gold? Given your answer, would you invest in gold at this time if gold cost $375 per ounce?

9. ***Four out of five.***  In Newcomb's Paradox, suppose that the psychologist predicts the choice of the subjects correctly four out of five times. What is the expected value of selecting both boxes? What is the expected value of selecting just the Zero-or-Million Dollar Box? What would you do?

10. ***Chevalier de Méré.***  Suppose that the Chevalier de Méré bets 1,000 francs that a 6 will appear at least once in four rolls of a fair die. If he wins, he receives 2,000 francs. What is the expected value for the Chevalier de Méré? (*Hint*: Return to the previous section, p. 558.)

### III. FURTHER CHALLENGES

1. ***The St. Petersburg paradox.***  Here is an interesting game. You pay a certain amount of money to play. Then you flip a fair coin. If you see tails, you flip again, and the game continues until you see a head, which ends the game. If you see heads on the first flip, you receive $2. If you see heads on the second flip, you receive $4. If you see heads on the third flip, you get $8, and so forth—the payoff is doubled every time. What is the expected payoff of this game? How much would you pay to play this game? Suppose you paid $1,000 to play. What is the probability that you will make money? Why is this game a paradoxical situation given the expected value?

2. ***Coin or god.***  In Newcomb's Paradox, first suppose that the psychologist just flips a coin to determine whether or not to place the million dollars in the box. What is the expected value of selecting both boxes? What is the expected value of selecting just the Zero-or-Million box? Suppose, instead, that the experiment was run by an all-knowing godlike being. What would you do?

3. ***An investment.***  You wish to invest $1,000, and you have two choices. One is a sure thing: You will make a 5% profit. The other is a riskier venture. If the venture pays off, you will make a 25% profit; otherwise, you lose your $1,000. What is the minimum

required probability of this riskier venture paying off in order for the expected value to exceed the value of the first investment?

4. *Pap test* (**H**). Assume that the insurance value of a life is $1,200,000. Suppose Pap smear tests will save one life in 3,000. A Pap smear costs about $30. What is the expected value of a Pap smear?

5. *Martingales.* A game is played with a fair coin. You bet some amount of money on heads. If you flip heads, then you receive even money. (You get your bet back, plus an extra amount equal to the bet.) If you flip tails, then you lose your bet. The *martingale strategy* is one where you continue to double your bet after each loss until you win. So, for example, if you first bet $1 and lose, then the next time you would bet $2. If you lose again, you would then bet $4, and then $8, and so on. How much would you earn if you used this strategy and lost seven times in a row before finally winning? How much money would you need to play eight times?

### IV. IN YOUR OWN WORDS

1. *Personal perspectives.* Write a short essay describing the most interesting or surprising discovery you made in exploring this section's material. If any material seemed puzzling or even unbelievable, address that as well. Explain why you chose the topics you did. Finally, comment on the aesthetics of the mathematics and ideas in this section.

2. *With a group of folks.* In a small group, discuss and work through the reasoning involved in Newcomb's Paradox. After your discussion, write a brief narrative describing your own opinion about the proper Newcomb choice.

3. *Creative writing.* Write an imaginative story (it can be comical, dramatic, whatever you like) that involves or evokes the ideas of this section.

4. *Power beyond the mathematics.* Provide several real-life situations—ideally, from your own experience—for which some of the strategies of thought presented in this section would provide effective methods for approaching and resolving them.

# 7.6 What the Average American Has:
## Peering into the Pitfalls of Statistics

*There are three kinds of lies:*
*lies, damned lies, and statistics.*

—Benjamin Disraeli

*Fact: The average American has one testicle and one ovary.* Aren't statistics great!? We find this perfectly valid statistical representation of the testicularity and ovarian endowments of humans funny because we know exactly what is actually going on. We know that the population is roughly equally divided between men and women, and we know some rudimentary facts about the physiology of each group. We also realize that the statistics about the average American's physiology are better understood through other analyses than simply averaging over the entire population.

Information drives our society. How can we collect it, understand it, analyze it, and use it effectively? Statistics provides a means for analyzing data in an attempt to uncover structure and explain pattern. Such results are powerful in that they can lead to anything from new insights to new public policies. However, unintentionally or not, sometimes statistics are used incorrectly to support conclusions that are invalid. We know two organs that the average American has one of, but, in reality, there is no American (or not many) who has one of each. The statistic is correct, but

**Doonesbury** BY GARRY TRUDEAU

is it meaningful? Statistics in newspapers and in 30-second soundbites often suffer from a similar defect. Unfortunately, the more meaningful analysis of the data required to understand the information is often omitted from the headlines. Thus we should be wary of the statistics we read or hear.

To avoid erroneous conclusions, one strategy is to face potential pitfalls and fallacies. Knowing what can go wrong helps us to see potential defects in presentations or interpretations of data.

## *Collecting Data*

Dealing with data involves two basic steps: collecting it and interpreting it. It may seem as though collecting data is no big deal. However, a plethora of pitfalls surrounds such a process, and it is into these pits that we now will peer.

## *Asking Embarrassing Questions*

One of the easiest ways for survey data to be misleading and inaccurate is if the surveyed people lie. Suppose we want to take a survey in which we wish to acquire accurate data on an embarrassing or controversial topic. We might expect many people to lie about the topic if we just asked them outright, so we would like to use a method of taking the survey in which no one feels any danger of their privacy being invaded. How can we proceed?

One of the most serious problems facing colleges and universities today is the problem of alcohol abuse among students. A first step in dealing with this problem is to get accurate data about alcohol use among college students. Although some students might forthrightly and honestly report about such activity, many would lie, and some would exaggerate for various purposes, thereby skewing the data. Let's suppose the question we want to ask is: Have you been drunk during the last week? The goal is to structure the survey such that we can deduce approximately what fraction of the students got drunk; however, we cannot tell definitely which individual students got drunk. An easy method might be to promise to keep the survey anonymous. However, people might not believe us, and they also wouldn't want us to be able to peek at their answers. Thus people may still lie.

Is a data-collecting method possible whereby, from public statements of the group, we are able to deduce what percentage of students got drunk, and yet no one (including us) knows whether any individual did or did not get drunk? Sounds like a pretty tall order, but the apparently impossible is not always impossible.

If we don't know the behavior of even one student, how can we hope to know the behavior of the aggregate? Let's think about this conundrum for a moment. As a first observation, we might note that, if we chose only half the students at random and asked them the question, we would be getting a representation of the behavior of the whole group. At first this fact seems to be of no use, since we would have the same problem of getting inaccurate answers from half the population we ask. Let's now add a twist. Suppose we ask only half the group, but no one knows which half we are asking. This strategy is actually successful, but what in the world does it even mean?

## A Secret Half—A Survey Method Guaranteeing Some Anonymity

Let each student privately flip a coin. If the coin lands heads up, then the student answers yes to the original question whether or not the student actually got drunk last week. If the coin turns up tails, then the student answers the question honestly. Thus, a yes answer could mean that the student actually did get drunk, or it could mean that the student flipped heads and either did or did not get drunk.

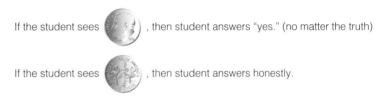

From the results of that experiment, how can we determine the actual percentage of students who got drunk? Well, if no one got drunk, we would expect 50% of the students to answer yes by virtue of flipping a heads. If 40% of the students got drunk, we would expect half of them to flip a tails and answer yes. Thus, 20% more than the half of the students who flipped a heads would answer yes. Thus, if 70% answered yes, we could deduce that about 40% got drunk.

For example, suppose we try this procedure in a class of 60 students, and 41 answer yes. We assume that about 30 people said yes because they saw heads, so the remaining 11 people said yes because they saw tails *and* got drunk. Thus, if 11 people who saw tails got drunk, it seems reasonable to assume that 11 people who saw heads also got drunk (we cut the class randomly into two groups of 30 and 30); so, we would predict that 22 students out of the 60 got drunk last week.

## Totally Secret—A Survey Method Guaranteeing Total Anonymity

One defect of this technique is that if a person answers no, then we do know that that person definitely did not get drunk. Think about the different scenarios in which a person would answer no and verify that, in all these cases, we know for sure that the person did not get drunk. Can we further refine this method so that we do not know *either* way? That is, a person who answers yes may or may not have gotten drunk, *and* a person who answers no may or may not have gotten drunk?

The possibly surprising answer is that we *can* devise a test so that we do not know anything about any particular person, no matter how the person answers. Suppose, for example, that we have each person secretly flip two coins. If the coins both land heads up, then that person reports the *opposite* of the truthful answer. Anyone who flips at least one tails answers truthfully. So let's analyze the situation if, in another class of 60 students, 24 people answer yes, and 36 people answer no. Notice first that we have no information about any individual. In fact, we could have the students answer by raising their hands: If someone said yes he or she could be telling the truth or not, and if another person answered no he or she could be telling the truth or not. So, how do we determine the number of students who got drunk?

If the student sees      , then the student lies.
Otherwise, the student answers truthfully.

There are four equally likely coin outcomes:

HH, HT, TH, TT.

We assume that these four outcomes occur randomly among the class so the number of students who got drunk who flipped HH is the same as the number of students who got drunk who flipped HT or TH or TT—and the same for the sober students. Let's suppose that the number of students who got drunk and who flipped HH is $D$, and the number of sober students who flipped HH is $S$. Then we expect that $D$ students who got drunk and $S$ sober students flipped HT, TH, and TT. So, how many students who got drunk are there in the class? Answer: $4D$. ($D$ of them flipped HH, another $D$ of them flipped HT, still other $D$ of them flipped TH, and finally yet another $D$ of them flipped TT.)

Now which people answered yes? Any student who flipped HH is to answer the opposite of the truth, so the $S$ sober students in that group answered yes. The students who see at least one tails answer truthfully, so, since there are three such groups (HT, TH, and TT), there are $D + D + D$ students, which is $3D$, who answer yes from those groups. So we see that

$3D + S = 24$ (the number who answered yes).

The rest answered no. So,

$D + 3S = 36$ (the number who answered no).

So, now we have two equations with two unknown values, and we can deal with them. We want to find how many students got drunk, so we need to figure out what $D$ equals. The first equation tells us that

$S = 24 - 3D$,

and we can substitute this fact into the second equation and get rid of that pesky S:

$$D + 3 \times (24 - 3D) = 36.$$

We now distribute that factor of 3,

$$D + 72 - 9D = 36,$$

which is the same as

$$-8D = 36 - 72 = -36.$$

So, we discover that

$$D = \frac{36}{8} = \frac{9}{2}.$$

How many students in the class got drunk? Remember that the number of students who got drunk is $4D$, so we would estimate that $4 \times 9/2 = 18$ students got drunk last week, and yet we do not know whether any particular student got drunk or not.

By devising this method for maintaining anonymity, we can overcome the potential pitfall of having people lie on a survey. This method becomes more accurate and reliable as our pool grows larger.

## Polluted Pools

Besides the problem of lying on surveys, there may be problems with the pool. We now illustrate the pitfall of a biased sample as we continue to discover the potential perils of data collection. Here we see that answers often depend on whom you ask.

### The Literary Digest—Story of a Statistical Fiasco

Before its demise in 1937, the *Literary Digest* no doubt provided its readers with well-digested literature; however, the publication is perhaps most famous for a monumental statistical fiasco caused by sampling bias. As you no doubt recall, the 1936 presidential election was contested between the Republican Alfred Landon and the incumbent Democrat Franklin Delano Roosevelt. Before the election, the *Literary Digest* conducted a poll and predicted that, in the electoral college, Landon would win by the substantial margin of 370 to Roosevelt's 161. When the election took place, the electoral college vote was in fact a landslide the other way: Roosevelt

523, Landon 8 (and some statisticians at the *Literary Digest* moved on to other careers). How can we account for such an incredibly mistaken survey? The answer: a biased sample.

The year 1936 was in the middle of the Great Depression. Most people were struggling financially and had eliminated luxuries and unnecessary expenses from their budgets. The sample of citizens polled by the *Literary Digest* came from several sources: *Literary Digest* subscribers, people with telephones, and automobile registration records. Those surveyed people did, in fact, prefer Landon and no doubt voted for him. The problem was that subscribing to the *Literary Digest* and owning a telephone or a car were among the unnecessary expenses that most people had eliminated from their personal budgets. So, the poll was obtaining honest information from a collection of voters whose views and experience did not proportionately reflect the opinions of the whole voting population.

## *Eliminating Sample Bias*

Eliminating sample bias is not always an easy proposition. Sometimes bias can be removed by changing one's claims about what the survey shows rather than changing the sample. For example, if the *Literary Digest* had reported their survey as an indication that the subscribers of their magazine preferred Landon, then for that purpose, the sample selection they chose may have been just fine. Thus, we see a need for ensuring that the pool in which we gather our data is clear of bias. But how big does that pool have to be?

## *Bad, Bad Biltong*

Biltong is a cured, air-dried beef product developed in South Africa and is a staple in many African countries. It is similar to beef jerky and is also comparable to prosciutto. Another popular South African dish is the meat pie. Recently the Onderstepoort Veterinary Institute, located in Onderstepoort, South Africa, made a shocking announcement. It found that three-quarters of the biltong and half of the meat pies tested by the institute in one month contained horse meat. There is great danger in eating old racehorses: Consumers might be exposed to steroids, antibiotics, and pesticides. This disturbing finding immediately prompted the local meat board to launch a full-scale investigation, which led to big headlines in the local papers: "Meat Pies, Biltong Made with Horse Meat."

It appeared that the Onderstepoort Veterinary Institute made an important discovery that might, in fact, save the lives of many people. The meat board was willing to spend as much money and time as needed to make sure this dangerous and illegal act of manufacturers was stopped and punished. Sounds like a happy ending. But not in the eyes of the meat pie manufacturers. They were not happy. In fact, they were downright outraged at the allegations, since they believed that essentially all meat pie manufacturers produced healthy and safe products free of horse meat.

## *A Case of Insufficient Data*

When pressed by the meat pie manufacturers, the Onderstepoort food hygiene division manager, Dr. Werner Giesecke, revealed that the institute's investigation consisted of testing *two* meat pies. Now, are we as outraged and scared as we were after reading the first paragraph? Well, probably not. The crucial point is that, for data and the interpretation of that data to have any real meaning, one must be convinced that the data is accurate, not biased, and sufficiently robust to be representative of the whole population. Often attaining such clean data is not an easy task. Dirty data often leads to dirty stats.

## *Interpreting Data*

We now realize the importance and difficulties involved in collecting accurate and representative data. Assuming that we have the data, we have to be able to analyze it in some reasonable manner to better enable us to discover trends, see patterns, and make predictions. Where do we begin?

The most basic and perhaps the best thing to do with data is to *look at it*. Visualizing data through graphics and pictures often reveals important structure. Suppose for example, you measure the height of every male student in your math class and record how many there are in each 3-inch height interval starting with intervals

Look at it.

◆ ◆ ◆ ◆ ◆ ◆

| | |
|---|---|
| 4′0″–4′2″; | 5′0″–5′2″; |
| 4′3″–4′5″; | . . .; |
| 4′6″–4′8″; | 6′3″–6′5″; |
| 4′9″–4′11″; | 6′6″–6′8.″ |

Now, if you were to make a plot of this data, putting the various height intervals along the horizontal axis and the frequency (the number of male students) of those heights on the vertical axis, then the plot would probably look like this:

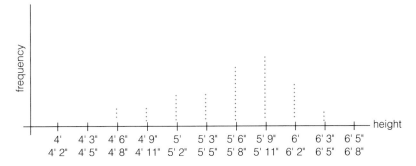

## Normal Distributions

If we connect the points, we get a nice bell-shaped curve that is characteristic of a *normal distribution*. Often examination grades from a class are normally distributed, but sometimes they are not. We caution

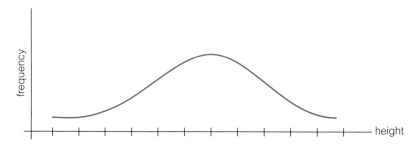

that many kinds of data, in fact, are not normally distributed. For example, if an examination covered a topic for which half the students had some prerequisite knowledge and half did not, we might see a *bimodal* distribution that looks like this:

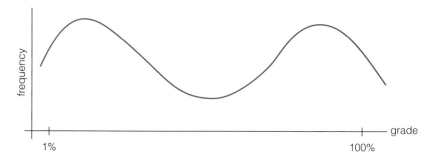

Often data about time and space, such as the times that cars arrive at a toll booth, results in a *Poisson distribution* that looks like this:

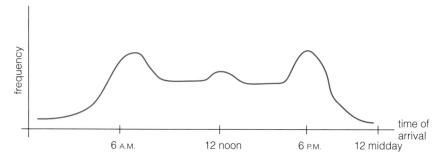

For normally distributed data it is easy to see the average: It's right in the center of the bell. Some normally distributed data is tightly bunched around the center, and other data is more spread out. By measuring how

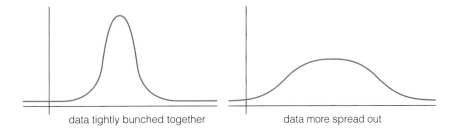

data tightly bunched together      data more spread out

steep or how shallow the bell curve is, we can tell how much the data is bunched up or spread out. Although we will not consider such measurements here, you may have heard of the best known one: the standard deviation. The smaller the standard deviation is, the tighter the data is bunched around the center.

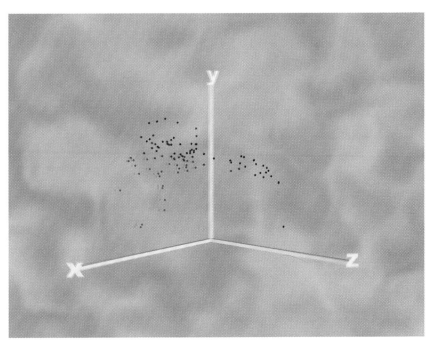

Looking at complicated data sets from different points of view sometimes leads to unexpected correlations—put on your 3D glasses and correlate away!

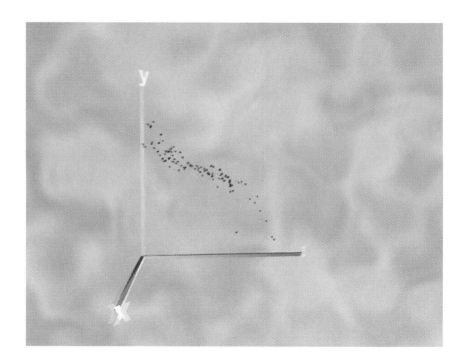

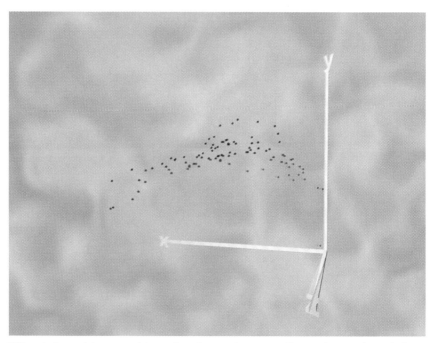

Different views of the same data set from the previous page. Connections and correlations become visible.

## Measures of Central Tendency

When the data is not normally distributed and does not produce a symmetric bell curve, then it is harder to locate the average of the data or alternatively the "center" of the associated graph. In fact, what does "average" even mean? We would like to get a sense of the overall tendency of the data. There are three basic measures that provide an overall sense of the data, and they are collectively referred to as *measures of central tendency*. The three basic measures of average are called the *mean, median,* and *mode.*

## The Mean

The mean is what we usually think of as an average. The *mean* equals the sum of all the numerical data divided by the number of pieces of data. For example, suppose someone has a handful of change consisting of three pennies, one nickel, one dime, and two quarters. What is the mean face value of the coins? The data in this case can be stated in cents as 1, 1, 1, 5, 10, 25, 25. To find the mean, we add all these numbers up and divide by the number of data points: 68/7 = 9.714 . . . cents. So, the mean is about 9.7 cents, which, looking at the data, makes sense. Suppose now that someone else has a handful of change. This person is holding three pennies, one nickel, one dime, two quarters, and one $50 gold eagle coin. What is the mean face value now? Try to compute the mean first in terms of cents.

Now our data looks like 1, 1, 1, 5, 10, 25, 25, 5,000, so the mean equals 5068/8 = 633.5, or roughly $6.34. That's the right answer, but does it accurately capture the aggregate of this handful of coins? On the one hand yes, but on the other hand, no: The average is extremely high because we had one coin, an *outlier*, that, in some sense, skewed the average value of the collection of coins. This possibility inspires another measure of average called the *median.*

## The Median

The *median* is just the middle data point when the data is lined up in numerical order. If there are an even number of data points, then we take the mean of the two middle values. In the first handful of coins, we would order them 1, 1, 1, 5, 10, 25, 25 and see the middle data point is 5, so the median is 5¢. In the second handful of coins, we see 1, 1, 1, 5, 10, 25, 25, 5,000, so the median will be the mean of the two middle numbers 5 and 10: (5 + 10)/2 = 7.5¢. Notice that 7.5¢ better captures the general size of most of the coins than the mean of $6.34. The median (think *middle value*) is a useful measure of average when there are outlying data points.

## The Mode

Another useful measure of average is the mode. The *mode* is simply the value that appears in the data most frequently. So, for the preceding examples, the mode would be 1 in each case, since the most frequent number appearing in either list is 1: It appears three times. The mode is a strange creature. Sometimes there may be more than one mode and sometimes no mode at all! For example, consider the data set 7, 9, 2, 1, 9, 3, 2. Notice that this set has two modes: 2 and 9. In contrast, 1, 6, 2, 3, 8 has no mode since there is no value that appears more often than any other. It turns out that the mode is often useful if there is a limited amount of data, or if the data is not numerical.

These measures of central tendency provide us with a sense of how the data looks, on average. But as we have already seen, sometimes averages can be misleading. *The average American has one testicle and one ovary.* If the population of males were exactly equal to the population of females, this statement would be correct for both the mean and the median. But what does the statement tell us? Really nothing, because we grouped two populations inappropriately with respect to our anatomical study: Sometimes analogous problems are less obvious to detect.

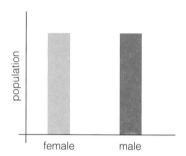

## School Daze

Surprising facts often have simple explanations.

◆ ◆ ◆ ◆ ◆ ◆

Suppose we wish to choose a high school for our children. There are many variables involved in such a choice, but one we might consider is the school's graduates' prospects for remunerative employment. We are statistically savvy, so we decide to compare the average earnings of graduates of various schools for the year 1997. On our list of schools to consider is a fine school in Seattle, Washington called Lakeside School. The Lakeside School has an excellent reputation in many respects, but when we look at the average income of its graduates last year, we are so excited that we instantly pack up the family and move to Seattle. We learn that last year's mean income, including salary and investments, of all the graduates of Lakeside School was more than $2,500,000. In fact, this average includes recent graduates who are still in college and graduates from the 1920s who are now dead. Wow, what a record!

When we read about or hear a fact that is almost unbelievable, it is a good idea to seek an explanation. The data about Lakeside School is readily explained: Namely, Bill Gates and Paul Allen, founders of Microsoft, are Lakeside School graduates. From 1996 to 1997, Bill Gates's net worth increased by about 18 billion dollars

The Lakeside School

and Paul Allen's increased about 7 billion dollars. Lakeside is an excellent school, so no doubt the typical Lakeside graduate does have an income higher than the country's average income. However, the mean salary we compute using Gates's and Allen's salaries in the mix totally misrepresents the facts. They are the ultimate outliers. For the purposes of computing a more meaningful description of the earning potential for Lakeside graduates, we would do better to consider the median income.

**A LOOK BACK**

Dealing with data involves collecting it and interpreting it. In collecting data, we should be certain that we are getting honest answers, that the sample is representative of the whole population, and that we have taken a large enough sample. When we are interpreting data, it is imperative that we understand which kind of summary is appropriate for that situation and whether the kind of summary we are looking at is what we really want to know.

A good way to begin to think about statistics is to look at potential pitfalls. Many fallacies are possible. Seeing some examples of startling errors promotes caution but also lets us devise some methods for avoiding mistakes.

Looking squarely at mistakes can be a fruitful way to develop insights. The more dramatic and pronounced the error, the easier it may be to see the underlying defect. Do not ignore errors—learn from them.

Don't blithely accept all assertions as true.

Analyze errors.

Learn from mistakes.

## MINDSCAPES  Invitations to Further Thought*

### I. SOLIDIFYING IDEAS

1. ***The questionnaire.*** Answer the following survey *anonymously* and *honestly*. DO NOT PUT YOUR NAME ON YOUR RESPONSE SHEET, AND DO NOT ANSWER ANY QUESTION YOU WOULD RATHER NOT ANSWER. The data from all the students in the class should be collected, compiled, and returned to you for further analysis.

---

*In the Mindscapes section, exercises marked (H) have hints for solutions at the back of the book. Exercises marked (S) have solutions.

**Class Survey: How Do You Measure Up?**

For each question, select the one answer that most closely reflects the correct answer in your case. Give only one answer per question.

1. What is your gender?   (1) male   (2) female
2. How much do you weigh?
    (1) below 90 lbs.   (2) between 90 and 119 lbs.   (3) between 120 and 149 lbs.   (4) between 150 and 179 lbs.   (5) 180 or more lbs.
3. What is your height?
    (1) below 4′6″   (2) between 4′6″ and 4′11″   (3) between 5′ and 5′5″   (4) between 5′6″ and 5′11″   (5) 6′ and over
4. What is your favorite part of your own body?
    (1) hair   (2) face   (3) arms   (4) stomach   (5) legs
5. How many siblings do you have?
    (1) none   (2) one or two   (3) three or four   (4) five or six   (5) more than six
6. How often do you call home on average?
    (1) once every two months   (2) once a month   (3) twice a month   (4) once a week   (5) several times a week
7. How often do you do your laundry?
    (1) every time you go home   (2) once a semester   (3) once a month   (4) twice a month   (5) once a week
8. How often do you drink alcohol?
    (1) never (2) once or twice a month   (3) once or twice a week   (4) three or four times a week   (5) more than four times a week (seek help)
9. How many hours do you sleep on average per night?
    (1) less than two hours   (2) two or three hours   (3) four or five hours   (4) six or seven hours   (5) eight or more hours
10. How many hours of athletic activities do you participate in per week?
    (1) less than one hour   (2) one or two hours   (3) three or four hours   (4) five or six hours   (5) seven or more hours
11. What is the average number of dates you go out on per month?
    (1) at most one   (2) two or three   (3) four or five   (4) six or seven   (5) more than seven
12. When did you see your last movie in a movie theater?
    (1) at least two months ago   (2) a month ago   (3) three weeks ago   (4) two weeks ago   (5) this past week or this week
13. How many hours do you surf the Web per week?
    (1) at most one hour   (2) between one and two hours   (3) between two and three hours   (4) between three and four hours   (5) more than four hours
14. What is the average number of e-mail messages you receive per week?
    (1) at most four   (2) between five and nine   (3) between 10 and 14   (4) between 15 and 19   (5) 20 or more
15. How many hours do you study per week?
    (1) at most nine hours   (2) between 10 and 19 hours   (3) between 20 and 29 hours   (4) between 30 and 39 hours   (5) 40 or more hours

16. *How many courses have you taken in college so far?*
    (1) five or less   (2) between six and 10   (3) between 11 and 15
    (4) between 16 and 20   (5) More than 20

17. *What year do you expect to graduate?*
    (1) this year   (2) next year   (3) in two years   (4) in three years
    (5) in four years

18. *What would you guess you will do after you graduate?*
    (1) pursue graduate studies in the arts and humanities   (2) pursue graduate studies in the sciences   (3) attend a professional school
    (4) find a job in business   (5) find a job not in business

19. *How many people in this class do you find attractive?*
    (1) at most two   (2) three or four   (3) five or six   (4) seven or eight
    (5) nine or more

20. *What is your favorite topic in the book from the choices below?*
    (1) infinity   (2) the fourth dimension   (3) chaos   (4) Fibonacci numbers   (5) the Monty Hall issue

2. **Class stats (H & S, I.2–8).** Make a graphical plot and then compute the mean, median, and mode (if the mode exists) for each of questions 2 and 3 of the class survey. State your results in complete, self-contained, declarative sentences (for example, "The median number of times the average student from this class does laundry is once a month").

3. **How often on average?** Make a graphical plot and then compute the mean, median, and mode (if the mode exists) for each of questions 6, 7, and 8 of the class survey. State your results in complete, self-contained, declarative sentences (see the sample in question 2).

4. **How many per month?** Make a graphical plot and then compute the mean, median, and mode (if the mode exists) for question 11 of the class survey. State your results in complete, self-contained, declarative sentences (see the sample in question 2).

5. **How many hours?** Make a graphical plot and then compute the mean, median, and mode (if the mode exists) for each of questions 9, 10, and 13 of the class survey. State your results in complete, self-contained, declarative sentences (see the sample in question 2).

6. **How much school?** Make a graphical plot and then compute the mean, median, and mode (if the mode exists) for each of questions 15, 16, and 17 of the class survey. State your results in complete, self-contained, declarative sentences (see the sample in question 2).

7. **Outside of school.** Make a graphical plot and then compute the mean, median, and mode (if the mode exists) for each of questions 4, 19, and 20 of the class survey. State your results in complete, self-contained, declarative sentences (see the sample in question 2).

8. **Life.** Make a graphical plot and then compute the mean, median, and mode (if the mode exists) for each of questions 5, 14, and 18 of the class survey. State your results in complete, self-contained, declarative sentences (see the sample in question 2).

9. ***Surprises.*** Examine your analyses from the previous exercises regarding the class surveys. What are, for you, the most surprising discoveries? The discoveries could come from the measures of central tendency, the discrepancy among the different measures, or the plots of data.

10. ***Plotting.*** Make a graphical plot and then compute the mean, median, and mode (if the mode exists) for the data found by referring to **www.heartofmath.com**.

11. ***Plotting 2.*** Make a graphical plot and then compute the mean, median, and mode (if the mode exists) for the data found by referring to **www.heartofmath.com**.

12. ***Plotting 3.*** Make a graphical plot and then compute the mean, median, and mode (if the mode exists) for the data found by referring to **www.heartofmath.com**.

13. ***Plotting 4.*** Make a graphical plot and then compute the mean, median, and mode (if the mode exists) for the data found by referring to **www.heartofmath.com**.

14. ***Embarrassing data*** (**H**). Suppose you asked 100 students to answer the question: Have you ever cheated on a test? You used the one-coin method described in this section, and 54 people answered yes. Estimate how many students in the group have ever cheated.

15. ***Hugging both parents*** (**S**). You ask 150 students to answer the question: Do you still hug your parents? You use the one-coin method described in this section, and 83 people answer yes. Estimate how many students still hug their parents.

16. ***Drug data.*** You ask 250 students to answer the question: Have you used an illegal drug in the past 72 hours? You use the two-coin method described in this section, and 78 people answer yes. Estimate how many students in the group have used drugs recently.

17. ***Kissing.*** You ask 180 students to answer the question: Did you kiss on your most recent first date? You use the two-coin method described in this section, and 50 people answer yes. Estimate how many students in the group kissed (but didn't necessarily tell).

18. ***Ask them.*** In a dining hall or at a gathering of friends (at least 20 or so), ask people an embarrassing question and use the two-coin method to extract information. Make sure you clearly explain the instructions, and make sure they understand that they will be revealing no information! Record the outcomes, estimate the answer to the question, and record their reactions.

19. ***More homework.*** In a recent survey of random college students, 50% believed that there should be more homework assigned in all classes. What is your guess as to the sample size of this survey? Explain your answer.

20. *Amazing stats.* Think of an appropriate question (political, fashion, music, and so on) and then ask your friends to answer the question. Try to construct the question so that the statistics you generate (the mean, say) is not accurate owing to either an extremely small sample size or a biased sample. Record the question, sample size, and mean. Explain the bias.

## II. CREATING NEW IDEAS

1. *News data.* Find some data in a recent newspaper. Describe the data and how it was analyzed. If an average was given, was it clear which measure of central tendency was used? Which do you think was used? Why? Compute the other measures of central tendency. What conclusions are deduced from the data? Do you agree with the conclusions? Are there other ways of interpreting the data? Try to give another explanation for the statistics given other than what was mentioned.

2. *9:00 A.M. versus 9:00 P.M.* You take a survey of students in the library at 9:00 A.M. on Saturday morning and ask: How many hours per week do they study? You ask the same question at 9:00 P.M. at a wild party. What would you expect to see in your data? Explain the biases.

3. *Four out of five.* Watch several TV commercials in which ordinary people are asked to comment about a product. Describe the ads and list at least three questions you would like to ask to determine the accuracy and significance of the claims made.

4. *Which half?* Half of the people in the United States are below average. Is this general statement true or false? Explain.

5. *Grades (S).* A student received the following quiz grades: 72, 84, 61, 95, 92, 98, 87, 84. Compute the student's mean quiz grade and the median quiz grade. If on the next quiz the student earns an 80, will the mean go up or down? What about the median?

6. *Raising scores.* A student received the following quiz grades: 90, 83, 80, 72, 78, 63, 79, 90. What is the lowest grade the student can receive on the next quiz that would raise the student's mean grade? Can you generalize your observations?

7. *Mean and mode.* Give an example of a data set in which the mean is greater than the mode.

8. *Mode and mean (H).* Give an example of a data set in which the mode is greater than the mean.

9. *Big mode.* Is it possible to have a data set in which the mode is greater than both the median and the mean? If so, give an example. If not, explain why.

10. *Big mean.* Is it possible to have a data set in which the mean is greater than both the median and the mode? If so, give an example. If not, explain why.

## III. FURTHER CHALLENGES

1. *Whoops.* A survey of 100 recent college graduates was made to determine the mean salary of recent college graduates. The mean salary found was $35,000. It turns out that one of the alums incorrectly answered the survey. He said he is earning $29,000 when, in fact, he is earning $42,000. What is the actual mean salary of the 100 graduates?

2. *Mean wins.* Suppose the state keeps 30% of the bets made on a lottery as profit and pays out the rest. What is the mean return for each $1.00 bet? Why does this mean mean very little? What example from the text does this scenario most closely resemble?

3. *M & M & M.* Is the median numerically always between the mean and the mode? If so, explain why, if not, provide several examples.

4. *Average the grades.* In a math class of 23 men and 25 women, the mean grade on the most recent exam for the women was 89%, and for the men it was 83%. Is it possible to compute the mean exam grade for the entire class of 48 students? If so, do it; if not, explain why. Is it possible to compute the median exam grade for the entire class? If so, do it; if not, explain why.

5. *Up and down* (H). The grade point average of the women at the college went up. The grade point average of the men at the college went up. The grade point average of all students at the college went down. Is this possible.? Explain.

## IV. IN YOUR OWN WORDS

1. *Personal perspectives.* Write a short essay describing the most interesting or surprising discovery you made in exploring this section's material. If any material seemed puzzling or even unbelievable, address that as well. Explain why you chose the topics you did. Finally, comment on the aesthetics of the mathematics and ideas in this section.

2. *With a group of folks.* In a small group, discuss and work through the reasoning involved in collecting accurate answers to embarrassing questions. After your discussion, write a brief narrative describing the methods in your own words.

3. *Creative writing.* Write an imaginative story (it can be comical, dramatic, whatever you like) that involves or evokes the ideas of this section.

4. *Power beyond the mathematics.* Provide several real-life situations—ideally, from your own experience—for which some of the strategies of thought presented in this section would provide effective methods for approaching and resolving them.

## 7.7 Navigating Through a Sea of Data:
### How Interpreting Data Reveals Surprising and Unintended Results

*It is a capital mistake to theorize before one has data.*

—Sherlock Holmes
via Sir Arthur Conan Doyle

Blue (Moby Dick) (1943) by Jackson Pollock. *A sea of unusual data.*

Peas, AIDS, air safety, and twins: In common is an issue of data. Data about peas played a prominent role in developing our understanding of genetics. Data about the devastating AIDS epidemic has implications about policies on testing. Data about flying can help us decide how to make travel safer. And interpreting data about twins raised apart brings up questions about randomness and coincidence. Our world is awash in data. When faced with a seemingly endless ocean of numbers, it is easy to feel dragged down by the undertow of information. But statistics and careful reasoning can save us from drowning.

Part of the strategy for dealing with data is to appreciate the relationship between randomness and forces of nature. When we are confronted with data, we can ask what data we would expect from random events and compare those results with the facts. That comparison can then lead us to find forces that must be at work designing interactions in the world.

### Pea Parenting

Statistics lead to insight.

✦ ✦ ✦ ✦ ✦ ✦

Examining data can have profound consequences. One of the most significant instances of drawing conclusions from data occurred in the 1860s when Gregor Mendel, an Augustinian monk, did some experiments concerning inherited characteristics of peas. His results led to our modern understanding of genetics and heredity. His analysis is a wonderful illustration of probability and statistics at work.

One of Mendel's experiments was to crossbreed two varieties of peas—those with green pods and those with yellow pods. Some green-pod peas,

when crossed with yellow-pod peas, will have only green-pod offspring and some will have green- and yellow-pod offspring. He began with green-pod peas whose offspring were pure green. Yellow-pod peas always had only yellow-pod offspring. Mendel's experiment was to breed the pure green-pod peas with the yellow-pod peas, which we refer to as the first generation, and see what happened. He discovered that the second-generation offspring had only green pods. However, when he bred the second-generation plants with one another, one quarter of the third-generation plants were yellow-pod peas. From this basic result, and with some appreciation for randomness and probability, he was able to deduce basic features of genetics.

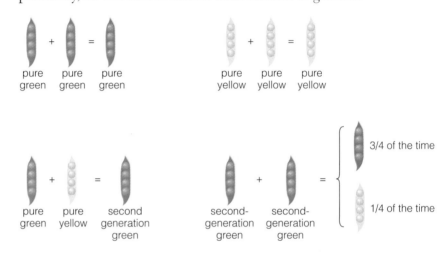

He realized that some of the yellow-pod genetic material must be in the second-generation offspring even though the pods were green, because in the third generation the yellow-pod variety reappeared. He also noticed that the color was not a yellow-green blend in any generation. From these observations he hypothesized that offspring of the breeding of two plants contained genes from each parent and that the green gene was dominant. *Dominant* means that, if the green gene is present, the pod will be green regardless of what the other gene is.

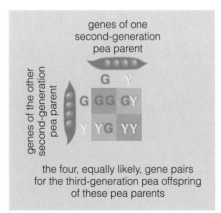

Finally, he realized the significance of the fact that one quarter of the third-generation peas were yellow-podded. The second-generation plants had two parents—one yellow pod, one green pod. So, Mendel assumed that the second-generation plants had one green-pod gene and one yellow-pod gene. If one gene from a yellow-green plant is randomly chosen and combined with a random gene from another yellow-green plant, what would we expect to see? By comparing expectations with reality, we can confirm or refute hypotheses.

Let's count the possibilities. From the first parent pea, we could get a green-pod gene or a yellow-pod gene. For each of those possibilities, we might get a green-pod gene or a yellow-pod gene from the other parent. Therefore, four possibilities occur: green-green, yellow-green, green-yellow, or yellow-yellow. In the first three of those four cases, the pods of the offspring will be green. In the fourth and last case, the pods will be yellow.

This simple example of counting the possible outcomes and counting how many of these result in yellow-yellow genes gave powerful evidence for the important theory of how genetic information from parents is used in their offspring. Mendel's seminal work was fundamental to our understanding of genetics.

## AIDS Testing

Analyzing data is important not only in the wonderful world of life and genetics but also in the devastating world of death and disease. AIDS (acquired immune deficiency syndrome) has been the most lethal disease for young people in the world for the last 10 years. A tremendous research effort has resulted in an understanding of some of the mechanisms by which the disease works, and clinical research has given some hope for an eventual cure. However, during the period 1980–1996, AIDS was viewed as inevitably fatal and, as such, was one of the most feared diseases in the world. AIDS is the active disease that results from a person being infected with HIV (human immunodeficiency virus). People carrying HIV will develop AIDS after some interval of time has passed, sometimes 10 to 15 years.

Until 1996, there were 27 million people who were HIV positive. Of those, 7.7 million had developed AIDS. By the middle of 1996, approximately 4.5 million adults and 1.3 million children had died as a result of HIV/AIDS. By the year 2000, 40 million people worldwide are expected to be HIV positive. In the United States, approximately 1 person in 1,000 is estimated to be HIV positive. These figures are painful to contemplate.

A person who is HIV positive can transmit the disease to others through sexual contact or contact with blood. Peoples' knowing who is HIV positive has the potential to curb the spread of this dreaded disease. Scientists have worked hard to develop reliable tests for HIV. One such test, the ELISA blood-screening test, is 95% accurate in that it tests positive for 95% of people who have the disease, and it tests negative for 99% of people who do not. The 5% error in detecting HIV in carriers may be due to low concentrations of the virus or antibodies to the virus among some people. The 1% false positive error may or may not be caused by mistakes in running the test. Some false positives might be due to some unknown idiosyncrasies in the body chemistries of that 1% of the population.

Potentially infected people should be warned to avoid the kind of contact that would spread the disease. But telling an uninfected person that he or she is infected has tremendous costs: It frightens the person half to death, and he or she must make lifestyle changes unnecessarily. A fundamental policy question is: Should we undertake universal testing for

*During a crisis, try not to panic. Instead, focus on clear and rational thoughts.*

◆ ◆ ◆ ◆ ◆ ◆

HIV/AIDS? Should every man, woman, and child in the country be tested?

Let us suppose for a moment that the policy of universal testing is instituted. You have had your test, and one day you receive the dreaded result: "We regret to inform you that your test was positive." When faced with this devastating situation, you need to control any instincts to panic and force yourself to think clearly and rationally.

Given the accuracy of the test and the data about what percentage of the population are actually HIV positive, what is the probability that you have the disease?

At first glance, this question seems to have been answered in the description of the reliability of the test. The test gives a positive result to an uninfected person only in 1% of the cases. It is 99% accurate, so you might conclude that you are 99% certain to have the disease—a most frightening prospect.

But let's think a bit more deeply about this case by actually looking at the numbers. Let's write down a few of the facts we know.

1. The population of the United States is about 250,000,000.
2. About 1 in 1,000, or 250,000, people in the United States are HIV positive.
3. Of the 250,000 who have the disease, the test will come out positive 95% of the time, which equals 237,500 cases.
4. There are 249,750,000 (250,000,000 − 250,000) people who do not have the disease.
5. Of the 249,750,000 people who do not have the disease, the test will come out falsely positive 1% of the time, which equals 2,497,500 cases.
6. So, the total number of people receiving a positive test result is 237,500 + 2,497,500 = 2,735,000.
7. Of the 2,735,000 who get positive test results, only 237,500 actually have the disease. Therefore, if you get a positive test result, your chance of having the disease is 237,500/2,735,000, which is less than 9%. So, when you receive a positive test result, although you are worried, you in fact have less than a 9% chance of actually having the disease.

How can we reconcile this apparent paradox? On the one hand, the test is 99% accurate. On the other hand, in the case of a positive result, the chance of having the disease is less than 9%.

*If the truth seems strange, we do not understand it well enough.*

◆ ◆ ◆ ◆ ◆ ◆

## The Hidden Test

This result is surprising at first or second glance. We need to understand this apparent paradox so that, instead of appearing counterintuitive, it becomes natural and reasonable.

*Look at things from new points of view.*

❖ ❖ ❖ ❖ ❖ ❖

How can we change our view of a result from surprising to obvious? We have encountered such situations before. We became accustomed to the ideas of infinity, irrational numbers, the fourth dimension, and distortions of objects. One way we retrained our intuition about those subjects was to look at the issues from different points of view and to look at examples that illustrated some feature in a clear way.

One way to look at the AIDS-testing situation differently is to recognize that the scenario really presents two tests for AIDS, but only one was called a test. The hidden "test" was the statistic given that 1 in 1,000 people are HIV positive. Implicitly this gives a test. That test is as follows: Take a person at random; with probability .999 that person is not HIV positive. It doesn't seem like much of a test, but we have to admit, it is pretty accurate. If we take a person at random and tell him or her, "You are not HIV positive," we will be correct 999 times out of 1,000. That's a pretty impressive success rate.

By comparison, the test explicitly mentioned in the scenario states that the blood test is accurate only 99 times out of 100. So, the blood test is actually far inferior to the hidden "test" of just looking at the proportion of people in the population who are HIV positive.

So, now let's change the presentation of the situation just slightly and see if we can get a different sense of the significance of the outcome. Instead of having said that 1 in 1,000 people are HIV positive, suppose we said that you would be taking two different tests and we did not mention what kind of tests they were: one test called *PoorerTest* that is accurate 99% of the time and another test called *BetterTest* that is accurate 99.9% of the time. Suppose according to PoorerTest you have the disease, and according to BetterTest you do not have the disease. Of course, you would be pleased if both tests reported that you are healthy; however, the result of the more accurate test is more important than the result of the less accurate test. The disparity in the accuracy of the two tests is what accounts for the rather low probability that you actually have the disease.

## *The Big Picture*

Another alternative view of the AIDS test issue is the following. There are roughly 250,000 people who have AIDS. Roughly 2,500,000 people receive positive test results. So, only about 1 in 10 positive test results can be correct.

The surprising part of the AIDS-testing scenario was the mismatch between the accuracy of the test (99%) and the confidence with which we could assert that a person has the disease if the test is positive (9%). We strove to understand this example better by rephrasing it. Although this result might now be a bit more intuitive and make more sense, let's be certain to understand that the 99% accurate test being positive makes a large difference in the probability of having the disease. Specifically, it changes the probability of being HIV positive from 1 in 1,000 to roughly 1 in 10.

In practice, the paradoxical situation in AIDS testing does occur and has resulted in methods of screening for diseases that use a preliminary, inexpensive test to reduce the initial population and then a more refined, conclusive test on the sub-population.

This powerful AIDS example shows how data and the understanding of probability can have a significant effect on our thinking about policy decisions. As a further illustration of how data and probability affect public policy issues, we now consider another serious topic: travel safety.

### *The Friendly Skies*

No one wants to be in an airplane crash, and, happily, our chances of being in one are extremely low. Each day U.S. commercial airlines log approximately 1.7 billion passenger miles, and yet accidents are so rare that we remember big crashes for years. U.S. commercial airplane crashes are so infrequent that, during the last several years, on average, there was only about one fatality per 3.4 billion passenger miles. We would have to fly 500 miles every day for more than 18,000 years between crashes. That's a long wait, and it doesn't even include waiting for our luggage.

Nevertheless, safe as flying is, when a plane does crash, people naturally turn their minds to air safety; there is always room for improvement.

Hey, wait! You forgot these bags!

### *Should We Support Increased Air Traffic Safety?*

Air safety could be improved, but at a price, and that price is real money. We could probably make the airlines 10 times as safe as they are now, but ticket prices would go up to pay for the safety improvements.

Move from the qualitative to the quantitative.

◆ ◆ ◆ ◆ ◆ ◆

We are responsible citizens and must decide whether regulations should require the airlines to institute better safety measures, and we are naturally interested in the consequences of various possibilities. So, if new air safety measures are instituted next year, what would be the effect on the loss of life? Let's take the trouble to figure out the quantitative consequences instead of relying entirely on a qualitative impression about safety.

From the facts given, we can estimate how many commercial airline fatalities are expected per year. So, we can estimate how many lives would be expected to be saved if airline fatalities were only a tenth as great.

Let's do the arithmetic and see how many people can be expected to live or die if we improve airline safety as suggested. First let's note that, since U.S. airlines fly about 1.7 billion passenger miles each day and there is an airline fatality about every 3.4 billion passenger miles, then, on average (the mean), there is one death every two days. Of course, what really happens is that many people are killed all at once instead of at a steady one-death-per-two-days rate. However, one fatality every two days means that, with current safety practices, the estimated number of fatalities expected due to U.S. commercial airline crashes is roughly 183 per year. Now suppose that air travel becomes 10 times safer. Then, instead of 183 people dying, only about 18 commercial air fatalities would be expected per year. That saves about 165 lives per year. This safety proposal looks pretty darn good.

## Consider Unintended Consequences

We must remember that this improved safety will cost money, and the cost of airplane tickets will go up. When ticket prices go up, some people will drive instead of fly. If planes were made 10 times safer and the airplane ticket prices went up to pay for that safety, let's suppose that 10% of travelers who now fly would choose to drive instead. Sadly, fatal automobile accidents are much more common than airplane crashes. In fact, our chances of dying in an automobile accident are about 34 times greater per mile than they are by traveling with commercial airlines. What will be the result of this extra driving?

If 10% of the people who now fly would drive instead, that is an extra 170 million miles of driving per day. As we mentioned, driving is 34 times more dangerous per mile than commercial flying. Since the fatality rate for flying was one death per 3.4 billion passenger miles, the fatality rate for driving is 34 deaths per 3.4 billion miles, or one death per 100 million miles. Since people are now driving 170 million more passenger miles per day (instead of flying), on average there would be 1.7 extra automobile deaths per day. Therefore, there would be an extra 620 people killed in automobile accidents each year. Uh oh.

So the consequences of our idea to increase air safety are somewhat counterproductive. It is true that 165 fewer people would die in airplane crashes, but then 620 more people would die in car crashes. Unfortunately, that means that approximately 455 additional people would be killed if airline safety were improved. Maybe air traffic safety is not such a great idea after all. In fact, maybe more lives would be saved by making airlines even less safe if that would make airline tickets cheaper as well and thus get more people flying instead of driving. The moral: *Beware of unintended consequences.*

This problem illustrates an important lesson. Even simple arithmetic combined with a rough analysis of data can lead to striking conclusions. In this case, we see that a side effect of increased safety measures, namely, higher prices for airline tickets, could well result in an effect opposite from what we intended. When we are making decisions, making approximations of all the likely consequences of our actions is extremely useful in visualizing the entire picture.

## Twin Studies: A True Story

Jim Lewis and Jim Springer were identical twins who were separated at birth and were brought up without any contact with each other. Thirty-nine years later, on February 9, 1979, they met for the first time. Elaborate tests

Jim Lewis and Jim Springer toast each other in 1979.

were administered to the two, and remarkable similarities were reported between their lives, tastes, and habits. Both were named James. Both lived on a block with only one house on it in towns in Ohio. Both had a white bench around a tree in their yards. Both liked Miller Lite beer. Both first married women named Linda, and both remarried women named Betty. Lewis had a son named James Alan, and Springer had a son named James Allen. Both, at one time, owned dogs named Toy. Both had workshops in their basements. Both drove Chevrolets. Both chain-smoked Salems. Both liked stock-car racing but disliked baseball. Both vacationed at the same three-block-long beach on the Florida gulf coast. Both were guests on the *Tonight Show Starring Johnny Carson* (well, okay, they appeared together after all the coincidences were revealed). Surely, then, there is a strong genetic component to personality, right?

What additional information would we have to know about the study to have confidence that it truly supported the theory that personality is significantly influenced by genetics? The answer is that we would have to compare the twin study rate of coincidence with a random rate of coincidence and make sure that the twin study rate is in fact greater.

> *"'Genetics may not be the main reason that identical twins raised apart seem to share so many tastes and habits,' said Richard Rose, a professor of medical genetics at Indiana University. 'You're comparing individuals who grew up in the same epoch, whether they're related or not. If you asked strangers born on the same day about their political views, food preferences, athletic heroes, clothing choices, you'd find lots of similarities. It has nothing to do with genetics.'"°*

Before being impressed with a correlation, we must first compare it to pure chance.

◆ ◆ ◆ ◆ ◆ ◆

Genetics appear to influence personality and tastes; however, we cannot evaluate the significance of a list of coincidences without additional information. In fact, note that some of the coincidences listed in the twin study cannot be more than coincidence. For example, their both being named Jim had nothing to do with their being twins—they didn't name themselves. Their both choosing wives with the same name is interesting, but what pertinence has it to the question of genetic influence on personality? Is the study suggesting that the preference for names is genetically based? It's a good thing one of the twins was not raised in Japan—it may be hard to find a Linda there. More pertinent would be statements about the personality types of the Lindas.

How many questions were asked of these twins? If they were asked thousands of questions, our sense of amazement at finding a list of surprising coincidences drops dramatically. If we ask any two people thousands of questions about their preferences, life experiences, and physical

---

°K. C. Cole, *Los Angeles Times*, 4 January 1995.

surroundings, we could easily pick out many eclectic examples that would appear to be amazing coincidences.

## Post Hoc Ergo Propter Hoc *(After This, Therefore Because of This)*

As we have seen, statistics can be misleading or inappropriate. Some of the potential abuses of statistics correspond to some classical logical fallacies. Statistics often illuminate correlations between two phenomena. Two collections of data are correlated if there is a relationship between the two. For example, one would expect a correlation between height and weight. Various types of correlations of pairs of data can be detected through the use of scatter plots. A point on a scatter plot represents the two data measurements.

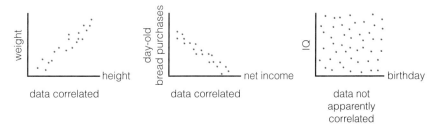

When examining correlated data, we often wonder if one collection is the cause of the other. It is a potential error to deduce cause and effect from correlation. Here are some examples of the *post hoc ergo propter hoc* fallacy.

- While writing our section on randomness, five famous people died. Therefore, writing a section on randomness results in the deaths of five famous celebrities.
- 94% of all CEOs watched cartoons on television in their youth. Therefore, cartoon watching produces corporate leaders.
- Almost two thirds of the inmates in prison have divorced parents. Thus, divorce leads children to criminal behavior.

Statistics is a method for helping us deal with the uncertain and the unknown. Our world is determined meaningfully by events that combine the unpredictable, the chaotic, and the random. We must make decisions and act on processes that are beyond our knowledge. In many cases, statistics give us a hint about what causes what. In other cases, statistics lead us to visit the never-never land of explaining the fundamentally random. Superstition, prejudice, and lucky socks are some of the consequences of finding meaning in the meaningless.

**A LOOK BACK**

Interpreting data can be insightful and valuable. Organizing the data may suggest an underlying structure to the situation we are exploring. Guided by statistical indications, we can seek underlying reasons. Mendel, in his experiments with peas, went from statistics to a theory that was later confirmed by other methods. Sometimes, our analysis of data surprises us by showing that the likely results of an action may be different from what we expect. The interpretation of universal AIDS-testing results or the possible results of improving airline safety demonstrate the potential for unintended consequences.

A principal strategy is to reason about data using probability. Mendel's pea experiment is explained as the natural result of an underlying genetic mechanism combined with probability. The AIDS-testing result combines data with chance, and the air safety issue combines data about miles traveled and the probability of accidents. The twin study is entirely an issue of trying to distinguish chance between random occurrences from a causal relationship.

Look quantitatively to interpret the world more clearly. Go from a vague impression of the way things are to a clear understanding that is confirmed quantitatively. Look for unintended consequences.

*Do not be content with a vague impression.*

*Take the step from a qualitative feeling of a situation to a quantitative understanding of it.*

*Look for unintended consequences.*

◆ ◆ ◆ ◆ ◆ ◆ ◆ ◆ ◆ ◆ ◆ ◆ ◆ ◆ ◆ ◆ ◆ ◆ ◆ ◆

## MINDSCAPES  Invitations to Further Thought*

### I. SOLIDIFYING IDEAS

1. *Mendel's snapdragons.* Another of Mendel's experiments involved red-flowered and white-flowered snapdragons. Through crossbreeding experiments he concluded that each flower had two genes. The red flowers had two red genes, and the white had two white genes. We will call those parent plants the first generation. When two flowers bred, the offspring received one gene from each parent. If the offspring inherited one red and one white gene, the genes combined to produce a pink flower. Suppose a red-flowered snapdragon breeds with a white-flowered snapdragon. What are the possible gene outcomes for the second generation of this crossbreeding?

---

*In the Mindscapes section, exercises marked (H) have hints for solutions at the back of the book. Exercises marked (S) have solutions.

2. ***More Mendel* (S).** Given Mendel's snapdragon experiment in question 1, suppose that two second-generation snapdragons are bred. Construct a 2 × 2 table showing all the possible gene pairs for the third generation of snapdragons. If a third-generation plant is equally likely to inherit a red or white gene from the second-generation parent, what is the probability that a third-generation plant has a white flower? Red flower? Pink flower? What proportions of the colors would you expect to see in a bed of such flowers?

3. ***Oedipus red.*** Given Mendel's work described in questions 1 and 2, suppose that a second-generation plant is bred with a red parent plant. What is the probability of the offspring having a white flower? A red flower? A pink flower?

4. ***Oedipus white.*** Given Mendel's work described in questions 1 and 2, suppose that a second-generation plant is bred with a white parent plant. What is the probability of the offspring having a white flower? A red flower? A pink flower? What proportions of the various colors of flowers would you expect to see in a bed of such flowers?

5. ***Pure white.*** Given Mendel's work described in questions 1 and 2, suppose that a second-generation plant breeds with a third-generation plant. Is it possible for these hybrids to produce an offspring with a pure white flower? If so, explain how; if not, explain why not.

6. ***Blonde, bleached blonde* (H).** You have high standards with respect to truth in advertising, particularly when it comes to hair color. One day at the Laundromat, you meet an attractive, blonde stranger named Chris and wonder if you should pursue a relationship. Unfortunately, you have the nagging belief that Chris's golden locks may have been the result of peroxide—presenting the specter of a dark (haired) future. However, you also know several facts about the incidence of dyed hair and about your ability to detect fraudulent follicles. You know that 90% of blonde people in the world are naturally blonde. You have done a personal survey and learned that you are 80% accurate in your ability to correctly categorize fake hair color as fake and real hair color as real. What is the probability that Chris's hair is fair and that your bleached beliefs were incorrect? Given these facts, should you pursue your relationship with Chris?

7. ***Blonde again* (S).** Given the scenario in question 6, suppose now that 70% of blonde people are naturally blonde and that you are able to accurately detect dyed hair 85% of the time. What is the probability that Chris's hair is fair and that your bleached beliefs were incorrect? Given these facts, should you pursue your relationship with Chris?

8. ***Bleached again.*** Given the scenario in question 6, suppose now that 80% of blonde people are naturally blonde and that you are able to accurately detect dyed hair 50% of the time. What is the probability that Chris's hair is fair and that your bleached beliefs

were incorrect? Given these facts, should you pursue your relationship with Chris?

9. **Pocket change.** Ask 10 people to count the number of coins in their pockets. Record their answers. How many matches are there? Are you shocked by the similarities? Explain.

10. **Random person.** Pick a person at random (in your dorm, your class, or a dining hall) whom you do not know well. Find seven unusual common features with that person (your mothers were both born in San Francisco, and so on). Devise a test of 10 questions that would make these observations seem remarkable.

11. **Another random person.** Using the 10-question test created in the previous question, survey another random person. How many answers do the two of you have in common? Are you surprised? Explain.

12. **Correlated issues.** From the class survey of Section 7.6 (Mindscape I.1), find five examples of pairs of questions for which you would expect to find some correlation of the answers. Explain what correlation you would expect. Include guesses as to how the scatter plots might look.

13. **College correlation.** Using the student survey data given below, create the five scatter plots for the number of parties per month versus the other five columns. Which pairs are correlated? For the correlated pairs, what type of correlation is there?

Data of a Student Survey

| Student | Hours on Web/Week | Number of E-Mails Received/Week | Number of Dates/Month | Number of Hours/Week Studies | Number of Hours/Week Sports | Number of Parties/Month |
|---|---|---|---|---|---|---|
| 1 | 16 | 45 | 3 | 25 | 10 | 3 |
| 2 | 5 | 17 | 7 | 50 | 6 | 1 |
| 3 | 10 | 26 | 11 | 21 | 15 | 6 |
| 4 | 13 | 30 | 9 | 25 | 8 | 2 |
| 5 | 3 | 10 | 7 | 30 | 18 | 4 |
| 6 | 7 | 19 | 9 | 14 | 5 | 8 |
| 7 | 9 | 24 | 8 | 35 | 5 | 1 |

14. **Correlation web.** Using the student survey data given in question 13, create the five scatter plots for the number of hours per week spent surfing the Web versus the other five columns. Which pairs are correlated? For the correlated pairs, what type of correlation is there?

15. **Date correlation.** Using the student survey data given in question 13, create the five scatter plots for the number of dates per week versus the other five columns. Which pairs are correlated? For the correlated pairs, what type of correlation is there?

16. *Safety first* (H). Suppose a car is widely believed to be the safest car made. You might expect people driving this car to be less severely injured in accidents, but why would you expect that model of car to be involved in fewer accidents per million miles than a sports car?

17–20. For Mindscapes 17–20, say whether or not each is an example of *post hoc ergo propter hoc*. Explain your reasoning.

17. *Lucky charms.* "I was doing lousy on the soccer field. Then my girlfriend gave me a necklace to wear. Ever since then, I've blocked every shot to our goal. That necklace bought me good luck. If I keep wearing it, we'll keep winning."

18. *Read my lips.* "Shortly after a Republican tax bill bringing tax breaks to the wealthy becomes law, the economy moves into a recession. The Democrats campaign that this new law was the cause of the country's economic crisis and thus want to repeal it."

19. *Penny luck.* "Right before going in to buy my lottery ticket, I saw a penny on the ground. Usually I don't bother to pick up pennies, but this time I did. That night I won $5,000! I'm always going to pick up any pennies I see on the ground from now on."

20. *The lick.* "I was visiting some friends and their dog Golden licked my hand. Two days later I felt achy and had a mild fever. I must have caught some bug from that darn dog lick."

## II. CREATING NEW IDEAS

1. *Martian genetics.* A certain alien species has three genes that determine the color of their eyes. There are three types of genes: Red, White, and Blue. If a creature has all three genes the same color, then the creature will have that color eyes. If a creature has R R W, then its eye color is red; if it has R W W, then its eye color is pink; if it has R W B, then its eye color is light purple; if it has R R B, then its eye color is purple; if it has R B B, then its eye color is dark purple; if it has B B W, then its eye color is blue, if it has B W W, then its eye color is sky blue. Three alien parents are required to produce one offspring. Each parent contributes one gene to the new alien, and the gene is randomly selected. Suppose an R R R and a W W W and a B B B mate and produce an offspring. What are the possible eye colors of this second-generation offspring, and, for each such color, what is the probability that the offspring will have that eye color?

2. *More martians.* Given the scenario from the previous question, suppose that a second-generation alien with genes R W B mates with a W W W and another W W W. What are the possible eye colors of their offspring, and, for each such color, what is the probability that the offspring will have that eye color?

3. **Scholarship winner (H).** You apply for a national scholarship. 100,000 students apply and 200 scholarships will be awarded. You call to find out if you are one of the winners. The absent-minded professor in charge of the program reports that the letters have just been sent out, but no list of all the winners is available. The professor also recalls that you were selected. The professor correctly recalls information 90% of the time. Assuming that the scholarships are awarded randomly (not a completely ridiculous assumption), what is the probability that you won one?

4. **Taxi blues.** An eyewitness observes a hit-and-run taxi cab accident in a city in which 85% of the cabs are green and 15% are blue. The witness is 100% certain that the cab was blue. Given all this information, how likely is it that the cab actually was blue?

5. **More taxi blues (S).** An eyewitness observes a hit-and-run taxi cab accident in a city in which 85% of the cabs are green and 15% are blue. The witness is 80% sure that the cab was blue. Given all this information, how likely is it that the cab actually was blue?

6. **Few blues.** An eyewitness observes a hit-and-run taxi cab accident in a city in which 95% of the cabs are green and 5% are blue. The witness is 80% sure that the cab was blue. Given all this information, how likely is it that the cab actually was blue?

7. **More safe.** Given the scenario of our air safety discussion, suppose now that planes could be made 10 times safer with government support, so that the airplane ticket prices would rise less and thus only 5% of travelers who now fly would choose to drive instead. Assuming all the other data still holds from our discussion, what is the net result on lives lost if we make the planes safer?

8. **Less safe.** Given the scenario of our air safety discussion, suppose now that, if planes were made only five times safer, then the airplane ticket prices would rise less than before, and thus only 1% of travelers who now fly would choose to drive instead. Assuming all the other data still holds from our discussion, what is the net result on lives lost if we make the planes safer?

9. **Politics as usual.** During the presidential campaign of 1968, it is reported that Richard Nixon made the following statement about his opponent Hubert Humphrey: "Hubert Humphrey defends the policies under which we have seen crime rising 10 times as fast as the population. If you want your president to continue a do-nothing policy toward crime, vote for Humphrey. Hubert Humphrey sat on his hands and watched the United States become a nation where 50% of the American women are frightened to walk the streets at night." Is this a logical reasoning fallacy? If so, what kind of fallacy is it, and what is the fallacy?

10. **Whoops.** Find an example in a recent newspaper or magazine story in which a *post hoc ergo propter hoc* error was stated. Analyze the error.

## III. FURTHER CHALLENGES

1. ***Mendel genealogy*** **(H).** We return once again to Mendel's snapdragon experiment from question I.1. Suppose two second-generation crossbred plants mate. What is the probability that the offspring will be white-flowered?

2. ***Modified Mendel.*** We again return to Mendel's snapdragon experiment from question I.1. Suppose that, in the crossbreeding of the second-generation pink flowers, the probability that a pink snapdragon passes on the white gene is 1/3, and the probability it passes on the red gene is 2/3. What would be the possible outcomes for the third-generation offspring if two second-generation flowers are bred? What would be the probabilities of the various colors? What color would you expect to see?

3. ***HIV tests.*** Recall that, in the United States, approximately 1 person in 1,000 is estimated to be HIV positive. Suppose a person decides to take two independent tests. Test A determines if a person is infected or not with HIV with 95% accuracy, whereas Test B is 99% accurate. Suppose this person learns that both tests came back positive—that is, both predicted that the person was infected with HIV. What is the probability that this person is actually infected?

4. ***More HIV tests.*** Given the tests described in the previous question, assume now that Test A came back positive, and Test B came back negative. What is the probability that this person is actually infected?

5. ***Reduced safety.*** Given the scenario of our air safety discussion, suppose now that the FAA reduced many of its safety requirements, and air travel safety dropped by 5% (that is, 5% more people would be killed in air travel). However, the airlines passed this savings onto the consumers, and lower ticket prices resulted in a 15% increase in air travel and a like reduction in travel by car. Assuming all the other data still holds from our discussion, what is the net result on lives lost if we make the planes less safe?

## IV. IN YOUR OWN WORDS

1. ***Personal perspectives.*** Write a short essay describing the most interesting or surprising discovery you made in exploring this section's material. If any material seemed puzzling or even unbelievable, address that as well. Explain why you chose the topics you did. Finally, comment on the aesthetics of the mathematics and ideas in this section.

2. ***With a group of folks.*** In a small group, discuss and work through the reasoning involved in AIDS testing and false positives. After your discussion, write a brief narrative describing the reasoning in your own words.

# Farewell

As you read these final pages, we wonder what you will take away from your journey. As you close this book for the last time, we hope that some parts of it will remain open in your mind. Perhaps a few specific concepts will become part of your permanent collection of ideas. Perhaps your sense of mathematics has changed. Maybe mathematics will never seem quite so dry; maybe it seems richer now. We hope the life lessons of this book expanded your repertoire of strategies and modes of thought. These lessons are simple yet profound. We hope that they help you strengthen your confidence to face challenging life issues and conquer them. Mathematics is an empowering force. If you can grapple with infinity and the fourth dimension, then what can't you do? We hope that in 20 years you will still be contemplating some of the ideas you encountered here.

We now invite you to look back over your experience. Perhaps some concepts were challenging and appeared always just out of reach; some you struggled with before making them your own; and some are now so solid in your mind that they are now a part of you. In any case, we hope that you will recall all of them with great satisfaction.

It all began, as it should, with fun and games. Our mission was to pose some entertaining and thought-provoking conundrums that would introduce certain modes of logical thought and problem solving. Simultaneously, we foreshadowed ideas to come.

> *... the primary question was not What do we know, but How do we know it.*
> —Aristotle

What do you now think of when you contemplate the notion of *number*? Perhaps you think about the power of counting, as in the Pigeonhole principle, or you recall intriguing patterns of natural numbers, such as the Fibonacci numbers or the prime numbers. Perhaps your mind drifts to mod clock arithmetic, and you remember that arithmetic serves our technological world in unexpected ways, from error-correcting digits to public key cryptography. Perhaps you now see and appreciate subtle distinctions between the rational and the more mysterious and less understood irrational numbers.

There are a handful of mathematical arguments that are so artful that they may be viewed as elegant. Three of these arguments are from the study of numbers: There are infinitely many prime numbers; the square root of 2 is irrational; and 1 equals .99999 . . . . We hope you made these proofs your own and that you will share them with others.

> *Wherever there is number, there is beauty.*
> —Proclus

What could be more empowering than understanding the nuances of infinity? That some infinities are actually larger than others is a profound, counterintuitive idea that has shaken the foundations of many people's beliefs. Certainly, when you think of the infinite, part of what we hope you will remember is

> *Oh moment, one and infinite!*
> —Robert Browning

the power of a simple and clear definition—that of a one-to-one correspondence—and how carefully following its consequences opened a new world of insight.

We hope that the Pythagorean Theorem and its jigsaw-puzzle proof are yours for life. Once the aesthetically appealing Golden Rectangle becomes part of your consciousness, you will see it often—in architecture, art, and snails. The jumbled pinwheel pattern with the same symmetry of scale as a regular checkerboard pattern, despite its chaotic appearance, is an enticing illustration of the beauty and the nuance of the aperiodic.

> *Might is geometry; joined with art, resistless.*
> —Euripides

Can you sit in a room and see in your mind's eye the outline of its dual octahedron? Maybe when you see a soccer ball, you will recognize the wedding of a Platonic solid and its dual. We hope you have developed a sense of the geometry of the fourth dimension—where being inside a box may not be an obstacle. Indeed, building some intuition about the fourth dimension is an outstanding example of the power of analogy and abstraction.

The amorphous universe of topology thrives on the unexpected. Whether you are removing a rubber vest without removing your outer coat, untying a knotted ring, or dissecting Möbius Bands, we hope you discover important lessons in visualization and the wonderful consequences of bending and twisting space.

> *The true spirit of delight . . . is to be found in mathematics as surely as in poetry.*
> —Bertrand Russell

Topology also has unexpected consequences in real life—including the life of the twisted DNA molecule. Perhaps you now understand why at any instant there must be two diametrically opposite places on the Earth where the temperatures are identical. We believe it is exciting to discover abstract mathematical issues that lead to interesting new facts about the real world. We certainly hope you share our appreciation for the wonderful interplay between the physical and the abstract.

Fractal pictures are literally infinitely intricate. They all arise from repeating a simple process infinitely often, and they seem to capture the complexity of nature. Repetition is at the heart of both chaos and fractals. As ordered as we wish our universe to be, in reality, chaos reigns.

> *God has put a secret art into the forces of Nature so as to enable it to fashion itself out of chaos into a perfect world system.*
> —Immanuel Kant

We hope that you now see that amazing coincidences are nearly certain to happen to you. After one thinks about chance and randomness, surprises appear in a whole new light. Randomness is a powerful force; it permits us to collect answers to embarrassing questions that people would normally never answer truthfully. We hope you have developed a feel for expectation and that you will look at lotteries and other games of chance with a critical eye.

> *Chance, too, which seems to rush along with slack reins, is bridled and governed by law.*
> —Boethius

You've seen that we must be cautious with statistics and not read more into an average than we rightfully should—remember what the average American has. We hope that you have a better sense of the significance of false positives in medical tests. Perhaps you are more sensitive to the possibilities of unintended consequences. If improved airline safety raises prices and therefore makes more people drive, maybe safer airplanes would lead to more accidental deaths—but not in the air.

As we approached the end of this *Farewell*, we considered copying it in a smaller font that would, of course, contain the entire *Farewell* in an even smaller font, which would, of course, contain another copy of the whole *Farewell*, and so on. That fractal font farce would be fun, but we resisted the temptation.

Nevertheless, we do end with a paradox, because we hope that this end is also a beginning. One of the great features about mathematics is that it has an endless frontier. The farther you travel, the more you see over the emerging horizon. The more you discover, the more you understand what you've already seen. How many more ideas are there for you to explore and enjoy? How long is your life?

*The Masterpiece or The Mysteries of The Horizon* (1955) by René Magritte.

**LESSONS FOR LIFE**

1. Just do it.
2. Make mistakes and fail, but never give up.
3. Keep an open mind.
4. Explore the consequences of new ideas.
5. Seek the essential.
6. Understand the issue.
7. Understand simple things deeply.
8. Break a difficult problem into easier ones.
9. Examine issues from several points of view.
10. Look for patterns and similarities.

# Acknowledgments

One theme of this book is that ideas arise from many sources, and certainly many sources were important in the creation of this book. Here we thank those who contributed to *The Heart of Mathematics: An invitation to effective thinking*. We began writing this book in February of 1995. Since then, hundreds of people have helped to inspire us and shape our thinking. First and foremost we would like to thank all our students who used various draft versions of this text (many without figures). Our students' input had an enormous impact on what we wrote and how we wrote it, but more fundamentally, they gave and continue to give us the motivation to make mathematical thought a pleasurable part of their lives.

We hope this textbook helps teachers bring mathematics to life for students. We are therefore especially grateful to the many professors of mathematics who reviewed drafts of the manuscript. Our reviewers include:

Thomas Banchoff
Brown University

Phillip Johnson
Appalachian State University

John Emert
Ball State University

Darrell Kent
Washington State University

Earl Fife
Calvin College

George Kertz
University of Toledo

Rich France
Millersville University

Joe Malkevitch
CUNY, York College

Kay Gura
Ramapo College of NJ

John Orr
University of Nebraska

Edwin Herman
St. Thomas University

Edward Thome
Murray State University

Fred Hoffman
Florida Atlantic University

Stan Wagon
Macalester College

We would especially like to thank Earl Fife and Joe Malkevitch, who read several versions of this text. Their detailed comments were of great value. We also wish to thank Mary L. Platt from Salem State College;

Bret A. Simon and Angela Vanlandingham, from the University of Colorado, who used a manuscript version of the text in their classes; and Stephen A. Kenton, from Eastern Connecticut State University, who used a draft of the infinity chapter in his course *Infinity: Math and Philosophy*. Their comments and their students' reactions were extremely valuable.

The Editorial Director at Springer-Verlag, Jeremiah Lyons, was a constant source of enthusiasm, inspiration, and insight. From the very beginning, Jerry supported our vision of an enticing book that would be truly enjoyable for real students to read, and he remained confident of its potential. He enabled us to turn our more unconventional ideas into reality. Throughout this process he has demonstrated a great sense of humor, and it is our hope that one day he will win the Superfecta at Saratoga. It has been a privilege to work with Jerry Lyons.

Our developmental editor, Jeanne Woodward, helped to focus us on our meta-message of effective thinking. Jeanne inspired us to reexamine the format and organization of the sections. Her comments and those of John Bergez resulted in a much more readable text. Eric Gerde accomplished the monstrous task of answering each and every Mindscape question. Larry Cannon, Robert Heal, and Richard Wellman from Utah State University allowed us to tap their creative talents for the CD-ROM component of the kit. Daniel Symmes and Jim Carbonetti created the wonderful 3D art. The production of a complicated project such as this text requires many editors, artists, designers, and others. Their imagination and talent helped bring this book to life. We thank everyone for their creativity, hard work, and dedication.

Finally, we would like to thank the Educational Advancement Foundation for their support. The EAF fosters methods of teaching that promote independent thinking and student creativity.

We close with our individual acknowledgments.

*Edward Burger:* The two greatest influences on my life have been my parents, Florence and Sandor Burger. They have been a constant source of love, support, joy, and inspiration.

My interest in this course started in 1988. I was intrigued by a course offered at Austin Community College entitled *Mathematics: Its Spirit and Use*. I want to thank Stephen Rodi from ACC for giving me the opportunity to teach liberal arts students. Also from those early years, I wish to thank Jeffrey Vaaler, my Ph.D. advisor, who has been and continues to be one of the greatest teachers, mentors, and friends I've ever had.

In 1993, I developed a similar course at Williams College. My colleagues and students at Williams have always been a great source of inspiration, and I thank them all. Two who deserve special recognition are Stephen Fix, from the English Department, and my student Eric Cohen, both of whom provided me with many valuable specific comments.

For me, the work on this book began and ended at the University of Colorado at Boulder, where I spent sabbatical leaves in 1995 and 1999. I wish to thank the Department of Mathematics and the Center for Number Theory for all their interest, encouragement, and support in my work.

The greatest aspect of this project for me was my association with Michael Starbird. Mike is a sheer joy and inspiration to be around. He has the wonderful ability to turn work into play and play into innovative ideas. Far beyond this project, Mike has taught me life lessons too numerous to mention. My respect and admiration for Mike is immense, and I am most fortunate and grateful to have him as a role model, mentor, and close friend. My collaboration with Michael Starbird has been one of the most valuable experiences of my life.

*Michael Starbird:* I am incredibly lucky to have a family whose support and encouragement have inspired me throughout my life. My parents gave me curiosity and independence. My wife, Roberta, and children, Talley and Bryn, are the joyful foundation for all I do. I want to thank all my family from the bottom of my heart.

I began teaching a mathematics course for liberal arts students because Betty Sue Flowers, the Director of the Plan II Honors Program at UT, needed a new mathematics course to enrich the technical side of the program. That request opened the door to a decade-long project that includes this book. Paul Woodruff, the subsequent Director of the Plan II Honors Program, continued to support the course and encouraged me to continue its development. Friends are often most significant for what they ask rather than what they give. I thank my friends for what they asked.

I thank all my friends and colleagues from the Department of Mathematics and across UT who have been uniformly supportive of the idea of making mathematics fun. They encouraged me by showing a real interest in the mathematical ideas that are at the heart of this book.

Finally, I want to thank my coauthor. Certainly the single greatest joy from this whole project was working with Ed. His imagination is boundless. His energy is endless. His insight and humor are uncontainable. He has a clear vision of the whole project and can see every aspect of what needs to be done. He does global thinking and attends to details. But best of all, the collaboration we created brought out the strengths of each of us. It is an honor and a privilege for me to learn from and work with my friend and coauthor, Edward Burger.

# Hints and Solutions
# (H = Hint; S = Solution)

## CHAPTER 2: NUMBER CONTEMPLATION

### 2.1 Counting

**I.1. (H)** Estimate the weight of the million dollars. How does the weight of a single bill compare to the weight of a single $8.5 \times 11$ piece of paper or a ream of paper?

**I.3. (H)** Don't worry about the total number of pieces on the board. Just estimate how many pieces are on the 64$^{\text{th}}$ square.

**I.6. (S)** Let's first estimate the width of an ordinary piece of paper by noting that packages of 200 sheets of paper are more than 1/2 inch thick. So, a single piece of paper is at least 1/400 inch thick. Now, after one folding, the paper is twice the original thickness. After two foldings, the paper is $4 = 2^2$ times as thick. After 50 foldings, the paper will be $2^{50}$ times as thick. The resulting paper is then $2^{50}/400$ inches thick. (That's more than $2.8 \times 10^{13}$ inches, which is more than 40 million miles!)

**II.2. (H)** How many numbers less than 10,000 have no 3 in them? How many choices are there for each of the four digits of such a number?

**II.5. (S)** Within the next 100 years, the 5.8 billion people currently populating the Earth will die. If less than 50 million people died each year, then, at the end of 100 years, only 5 billion people would have died. This contradiction shows that at some point more than 50 million people will die in a year. Alternatively, note that the average number of people who will die each year is $5{,}800{,}000{,}000/100 = 58$ million. Since this is the average, there must be at least some year in which more than 58 million people will die (otherwise the average would be *lower*).

### 2.2 Numerical Patterns in Nature

**I.2. (S)**

| $n$ | 1 | 2 | 3 | 4 | 5 | 6 | ... |
|---|---|---|---|---|---|---|---|
| $(F_n)^2$ | 1 | 1 | 4 | 9 | 25 | 64 | ... |
| $(F_{n+1})^2$ | 1 | 4 | 9 | 25 | 64 | 169 | ... |
| sum | 2 | 5 | 13 | 34 | 89 | 233 | ... |
|  | $F_3$ | $F_5$ | $F_7$ | $F_9$ | $F_{11}$ | $F_{13}$ | ... |

Note that we are getting all the odd Fibonacci numbers. This leads to the formula, $(F_n)^2 + (F_{n+1})^2 = F_{2n+1}$.

**I.4. (H)** Make a grid with five rows: month number, number of mature pairs, number of new babies, number of older babies, and total number of rabbits. Start with one mature pair and no babies and fill out the rest of the grid.

**I.10. (S)** If you look closely, you'll notice that the pieces don't line up exactly. The triangle with sides 3 and 8 appears to be similar to the big "triangle" with sides 5 and 13. If this were true, then the corresponding ratios would be equal. But 8/3 isn't 13/5. Since these are ratios of consecutive Fibonacci numbers, the ratios are close, and that's why this is a convincing trick!

**I.11. (H)** Start with the largest Fibonacci number smaller than the given number and work your way backward. $52 = 34 + 18$. ($18 = 13 + 5$), so $52 = 34 + 13 + 5$.

**II.2. (S)** Question I.3. showed us that $F_{2n} = (F_{n+1})^2 - (F_{n-1})^2$, which can be factored as the product of $(F_{n+1} - F_{n-1}) \times (F_{n+1} + F_{n+1})$. This means that besides 3, none of the even Fibonacci numbers are prime!

**II.10. (H)** Start with a natural number $N$. If $N$ is a Fibonacci number, then we're done; if it isn't, then grab the largest Fibonacci number smaller than $N$, and call it $F$. Now repeat with $N - F$.

**III.5. (H)** Rewrite the fraction under the square root as $F_{n+2}/F_n = (F_{n+2}/F_{n+1}) \times (F_{n+1}/F_n)$.

## 2.3 Prime Cuts of Numbers

**I.10. (H)** One of the two primes in the sum will have to be 2.

**I.11. (S)** The harder question is, Are any of these prime? We can describe each element in the list by its number of digits. If it has an even number of digits, then the number is divisible by 11. If it has 3,6,9, … digits, then the number is divisible by 3. By computer search, the first three primes in the sequence have 19, 23, and 317 digits.

**I.15. (H)** Try to factor the $n = 16$ and $n = 17$ cases. (Factor the $n = 40$ and $n = 41$ cases for the bonus.)

**I.19. (S)** Write the original number as $X = 13A + 7$. Adding 22 yields, $X + 22 = 13A + 7 + 22 = 13A + 29 = 13A + 13 \times 2 + 3 = 13 \times (A + 2) + 3$. So 13 goes into our new number $(A + 2)$ times with a remainder of 3. Alternative, just add $7 + 22$ and find its remainder.

**II.1. (H)** Write the two numbers, $X$ and $Y$, in the following way: $X = 57A + r$, and $Y = 57B + r$. The difference is $(X - Y) = 57A - 57B = 57 \times (A - B)$.

**II.10. (S)** By exhaustively looking at differences between successive primes you will find that the first string of six nonprimes appears between 89 and 97. But it's more interesting to consider the hint. Let $M = 2 \times 3 \times 4 \times 5 \times 6 \times 7$. Both $M$ and $M + 2$ are divisible by 2, so $M + 2$ isn't prime. Similarly, $M + 3$ is divisible by 3, $M + 4$ is divisible by 4, and so on up to $M + 7$, which is divisible by 7. $(7! + 1) = 5041$.

**III.5. (H)** Suppose there were only finitely many primes, $p_1, p_2, p_3, \ldots, p_L$. Consider $N = p_1 p_2 p_3 \ldots p_L + 1$. Think about the prime factors of $N$.

## 2.4 Crazy Clocks and Checking Out Bars

**I.3. (H)** Express each number as a simpler number mod 12.

**I.4. (S)** Note that $(3 \times 0) + (1 \times 7) + (3 \times 1) + (1 \times 7) + (3 \times 3) + (1 \times 4) + (3 \times 0) + (1 \times 0) + (3 \times 0) + (1 \times 2) + (3 \times 1) + (1 \times 8) = 43$. Since the sum is not evenly divisible by 10, it's not a correct UPC. The corresponding sums for the next two codes are 40 and 42, respectively. So the second code is the correct one.

**I.8. (H)** If the covered digit were $D$, then the sum would be $55 + 3D$. Now find a value for $D$ that will make the sum divisible by 10.

**I.14. (S)** There are three unknown digits, the 9, the 1, and the 7. Since each digit could be one of two different numbers, we have eight possible combinations in all to try: 903068823517, 903068823511, 903068823577, 903068823571, 403068823517, 403068823511, 403068823577, 403068823571. Of all these numbers, only 903068823577 is a valid code. This is your best guess.

**II.1. (S)** $129 = (9 \times 13) + 12$, so 12 is the remainder when 129 is divided by 13. We can also say $129 \equiv 12 \pmod{13}$. You would spin around 9 times and then move the clock hand ahead 12 hours more.

**II.8. (H)** Let $D$ represent the check digit, and compute the resulting sums. Now find the values of $D$ between 0 and 9 that will make the sum divisible by 11.

**III.2. (H)** Look at numbers of the form $x\,y\,0000000\,z$ and $y\,x\,0000000\,z$ whose sum is $3x + y + z \pmod{10}$, and $3y + x + z \pmod{10}$, respectively. When does $3x + y \equiv 3y + x \pmod{10}$?

## 2.5 Secret Codes and How to Become a Spy

**I.4. (H)** Encrypting a message involves raising it to the public power 7 mod the public number 143. Decoding an encrypted message involves raising the encrypted message to the secret power 103 mod 143.

**I.6. (S)** Note first $m = (3-1) \times (5-1) = 8$. Since $e$ must be relatively prime to $m$, we need only consider the values $e = 1, 3, 5,$ and $7$. For each possible value of $e$, find $d$ and $y$ that satisfy $de - 8y = 1$. For example, for $e = 1$, fill in the following blanks: \_\_\_ $\times 1 - 8 \times$ \_\_\_ $= 1$. Since $1 \times 1 - 8 \times 0 = 1$, $(e = 1, d = 1)$ is a pair. Similarly, since $3 \times 3 - 8 \times 1 = 1$, $5 \times 5 - 8 \times 2 = 1$, and $7 \times 7 - 8 \times 6 = 1$, $(e = 3, d = 3)$, $(e = 5, d = 5)$, and $(e = 7, d = 7)$ are possible pairs.

**I.10. (H)** Remember that decoding an encrypted message involves raising the encrypted message to the secret power.

**II.1. (S)** The hint asks us to recall that $5^6 \equiv 1 \pmod{7}$. This means that $(5^6)^k \equiv 1^k \equiv 1 \pmod{7}$ for any integer $k$. In particular, since $600 = 6 \times 100$, it is convenient to choose $k = 100$, giving us $5^{600} \equiv (5^{6 \times 120}) \equiv (5^5)^{120} \equiv 1^{120} \equiv 1 \pmod{7}$. Similarly, since $1000000 = 10 \times 100000$, $8^{1000000} \equiv 1 \pmod{11}$.

**II.2. (H)** Remember that $5^6 \equiv 1 \pmod 7$. So, $5^{6k}$ will also be congruent to 1 (mod 7) for any natural number $k$.

## 2.6 The Irrational Side of Numbers

**I.5. (H)** Imitate the proof of the irrationality of $\sqrt{2}$, except replace the notions *even* and *odd* with *divisible by 5* and *not divisible by 5*.

**I.10. (S)** We need a modification of the proof in problem I.5. Assume $\sqrt{10} = c/d$ with $c$ and $d$ having no common factors. Squaring gives $c^2 = 10d^2$. Since the right-hand side is divisible by 5, the left-hand side is also divisible by 5, so that $c$ is divisible by 5. (If $c$ weren't divisible by 5, then $c^2$ wouldn't be divisible by 5 either.) We can replace $c$ with $5n$ to get $25n^2 = 10d^2$, or $5n^2 = 2d^2$. We want to show that 5 divides $d$. Imagine writing out the prime factorization for the left and right sides of the equation. On the left we have all the prime factors of $n$ (listed twice since we are squaring $n$) and 5. On the right we have 2 and all the prime factors of $d$ (listed twice). Since numbers are equal, we call on the uniqueness of prime factorizations to argue that the list of primes for both numbers are the same. Since the prime 5 appears on the left side, it must also appear on the right side. And since it can only come from the prime factorization of $d$, we must have that 5 is a prime factor of $d$. So $d$ is divisible by 5, and we have our contradiction.

**I.12. (H)** Assume that $E$ is rational—that is, $E = n/m$, so that $12^{(n/m)} = 7$. Raise both sides to the $m^{\text{th}}$ power; $12^n = 7^m$ and look at the prime factorizations of each side.

**I.20. (S)** Assume $\sqrt{(2/3)} = a/b$ (with no common factors). Squaring gives $2b^2 = 3a^2$. At this point it doesn't matter whether you choose 2 or 3, but you must stick with it! (We'll choose 3.) The right side is divisible by 3, and so $2b^2$ is divisible by 3. Since 3 is prime and 2 isn't divisible by 3, $b^2$ must be divisible by 3. Again, since 3 is prime we conclude that $b$ is divisible by 3. Writing $b = 3n$, and substituting it into the equation gives $18n^2 = 3a^2$, or $6n^2 = a^2$. Using the same reasoning, we conclude that 3 divides $a$. So 3 divides both $a$ and $b$ contradicting the fact that $a$ and $b$ had no common factors. We conclude that our original assumption was wrong, and therefore $\sqrt{(2/3)}$ is irrational.

**II.5. (H)** Use the $\sqrt{2}$-is-irrational proof as a template, but begin by *cubing* both sides instead of squaring both sides.

**II.8. (S)** Not always. The not-so-satisfying counterexample: $\pi$ and $(-\pi)$ are both irrationals, yet their sum is zero, which is rational. The numbers $1.01001000100001\ldots$ and $0.10110111011110\ldots$ are both irrational because their decimal expansion doesn't repeat. Their sum is $1.11111111111111\ldots = 10/9$, a rational number. Keep in mind that sometimes the sum is irrational, for example, $\pi + \pi = 2\pi$.

**III.5. (H)** Assume that they are both rational—that is, $(a + b) = m/n$ and $(a - b) = r/s$, and solve for $b$.

## 2.7 Get Real

**I.5. (H)** Take an irrational number that you know and stick its nonrepeating decimals after 5.70.

**I.9. (S)** Use long division (or a calculator!). 7 goes into 60 eight times with a remainder of 4; bring down the zero. 7 goes into 40 five times with a remainder of 5; and so on. $6/7 = 0.857142857142857142857143\ldots$

**I.16. (S)** Call our elusive number $E$, so that $E = 43.121212\ldots$ Since there are two digits in our repeating segment, multiply $E$ by 100 to shift the decimals by two digits: $100E = 4312.121212\ldots$. Subtracting gives:

$$100E = 4312.121212\ldots$$
$$-E = 43.121212\ldots$$
$$99E = 4269.$$

Thus we see that $E = 4269/99$.

**I.20. (H)** This number has arbitrarily long sequences of zeros. Why does that imply that there is no repeating sequence and that the number is irrational?

**II.3. (H)** $(10^n + 1)$ has $n - 1$ adjacent zeros. So, the decimal expansion contains arbitrarily long sequences of zeros. Could this number have a repeating sequence of length $N$?

**II.6. (S)** Using the same argument used in question I.20, let's show that $y = 0.21221222122221\ldots$ is also irrational. Suppose there were a repeating sequence of length $N$. If the repeating sequence were all 2's, then we'd end up with a rational number. If the repeating sequence were not all 2's, then eventually we would see a non-2 digit after every $N$ digits. But this isn't the case.

**III.2. (H)** If the two real numbers were farther than 1 apart, then there would be an integer between them. Can you find a sequence of rationals that are evenly spaced over the whole line (like the integers) where the spacing is less than the difference between the two given reals?

# CHAPTER 3: INFINITY

## 3.1 Beyond Numbers

**I.1. (H)** Weighing two jars of pennies against each other indicates which jar has more pennies without giving any clue as to the number of pennies in each jar. The process of passing out tests to students quickly tells you whether you have more tests, more students, or exactly the same number of tests as students.

**I.2. (S)** If the correspondence were not one-to-one, we would have one of the following situations:
a. A symbol representing no company: No one would buy stock in a company that didn't exist. Conversely, if a company has no symbol, then no one will ever purchase its stock.
b. A symbol representing two different companies: This poses a problem if you want to invest in one of the companies but not in the other.

**I.8. (H)** Does each resident of the United States have a Social Security number? Are they different?

**II.2. (S)** The number of possible social security numbers is $10^9$, but the population is only about 250 million. If a one-to-one correspondence existed, then there would be the same number of elements in each set.

**II.5. (H)** Given the information in the question, there may be more rooms than students, more students than rooms, or the same number of students and rooms. Why? Ask your roommate.

## 3.2 Comparing the Infinite

**I.6. (H)** Make a chart with two columns: one containing the list of natural numbers and the other containing the elements of **TIM**. The correspondence can't be given by a nice formula.

**I.11. (H)** What if the hotel manager instructs all guests to move one room down the hall?

**I.20. (S)** If the sets had the same cardinality, then there would exist a correspondence that matched every element from one set with every element of the other, something the question states specifically cannot happen. Therefore, the sets have different cardinalities, and the set that intuitively seems larger in fact is.

**II.1. (H)** First list all fractions with 1 in the denominator. Then list all new fractions that can be obtained by using 2 for the denominator. Continue increasing the denominator and listing in order all the new fractions that can be created.

**II.7. (S)** Let's suppose that the streets are laid out as in the figure below. Start at the lower left intersection, label it 1, and start spiraling outward. As we move outward, label each intersection with the next natural number. Since we will eventually "hit" any single intersection, we've provided a valid one-to-one correspondence between intersections and natural numbers.

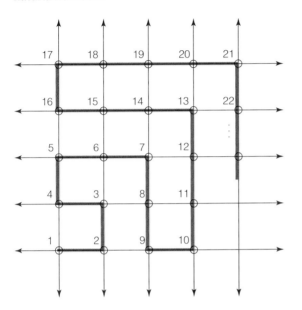

**III.1. (H)** Is ball number 1 in the barrel? No, it was gone after the first step. How about number 2? Number 1294959?

## 3.3 The Missing Member

**I.2. (S)** Instead of a six by six Dodge Ball game, Cantor's game has infinitely many rows and infinitely many columns. Since Cantor starts with a listing of the reals, it is as if Player One has filled out the entire Dodge Ball chart before Player Two begins. Finally, instead of filling a table with X's and O's, Cantor creates his new number using two digits (say, 2's and 4's) and places them in such a way that the new number differs from the $n$th number on the list in at least the $n$th decimal place.

**I.8. (H)** Assume that the set of colorings were the same size as the naturals. Now use the diagonalization argument to create a new coloring that isn't on the list.

**I.10. (H)** Make a list of decimal numbers in which the first number begins with a 7 after the decimal point. Now write down the remaining digits of that first number (just use 2's and 4's), and then complete the list so that our first number turns out to be the same as M.

**II.1. (H)** Modify Cantor's diagonalization argument. Note the connection between this question and both Coloring Revisited (I.8.) and A Penny For Your Thoughts (I.9.)

**II.2. (S)** This procedure does work; an example is given below.

$$0.24434322234\ldots$$
$$0.43223253242\ldots$$
$$0.23424432133\ldots$$
$$0.23435442432\ldots$$
$$\ldots$$
$$M = 0.22442222\ldots$$

Note that each pair of digits in M is determined by a single number on the list. Also, M is different from each number on the list. For example, it differs from the third number in the third pair, the fourth number in the fourth pair, and so on.

## 3.4 Travels Toward the Stratosphere of Infinities

**I.2. (H)** Each agenda item either appears or does not appear on the final agenda. The number of different agendas is equal to the number of different ways that eight named boxes could be checked with either Y's (yes's) or N's (no's).

**I.3. (S)**

@ and ! are in S,

{!, #, %} and {#} are in $\mathcal{P}(S)$,

{ {!}, {@}, {#}, {%}, {&} } and { {@}, {$, !} } are in $\mathcal{P}(\mathcal{P}(S))$

{ {{@, !}}, {{$}} } and { {{@}}, {{#, $}}, {{1}, {%, &}} } and {{{!}}} are in $\mathcal{P}(\mathcal{P}(\mathcal{P}(S)))$

Note that the set { @, {#} } isn't in any of the sets on the left!!

**I.9. (H)** Is @ in { @, !, $}? Yes, so leave it out. Is & in { }? No, so put it in, and so on.

**II.3. (S)** This question is equivalent to "Is there a largest cardinality?" If there were a largest set, how would it compare in size to its power set? Consider the following proof-by-contradiction argument. Assume that there is a largest set, $S$. Cantor's theorem states that the power set of $S$ is even larger, contradicting the assumption that $S$ is the largest set. So our assumption is wrong; there is no largest set.

**II.4. (H)** Classify all men as either self-shavers or non-self-shavers. If the barber belongs to the self-shavers group, then would he shave himself? If he belongs to the non-self-shavers group, would he shave himself? Similarly, if we classify all sets as either self-included or non-self-included, could **No Way** be a self-included set? A non-self-included set?

## 3.5 Straightening Up the Circle

**I.1. (H)** Place the circles inside one another, and pair points by drawing rays emanating from the center of the smaller circle.

**I.6. (H)** Un-shuffle the digits of the given number, and then recall that we decided to express all numbers, whenever possible, with trailing 9's instead of trailing 0's.

**I.10. (S)** Write the digits of **p** as 0.1 2 0001 0001 0001 0001 0001 ...

$x = 0.1\ 0001\ 0001\ 0001\ 0001\ \ldots$

$y = 0.2\ 0001\ 0001\ 0001\ 0001\ \ldots$

**p** gets paired with the point: (0.100010001 ..., 0.200010001 ...).

**II.2. (H)** Cut the two letters up as in the figures in the next column. Note that both figures are made up of a single solid line segment (including the two endpoints) and one half-open line segment, where one of the endpoints is missing.

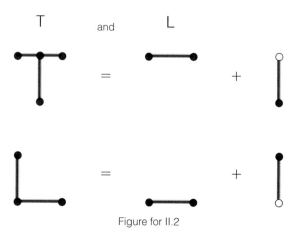

Figure for II.2

**II.5. (S)** We need to extend the stereographic projection to three dimensions. Place the punctured sphere on the plane, and identify points on the plane with points on the sphere by drawing lines through the missing North Pole. Given an arbitrary point on the sphere, we find its mate by constructing the line from this point to the North Pole.

## CHAPTER 4: GEOMETRIC GEMS

## 4.1 Pythagoras and His Hypotenuse

**I.4. (S)** "Directly above her" means that the triangle formed by the student, the kite, and the spectator is a right triangle. If the height is $A$, then the Pythagorean Theorem says that $A^2 + 90^2 = 150^2$. (If you make the 3-4-5 right triangle 30 times bigger, you get a 90-120-150 right triangle, and that makes solving for $H$ easier!) The kite is 120 feet high.

**I.5. (H)** The tip of the mast, the base of the mast, and the stern of the sailboat form a right triangle, and so the height ($L$) satisfies the equation $L^2 + 50^2 = 130^2$.

**I.9. (H)** In a right triangle, the lengths of the sides would satisfy $a^2 + b^2 = c^2$. Do the numbers given satisfy that relationship? If $a^2 + b^2 < c^2$, is the angle opposite $c$ larger or smaller than $90°$?

**II.4. (H)** Make a triangle whose sides are as long as the three pizza diameters. What does it mean if the resulting triangle is a right triangle?

**II.5. (S)** The longest side has length $r^2 + s^2$, so we need to check the following equation: $(2rs)^2 + (r^2 + s^2)^2 = (r^2 + s^2)^2$. Expanding gives $(4r^2s^2) + (r^4 - 2r^2s^2 + s^4) = (r^4 - r^2s^2 + s^4)$, which is true. So yes, there are infinitely many integer-valued Pythagorean triplets.

## 4.2 A View of an Art Gallery

**I.4. (H)** There are numerous correct triangulations. Here is one.

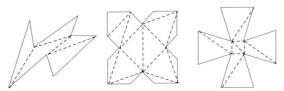

**I.6. (H)** You are free to pick the colors of two adjacent vertices, but, after that, the colors of the rest of the vertices are determined.

**I.10. (S)** See the zig-zag museum in the figure.

**II.4. (S)** Here are some possible guard placements.

There are several ways to do this. The museums that look like 3's need at least three guards; the museum that resembles the number 2 needs only two guards!

**II.5. (H)** Modify the comb-shaped museum from question II.2.

## 4.3 The Sexiest Rectangle

**I.6. (H)** If our original rectangle had lengths 2 and $(1 + \sqrt{5})$, then our new rectangles would have dimensions $2 \times 1$ and $2 \times \sqrt{5}$. Why? Are either of these Golden Rectangles?

**I.7. (H)** If the original lengths are 1 and $(1 + \sqrt{5})/2$, then the new side will have length $(3 + \sqrt{5})/2$. Simplify the ratio $(3 + \sqrt{5})/(1 + \sqrt{5})$ by multiplying both top and bottom by $(1 - \sqrt{5})$.

**I.9. (S)** Consecutive Fibonacci numbers (1, 1, 2, 3, 5, 8, 13, ...) are wonderful approximators of the Golden ratio. In a $10 \times 10$ grid, an $8 \times 5$ rectangle is your best bet. You can convince yourself by computing the 100 possible ratios and seeing which is closest to $(1 + \sqrt{5})/2$. Surprisingly, it doesn't help to consider nonhorizontal right triangles! (Write a small computer program to check this ... but don't do it by hand!)

**II.3. (S)** Let the starting rectangle be two units wide and one unit high.

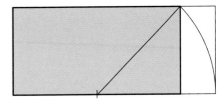

The new longer rectangle has length $(1 + \sqrt{2})$, width 1, for a corresponding ratio of $\sqrt{2} + 1 = 2.141\ldots$ The smaller rectangle has width $(\sqrt{2} - 1)$ and a ratio of $1/(\sqrt{2} - 1) = 2.141\ldots$ It has the same proportions! To see that these two expressions represent the same number, multiply both numerator and denominator by $(\sqrt{2} + 1)$.

**II.4. (H)** There are two similar triangles in the drawing, both having a leg that corresponds to a long side of a Golden Rectangle. Use the relationships between sides of Golden Rectangles and properties of similar triangles.

## 4.4 Soothing Symmetry and Spinning Pinwheels

**I.4. (H)** Fix an interior triangle in the center of the floor of a large warehouse and argue that, eventually, the whole warehouse floor will be covered.

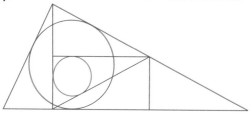

The 5-unit super-tile doesn't completely surround the original triangle; how about the 25-unit super-tile?

**I.8. (S)** When tile ★ covers tile 1, two 5-unit super-tiles line up, for a total of 10 matching triangles. When tile ★ covers tile 2, only one super-tile lines up, for a total of five triangles. Using the following tile-numbering scheme, all three tiles are interior children of their parents. ★'s parent is a 4th child, 1's parent is an interior child, and 2's parent is the 2nd child. Since neither tile 2's parent nor tile 1's parent is the 4th child, the 25-unit super-tiles will not line up.

**I.9. (H)** Yes, but it won't look anything like the pinwheel construction—it will look regular.

**II.1. (S)** Consider

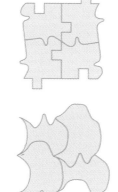
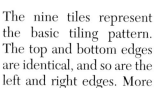

The nine tiles represent the basic tiling pattern. The top and bottom edges are identical, and so are the left and right edges. More complicated patterns with one or two tiles can be made by grouping four of them together (see figure) to make one standard tile.

**I.2. (H)** The angle at the vertex of the regular polygons must evenly divide 360°.

**II.5. (S)** To see that the order matters, it is easier to look at the end results of the modified square pattern of question II.1. Flipping diagonally and rotating by 90 is equivalent to flipping about a horizontal line. Reversing the steps is equivalent to flipping about a vertical line.

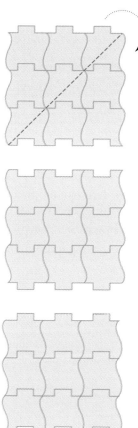

## 4.5 The Platonic Solids Turn Amorous

**I.3. (H)** Visit our Web site **www.heartofmath.com** to find java applets for displaying these stereographic projections.

**I.7. (H)** View the soccer ball as an icosahedron with its vertices sliced off.

**I.9. (S)** Cubes, tetrahedrons, and dodecahedrons yield triangles when their vertices are sliced. Octagons yield squares, and icosahedrons yield pentagons. The number of sides of the boundary correspond to the number of faces that meet at a vertex.

**I I.1. (H)** Try to find a plane that intersects all six sides of the cube! Can you find other planes that intersect 3, 4, and 5 sides?

**I I.2. (S)** The edges of the octagon connect the centers of the square faces. Each edge is the hypotenuse of a right triangle with legs of length 1/2. Therefore, by the Pythagorean Theorem, the edges of the octagon have length $\sqrt{2}/2$.

## *4.6 The Shape of Reality?*

**I.4–6. (H)** If you don't have a globe, then visit **www.heartofmath.com**.

**I.8. (S)** The triangle formed has angles 76°, 88°, 63°, for a total of 227°.

**I.16. (S)** The figure shows the optimal unfolding of the box. The minimum distance (using the Pythagorean Theorem) is $D = \sqrt{(2.5^2 + 1.5^2)} = 2.19\ldots$.

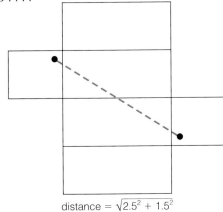

distance = $\sqrt{2.5^2 + 1.5^2}$

**I.18. (H)** On the plane, a triangle's angles add to 180; on the sphere, they add to more than 180; on a hyperbolic surface, they add to less than 180.

**I I.1. (H)** Every triangle divides the sphere into two triangular regions that cover the entire sphere. We can use either of those two triangles—one may have a much bigger sum of its angles.

**I I.2. (S)** If all three great circles intersect at a point you get *no* triangles, but, otherwise, you will get eight triangles whose total sum is (8 × 180) + 720 = 2160 (the sum of the angles of eight planar triangles plus an excess of 720°). The simplest example is three perpendicular circles that cut the sphere into eight equilateral triangles, all of whose angles are 90°.

**I I I.1–4. (H)** When we create a cone by removing a thin sliver, the cone is almost like a flat piece of paper. When we remove a bigger sliver, the cone becomes more pointed, like a wizard's hat. The sums of angles of triangles that go around the cone point will differ more from the sums of the angles of a triangle on the plane when the cone is more pointed.

## *4.7 The Fourth Dimension*

**I.2. (S)** Object 1: Either a sphere or two cones—a diamond spun about its axis. Object 2: A tetrahedron. Object 3: A vaselike object—a capped-off cylinder with a bulge in the middle. Object 4: A doughnut.

**I.6. (H)** Take a small portion of one of the rings and push it into the fourth dimension.

**I.9. (H)** It's an artistic feat to make a 2-dimensional drawing of a 6-dimensional cube. Here are 4-D and 5-D edge drawings: A 3-D cube has 4 + 4 + 4 = 12 edges. A 4-D cube has 12 + 12 + 8 = 32 edges. A 5-D cube has 32 + 32 + 16 = 80 edges, and a 6-D cube has 80 + 80 + 32 = 192 edges!

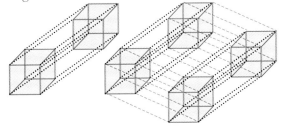

**I I.1. (H)** Since all the vertices are connected to each other, we have an edge for any pair of vertices and a face for any triplet of vertices.

**I I.3. (S)** There are 24 square faces: 6 around the center cube (hidden) and 12 next to the 12 edges of the center cube. (See the faces labeled 1, 2, and 3 in the figure.) The remaining 6 faces are labeled, A, B, C, D, E, and F.

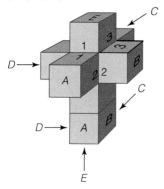

## CHAPTER 5: CONTORTIONS OF SPACE

### 5.1 Rubber Sheet Geometry

**I.1. (S)** Removing the two intersection points breaks the theta curve in three pieces, while removing *any* two points on the circle always leaves exactly two pieces. Suppose there were a deformation that turned the theta curve into a circle. Mark the two intersection points on the theta curve in red and follow the deformation process. At any stage, the removal of the two red points should break the distorted theta curve in three pieces; yet when we get to the final stage, the circle, we find that the object falls into just two pieces. This contradiction shows that our assumption is wrong: The theta curve is not equivalent to the circle.

**I.2. (H)** Imagine each letter being made out of rubber-bandlike material. Any letter with a loop cannot be distorted into a letter without a loop. What other characteristics distinguish one type of letter from another?

**I.6. (H)** Take a Gumbylike coffee mug and flatten out the part that holds coffee so that it looks like a flat disk connected to a ring (the handle).

**I.8. (S)** Yes. First rotate your hand 180° by bringing the glass under your armpit. Roll your shoulder forward and continue rotating your hand (while straightening out your arm) until you've rotated the palm of your hand 360°. You might feel close to dislocation, but don't stop here. Swing your arm in front of you and bend your elbow so that the glass is just above your head (540°). Continue swinging your arm above your head until the glass is back in its original position. In fact, you've rotated the glass 720° with your arm intact!

**II.2. (H)** Grab the four corners of the large squarish hole and pull them toward the center.

**II.8. (S)** Fold the sheet into a cylinder by gluing two opposite sides together. Now bend the tube around so that the top circle is right up against what was formerly the bottom circle, and glue these circles together.

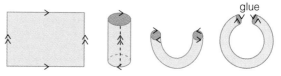

**III.3. (H)** Consider stretching out the knot so that you can pull parts of it around the sphere.

### 5.2 The Band That Wouldn't Stop Playing

**I.5. (H)** The original strip has only one edge. Cutting along the center line will not cut through that edge. So, the result can have only one piece.

**I.6. (S)** With the left and right ends identified appropriately, a rectangular strip of paper of length $L$ represents a Möbius band whose boundary edge has length $2L$. The centerline of the Möbius band still has length $L$, though. Cutting along this centerline produces a longer two-sided strip whose edges have length $2L$, but the centerline of this new strip is $2L$ as well.

**I.17. (H)** The two edges were not connected. How about the sides?

**I.20. (S)** Compare with question I.8 in section 5.1. Glue the left and right sides to make a cylinder. Now bend the cylinder into a U shape, pass one end through the side of the cylinder (in the fourth dimension this can be done without cutting into the surface), and then join the two ends.

**II.2. (S)** The half twist of the Möbius band arises from identifying two edges with arrows pointing in different directions. The centerline of the Möbius band crosses this identification edge one time—corresponding to the one half twist. The centerline of the cut Möbius band is twice as long and crosses the identification band *two* times in the same direction.

**II.6. (H)** Start drawing lengthwise anywhere in the strip and continue until you get back to where you started.

**III.2. (H)** In fact, you can draw a map on the Möbius band with six countries all bordering one another (showing that you need at least six colors to color this map). See www.heartofmath.com for a discussion of how many colors are needed for maps on other surfaces.

## 5.3 Feeling Edgy?

**I.4. (H)** Either both ends of the new edge are attached to existing vertices or one end of the new edge is attached to an existing vertex and the other end is a new vertex.

**I.7. (S)** Connected graphs on the plane satisfy the formula $V - E + F = 2$, where $F$ represents the number of pieces of the plane after removing the edges. If the plane isn't separated into different pieces, we have $F = 1$, so $V - E = 1$.

**I.9. (H)** Substituting $E = 151$ into the formula $V - E + F = 2$ gives $V + F = 153$. Could $F = 153$?

**I.14. (S)** Three faces, two edges, and two vertices. $V - E + F = 3$, not the expected number 2. The Euler Characteristic Theorem makes statements about connected graphs on the sphere, and the graph described in this question is not connected.

**II.2. (S)** Let $V_1, E_1, F_1$ represent the vertices, edges, and faces of the first piece and $V_2, E_2, F_2$ represent the analogous components of the second piece. The total number of vertices is $V = V_1 + V_2$; edges: $E = E_1 + E_2$; faces $F = F_1 + F_2 - 1$, where the $-1$ counteracts the double counting of the outside face. In all we have $V - E + F = (V_1 + V_2) - (E_1 + E_2) + (F_1 + F_2 - 1) = (V_1 - E_1 + F_1) + (V_2 - E_2 + F_2) - 1 = 2 + 2 - 1 = 3$. The simplest example of such a graph is two dots: $V - E + F = 2 - 0 + 1 = 3$.

**II.5. (H)** When pasting the two rectangular faces, we identify four vertices and four edges, and lose two faces entirely. Remember this when you count vertices, edges, and faces of the glued object.

**III.4. (H)** Calculate $E$ and then use Euler's formula to compute what $F$ must be. Now recompute what $E$ must be by reasoning that every face has at least three edges on its boundary. Find a contradiction.

## 5.4 Knots and Links

**I.9. (S)** No, it isn't a knot. It only seems like a knot when you hold the ends. As long as you don't cut the rope, an unknot will remain an unknot, because you can always reverse your steps!

**I.14. (H)** Switching any crossing in either figure will work.

**I.18. (H)** How can you remove the one crossing?

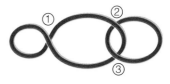

**I.19. (S)** Yes. However, if the two loops are painted red and blue, the number of red-blue crossings will always be even.

**II.2. (S)** Pick any point and start moving along the knot. Flip crossings when necessary so that it is as if you are constantly traveling under crossings. When you visit a crossing for the second time, do not flip it. This crossing is part of a loop that lies under the rest of the knot. (Imagine shrinking this part of the loop so that all the corresponding crossings vanish. This leaves you with a smaller problem that you can simplify in the same manner.) Continue the process.

**II.7. (H)** Yes! Try to draw a curve on the torus that goes three times around the long way while it goes twice around the short way.

**III.2. (H)** After you have made your rubber band set of Borromean rings, move them around and lay the ensemble on a table so they are roughly in a line with the left one round, the right one round and completely separate from the left one, and the middle one somewhat jumbled. Look for a way to put in more loops.

## 5.5 Fixed Points, Hot Loops, and Rainy Days

**I.3. (H)** For each figure, place a compass on the green circle that continually points to the corresponding point on the yellow curve. Now move the compass clockwise around the green circle (one time) and notice how many times the compass hand spins around (and in what direction). The winding number for the first figure is 1 (clockwise).

**I.5. (S)** Yes, for exactly the same reason given in the text as to why there exist two antipodal points having the same temperature. Let Diff($p$) equal the difference between the pressure at a point $p$ and the pressure at $p$'s antipodal cousin. If $A$ and $B$ are antipodal points, then Diff($A$) = $-$Diff($B$). Since the difference (Diff) varies continuously from a positive number to a negative number, there exists a point $C$ (between $A$ and $B$) where Diff($C$) is 0. Therefore, $C$ and its antipodal point have the same pressure.

**I.7. (H)** Draw a circle with a 5-foot diameter somewhere in your room. Must some pair of opposite points on that circle have the same temperature?

**II.1. (H)** Yes. Measure how far each red point is moved left or right. The right end is either fixed or moves left. The right end is either fixed or moves right. If the left end moves right and the right end moves left, why must there be a point between them that doesn't move at all? The Brouwer Fixed Point Theorem is true for line segments as well as disks.

**II.5. (S)** You drove 140 miles in 2 hours, for an average speed of 70 miles per hour. At some point, you must have been traveling at least 70 miles per hour (otherwise your average speed would be lower). So, you will get a ticket for driving at least 5 m.p.h. over the speed limit sometime while you were on the tollway.

# CHAPTER 6: CHAOS AND FRACTALS

## 6.1 Images

**I.2. (S)** There are a variety of answers. The largest such sub-figure is half the size of the original picture. Note that the original picture can be subdivided into three of these sub-figures. Also find sub-figures that are 1/4, 1/8, 1/16, . . . as large. (Any power of 1/2 will do.)

**I.10. (H)** You could start with a border around your picture. Then you could draw four reduced copies of the border, making a picture of four rectangles. In each of those four, draw four smaller rectangles. Continue forever.

## 6.2 The Dynamics of Change

**I.4. (S)** Let's estimate first: The rule of thumb from the previous question says that the population will double roughly every 70 years. So, in 700 years the population will double 10 times. $2^{10} = 2 \times 2 \times 2 \times 2 \times 2 \times 2 \times 2 \times 2 \times 2 \times 2 = 1024$, or about 1,000. So, in 700 years our population will grow by about a factor of 1,000, to about $5.7 \times 10^{12}$. In another 700 years, we get another factor of 1,000 for a total of $5.7 \times 10^{15}$, which already exceeds the total square yardage! So, in far less than 1,400 years, the population growth will decline. The precise solution to $(5.7 \times 10^9)1.01^y = (2.5 \times 10^{15})$ is $y = \log(2.5 \times 10^{15}/5.7 \times 10^9)/\log(1.01) = 1305.6 \ldots$ years.

**I.5. (H)** Don't focus on the total number of grains of rice on the checkerboard. Instead, just estimate the total number of grains on the last square! Since each square has twice the number of grains as the previous one, the last square will have $2^{64}$ grains of rice. That's more than $1.8 \times 10^{19}$ grains of rice! (For comparison, the Earth has a volume of only $10^{21}$ cubic millimeters.)

**I.7. (S)**

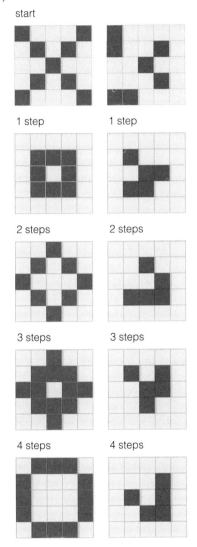

**I.18. (H)** Combining a figure having period 2 with one having period 3 is one way to generate a pattern that repeats only after six generations!

**II.6. (S)** Periodic. The $x$−coordinate oscillates between 0.79 and 0.52.

**II.9. (H)** Let $x$ represent the distance from the left end. For $x < 1/2$, $x \to 2x$; for $x > 1/2$, $x \to 2 - 2x$. To find points having period 2, assume $x$ first moves right (so $x \to 2x$), and then that result moves back left (so $2x \to 2 - 2(2x)$). Now solve the equation $x = 2 - 2(2x)$.

**III.1. (H)** Traditional cobweb iteration takes the form "go vertically to the function and horizontally to the diagonal, repeat." Reverse this to find the intervals of points that are moved into the interval $[1/3, 2/3]$.

## 6.3 The Infinitely Detailed Beauty of Fractals

**I.8. (S)** The edge of every triangle at each stage is eventually replaced with a scaled version of the standard Cantor Set. In fact, each point of the Sierpinski Dust can be associated with a pair of numbers in the Cantor Set on the base of the triangle. Specifically, each point of the Sierpinski Dust above the $x$-axis is the apex of an equilateral triangle whose other two vertices are points of the Cantor Set on the base of the original triangle.

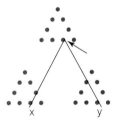

**I.10. (H)** Your third stage will look like 64 dots (each dot representing a tiny copy of the original picture) arranged in four groups of four groups of four.

**I.11. (S)** The third stages looks like a flock of birds flying in V formation. After infinitely many stages, the burgers vanish, leaving instead three copies of the entire picture at different scales.

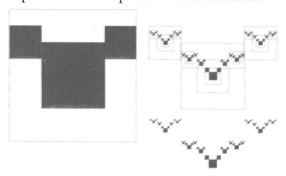

**I.19. (H)** Work backward; in order to get to triangle $b$, you must have been in the top corner of the left triangle and then rolled a 3. To get to the top corner of the left triangle, you must have been in the top triangle and then rolled a 1. To get to the top triangle, you need only roll a 2.

**II.1. (H)** *Moving 2/3 the way toward zero is equivalent to divide your current position by 3.* You can move all the points in the Cantor Set at once by reducing the set by a factor of 3 and lining it up with the origin. Notice that the Cantor Set neatly lines up with itself, because the left and right thirds are each 1/3 replicas of the entire set. Moving toward 1 is essentially the same.

**II.2. (S)** Like the Sierpinski Dust question above, the edges at each stage will be replaced with Cantor Sets after infinitely many iterations. The coordinates of each point in the Cantor Square have the form $(x, y)$ where both $x$ and $y$ are members of the Cantor Set.

**III.3. (H)** Start with any picture. Reduce it by 50%, make six copies, and arrange the centerlines of the six pictures to form a hexagon. Experiment with rotating and/or slightly shifting the positions of the six reduced copies.

## 6.4 The Mysterious Art of Imaginary Fractals

**I.2. (H)** $(1/2 + 3i) + (5/2 - i) = 3 + 2i$.

**I.5. (H)** $(1 + i)$: angle 45, length $\sqrt{2}$. $(1 + i)^2 = (2i)$: angle 90, length 2.

**I.7. (S)** $0 \to i \to -1 + i \to -i \to$ (periodic with period 2).

**I.13. (S)** $2 + 3i \to -4.73 + 12i \to -121.4 - 113.5i \to 1841.0 + 27552.9i \to$ (blows up!)

**II.4–5. (H)** Those numbers corresponding to connected Julia Sets will lie *inside* the Mandelbrot Set; otherwise, the point lies outside the Mandlebrot Set.

**II.6–9. (S)** If the Julia Set is connected, then zero will lie inside the Julia Set.

**III.1. (H)** Consider the geometric method of squaring and adding complex numbers. If $z_1$ has length greater than 2, then $z_1^2$ has length greater than 4, and $z_1^2 + a$ has length greater than 3. How long must the next iterate be?

## 6.5 Predetermined Chaos

**I.4. (H)** Work backward from one of the two fixed points (0 and 2/3). For example, what goes to 0? 0 and 1. What goes to 1? 1/2. What goes to 1/2? 1/4 and 3/4. Therefore, both 1/4 and 3/4 are two starting points that will get fixed at 0 after three iterations. For the fixed point 2/3 we have: 1/3 goes to 2/3; both 1/6 and 5/6 got to 1/3; both 1/12 and 11/12 go to 1/6; and both 5/12 and 7/12 go to 5/6. So 1/12, 11/12, 5/12, and 7/12 all get fixed at 2/3 after three iterations.

**I.6. (S)** The diagonal line intersects the inverted V at two points: 0 and 2/3. The equation of the right line is $y = 2 - 2x$, and the equation of the diagonal is $y = x$. Solving this system gives $x = 2/3$.

**I.7. (S)** $x = 0.4$. This is the cobweb plot of question I.1. Aside from guessing and trial and error, you can find the solution via algebra. Let's assume that $x$ goes up to the left curve ($y = 2x$), over, and then down to the right curve ($y = 2 - 2x$). So, $x \to 2x \to 2 - 2(2x)$. Solving the equation $2 - 2(2x) = x$ gives the answer.

**I.12. (H)** 0 stays at 0 while 0.01 → 0.02 → 0.04, and so on. How many steps before the whole interval is covered?

**II.5. (S)** The iterates are 0.48, .8077225673, .5025834321, .8089953967, .5000431939,

.8090169884 . . . , which get closer and closer to the repeated values starting with 0.5 → .8090169945 . . . → 0.5 → . . . .

**II.7. (H)** Try about a dozen iterates and see how every fourth value compares.

**III.4. (H)** We can get a precise answer by numerically solving the $8^{th}$ degree equation, $F(F(F(x))) = x$, where $F(x) = 4x(1 - x)$. There are eight solutions corresponding to the two fixed points, and two 3-periodic cycles.

## 6.6 Between Dimensions

**I.2. (H)** Four copies of the rectangle can make a rectangle twice as large. So $S = 2$ (scale), and $N = 4$ (number of copies). $2^d = 4$ implies $d = 2$, which agrees with our expectation.

**I.5. (S)** The eight squares surrounding the center are a scaled version of the larger square. Therefore, eight copies make a square three times as large. $S^d = N$ becomes $3^d = 8$ implying a fractal dimension of $d = \log(8)/\log(3) = 1.89 \ldots$.

**I.7. (H)** Two copies of the Cantor Set can be arranged to make a Cantor Set that is three times as large.

**II.1. (S)** The collage-making instructions imply that three copies of this object (Sierpinski Dust) can be formed to make a new object three times larger. $3^d = 3$ implies that $d = 1$. We have a totally disconnected set of points whose fractal dimension is 1.

**II.5. (H)** Find a way to replace each segment with five segments that are each a fourth as large.

## CHAPTER 7: RISKY BUSINESS

## 7.1 Chance Surprises

**I.6. (S)** You've seen one black card, and there are 25 other black cards in the remaining deck of 51. So, the probability of seeing another black card is $25/51 = 0.490 \ldots$. You should find that roughly half of the black-first pairs had a second black card.

**I.7. (H)** There are four ways to pick two cards: BB, BR, RB, RR. Since we are concerned only with pairs containing a black card, we have only three allowable outcomes: BB, BR, and RB. What fraction of these cases has two black cards?

## 7.2 Predicting the Future in an Uncertain World

**I.7. (S)** Answer: 2/9. There are six ways to roll a 7, (1 & 6, 2 & 5, 3 & 4, 4 & 3, 5 & 2, 6 & 1) and two ways to roll an eleven (5 & 6, 6 & 5). Since there are 36 equally likely ordered rolls, the chances of rolling 7 or 11 are 8/36, or 2/9.

**I.10. (H)** How many equally likely outcomes are there? How many equally likely outcomes contain a head? How many equally likely outcomes include seeing Lincoln?

**I.15. (H)** It is easier to find the probability that the sum *doesn't* exceed 4. The only such sums are (1 & 1), (1 & 2), (2 & 1), (1 & 3), (3 & 1), (2 & 2).

**I.19. (s)** Answer: 0.1087 . . . . The probability that the second person doesn't match the first is 365/366. Now two birthdays are used up, so the chance that the third doesn't match either of the first two is 364/366. The probability that the first doesn't match AND the second doesn't match is (365/366) × (364/366). Continuing in this way, the probability that the first 40 people don't match is (365/366) × (364/366) × . . . × (327/366) = 0.1087681902 . . . .

**II.1. (S)** Probabilities are easiest to calculate when we can start with a list of equally likely outcomes. The simplest way to do this is to view the coins as different and order the outcomes. With this in mind, there are four equally likely outcomes that contain at least two heads: (T, H, H), (H, T, H), (H, H, T), and (H, H, H). So, the chance of three heads is 1/4.

**II.5. (H)** To help your analysis, assume that each side of each card is a different shade of the colors given—for example, (red & hot pink), (blue & deep blue), and (ruby-red & lagoon-blue). You are shown a side of a card with one of the red hues. It is equally likely to be the hot-pink side, the red side, or the ruby-red side.

**II.9. (S)** Matching at least one number is the opposite of not matching any numbers, an event whose probability is easier to determine. The chance that the first number isn't a match with 2 or 9 is 8/10, whereas the chance that the second number also isn't a match is 7/9 because, of the nine remaining balls, only seven are different from 2 and 9. The probability of neither matching is $(8/10) \times (7/9) = 56/90 = 28/45$. So, the probability that you have at least one match is $1 - 28/45 = 17/45$. There are 10 possible balls selected first and nine selected second for a total of $10 \times 9$ total ways of selecting two balls in order. Of those 90 possibilities only 2-9 or 9-2 will match both numbers. So, your probability of matching both numbers is $2/90 = 1/45$.

**III.5. (H)** Think of the coins as different—penny, nickel, dime. Saying that two will be the same is different from saying the penny and dime are the same. Always think about the eight equally likely outcomes.

## 7.3 Random Thoughts

**I.8. (S)** Suppose only 39 people died during every two-week period. This accounts for only $39 \times 27 = 1053$ deaths. So, 39 deaths per two weeks are not enough.

**I.13. (H)** After the first week, 300 people see a perfect record; second week, 150; third week, 75. Continue the pattern.

**I.16. (S)** With a 1/4 chance of correctly answering each question, the chance of a perfect score is $(1/4)^{100}$, or approximately $10^{-60}$. With educated guessing, the chances go up to $(1/2)^{100}$, or roughly $10^{-30}$, still unfortunately small.

**I.18. (H)** Forever is a *long* time. The odds of rolling any seven-digit number right off the bat is $(1/10)^7$. How many chances do you have?

**II.1. (H)** The chance of typing "c" first is 1/26. The chance of then typing an "a" is also 1/26. What is a chance of doing both those things?

**II.7. (S)** The probability of producing any particular sequence is $1/2^8 = 1/256 = 0.004\ldots$.

**III.1. (H)** List all the pairs of days on which Celebrity 1 and Celebrity 2 could die—for example, Mon-Mon, Wed-Mon, Mon-Wed, . . . .

## 7.4 Down for the Count

**I.7. (H)** How many different ways can you choose six numbers from 36? There are 36 ways to pick the first number, 35 ways to pick the second number, and so on, so there are $36 \times 35 \times 34 \times 33 \times 32 \times 31$ ways to pick six numbers in an *ordered* way. But we have each combination several times; for example, (123456), (263415), (354162), and so forth are all the same combination. How many different times have you counted each set of six numbers?

**I.8. (H)** This is identical to the previous question, but we'll answer it using a different method. Imagine that we are watching the lottery balls come out one by one. The chance that the first lottery number matches one of your six numbers is 6/40. The chance that the second lottery number matches one of the remaining five numbers is 5/39 (since there are only 39 balls left in the bin). What is the probability that all six events happen?

**I.10. (S)** There are eight questions in all: Do I want cheese? Do I want lettuce? and so on, and each answer has two possibilities (yes or no). There are $2 \times 2 \times 2 \times 2 \times 2 \times 2 \times 2 \times 2$ different ways to answer all the questions corresponding to $2^8 = 256$ different possible hamburgers.

**I.13. (S)** $250/800 = 0.3125$, so that 31% of all the cars in the area are Hondas. If $C$ represents the total number of cars, then $0.31 \times C = 10{,}000$. Dividing by 0.31 to solve for $C$ yields $C = 10{,}000/0.31 =$ roughly 32,300 cars.

**II.1. (S)** The question is really, How many different ways can you choose three things out of five? First imagine that order counts. There are five ways to pick the first item, four ways to pick the second, and three ways to pick the third. There are $5 \times 4 \times 3$ different ways to pick three items (where [eggs, bagel, coffee] represents a different choice than [bagel, coffee, eggs]). We overcounted each combo by exactly the number of orderings of three things. There are $3 \times 2 \times 1$ orderings of a three-item menu, and so the total number of distinct combinations is $(5 \times 4 \times 3)/(3 \times 2 \times 1) = 10$.

**II.4. (H)** The number fits the form $(4xxx)$ $(x4xx)$ $(xx4x)$ or $(xxx4)$, where each $x$ represents one of nine possible numbers (not a 4). How many total possibilities like that are there?

**III.2. (H)** Either you roll an 11 at least once, or you never roll it, and the odds of the second case are easier to compute. Of the 36 equally likely ordered outcomes for a pair of dice, only two total 11.

## 7.5 Great Expectations

**I.1. (S)** The two green slots give the house an edge. The chance of winning on red is 18/38, so that the expected value is $2,000(18/38) = \$947$.

**I.10. (S)** Suppose you get $D$ dollars if you flip tails. The value of the game is then $(-5) \times (2/3) + (D - 5) \times (1/3) = D/3 - 5$. Since the value of a fair game is zero, set $D/3 - 5 = 0$, and solve for $D$. So, $D = \$15$.

**I.11. (H)** The best way to approach insurance questions is from the point of view of the insurance company. What is their expected value from this deal? If the policy costs $\$P$, then 80% of the time they will make $\$P$ and 20% of the time they will make $\$P - \$700$ (here, making negative money means losing money!)

**I.15. (H)** The expected value (a.k.a. average return) is the sum of the individual probabilities of each roll multiplied by the corresponding payback.

**II.3. (S)** Look at this game through the eyes of the insurance company. The value of the car insurance to the company is $0.02 \times (\$200 - \$9,000) + 0.98 \times (\$200) = \$20$. The expected value of the bike insurance is $0.1 \times (\$75 - \$850) + 0.9 \times (\$75) = -\$10$. The company makes enough money on the car insurance to cover its losses on the bike insurance.

**II.4. (H)** Consider an equivalent question: Suppose you knew 100 people who were in a similar situation. After all the offers were accepted, what would their average salary be?

**III.4. (H)** View these questions from the point of view of an insurance company that has to shell out $1.2 million for the death of each customer. Is it worth it for the company to pay for all the women to have a Pap smear? How much does it cost the company to pay for 3,000 Pap smears compared to paying a death claim? What is the expected value to the insurance company of paying for a Pap smear?

## 7.6 What the Average American Has

**I.2–8. (H & S)** The figure shows a sample graph for the following data from question 2 (total of 30 responses: [a] $< 90, 0$]; [b] $90 - 119, 5$]; [c] $120 - 149, 10$]; [d] $150 - 179, 11$]; [e] $> 180, 4$]. No one weighed less than 90 lbs, five weighed between 90 and 119 lbs, and so on.

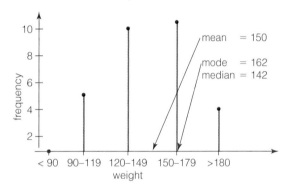

Make five tick marks along the horizontal axis representing the five weight categories, and draw tick marks along the vertical axis to represent the number of people in each category. Draw a line (or a number of dots) whose height is proportional to the number of people in that category.

To calculate the mean, let the letters (a)–(e) correspond with the five numbers 1, 2, 3, 4, and 5. If a person weighed 155 lbs, then he would correspond to category (d) and position 4 on the number line. Where will an average person appear on the number line? What is the expected value of the associated numbers?

**I.14. (H)** Suppose there were $H$ noncheaters in your group. The only no answers will come from these $H$ people when they flip tails. Roughly half of them will flip tails. So, how many no answers do you expect expressed in terms of $H$?

**I.15. (S)** Answer: 16. Only the nonhuggers will contribute to the no answers in the survey. If there are $X$ nonhuggers, then we'll see roughly $0.5X$ no answers. Since 83 answered yes, there are 67 no answers, and solving $0.5X = 67$ gives $X = 2 \times 67 = 134$. This leaves an estimated 16 people that still hug both parents.

**II.5. (S)** Mean = 84.125. Median is halfway between 84 and 87 = 85.5. Both statistics move a little in the direction of the added score, so they will both go down. With the extra score, the mean = 83.66..., and the median = 84.

**II.8. (H)** Suppose a professor gives an easy exam and nearly all of the A students as well as many of the B students get a perfect score.

**III.5. (H)** Yes, it is possible; however, if the ratio of men to women stays the same, then no, this isn't possible.

## 7.7 Navigating Through a Sea of Data

**I.2. (S)**

|  | Dad gives white gene | Dad gives red gene |
|---|---|---|
| Mom gives red gene | (red, white) → pink | (red, red) → red |
| Mom gives white gene | (white, white) → white | (white, red) → pink |

The third-generation flower is red with probability 1/4, white with probability 1/4 and pink with probability 1/2. You should see twice as many pink flowers as either red or white flowers.

**I.6. (H)** Suppose there are $B$ blondes in the world (fake or real). How many will you suspect of being fake? You'll mistake 20% of the 90% that are real blondes, and you'll correctly identify 80% of the 10% that are fake. Of those, how many are actually fakes?

**I.7. (S)** Answer: 29%. Use the reasoning in the previous question. There are $B$ blondes. You'll correctly identify 85% of the bleached blondes $(0.30B)$ and mistake 15% of the real blondes $(0.70B)$. If you judged every blonde, you would be suspicious of $[0.85 \times (0.30B) + 0.15 \times (0.70B) = .36B]$ 36% of all the blondes. Since you correctly suspected only $0.255B$ people, the odds of your suspicion being correct is $0.255B/0.36B = 0.708\ldots$, or 71%. So suspecting that Chris's hair color is fake, the probability that Chris is a true blonde is 29%. Keep waiting.

**I.16. (H)** Is the accident rate determined more by the kind of car or the kind of driver?

**II.3. (H)** Suppose every student calls about the scholarship. How many will be told they won? How many will actually have won?

**II.5. (S)** Answer: 41%. Let's interpret the statement, "The witness is 80% sure..." as, "With 80% accuracy, the witness sees a blue cab as blue and a green cab as green." Suppose that the witness were to look at all the cabs in the city—say, a total of $C$ cabs. Of the $0.85C$ green cabs, she'll identify 20% of them as blue, and of the $0.15C$ blue cabs, she'll identify 80% correctly. She'll claim that $0.20 \times 0.85C + 0.80 \times 0.15C$ cabs are blue, when in fact only $0.80 \times 0.15C$ of those cabs were blue. Her chances of correctly identifying a blue cab in this city is $0.80 \times 0.15C/(0.20 \times 0.85C + 0.80 \times 0.15C) = 12/29 = 0.41\ldots = 41\%$.

**III.1. (H)** To help visualize the situation, you could think of starting with 64 plants: 32 red, 32 pink. You could pair them in the correct proportions: eight red-red pairs (accounting for 16 red plants), 16 red-pink pairs, and eight pink-pink pairs. What colors result in what proportion?

# *Index*

Abbott, Edwin, 313
Adams, Colin, 262
Adams, Douglas, 542
Adleman, Leonard, 98, 99
Afghanistan, 223
AIDS testing, 603–606
Air traffic safety, 606
Alexander the Great, 373
*Alice in Wonderland*, 251
Alleles, in DNA, 560
Allen, Paul, 593
Allen, Steve, 542
Alternating picture of a knot, 381
American, average, 583
Amorous solids, 282
Anaxagoras, 174
And, 556–558
Animal husbandry, 27
Antoine's necklace, 454
Aristotle, 116, 249, 616
Arnold, Sir Edward, xii
Art gallery theorem, 223
Asking embarrassing questions, 584
Attractor, 497
Austin, TX, 299
Average face, 250
Average value, 570

Bacon, Francis, x, 5, 344
Bagel, 355
Balloon, 360
Bar codes, 87
Barnsley's Fern, 438, 441
Barnsley, Michael, 438
Base 3 expansion, 426
Beckmann, 514
Belly button, 237
Berger, Robert, 253
Bias, 588
Big Ten Conference, 39
Biltong, 588
Bimodal distribution, 590
Binary expansion, 127
Bing links, 383
Birthday question, 517, 528
Blah, blah, blah, 177
Blake, William, 138, 175
Boethius, 514, 618

*Bon Appetit!*, 431
Bond, James, 568
Book, this, value of, 572
Borromean rings, 377
Bowling, 365
Box, Thousand Dollar, 574
Box, Zero-or-Million Dollar, 574
Boxer shorts, 8
Bread sticks, tossed, 543
Broccoli, 402, 430
Brocken, 297
Brouwer Fixed Point Theorem, 387, 388
Browne, Sir Thomas, 371
Brownian motion, 545
Browning, Robert, 617
Brunnian links, 383
Buffon, Georges Louis Leclerc
    Comte de, 543
Bug, 291, 333
Bug, food, 291
Bug, in a proof, 347
Bunnies, 56
BURGER and STARBIRD, 534
Burger, Edward, Acknowledgments, 621
    in collage, 435
    in 2 dimensions, 309–312
    pants down, xiv, 8
    signature, xii
Burger heaven, 450
Butler, Bishop Joseph, 512, 518
Butterfly from Brazil, 490, 491

Caillebotte, 245
Calculator, chaos, 482
Candles, scented, 54
Cantor Set, 426, 452, 453, 509
Cantor, Georg, 164, 170, 180
Cantor, Georg, Power Set Theorem,
    180, 181
Cantor, Theorem, 169
Cardinality, 150
Carroll, Lewis, xi, 26, 213
Carson City Kid, 98
Carson, Johnny, 608
Central tendency, 593
    mean, 593
    median, 593
    mode, 593

Cervantes, 4
Chaos, 481, 484, 488, 489
Chaos game, 445
Check digits, 88
Cheers!, 350
Chicago, 290
Chickens, squawking, 223
Churchill, Sir Winston, 146
Chvatal, Vasek, 223
Circle, area of, 231
Class survey, 596
Clock arithmetic, 84
Cobweb plot, 423, 425, 492–496
Codes, 97
Coffee cup and doughnuts, 338
Cohen, Paul, 186
Coincidence, 276
Coincidence Happens, 540
Coincidence, amazing, 538
Coincidence, personal, 540
Cold War, 96
Cole, K.C., 608
Collage-making instructions, 435
Collecting data, 584, 587
Collections, ordered, 554
Combinatorists, 552
Complex addition, 461
Complex multiplication, 462
Complex plane, 463
Comte, Auguste, 47
Cone geometry, 304
Coincidence, deadly, 538
Connected graph, 358
Connected sum, 368
Continuum, 124
Continuum hypothesis, 185
Conway, John, 253, 411
Counting, generalized, 555
Cryptography, 96
Crystal ball, 520
Cube, 271, 274
Cubes, higher dimensional, 313
Cubes, unfolded, 316
cummings, e.e., 306

Da Vinci, Leonardo, 236
Dali, Salvador, 317, 328
Damn, I don't give a, 569

Darwin, Charles, x
Data, collecting, 584
　dirty, 589
　insufficient, 589
　interpreting, 589
　3D, 591, 592
De Divina Proportione, 236, 282
De la Tour, Georges, 520
De Morgan, Augustus, 328
Debussy, Claude, 238
Decimal expansion, 126
Del Vaga, Perino, 371
Demoth, Charles, 64
Descartes, René, 359
Devil, 250
Diagonalization argument, 168, 171
Dice, cool, 536
　fair, 521
　rolls, 526
Die, fair, 521
Dimension, 501, 503, 505
　Koch curve, 506
　Sierpinski triangle, 506
Diophantus, 74
Disreali, Benjamin, 583
Distribution, bimodal, 590
　normal, 590
　Poisson, 590
Divine Proportion, 236
Division algorithm, 65
DNA, 371, 373, 560
DNA evidence, 561
Dodecahedron, 272, 274
Dodge Ball, 8, 170, 175
Dominant gene, 602
Doonesbury, 583
Double helix, 371
　3D, 372
Doughnuts, holy, 335
Doyle, Sir Arthur Conan, 601
Duality, 279
　figures, 280–281
Duchamp, Marcel, 318
Duell, Charles, 76
Dynamic symmetry, 237
Dynamics, 418

Eat Me, 251
Edge, life on the, 345
Edison, Thomas, 76
Eerie, 306
Einstein, Albert, 25, 41, 326
Embarrassing questions, asking, 584
Equivalent by distortion, 331, 332
Escher, M.C., 250, 344, 481
Euclid, 68, 116
Euler Characteristic, 359
Euler, Leonhard, 359

Eureka, 213
Euripides, 617
Expected value, 570

Fair dice, 521
Fair die, 521
Fair game, 571
Farewell, 616
Farey factions, 135
Fechner, Gustav, 234
Female, 340
Ferguson, Helaman, 350
Fermat's Last Theorem, 75
Fermat's Little Theorem, 100
Fermat, Pierre de, 74, 524
Fibonacci nim, 55
Fibonacci numbers, 49, 57, 421
Fibonacci, Leonardo, 49
Figure-eight knot, 374
Fish, 414
Fisk, Steve, 223
Five-dimensional cube, 315
Flat Earth Society, 289
Flat tire, 329
Four-dimensional space, 309
Fourth dimension, 309
Fox, 544
　droppings of, 544
Fractal, 431, 460
Fractal dimension, 505
　Koch curve, 506
　Sierpinski triangle, 506
Fractals, 400–406
Franklin, Benjamin, 385
Fredddd, 310
French Impressionism, 237
Frieze patterns, 266
Fun, 1–622

Galileo, 2, 165, 513
Gambling, 569
Game of Life, 411
Game, fair, 571
Gates, Bill, 593
Gauss, Karl F., 64, 72, 297
Generalized counting, 555
Genes, 250
Geometric probability, 543
Giesecke, Werner, 589
Gilbert, William S., 210
Gödel, Kurt, 186
Goldbach Question, 77
Golden ratio, 53, 234, 283
Golden rectangle, 54, 234, 282
　minus square, 240
　construction of, 240
　linked, 378
Goldenheim, David, 213

Gombauld, Antoine, 524
Gordian Knot, 371, 373
Granny knot, 374
Graph, connected, 358
Great circle, 290, 291
Grecian eye cup, 235
Guggenheim Museum, 219
Gustafson, Pier, 262

Hairy bodies, 39
Hall, Monty, 10
Handcuffs, 339
Hardy, G.H., 42
Heraclitus, xi
Hilbert, David, 164
Hints, 623
HIV, 603
Hoech, Hannah, 434
Hohenhagen, 297
Hokusai, 428
Holmes, Oliver Wendell, 551
Holmes, Sherlock, 601
Homogeneous, 298
Hors d'oeuvre, 516
Hot Loop Theorem, 391
Hotel Cardinality, 159, 172
Human knots, 380
Huxley, Thomas H., 140
Hyperbolic geometry, 297
Hypercube, 314
Hypotenuse, 210

Icosahedron, 272, 274
　with Borromean rings, 378
Imaginary numbers, 461
Infinite Monkey Theorem, 542
Infinite set, 147
Inselsberg, 297
Insurance, bicycle, 578
　life, 572
Iran, 223
Irrational numbers, 116
Irrationals, decimal characterization of, 129
ISBN, 93
Isotropic, 298
Iterative process, 48

Jacks, Menger, 454
Jai alai, 552
Jawbreakers, 561
Jell-O cubes, 336, 342
Johns, Jasper, 38
Julia Sets, 404–406, 457–459, 468–471
Julia, Gaston, 457, 458

Kansas, 289
Kant, Immanuel, 398, 492, 618
Keats, John, 457

Kepler, Johannes, 233, 236, 243, 273, 288, 364
Khashanah, Khaldoun, 433
Kipling, Rudyard, 83
Klee art gallery question, 221
Klee, Paul, 219, 221
Klee, Victor, 221
Klein bottle, 349
   glass, 350
   stone, 350
Knot theory, 379
Koch Curve, 432, 506
   fractal dimension, 506
Koch Stool, 451
Koch, Helge von, 432
Kronecker, Leopold, 112, 147

Lakeside School, 593
Langlois, Judith, 250
Larson, Gary, ix
Law of Large Numbers, 524
Le Corbusier, 237, 238, 409
Legendre, A.-M., 72
Léger, Fernand, 4, 96, 568
Let's Make a Deal, 515
Letterman, David, 26
Lewis, Jim, 607–608
Li, Tien-Lien, 489
Life insurance, 572
Life, top ten lessons for, 25, 619
Likelihood, 522
Lincoln, dizzy, 515
   on edge, 514
   takes a hit, 533
Litchfield, Daniel, 213
Literary Digest, 587
Locke, John, 38
Logarithmic spiral, 243
Lorenz attractor, 489
Lorenz, Edward N., 488
Loser, 560
Lottery, 553
Lotto, 551
Lovelace, Augusta, 124
Lucas numbers, 58
Luggage, lost, 606

M&M's, 139
Magritte, René, ix, xiii, 618
Male, 340
Mandelbrot Set, 400, 459, 472–474
Mandelbrot, Benoit, 428, 443, 459
Map coloring, 366
Marsaglia, George, 543
Martingales, 582
Math rocks, 376
Math student, anonymous, ix
Mathematics, cliff of, 186

Matisse, Henri, 244, 386
May, Robert, 418
McGwire, Mark, 538
Mean, 593
Mean and Extreme Ratio, 237
Median, 593
Mendel, Gregor, 601–602
Menger jacks, 454
Menger Sponge, 433
Menger, Karl, 433
Meteorology Theorem, 391
Michelangelo, 146
Milton, John, 481, 501
Mindscapes, 25
Mirrored gallery, 227
Missing member, 164
Mission Impossible, 544
Möbius band, 345
Möbius resistors, 348
Mode, 593
Modular arithmetic, 85
Mondrian, Piet, 237
Money, 410
Monkeys typing, 541
Mooing, 8
Musgrave, Ken, 443
Mysterahedron, 361

Napoleon I, x
NASA, 298
Natural numbers, 38, 147
Nautilus sea shell, 243
Negative curvature, 295
Neufchatel, Nicolas, 268
New York City, 289
Newcomb's Paradox, 574
Nietzsche, Friedrich, 96
Nonexistahedron, 361
Normal distribution, 590
Number line, 124, 126

Octahedron, 271, 274
Olympic rings, 378
1 = 0.99999…, 132
One-dimensional space, 307
One-to-one correspondence, 142, 146
Or, 556–558
Ordered collections, 554
Ovary, 583

Pacioli, Luca, 236, 282
Pap test, 582
Parallel lines, 303
Parallel mirrors, 429, 449
Parthenon, 235
Pascal, Blaise, 524
Pasteur, Louis, 489, 568
Peas, 601–603

Penrose Patterns, 253, 257
Penrose tiles, 267
Penrose, Roger, 253, 267
Pereira, I. Rice, 210
Perfect numbers, 44
Phelps, Jim, 544
Phi ($\varphi$), Golden Ratio, 53, 234
Phidias, 234
Pi ($\pi$), 119
   random approximation, 543
Picasso, 357
Piece, Warren, 22
Pigeonhole principle, 41, 145
Ping-Pong balls, infinitely many, 150
Pinwheel pattern, 253
   no rigid symmetry, 260
   scalable, 255
Pinwheel triangle, 254, 451, 453
   5-unit, 254
Pizza, 217
Planetary positions, 410, 488
Plato, x, 36, 208, 268
Platonic solids, 269, 359
   only five, 361
   swiss cheeselike, 501
Poe, Edgar Allan, 43
Points without partners, 386
Poisson distribution, 590
Polio virus, 364
Pollock, Jackson, 601
Polygonal closed curve, 220
Pope, Alexander, 400
Population chaos, 486
Population models, 410, 486
Positive curvature, 295
Post Hoc Ergo Propter Hoc, 609
Power set, 179
Prime Number Theorem, 73
Prime numbers, 66
Primes, infinitude of, 68
Primes, prime factorization, 66
Probability, 523
   birth of, 524
   event doesn't happen, 527
Problems, problems, problems, ix
Proclus, x, 616
Projective plane, 349
Public key codes, 97
Purple cows, 8
Pushkin, Aleksandr, 356
Puzzles, 2–12, 26–30
Pythagoras, 114
Pythagorean School, 114
Pythagorean Theorem, 210
Pythagorean Theorem, puzzle proof, 211
Pythagoreans, 273

QED, 184

Quaker Oats, 403, 429, 449
Quasicrystals, 261
Quaver units, 238

Radin, Charles, 253
Radiolaria, 364
Ramanujan, Srinivasa, 41
Random walk, 544
Rational numbers, 113
Rationals, decimal characterization of, 129
Real numbers, 130
Reality, shape of, 288
Record player, 387
Recycling logo, 348
Regular polygons, 268
Regular solids, 269
    unfolded, 367
Relative frequency, 523
Repeller, 497
Reunion, 516
Rigid symmetry, 251
Ring, 330, 332
Rivest, Ronald, 98, 99
Romantic, 306
Rome, Italy, 290
Roommate, body hair count, 39
Roosevelt, Franklin Delano, 587
Rose, Richard, 608
Roulette, 569
RSA coding scheme, 98, 102, 104
Rubber geometry, 328
Rubber, fun with, 329
Russell's Paradox, 186, 190
Russell, Bertrand, 538, 617

Saddle geometry, 295
Safe sex, 521
Sandia Laboratories, 348
Scalable, 252
Scarlett, Ms., 569
Scream, blood-curdling, 560
Scrod, 414
Self-destruct, tape recorder, 544
Self-similarity, 430
Sense of humor, cutting, 347
Set, 147, 186
Seurat, Georges, 237
Sexiest rectangle, 233
Shafer, Paul, 26
Shakespeare, William, 85, 192, 520, 541
Shamir, Adi, 98, 99
Sierpinski carpet, 10, 21
Sierpinski triangle, 402, 433, 439, 506
    fractal dimension, 506
Sierpinski, Waclaw, 433
Sinatra, Frank, 54
Sloppy joes, 124
Soap film, 365

Soccer ball, 369
Solids, stellated, 286
Solids, truncated, 286
Solutions, 624
Sosa, Sammy, 538
Spanning arc, 225
Sphere, 334
Spherical geometry, 293
Spleen of Mathematics, 311
Springer, Jim, 607–608
Square knot, 374
Starbird, Michael, Acknowledgments, 622
    grandmother, 32
    in collage, 435
    in 2 dimensions, 309–312
    signature, xii
Stellated solids, 286, 367
Stereographic projection, 198, 361
    in 3 dimensions, 206
Stick number, 380
Stifel, Michael, 134
Stock prediction, 541
Subset, 176
Sumners, Dewitt, 373
Super-super-super-tiles, 259
Super-tiles, 252
Surface, 334
Surfing, xiii
Symmetry, 250
Symmetry of scale, 252
Symmetry, rigid, 251

T-arrangement, 254
Tangent line, 244
Tasty treat, 430
Tattoos, 375, 376
Teller, Edward, 490
Testicle, 583
Tetrahedron, 270, 273
That leg, 210
Theta curve, 333
Think, xi
This leg, 210
This page, 644
3D art, 84, 126, 270–272, 277, 278,
    280–281, 296, 315, 372, 378, 591, 592
Three-dimensional space, 308
Thurston, William, 301
Tigers, counting, 561
Tire iron, 330
Toilet paper, 536
Tonight Show, 542
Tonight Show Starring Johnny Carson, 608
Topology, 329
Torus, 334
Torus knots, 382
Torus, two-holed, 335
Toto, 491

Trash, toss that idea into, ix
Travel tips, xi
Trefoil knot, 374
Triangles, higher dimensional, 322
    hyperbolic, 297
    spherical, 293
Trou, dropping, 8, 19
Trudeau, Garry, 583
Truncated solids, 286, 367
Twin Prime Question, 77
Twin studies, 607–609
Two-dimensional space, 307
Typing monkeys, 541

Underwear, 143, 338, 564
    clean, 39, 565
Undies, 564
Uniqueness of scaling, 258
United Nations, 238
Universal Product Code (UPC), 87
Unknot, 374
UPC (Universal Product Code), 87
Uranus, 273

Vajradhatu Mandala, 430
Van Gogh, Vincent, 385
Verhulst logistic equation, 421, 486
Verhulst Model, 414
Verhulst, Pierre, 414
Vertex, 358
Vertices, 220
Video feedback, 429, 449
Vietnam, 105
Voltaire, Francois, 13, 553
Volvo, 141
Von Neumann, John, 33

Wales, coastline, 431
Weber, Max, 306
Welcome!, ix
White, Vanna, 11
Wiggles, 431
Wild side, walk on the, 345
Wiles, Andrew, 75
Williams mascot, 8
Williamstown, MA, 299
Winding number, 389
Wizard of Oz, 215
Woolums, Scott, 443

XX page!, xx

Yorke, James, 489

Zaman, Arif, 543
Zero curvature, 295
Zero-dimensional space, 307
Zinfenblatte, 542

# Credits

**Page ix**, *left: The False Mirror,* {*Le Faux Miroir*}. (1928), René Magritte. Oil on canvas, 21¼ × 31⅞″. The Museum of Modern Art, New York. Purchase. Photograph ©1999 The Museum of Modern Art, New York. **ix**, *right:* THE FAR SIDE by Gary Larson, ©THE FAR SIDE, ©1987 FARWORKS, INC. Used by permission. All rights reserved. **xiii:** *The Masterpiece or the Mysteries of the Horizon,* 1955, René Magritte. ©ARS, New York. Giraudon/Art Resource. **xiv:** Ed Burger, Photograph ©Williams Record. **xix:** Michael Dalton/Fundamental Photographs. **xviii:** Celestial Image Co./Science Photo Library/Photo Researchers Inc. **4:** *The Great Parade, (definitive state),* Leger, 1954. Oil on canvas, 117.75 × 157.5″. The Solomon R. Guggenheim Museum, New York 62.1619. David Heald/The Solomon R. Guggenheim Foundation, NY. FN. **8:** photograph © Williams Record. **37,** *middle: The Thinker,* Rodin, Musée Rodin, Paris, France. Vanni/Art Resource, NY. **38:** *Numbers in Color,* Jasper Johns, 1958–59. Encuastic & newspaper on canvas, 66.5 × 49.5″. Albright-Knox Art Gallery, Buffalo, NY. Gift of Seymour H. Knox, 1957. **41:** The Granger Collection. **42:** Courtesy of History of Mathematics Archive, St. Andrews University. **45,** *middle left:* Townsend P. Dickinson/The Image Works. **47:** Townsend P. Dickinson/The Image Works. **48,** *bottom:* Bill Aron/PhotoEdit. **48,** *middle:* James Fisher/Photo Researchers Inc. **48,** *top:* Townsend P. Dickinson/The Image Works. **49,** *bottom:* From *Fibonnaci Numbers in Nature* by Trudi Garland and Edith Allgood.© 1988 by Dale Seymour Publications. Reprinted by permission. **49,** *middle:* Portrait of Leonardo Fibonacci (c.1170–c.1250) The Granger Collection. **49,** *top left:* R. Lanaud/Jacana/Photo Researchers Inc. **49,** *top right:* Photo Researchers Inc. **64:** *I Saw the Figure 5 in Gold,* Charles Demuth, 1928. Oil on composition board, 36 × 29¾″. Signed: C.D. Inscribed: W.C.W. (William Carlos Williams). The Metropolitan Museum of Art, New York. The Alfred Stieglitz Collection, 1949. Photograph ©The Metropolitan Museum of Art, New York. **68:** Euclid Joos van Ghent, Palazzo Ducale, Urbino, Italy Scala/Art Resource, NY. **72,** *left:* Photo Researchers Inc. **72,** *right:* Mary Evans Picture Library. **74:** Science Photo Library/Photo Researchers Inc. **75:** Denise Applewhite/Princeton University. **96:** *Mona Lisa with Keys,* 1954, Leger (ADAGP). Biot, musée Fernand Léger ©ARS, New York ©Photo Reunion des Musées Nationaux/Gerard Blot. **98,** *left:* Courtesy of Ronald Rivest. **98,** *right:* Courtesy of Leonard Adleman. **113:** *The Thinker,* Rodin, Musée Rodin, Paris, France. Vanni/Art Resource, NY. **138,** *left:* Celestial Image Co./Science Photo Library/ Photo Researchers Inc. **141:** Celestial Image Co./Science Photo Library/ Photo Researchers Inc. **146:** *The Creation of Adam,* Michelangelo, Sistine Chapel, Vatican State. Scala/Art Resource, NY. **150:** Rob Crandall/The Image Works. **164:** Corbis. **175:** *The Ancient of Days,* William Blake Leonard de Selva/Corbis. **177:** Richard Magna/Fundamental Photographs. **208,** *middle left:* Louis Goldman/Photo Researchers Inc. **208,** *middle right:* Edward Miller/Stock Boston. **209,** *left:* North Wind Pictures. **209,** *right:* Michael Dalton/Fundamental Photographs. **210:** *Oblique Progression,* 1948, Rice Pereira. Oil on canvas, 127 × 101.6 cm Collection of Whitney Museum of American Art Purchase 4822 Photograph ©Whitney Museum of American Art. **219:** Louis Goldman/Photo Researchers Inc. **221,** *left: Open Book,* Paul Klee, 1930. Water-based paint & varnish over white lacquer on canvas, 45.7 × 42.5 cm The Solomon R. Guggenheim Museum, New York 48.1172 × 526, David Heald/The Solomon R. Guggenheim Foundation, NY. FN. **221,** *right: Glas-Fassade,* 1940, Paul Klee. 288(K8) Wachsfarben auf Jute Über Keilrahmen; 71.3 × 97.7 cm Paul-Klee-Stiftung, Kunstmuseum Bern, Switzerland. **222:** *Construction under a Waterfall, a Crossing,* 1924, Paul Klee. Oil on canvas. Private Collection, Milano, Italy Erich Lessing/Art Resource. **223:** Courtesy of Steve Fisk. **233:** Edward Miller/Stock Boston. **235,** *bottom:* Bohams, London, UK/The Bridgeman Art Library. **235,** *top:* Archivision **236,** *bottom: The Vitruvian Man,* da Vinci, Accademia, Venice, Italy Scala/Art Resource, NY. **236,** *top: Unfinished Portrait of St.Jerome,* da Vinci, Vatican Museums Scala/Art Resource, NY. **237,** *bottom: Place de la Concorde, 1938–1943,* Piet Mondrian, Dutch, 1872–1944. Dallas Museum of Art, Foundation for the Arts Collection, gift of the James H. Lillian Clark Foundation. **237,** *middle: Invitation to the Side Show,* Georges Pierre Seurat. Oil on canvas, 39.25 × 59″ Bequest of Stephen C. Clark, 1960. The Metropolitan Museum of Art. (61.101.17). **237,** *top:* Le Corbusier's *Modular Man* Courtesy of Corbusier Foundation and Le Corbusier by Robert Furneaux Jordan. **238,** *middle left:* Archivision **238,** *middle right:* From *Debussy in Proportion: A Musical Analysis* by Roy Howat, Cambridge University Press. **238,** *top,* **239:** Courtesy of Corbusier Foundation and *The Mathematics of the Ideal Villa and other Essays* by Rowe. **243:** Edward Miller/Stock Boston. **244:** *L'Escargot,* Matisse, 1953. Gouache on cut & pasted paper. ©ARS, NY. Tate Gallery, London, Great Britain Tate Gallery/Art Resource, NY. **245:** *Paris, A Rainy Day,* Caillebotte, 1877. Oil on canvas, 212.2 × 276.2 cm, Art Institute of Chicago. Erich Lessing/Art Resource, NY. **249,** *bottom: Strawberry Thief,* William Morris. Indigo discharged block printed cotton, 1883. Victoria & Albert Museum, London/Art Resource, NY. **249,** *top left:* Paolo Koch/Photo Researchers Inc. **249,** *top right:* Paul Almasy/Corbis. **250,** *top:* Judith Langlois **250,** *bottom: Symmetry Drawing E85* ©1999, M.C. Escher/Cordon Art, Baarn, Holland. **262,** *top left:* Property of AT&T Archives. Reprinted with permission of AT&T. **266,** *bottom left:* John Heseltine/Corbis. **266,** *bottom middle:* Scala/Art Resource. **266,** *bottom right:* Kelly-Mooney Photography/Corbis. **266,** *top:* Michael Newman/PhotoEdit. **267,** *left:* Historical Picture Archive/Corbis. **267,** *middle:* Historical Picture Archive/Corbis. **267,** *middle left:* Historical Picture Archive/Corbis. **267,** *middle right:* Historical Picture Archive/Corbis. **267,** *right:* Historical Picture Archive/Corbis. **268:** *Portrait of Johannes Neudorfer and His Son*, Nicolaus Neufchatel, 1561. Musée des Beaux-Arts, Lille, France/Giraudon/Art Resource. **273:** North Wind Pictures. **288:** NASA/Science Photo Library/ Photo Researchers Inc. **295:** Benson Photography. **306:** *Interior in the Fourth Dimension*, 1913, Max Weber. Brush, watercolor and gouache over black crayon on heavy watercolor paper, $18\frac{1}{2} \times 24\frac{1}{2}''$. Baltimore Museum of Art: Bequest of Saidie A. May BMA 1970.42. **315:** Michael Dalton/Fundamental Photographs. **317:** *The Crucifixion*, Dali, 1954. Oil on canvas, $76\frac{1}{2} \times 48\frac{3}{4}''$. Gift of the Chester Dale Collection, 1955. The Metropolitan Museum of Art, New York. Photograph ©The Metropolitan Museum of Art, New York. **318:** *Nude Descending a Staircase*, #2, Duchamp, 1912. Oil on canvas. Acc.#50-134-59 The Louise & Walter Arensberg Collection. Philadelphia Museum of Art. **328:** *Persistence of Memory*, {Persistance de la mémoire}, 1931, Salvador Dali. Oil on canvas, $9\frac{1}{2} \times 13''$. The Museum of Modern Art, New York. Given anonymously. Photograph ©1999 The Museum of Modern Art. **339:** Dion Ogust/The Image Works. **341:** Erv Schowengerdt. **344:** *Moebius Strip II*, M.C. Escher, ©1999 Cordon Art, Baarn, Holland. All rights reserved. **348:** Courtesy of the Sandia Corporation. **350,** *bottom: Eine Kleine Rock Musik III*, 1986. Helaman Ferguson *Mathematics in Stone and Bronze* Courtesy of Ed Bernik and Meridian Creative Group. **350,** *top:* Photo by Erv Schowengerdt. **364,** *right:* Oxford Science Films/Animals Animals. **371:** *Alexander the Great Cutting the Gordian knot,* del Vaga. Fresco. Castel Sant'Angelo. Scala/Art Resource, NY. **373:** Courtesy of Dr. Nicholas R. Cozzarilli/U.C.Berkeley. **385:** *Starry Night*, 1889. Vincent Van Gogh. Oil on canvas, $29 \times 36\frac{1}{4}''$. The Museum of Modern Art, New York. Acquired through the Lillie P. Bliss Bequest. Photograph ©1999 The Museum of Modern Art. **386:** *Dance*, Matisse, 1910. Oil on canvas, 260 × 391 cm, ©ARS, NY. The Hermitage Art Resource, NY. **398,** *left:* Benoit Mandelbrot. **398,** *middle right:* Richard F. Voss/IBM Research. **399,** *left:* Dr. F. Espenak/Science Photo Library/Photo Researchers Inc. **399,** *right:* ©1991 John C. Hart, Electronic Visualization Laboratory, University of Illinois at Chicago. **401,** *top:* Benoit Mandelbrot. **401,** *bottom:* Ken Musgrave and Benoit Mandelbrot. **402:** Courtesy of the author. **409:** Science Photo Library/Photo Researchers Inc. **428:** *The Great Wave*, from *Thirty-Six Views of Mount Fuji*, Hokusai, 1823–29, Private Collection. Art Resource, NY. **429,** *top:* Richard Megna/Fundamental Photographs. **430,** *bottom left:* Courtesy of Edward Burger. **430,** *bottom right:* Courtesy of Edward Burger. **430,** *top:* The Vajradhatu Mandala, Central Tibet, c.14th cent. **433:** *Sierpinski Garden*, 1997, Khaldoun Khashanah. Courtesy of the artist. **434,** *bottom: Kitchen knife cuts through Germany's first Weimar beer-belly culture*, photomontage, 1919, Hannah Hoech. National Gallery. ©ARS,NY Erich Lessing/Art Resource. **443,** *left: Carolina*, Ken Musgrave and Benoit Mandelbrot. **443,** *right: Sunset over Mount Annapurna*, Scott Woolums/Northwest Images. **444,** *bottom right: Blessed State*, Ken Musgrave and Benoit Mandelbrot. **444,** *left:* All fractals on left: Richard F.Voss/IBM Research. **444,** *middle right: Lethe*, Ken Musgrave and Benoit Mandelbrot. **457:** *God as the Architect of the Universe*, parchment, 344 × 260 mm, miniature from a *Bible moralisée*, c. 1230. Vienna, Östereichische Nationalbibliothek, cod.2554, fol.1v. **458,** *top:* Courtesy of the History of Mathematics Archive, St. Andrews University. **459,** *bottom left:* Hank Morgan/Science Source/ Photo Researchers Inc. **481:** *Order and Chaos*, M.C. Escher, 1950 ©1999 Cordon Art, Baarn, Holland. **491:** Townsend P. Dickinson/The Image Works. **501:** ©1991 John C. Hart, Electronic Visualization Laboratory, University of Illinois at Chicago. **512,** *middle right:* AFP/Corbis-Bettmann. **512,** *right:* Stacy Pick/Stock Boston. **513,** *left:* David Sams/Stock Boston. **514:** *The Dream*, 1921, Max Beckmann. Bequest of Morton D. May. The Saint Louis Art Museum. **520:** *Le Tricheur à l'As de Carreau*, Georges de la Tour, The Louvre, Paris. Erich Lessing/Art Resource,NY. **538:** AFP/Corbis-Bettmann. **541:** Joe Sohm/ Stock Boston. **543:** Courtesy of Esquire Magazine. **549:** Wide World Photos. **551,** *bottom:* Erv Schowengerdt. **551,** *top:* Richard B. Levine. **552,** *bottom:* Scott Halleran/Allsport USA. **552,** *top:* Stacy Pick/Stock Boston. **560:** Courtesy of Dr. Nicholas R. Cozzarilli/U.C. Berkeley. **568:** *Soldiers Playing at Cards*, Leger. Collection Kroller-Muller Museum, Otterlo, The Netherlands. Reproduced with permission. **569:** David Sams/Stock Boston. **583:** DOONESBURY ©1993 G.B. Trudeau. Reprinted with permission of UNIVERSAL PRESS SYNDICATE. All rights reserved. **587:** Culver Pictures, Inc. **588:** ©Mark Blumberg and Lisa Singer, 1998. **594:** Courtesy of the Lakeside School. **601:** *Blue (Moby Dick)*, Jackson Pollock, 1943. Courtesy of Ohara Museum of Art, Kurashiki, Japan. **608:** AP/Wide World Photos. **618:** *The Masterpiece or the Mysteries of the Horizon*, 1955, René Magritte. ©ARS, New York. Giraudon/Art Resource.